Red Beet Biotechnology

Bhagyalakshmi Neelwarne
Editor

Red Beet Biotechnology

Food and Pharmaceutical Applications

Springer

Editor
Bhagyalakshmi Neelwarne
Plant Cell Biotechnology Department
Central Food Technological Research Institute
Mysore, India

ISBN 978-1-4614-3457-3 ISBN 978-1-4614-3458-0 (eBook)
DOI 10.1007/978-1-4614-3458-0
Springer New York Heidelberg Dordrecht London

Library of Congress Control Number: 2012940416

Printed on acid-free paper

Springer is part of Springer Science+Business Media (www.springer.com)

Preface

Desire for living longer with good health has made people wise in shifting their food choices; and hence fruit, vegetables, pulses (legumes) and fish products occupy more space in their diet plates than starchy cereals, fries and red meats. Red beet - a vegetable, includes all sorts of colored beets belonging to the plant species *Beta vulgaris*. Anciently, beets were indeed considered as powerful food because of the rapid nourishment imparted to cattle and horse, and energy to humans to thrive in cold winter. However, their popularity declined with changing fashions in food, where more preference for starchy cereal products, red meats and fries dominated. Such fads, fuelled by the desperate requirement for sugar those days, made people concentrate their research efforts on white sugar beets, rather than the life elixir—the red beet. Recent health concerns and enormous developments made in research methodologies channelled more organized research towards finding the key health-deteriorating factors, with a main emphasis on oxidative damage. To combat such deleterious effects, researchers took a re-look at several food items consumed by humans, as well as animals sourced for human food. Stream-lined research efforts resulted in placing red beet among the top ten health-promoting vegetables because of its rich content of nutrients and enormous nutraceutical properties, which are expected to benefit children to build a strong foundation towards good health and elderly humans to tide-over geriatric problems.

Although recent research developments made on red beet have been chronicled in research journals and reports, a unified summary of this exciting crop was made available only in 2004, through an online publication by Nottingham. Research on red beet has increased rapidly in the last decade owing to the presence of brightly colored water-soluble pigments complemented by its richness of antioxidants, neuro-stimulators and strong anti-hypertensive and anti-cancer properties. This book has a special focus on biotechnological research done in red beet, the topics oriented towards the industrial development of food and pharmaceutical formulations. This treatise will hopefully substantiate its usefulness to all, particularly those in horticulture, corporate and academia, in keeping abreast with recent developments in red beet material per se, the science associated with it as well as processing technologies, all with projections towards future perspectives.

With an overview about the red beet crop and its chemical composition, Chap. 1 is a compilation of both biotechnological and other complementing breakthroughs made in science using red beet as a model plant. The presence of large quantities of bright pigments in colored beets have attracted plant physiologists to unravel the pigment biosynthesis pathways, as discussed in Chap. 2, whereas, the stability of beet pigments during their extraction process as well as in food/pharmaceutical products are important for their sustained applications as nutraceuticals. Characterizations of the stability of betalains, from their biosynthesis all the way to products that receive them, make up the content of Chap. 3. Indeed, red beet has always been a material of choice for pigment research, owing to the large cellular pigment-laden vacuoles that can be isolated from beet tuber quite easily. Once separated from the cell, a large body of biochemical information is discernible from such vacuoles. Chapter 4 is a compilation of research developments made in unravelling physiologies governing the transport of nutrients and metabolites into and out of vacuoles and the enzymes/energetics involved in such transport mechanisms.

Mitochondria are another set of subcellular oragnelles with a main function of orchestrating the energy balance of the cell. Mitochondria also involve themselves in biosynthesis of vitamins and lipids and generate reactive oxygen/nitrogen species. Respiration, the chief function in mitochondria, is much different in bulky tuberous underground organs such as beet roots. Chapter 5 provides interesting insights into the physiological events involved in respiratory metabolism and alternative pathways evolved in tubers. For such studies, red beet has served as a model system, especially because of its very long shelf life after harvest. In nature, plant organs that generate oxidative molecules are invariably equipped with powerful radical quenching mechanisms. In red beet, apart from several antioxidant enzymes, betalains serve as strong antioxidants. Therefore, understanding the kinetics of the radical-scavenging process rendered by the antioxidants of red beet has interested scientists. Chapter 6 deals with kinetics of radical scavenging by betalains in comparison with standard antioxidants. Betalains have shown stronger lipoperoxyl radical-scavenging and other anti-oxidative effects in both chemical and biological models.

The strong antioxidant effects are further translated into the larger effects such as anti-cancer (Chap. 7) and anti-diabetic properties of red beet (Chap. 8). Cancer and diabetes have been chronic problems in modern societies, probably due to higher stress faced by people in their ever-increasing daily challenges. Such stress has culminated in life-threatening health problems such as cancer, cardiac/circulatory diseases, diabetes and neurodegeneration. The chapters dealing with these aspects in the book describe, with appropriate and enormous number of examples, how red beet helps in preventing, and often reversing such pathologies.

The growing realization of the application of plant cell/organ cultures for propagation and breeding has provided great research impetus to horticultural biotechnologists. Densely pigmented nature of red beet has attracted scientists to opt for this plant material as a research model for generating basic information about cellular nutrition, their growth, pigment biosynthesis and morphogenesis towards organ formation. On the other hand, cell cultures have also paved ways for technological developments such as designing bioreactors for cell culture, product enhancement and fine-tuning recovery mechanisms. Chapter 9 provides a summary of the progress made in the culture of red beet cells and tissues from test-tube level to bioreactor level cultivation.

Genetic engineering techniques have resulted in the establishment of transformed cells/organs as well as whole plants, which would not have been existed through natural breeding process. One area which has emerged recently is the use of genetically transformed cultures, where "hairy root" cultures obtained after genetic transformation with *Agrobacterium rhizogenes* are of tremendous technological importance. Hairy roots per se can synthesize and consistently produce novel biochemicals akin to the root system of the plant from which they originate, but they can also be engineered to produce new proteins, including human therapeutic proteins. Red beet hairy roots have been extensively researched as model systems for engineering technology development on one hand and provide opportunity for new products on the other hand. Their high sensitivities to sugars and their ability to undergo re-engineering into double transformants have made them very remarkable candidates for basic research. Chapter 10 has interesting information on the recent upsurge of research conducted using red beet hairy roots. While Chap. 11 provides different bioreactor designs and several gadgets developed through ingenious methods, using red beet hairy roots as model systems, Chap. 12 encompasses research observations made on enzymes produced in red beet hairy root cultures, their characterization and presumptive applications. Product recovery from live materials, particularly from cultured cells and hairy roots, has been a challenging strategy where a great level of intricate cellular mechanisms and their monitoring without causing damage to cell and the product needs to be understood. Chapter 13 provides information on innate and active modes of product release and recovery, where safe and novel permeabilization techniques are discussed. Subsequently, applications of current and newer technologies for product concentration and separation for integration to form continuous processes are presented. Bioengineers have also been successful in applying mechanotronics for in situ product recovery and on-line product separations. Other advanced smart technologies for the extraction of betalains at industrial scales are explained in Chaps. 14 and 15.

Opportunities are enormous for benefiting from the innumerable health potentials of red beet. To reap the health benefits of red beet, it is essential to supplement current research with the new innovative biotechnologies, particularly chemogenomics for the in vivo identification of new biochemical pathways, combinatorial chemistry for pathway engineering and computational chemistry for newer ingredients/processing parameters. Such and other technologies, as well as legal requirements of new biotechnology products and their environmental/economic impacts are the components of Chap. 16.

I remain immensely grateful to 24 scientists from 9 countries for their great efforts and elegant preparation of the chapters with lucid illustrations, making this compendium attractive to readers. I am thankful to all reviewers who have contributed tremendously towards improving the quality of this treatise.

I also acknowledge the encouragement of the Director, CFTRI, who enthused me to imbibe the task of compiling this book and I thank all my colleagues and other scientists of my beloved institute for their interactions and dialogues during editing. I am very thankful to editorial staff of Springer, particularly Susan Safren and Rita Beck, who were kind and very supportive from inception till the publication of this book. I am indebted to my family members for their continuos support.

Suggestions are solicited from all around the world, for the future improvement of the content and approach of this book.

Mysore, India Bhagyalakshmi Neelwarne

Contents

Contributors

Thomas Bley Institute of Food Technology and Bioprocess Engineering, Technische Universität Dresden, Dresden 01062, Germany

Kotamballi N. Chidambara Murthy Triesta Science, Health Care Global Oncology Hospital, Sampangiram nagar, Bengaluru 560 027, India

Fruit and Vegetable Improvement Centre, Texas A&M University, College Station, TX 77843-2119, USA

Patrick J. Cullen School of Food and Environmental Health, Dublin Institute of Technology, Dublin, Dublin 1, Ireland

Vasil G. Georgiev The Stephan Angeloff Institute of Microbiology, Bulgarian Academy of Sciences, Laboratory of Applied Biotechnologies, Plovdiv 4000, Bulgaria

Kapuganti J. Gupta Department of Plant Physiology, University of Rostock, Rostock D-10859, Germany

Sowbhagya B. Halagur Plantation Products, Spice and Flavour Technology Department, Central Food Technological Research Institute (Council of Scientific and Industrial Research), Mysore 570 020, India

Kirsten Jørgensen Department of Plant Biology and Biotechnology, University of Copenhagen, Frederiksberg C DK-1871, Denmark

Govind J. Kapadia Department of Pharmaceutical Sciences, College of Pharmacy, Howard University, Washington, DC 20059, USA

Maria A. Livrea Dipartimento STEMBIO, Università di Palermo, Palermo 90123, Italy

Hans J. Lyngs Jørgensen Department of Plant Biology and Biotechnology, University of Copenhagen, Frederiksberg C DK-1871, Denmark

M.C. Madhusudhan Department of Food Engineering, Central Food Technological Research Institute, (Council of Scientific and Industrial Research), Mysore 570 020, India

Shivapriya Manchali Department of Biotechnology and Crop Improvement, University of Horticultural Sciences, Bengaluru 560 065, India

Shruthi Nagaraju Triesta Science, Health Care Global Oncology Hospital, Sampangiram nagar, Bengaluru 560 027, India

Bhagyalakshmi Neelwarne Plant Cell Biotechnology Department, Central Food Technological Research Institute (Council of Scientific and Industrial Research), Mysore 570 020, India

Yoshihiro Ozeki Department of Biotechnology and Life Science, Tokyo University of Agriculture and Technology, Koganei, Tokyo 184-8588, Japan

Atanas I. Pavlov The Stephan Angeloff Institute of Microbiology, Bulgarian Academy of Sciences, Laboratory of Applied Biotechnologies, Plovdiv 4000, Bulgaria

Department of Organic Chemistry and Microbiology, University of Food Technologies, 26 Maritza Blvd, Plovdiv 4002, Bulgaria

K.S.M.S. Raghavarao Department of Food Engineering, Central Food Technological Research Institute, (Council of Scientific and Industrial Research), Mysore 570 020, India

Hardy Rolletschek Leibniz-Institut für Pflanzengenetik und Kulturpflanzenforschung (IPK), Gatersleben 06466, Germany

Thimmaraju Rudrappa DABT, Dupont Knowledge Center, E.I. DuPont India Pvt. Ltd., ICICI Knowledge Park, Genome Valley, Turkapally, Shamirpet (M), R. R. District, Hyderabad 500 078, India

Nobuhiro Sasaki Department of Biotechnology and Life Science, Tokyo University of Agriculture and Technology, Koganei, Tokyo 184-8588, Japan

Hiroshi Sekiguchi Department of Biotechnology and Life Science, Tokyo University of Agriculture and Technology, Koganei, Tokyo 184-8588, Japan

Nandini P. Shetty Plant Cell Biotechnology Department, Central Food Technological Research Institute (Council of Scientific and Industrial Research), Mysore 570 020, India

G. Subba Rao Global Technology Resource Center, Streamwood, IL 60107, USA

Luisa Tesoriere Dipartimento STEMBIO, Università di Palermo, Palermo 90123, Italy

Brijesh K. Tiwari Department of Food and Consumer Technology, Manchester Metropolitan University, Manchester M14 6HR, UK

Chapter 1
Red Beet: An Overview

Bhagyalakshmi Neelwarne and Sowbhagya B. Halagur

Abstract Research on red beet has increased rapidly in the last decade owing to the presence of brightly colored water-soluble pigments complemented by its richness of antioxidants, neuro-stimulators and strong anti-hypertensive and anti-cancer effects. Apart from their increasing traditional applications in food and pharmaceutical products, betalains have found newer applications such as for developing solar cells and anti-ageing formulations. While the visible presence of vacuolar pigments has made the tissues of red beet a model system for studying several vacuole-related cellular physiologies, the material is ideal for developing color-extraction techniques and related engineering aspects. Although its derivative variety – the sugar beet – has attracted great attention because of its capacity to accumulate high sugar content, similar emphasis has never been placed on red beet. The modern knowledge and molecular techniques of genomics, proteomics and metabolomics have not been applied efficiently for improving existing nutraceuticals within the plant or for identifying hitherto unexplored biomolecules. The present chapter provides an overview of recent biotechnological research developments on red beet, highlighting the important research areas worth addressing in the near future.

B. Neelwarne (✉)
Plant Cell Biotechnology Department, Central Food Technological Research Institute (Council of Scientific and Industrial Research), Mysore 570 020, India
e-mail: bneelwarne@cftri.res.in

S.B. Halagur
Plantation Products, Spice and Flavour Technology Department, Central Food Technological Research Institute (Council of Scientific and Industrial Research), Mysore 570 020, India
e-mail: sowbhagyahalagur@yahoo.com; ppsft@cftri.res.in

B. Neelwarne (ed.), *Red Beet Biotechnology: Food and Pharmaceutical Applications*,
DOI 10.1007/978-1-4614-3458-0_1,

1.1 Introduction

Beet is said to have got its name from the Greek letter *beta* owing to the presence of the swollen root resembling a Greek *B,* although it was the foliage beet variety, namely chard, that was domesticated by Greeks and Romans in 2000 BC. Natural diversity probably first occurred in the fleshy (swollen) taproot varieties, known to have existed in the second century AD, which probably were the first to serve as dietary sugar in winter times (McGrath et al. 2007). Italians made selections from wild beets of the Mediterranean seacoast, developing both red and white varieties, which became popular as the Roman beet. Beet was used as a vegetable boiled in stews, baked in tarts, roasted as a whole and cut into salads. Various other applications such as in drinks and for medicines emerged based on individual experience and their popularities.

1.1.1 Origin of the Genus Beta and Production of New Beet Cultivars

The plants of genus *Beta* have been traced to have originated in North Africa and spread through the Mediterranean Sea route, occupying the seashores of Asia and Europe. Many leafy types would have emerged while adapting to newer agro-climatic conditions where they were grazed by animals. Thus earliest usage of beets by humans was only the foliage. Storage tap roots probably started increasing their sizes and pigmentation while adapting to rich soils, and people selected the roots for food and found many uses for them. Thus Romans first cultivated beets to use their roots in food while the foliage was fed to domestic animals. Tribal people, who frequently changed their living locations widely in Europe, were probably responsible for spreading this plant in the continent. During the eighteenth century, the large-rooted beets known as Mangel wurzel, recognized as nutritious, were being fed to cattle presumably to reap high meat and milk yields. They were introduced into England in the 1770s for use as livestock feed after being developed from early fodder beets in Germany and Holland. Beet was then found throughout Europe, chiefly for hybridization purpose (Eastwood and Nyhlin 1995) to produce the vast range of colors and shapes found in table beets today. In those days, white beets were perhaps more common, but less desirable than red beets. More details of the history of origin and development of beet crops are described elsewhere (Nottingham 2004).

In European countries beet was first used as animal fodder in large scale and became a popular vegetable for human consumption in the sixteenth century. Analytical intervention by alchemists, and then by modern chemists of the nineteenth century, resulted in the discovery that beet root was a concentrated source of sugar. While sugar supply from cane industries of Britain was curtailed to other parts, Napoleon was the first to pronounce that beet be used as an alternative for sugar production, which not only catalyzed beet's popularity but also encouraged breeding and selection, resulting in many cultivars. Taxonomically, Beet is classified under Kingdom: Plantae; Order: Caryophyllales; Family: Chenopodiaceae; Genus: *Beta* and Species: *Beta vulgaris* by Linnaeus (C. Linnaeus, 1753, 1762, *Specia Plantarum* vol. 1 and 2).

Diversifying from red beets, modern sugar beets emerged after several series of selections, followed by repeated breeding and re-selection (Hanelt et al 2001) for enhancing sugar level from 6% to 20%. A large support for its commercialization emerged from the support by the King of Prussia in the eighteenth century. Andreas Marggraf (1749) first demonstrated in 1747 that sugar could be isolated from beet roots, which was almost similar to that produced from sugarcane at concentrations of 1.3–1.6%. From then on, beet was recognized for commercial sugar production. Beet cultivation in Europe reached a commercial scale in the nineteenth century following the development of the sugar beet in Germany as an alternative to tropical sugar cane. As a response to blockades of cane sugar by British during war, Napoleon opened schools in France specifically for studying sugar production at commercial scale. This encouragement came with an additional package of devoting 28,000 ha (69,200 acres) for growing the newly selected beet varieties, pushing the rapid growth of the European sugar beet industry (Zohary and Hopf 2000). In North America the sugar beet was introduced in 1830, where the first commercial production started at a farm in Alvarado, California in 1879 (Zohary and Hopf 2000). Meanwhile, in 1850, the sugar beet had been introduced also to Chile via German settlers. It now remains a widely cultivated commercial crop in Europe, America and Canada for table sugar production, accounting for 30% of the world's sugar.

Generally growing as a herbaceous biennial and sometimes as a perennial plant, *Beta vulgaris* is classified under the family Chenopodiaceae (Lange et al. 1999). However, in America this genus was recently classified under Amaranthaceae. Plants bearing red or purple tuberous root vegetable are known as beetroot or garden beet. Beetroot is a firm, clean, globe-shaped vegetable with no mucilaginous or watery tissues, and the tubers available in the market often contain freshly emerged young leaves. The characteristic feature of red beet beetroot is the presence of distinctive bright red, purple of yellowish–orange flesh. The sugar bearing varieties do not display many colors, although pigments are formed under stress and adverse conditions and can be induced when cultured in vitro (Pavoković et al. 2009). The two most common varieties of red beet are: (a) the Globe beetroot, having a round and smooth tuber with a dark red flesh; (b) the Egyptian beetroot, which is spherical with a reddish smooth surface. The Crapaudine beetroot is another variety of red beet with a slightly wrinkled, rough skin.

While many varieties of beetroot are available (Table 1.1), the traditional dark red globe beets are still favorites with growers. Among the dark red globe beetroots, the varieties Derwent Globe, Darkest Globe, and Detroit Dark Red are the most popular ones, while Rapid Red, Early Wonder, Early Egyptian, and Early Market are recommended for fast growth and maturation. Several globular varieties display unusual concentric circles of pink or red with white, similar to growth rings of tree trunks. Golden Beets are known for their non-bleeding qualities and the White Albina is said to be sweeter than the standard red varieties. Both are globular shaped

Table 1.1 List of some original and bred beet varieties

Genus, species and variety	History, appearance and use
Original varieties	
B.v. ssp. *vulgaris* convar. *cicla* (leaf beets)	The leaf beet group has a long history dating to the second millennium BC. These were used as medicinal plants in Ancient Greece and Medieval Europe. Their popularity declined in Europe following the introduction of spinach
B. v. ssp. *v.* convar. *cicla.* var. *cicla* (spinach beet)	This variety is widely cultivated for its leaves, which are usually cooked like spinach and popular around the world
B. v. ssp. *v.* convar. *cicla.* var. *flaviscens* (chard)	Chard, thought to have arisen from the spinach beet by mutation is also a leaf beet, having thick and fleshy midribs used as a vegetable. Some cultivars are also grown ornamentally for their colored midribs
B. v. ssp. *v.* convar. *vulgaris* var. *crassa* (Mangel wurzel)	This variety was developed in the eighteenth century for its tubers for use as a fodder crop
B. v. ssp. *v.*convar. *vulgaris* var. *altissima* (sugar beet)	The sugar beet is a major commercial crop owing to its high content of sucrose, which is processed to produce table sugar. Since the origin of this beet is from red beets, the genes for betalains remain redundant and often expressed under stress conditions and in vitro cultures
B. v. ssp. *v.*convar. *vulgaris* var. *vulgaris* (garden beet)	This is the red root vegetable that is most typically associated with the word "beet". It is especially popular in Eastern Europe where it is the main ingredient of borscht
Bred varieties	
Albina Vereduna	A white variety
Burpee's Golden	Beet with orange–red skin and yellow flesh
Chioggia, an open-pollinated variety originally grown in Italy	The concentric rings of its red and white roots are visually striking when sliced. As a heritage variety, Chioggia is largely unimproved and has relatively high concentrations of geosmin
Detroit Dark Red	Relatively low concentrations of geosmin, and is therefore a popular commercial cultivar in the United States
Lutz Greenleaf	A variety with a red root and green leaves and a reputation for maintaining its quality well in storage
India Beet	India beet is more nutritious than Western beet, and is not as sweet as Western beet
Red Ace	Principal variety of beet found in the United States, typical for its bright red root and red-veined green foliage
B. v. ssp. *v.* convar. *vulgaris* var. *altissima* (sugar beet) Many bred lines of sugar beet	The sugar beet is a major commercial crop due to its high concentrations of sucrose, which is extracted to produce table sugar. It was developed in Germany in the late eighteenth century after the roots of beets were found to contain sugar in 1747

B. v. Beta vulgaris

Fig. 1.1 Field-grown chards showing high diversity (**a**) and petioles of chards (**b**) showing the absence of pigments (***b**-1*), dominant presence of betaxanthin (***b**-2*), high betaxanthin + low betacyanin (***b**-3*), low betaxanthin + high betacyanin (***b**-4*), and dominance of betacyanin (***b**-5*)

and produce tops that can be harvested and cooked like spinach. The other cultivated varieties include the leafy vegetables chard and spinach beet, as well as the root vegetable sugar beet, which is important in the production of table sugar; the var. Mangel wurzel is a fodder crop. Leaf beets, generally known as chard, display high diversity (Fig. 1.1a), where the petioles exhibit an array of pigments from zero pigment to intense red (Fig. 1.1b).

1.1.2 Crop and Production

Beetroot is grown all over the world in temperate areas, with main production in North America, Europe and USA. In Europe, the production of beetroot is mostly concentrated in the UK and France but Italy, Netherlands, Germany, Greece, Spain and Denmark also contribute significant productions. The total production of beet in different parts of the world is presented in Table 1.2. Although the season for beet is June to October, it is available throughout the year, owing mainly to improved horticultural practices. Among largely cultivated varieties, mainly three subspecies are typically recognized, with all cultivated varieties grouped under *Beta vulgaris* subsp. *vulgaris*, while *Beta vulgaris* subsp. *maritima*, commonly known as the sea beet, is the wild ancestor of these and is found throughout the Mediterranean, the Atlantic coast of Europe, the Near East, and India. A second wild subspecies, *Beta vulgaris* subsp. *adanensis*, occurs from Greece to Syria.

Economic satisfaction is realized in red beet cultivation since the annual crop yield ranges from 50 to 70 t per hectare. This crop is entering into the economic

Table 1.2 World-wide beetroot production in 2009

Country	Production (t)
Asia (total)	31,709,126
Europe (total)	158,852,215
Africa	8,063,010
South America (total)	1,096,863
North America (total)	2,743,690
Germany	25,919,000
China	7,178,960
Japan	3,649,000
Canada	657,700
Italy	3,307,700
United States of America	26,779,200
Poland	10,849,200
Russian Federation	24,892,000
World	227,158,114

Source: FAOSTAT, Sept. 2011

zone because of the presence of a high content of natural red pigments, betalains, which display intense hue properties and have many-fold health benefits, as discussed in other chapters.

1.2 Cytogenetic Analysis

Diploid (2X) red beet has 18 chromosomes. The first tetraploids having 36 chromosomes were obtained as early as 1940 by colchicine treatment. The occasionally formed auto-tripoids (3X = 27) in Europe were found more resistant to the fungus Cercospora, and have 10% higher yields than diploids (Skaracis 1994). When excised ovules pre-treated with colchicine were cultured in vitro, polyploidization and chromosome doubling were observed. After further re-culturing, ovaries formed embryos, which subsequently regenerated, showing an average of 8.1% spontaneous chromosome doubling (Hansen et al. 1994). In *Beta vulgaris,* DNA content (C-value) has been reported to be 714–758 million base pairs per haploid genome. However, variations from the above numbers have been recorded among sub-species (Bennett and Smith 1976; Armuganathan and Earle 1991). About 60% of the beet genome is constituted of highly repetitive DNA sequences (Flavell et al. 1974). Many families of short (140–160 nucleotide) repeats with high copy numbers ($>10^5$ per genome), not considering ribosomal RNA repeats, have been recorded (Schmidt and Heslop-Harrison 1996). Several sequences, like those of transposable elements, have also been reported (Schmidt et al. 1995; Staginnus et al. 2001). Each chromosome in beet has characteristic repetitive sequences. All together, the repetitive sequence diversity is very high in the genus *Beta*, which has proven advantageous for their characterisation (Desel et al. 2002).

Molecular analyzes of chromosomes are recognized as highly effective tools for understanding genomic evolution, meiotic recombination and chromosome stability (Heslop-Harrison and Schwarzacher 1993), where fluorescence in situ hybridization (FISH) and genomic in situ hybridization (GISH) have been widely applied for identifying chromosomes, and detecting inserts of T-DNA and fragment repeat distribution. In the genus *Beta*, the FISH technique was applied to visualize mitotic metaphase chromosomes of *B. procumbens*, using one family of highly repeated DNA and two *Procumbens*-specific repeats, where probes bound mainly to positions in the centrometric regions, although not always on all chromosomes (Schmidt and Heslop-Harrison 1996). Subsequently, GISH and FISH techniques with various probes were applied to screen foreign chromosomes or chromosome fragments for selecting nematode-resistant genotypes (Schmidt et al. 1997). The foreign chromosome material could be visibly detected, and the fragments were classified into discretely distinguishable groups (Mesbah et al. 2000). High-resolution FISH techniques for application in plants, with enhanced resolution, was also obtained using pachytene chromosomes and extended DNA probes, which allowed chromosome studies at the molecular level (Fransz et al. 1996) (Zhong et al. 1996a, b; Shen et al. 1987). Using FISH, monosomic addition of *B. procumbens* genes in *B. vulgaris* (2n = 19) could also be established (Mesbah et al. 2000).

1.2.1 Genetic Markers

Similar to many other crops, the major success in improving beets was accomplished by traditional breeding methods. However, great contributions have been made utilizing genetic markers for rapid characterization of beet germplasm (McGrath et al. 2007). Genome fingerprinting was used for the characterization of individual alien chromosomes of both *B. patellaris* and *B. procumbens* (Mesbah et al. 1997). Diversity analysis using molecular markers suggested the occurrence of a great diversity among plants of *B. vulgaris*, because it comprises chards (foliage beets), red beets and sugar beets. Moreover, a greater diversity was found in *B. vulgaris, sub-species maritima*, which comprises wild beets. Ruderal beets emerge as an introgression of wild beets with cultivated varieties. When molecular and morpho-physiological characteristics of ruderal beets were compared with wild beets and cultivated sugar beets from coastal regions of Italy using amplified fragment length polymorphism (AFLP) markers, the ruderal beets were genetically distinctly different from both wild sea beets and cultivars. Such differences were manifested even at physiological levels, displaying significant differences in root structure, growth rate and branching patterns (Saccomani et al. 2009).

The alleles for betalain pigment expression were found to be *loci Y* and *R* when dominant pigment is expressed. Recessive *rr-yy* plants do not synthesize betalains and their seedlings express some amount of betaxanthins, and tubers are pigmentless, a characteristic of sugar beets (McGrath et al 2007). In an effort to screen for diversity of betalain content, recurrent half-sib family selections were practiced for seven cycles, each time selecting for high pigment and solids. Several random amplification of polymorphic DNA (RAPD) markers were found linked to genes involved in betalain synthesis (Eagen and Goldman 1996). Thus, cultivars with high pigment levels (>310 mg/100 g fresh weight) have also been developed (Gabelman et al. 2002).

1.2.2 Male Sterility Genes

Male sterility has played a major role in crop improvements. In the mitochondrial genome of *B. vulgaris*, specific markers were found to be associated with male sterility (Ivanov et al. 2004), and hence such organelle-associated genetic characteristic is often termed a cytoplasmic trait. Cytoplasmic male sterility (CMS) results from disrupted microsporogenesis, leading to abortive pollens, first described by Owen (1942) in *Beta vulgaris* L. The interaction of recessive alleles of at least two nuclear fertility restorer genes (*Fr-fr*) with the cytoplasm of a specific S type, characterized by an altered structure and expression of the mitochondrial genome, has been found mainly responsible for CMS in *B. vulgaris*. This was confirmed by partial restoration of fertility in CMS plants after introduction of the dominant *Fr* alleles. CMS has often been found accompanied by changes in a set of plasmid-like genes in mitochondrial DNA (mtDNA) (Dudareva et al. 1988), where the structure of the mitochondrial genome remains unchanged, with changes taking place in its transcription profile (Kubo et al. 1999). When several 5′-degenerate primers, designed and selected after computation, were used for typing mtDNA of S-type (typical for cytoplasmic male sterility) and N (normal)-type cultivars, a number of N- or S-specific markers corresponded to transcribed mitochondrial genes. One of these was from the *orf215* region of the N-type mtDNA. When a physical map of the corresponding region from S-type mtDNA was constructed, a substantial difference was observed for the two genome types. One N-specific marker was found to contain a truncated copy of *atp9* copy and a rearranged *rps3* region. After nucleotide sequencing, PCR primers were designed that showed that both variations occur simultaneously in the *rps3* region of the mtDNA pool (Ivanov et al. 2004).

1.2.3 Current Gene Flow Problems in Beets

Since most of the domesticated crops retain their genetic compatibility with their ancestors, they are often able to hybridize with their wild relatives. Therefore one may expect gene flow from cultivated populations into the gene pool of wild populations, as well as among different cultivated populations inhabiting the same vicinity. While the inter-specific gene flow may lead to evolutionary change of the receiving population, the small populations that receive genes may also be endangered (Levin et al. 1996), because they lose their original gene constitution. This is true in the case of *B. vulgaris*, where the wide varieties easily cross with each other (McGinnis et al. 2010), although such out-crosses segregate fast because of the high genetic load on them. When wild beet populations growing in the vicinity of cultivated populations were analyzed by measuring isozyme allele frequencies of cultivated beet, one allele specific to sugar beets and another allele with a much higher frequency in Swiss chard and red beet than in sugar beet were found (Bartsch et al. 1999). Analyzed through microsatellite markers, clear genetic cleavage between wild individuals and their weedy relatives could be depicted (Levigne et al. 2002). Also, using a chloroplast genome-based genetic marker and a set of nuclear microsatellite loci, the occurrence gene flow from beet crop to wild beets was investigated (Arnaud et al. 2003). Although the results did not reveal a large degree of pollen dispersal from weed to wild beets, several pieces of evidence clearly indicated an escape of weedy lineages from cultivated fields via seed flow. Although gene flow from a crop to a wild taxon does not necessarily result in a decrease in the genetic diversity of the native plants, such a gene flow may lead to significant evolutionary changes in the recipient populations. Such events are of concern when the cultivated crops are genetically engineered and capable of flowering profusely. Probably for these reasons, sugar beet lines genetically modified for glyphosate herbicide were allowed to be planted in 2008, but were later withdrawn due to environmental and socio–economic impacts (McGinnis et al. 2010). Thus concerns with biotechnology–derived crops remain to be sorted out, particularly when wind cross-pollinated species, such as beets, Swiss chard and spinach, require greater stringency and isolation than most other insect-cross-pollinated crops. Such isolation from genetically engineered crops is also important to organic producers and in other markets with low or no tolerance for biotechnology-derived material.

1.3 Beet Nutrients

The composition of the various nutrients of red beet presented in Table 1.3 shows its rich mineral and vitamin content, with distinct differences between foliage and tuberous root. tuberous root. An unusual compound, 5,5,6,6-Tetrahydroxy-3,3-biindolyl was identified in red beetroot peel (Kujala et al (2001a)), although no further information on its nutritional benefits are identified.

1.3.1 Beet Foliage

The foliage of red beet makes a delicious green vegetable used in a similar manner to spinach, with higher concentrations of various nutrients than in roots. Beet foliage is an excellent source of carotenoids viz., beta carotene, lutein and zeaxanthin; and flavonoids, which are strong antioxidants and dietary sources for the biosynthesis of vitamin A in mammals. Beet greens are a source of folate, a component of vitamin B, which is needed in the body to release energy and donate organic carbon for various cellular functions. Folate is instrumental in the functioning of the nervous

Table 1.3 Major nutrients found in foliage and tubers of red beet (100 g raw)

	Nutrient	Units	Greens	Tubers
1.	Water	g	91.02	87.58
2.	Energy	kcal	22	43
3.	Energy	kJ	92	180
4.	Protein	g	2.20	1.61
5.	Total lipid (fat)	g	0.13	0.17
6.	Ash	g	2.33	1.08
7.	Carbohydrate, by difference	g	4.33	9.56
8.	Fiber, total dietary	g	3.7	2.8
9.	**Sugars, total**	**g**	**0.50**	**6.76**
Minerals				
10.	Calcium, Ca	mg	117	16
11.	**Iron, Fe**	**mg**	**2.57**	**0.80**
12.	**Magnesium, Mg**	**mg**	**70**	**23**
13.	Phosphorus, P	mg	41	40
14.	Potassium, K	mg	762	325
15.	Sodium, Na	mg	226	78
16.	Zinc, Zn	mg	0.38	0.35
17.	Copper, Cu	mg	0.191	0.075
18.	Manganese, Mn	mg	0.391	0.329
19.	**Selenium, Se**	**μg**	**0.9**	**0.7**
Vitamins				
20.	Vitamin C, total ascorbic acid	mg	30.0	4.9
21.	Thiamin	mg	0.100	0.031
22.	Riboflavin	mg	0.220	0.040
23.	Niacin	mg	0.400	0.334
24.	Pantothenic acid	mg	0.250	0.155
25.	Vitamin B-6	mg	0.106	0.067
26.	**Folate, total**	**μg**	**15**	**109**
27.	Folic acid	μg	0	0
28.	Folate, food	μg	15	109
29.	**Choline, total**	**mg**	**0.4**	**6.0**
30.	**Betaine**	**μg**	**0.00**	**128.7**
31.	Retinol	μg	0	2
32.	**Carotene, beta**	μg	**3,794**	0
33.	Carotene, alpha	μg	3	20
34.	Vitamin A, IU	IU	6,326	0
35.	Lycopene	μg	0	33
36.	**Lutein + zeaxanthin**	**μg**	**1,503**	0
37.	Vitamin E (alpha-tocopherol)	mg	1.50	0
38.	**Vitamin K (phylloquinone)**	**μg**	**400.0**	**0**
Lipids				
39.	Fatty acids, total saturated	g	0.020	0.027
40.	Fatty acids, total monounsaturated	g	0.026	0.032
41.	16:1 undifferentiated	g	0.000	0.000
42.	18:1 undifferentiated	g	0.026	0.032
43.	Fatty acids, total polyunsaturated	g	0.046	0.060

(continued)

Table 1.3 (continued)

	Nutrient	Units	Greens	Tubers
44.	18:2 undifferentiated	g	0.041	0.055
45.	18:3 undifferentiated	g	0.004	0.005
46.	**Phytosterols**	**mg**	**21**	**25**
Amino acids				
47.	Tryptophan	g	0.035	0.019
48.	Threonine	g	0.065	0.047
49.	Isoleucine	g	0.046	0.048
50.	Leucine	g	0.098	0.068
51.	Lysine	g	0.064	0.058
52.	Methionine	g	0.018	0.018
53.	Cystine	g	0.021	0.019
54.	Phenylalanine	g	0.058	0.046
55.	Tyrosine	g	0.052	0.038
56.	Valine	g	0.065	0.056
57.	Arginine	g	0.063	0.042
58.	Histidine	g	0.034	0.021
59.	Alanine	g	0.081	0.060
60.	Aspartic acid	g	0.129	0.116
61.	Glutamic acid	g	0.267	0.428
62.	Glycine	g	0.081	0.031
63.	Proline	g	0.052	0.042
64.	Serine	g	0.070	0.059

Some of the rich components indicated with bold letters are deficient among populations of developing countries

Compiled based on data from USDA National Nutrient Database (Release No. 24, 2011); Accessed on 20-10-2011: http://www.nal.usda.gov/fnic/foodcomp/cgi-bin/list_nut_edit.pl

and immune systems and in the synthesis of red blood cells. Beet greens are also a good source of vitamin C, which is also a strong antioxidant, and hence an anti-ageing nutrient.

1.3.2 Beet Tuber

Raw beetroot is a good source of minerals. When cooked, based on the method, the nutrient composition can be markedly different. Boiled beetroot is particularly high in potassium, carbohydrate and protein content. Pickling boiled beetroot decreases the carbohydrate and protein content to a level below raw beetroot, with a corresponding decrease in energy value (Nottingham 2004). Beetroot is often recommended in calorie controlled diets because of its low calorific value. Apart from rich pigmentation, table beets are rich in vitamin B folate, which is essential for various cellular processes from carbon donation to normal tissue growth and cognitive functions. Dietary folate and its synthetic counterpart, folic acid, play important roles in cardiovascular diseases, cancer, in the prevention of neural tube defects in infants

(Scott et al. 2000) and exhibit antioxidant activity (Joshi et al. 2001; Asensi-Fabado and Munne-Bosch 2010). Under typical commercial processing conditions, 8% loss of vitamin C, 60% loss of color and 30% loss of dietary folate were observed. There was a significant 5% increase in the phenolic content of processed beets (Jiratanan and Liu 2004), which may be the reason for their unchanged antioxidant property even after processing.

1.4 Beet Pigments

The importance of red beet continues because of its high red pigment content, namely betalain, which display excellent hue values suitable for applications in food and pharmaceutical products. Although many plants accumulate betalains, only red beet and prickly pear (*Opuntia ficus-indica*) are approved for food and pharmaceutical applications (Jackman and Smith 1996; Mabry et al. 1963). Use of beet extract to color food is approved by the US Food and Drug Administration (FDA). Crop yield of red beet ranges from 50 to 70 t/ha, with betanin content ranging from 40 to 200 mg betanin/100 g fresh (0.4–20 mg/g of dry) beet root; a level that not has been reached with any other betalain-producing crop. Table 1.4 lists important beet cultivars and their pigment contents.

Table 1.4 Pigment content in tubers of different red beet cultivars

Cultivar	Source and accession no.	Betacyanin (mg/100 g)	Betaxanthin (mg/100 g)
Uniball	Burpee (Netherlands)	38.7	20.6
Slowbolt R-2289	Burpee (Denmark)	35.8	15.1
Red E 403	Burpee (Denmark)	34.9	19.3
Bordo 237	USSR	34.6	21.9
Detroit Nero RS	Burpee (Netherlands)	33.2	16.2
Detroit Dark Red MT	Burpee (USA)	32.0	11.5
Detroit Dark Red ST	Burpee (USA)	31.6	13.5
Podzimniaja 0474	USSR	31.3	15.0
Detriotsluis	Burpee (Netherlands)	30.7	17.3
Early Wonder	Burpee (USA)	30.4	13.2
Little Ball	Burpee (Netherlands)	30.3	11.1
Choghundur	P11631179 (India)	30.2	18.2
Iowa	IA	28.3	14.9
Gladiator	Burpee (USA)	27.9	12.0
Asmer Beethoven	Burpee (England)	27.8	10.8
Crveno	P1357355 (Yugoslavia)	27.1	15.2
Polsko	P1357351 (Yugoslavia)	26.9	16.7
Okragly Ciemnozerwony	P1285591 (Poland)	25.8	11.4
Spangsbjerg	P1269310 (Sweden)	25.2	16.7
Rubidus	Burpee (Netherlands)	23.5	9.6

Source: Sapers and Hornstein, 1979, J.F.S. 1245–1248

Table 1.5 Betaxanthin and betacyanin

Betaxanthin		Betacyanin	
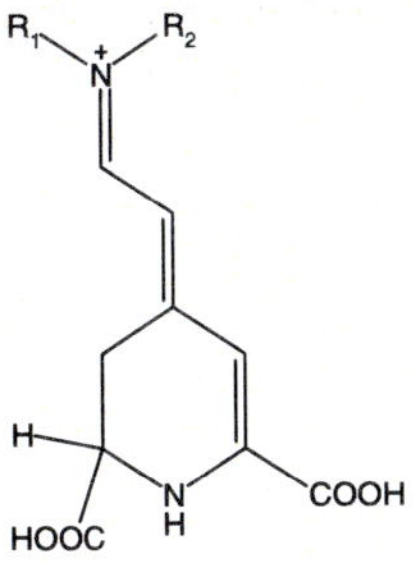		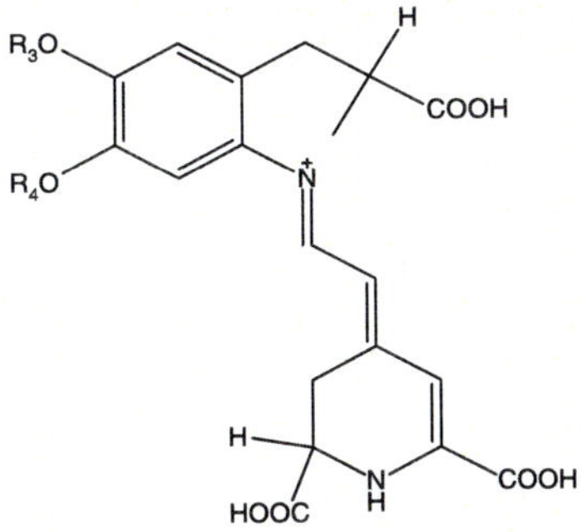	
	R1	R2	Botanical source
Vulgaxanthin-I	H	Glutamine	*Beta vulgaris*
Vulgaxanthin-II	H	Glutamic acid	*Beta vulgaris*
Indicaxanthin	Both groups together form proline		*Opuntia ficus-indica*
	R3	R4	
Betanin	β- glucose	H	*Beta vulgaris*

In addition to red beet, plants from 9 of 11 families of the Caryophyllales order contain betalains. The recent addition to the list of betalain families is Didieraceae, a small family from Madagascar. Evolutionarily, betalains are interesting because the presence of betalains or anthocyanis is mutually exclusive in Angiosperms, i.e., betalains and anthocyanins have never been reported in the same tissues of plants (Mabry and Dreiding 1968; Strack et al. 2003; Cai et al. 2005; Grotewold 2006). Betalains are also synthesized in a class of fungi: the mushrooms *Amanita*, *Hygrocybe*, and *Hygrosporu* (Delgado-Vargas et al. 2000).

1.4.1 *Characterization of Beet Pigments*

The visible red pigment of red beet has two major groups of water-soluble nitrogenous pigments: betacyanins, which display red to purple color, and the yellow water-soluble betaxanthins (Table 1.5). The ratio between the red and yellow pigments determines the hue of the pigment extract. The mechanism of *in planta* betalains production is proposed to occur as a defense response, since betalains accumulate when tissues are injured. For example, red beet leaves, which are normally not the major sites of pigment accumulation, showed betalain accumulation in wounded and infected areas, which also undergo oxidative bursts (Sepulveda-Jimenez et al. 2004). The high content of nutritious material in red beet root, along with the high sugar content (Table 1.3), probably calls for a strong defense mechanism in these tissues that is fulfilled by betalains. Unlike anthocyanins, which display a wide array of sugar moieties, betalains are condensed with either glucose (betacyanin) or one of eight amino acids in the case of betaxanthin (Table 1.5). The glucose of betacyanin (betanin) could also have other organic molecules attached at its R-marked positions, this causes color shifts. While the content of total betalains vary from 35 to 120 mg/100 g fresh weight, the content of betacyanin ranges between 0.04% and 0.21% and betaxanthin between 0.02% and 0.14%, depending on the variety (Nilsson 1970). However, under tissue culture conditions, wide variations in the ratios have been reported, sometimes with a higher content of betaxanthins in the Detroit dark red variety (Pavlov et al. 2002). The major pigment, betanin, has a tinctorial strength equal to or better than that of artificial dyes.

1.4.2 Attempts to Enhance Pigments in Beets

Since betalains are nitrogenous pigments, attempts to increase the synthesis of this pigment *in planta* comprised of the addition of either ammonium nitrate or ammonium sulfate to the soil, and/or foliar sprays of Fe and B (El-Tantawy and Eisa 2009). Despite its high betanin content, *B. vulgaris* root has several drawbacks. It has a limited pigment spectrum and adverse flavor due to geosmin (4,8a-dimethyldecalin-4a-ol) and various pyrazines. It contains a high nitrate level, which may form carcinogenic nitrosamines in the human body when ingested in large quantities. Common names of betalains are assigned in relation to plant source from where they were first isolated, and accordingly more than 50 betalain structures have been elucidated from various plants and fungi (Delgado-Vargas et al. 2000; Francis 1986). The chief pigment molecules of red beet are betacyanin or betanin and the betaxanthins, vulgaxanthin I and II (Table 1.5). The presence of other minor pigments and intermediary compounds presented in Tables 1.4 and 1.6 vary depending on the variety as well as the geographical status.

1.4.3 Betalains in Different Beet Varieties and Cultivars

The color of beetroot differs depending on the cultivars, varieties, growing conditions, age and size. The pigment content is influenced by the time of harvest; late harvest results in better color (Nilsson 1973). Pigment composition in five different varieties of red beet was studied by (Gasztonyi et al. 2001). A range of red–violet pigments betanin, isobetanin, betanidin and isobetanidin was observed, of which the chief pigment betanin ranged from 0.4–0.50 g/kg and yellow pigment vulgaxanthin I content ranged from 0.32 to 0.42 g/kg (Table 1.6). There was not much variation among the varieties.

Nilsson (1970) in Sweden studied the pattern of pigment accumulation in red beet from sowing to harvesting time, also recording the size of beets in three varieties viz., Banco, Egyptische-Platronde and Rubia. It was observed that pigment content was influenced both by sowing time and harvesting time. In all the varieties the amount of betacyanin and betaxanthin in June was lower than in September/October.

Table 1.6 Pigment content (g/kg) in Dutch (D) and Hungarian (H) varieties of red beet

	Betacyanins				Betaxanthins	
Variety	Betanin	Isobetanin	Betanidin	Isobetanidin	Vulgaxanthin 1	Vulgaxanthin II
Bonel (D)	0.50	0.27	0.04	0.01	0.42	0.06
Nero (D)	0.41	0.13	0.03	0.01	0.32	0.03
Favorit (H)	0.49	0.24	0.05	0.02	0.41	0.02
Rubin (H)	0.46	0.25	0.07	0.03	0.37	0.03
Detroit (H)	0.44	0.21	0.05	0.01	0.37	0.04

Gasztonyi et al. 2001

Table 1.7 Pigment content (mg/g dry weight) in peel and flesh of tubers of red beet cultivars

Cultivars		Vulgaxanthins I and II	Betanin	Isobetanin
Egyptische Platronde	Peel	1.4±0.3	3.8±0.4	1.2±0.2
	Flesh	1.5±0.2	2.9±0.2	0.03±0.02
Forono	Peel	1.8±0.1	7.6±0.1	3.1±0.1
	Flesh	4.0±0.2	5.2±0.2	0.4±0.03
Little Ball	Peel	4.3±0.4	7.5±0.5	2.1±0.2
	Flesh	1.9±0.1	3.6±0.2	0.02±0.01
Rubia	Peel	2.2±0.2	5.4±0.1	2.2±0.1
	Flesh	2.3±0.2	4.1±0.2	0.3±0.03

Kujala et al. 2002

Betacyanin content increased and reached a maximum in August and betaxanthin content continued to increase during the autumn, resulting in higher concentrations of yellow pigment in later harvests. The pigment content decreased with increasing beet diameter and the ratio between red and yellow content was affected by the beet diameter. Small beets had a higher pigment content than large beets. The pigment content in different cultivars of beet reported by different authors is presented in Tables 1.6 and 1.7.

1.4.4 Biotechnology for the Production of Betalains

Because the red beet tubers are loaded with pigments, this material has attracted the interest of many in the early years of plant tissue culture research for the production of secondary metabolites, since the presence of pigments is visible (Akita et al. 2000, 2001, 2002) and can be easily quantified by spectrophotometry. With the development of biotechnology, genetically transformed hairy root cultures find great use for the commercial production of red beet pigments (Pavlov et al. 2002; Georgiev et al. 2010a; Thimmaraju et al. 2003a, b, 2004, 2006; Pavoković et al. 2009). In addition to enhancing yields (e.g., up to 250% increase in betanin content even from sugar beets) (Pavoković et al. 2009), these technologies appear to provide better control over the quality of pigment, since the production is independent of environmental factors, which normally cause great variations in field-grown beets. However, there are reports of significant differences in the chemical composition of

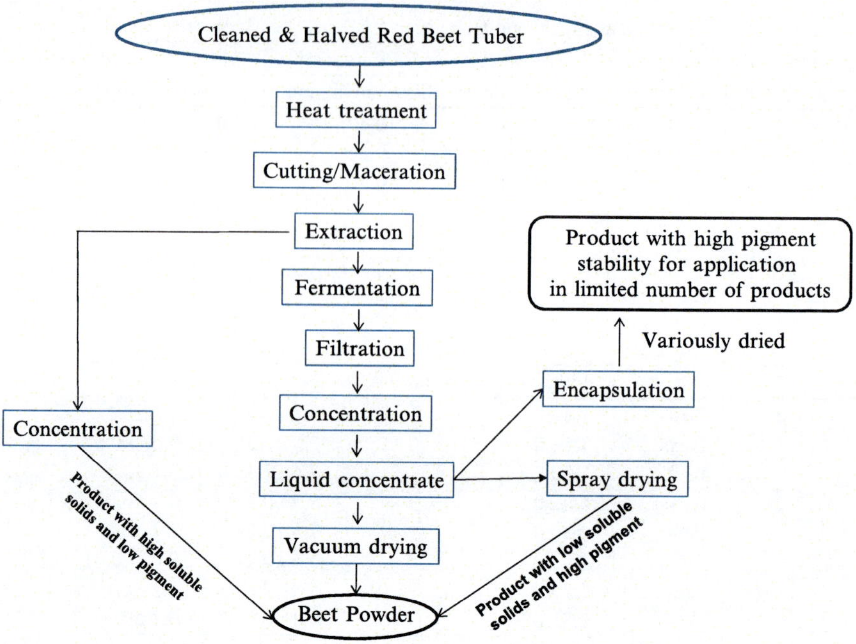

Fig. 1.2 Steps involved in the pigment extraction from red beetroots

the two sources. For example, rutin was present only in the extracts from hairy root cultures, whereas chlorogenic acid was found only in the extracts of field-grown plants (Georgiev et al. 2008). Also, the observed higher antioxidant activity of the hairy root extract compared with that of the field-grown beetroot extract (Georgiev et al. 2010b) was attributable to its increased concentrations (≈20-fold) of total phenol compounds, which may have a synergistic effect with the beetroot betalain pigments.

1.4.5 Extraction of Red Beetroot Pigments

Invariably, the roots of red beet have been utilized in all studies for pigment extraction, although there are rare occasions where beet leaves, stems and seeds were also researched (Gennari et al. 2011; Lee et al. 2009; Ninfali et al. 2007; Pyo et al. 2004; Križnik and Pavoković 2010). The steps involved in pigment extraction from red beet roots (Fig. 1.2) starts with chopping beet tubers into small pieces or quickly grinding them. Further steps are added to achieve the maximum yield of the betalain pigments, while keeping the losses to the minimum; the aim is to obtain a stable pigment extract with a long shelf life. In general, pigments are extracted with water, although methanol or ethanol solutions (20–50%) may be required for complete

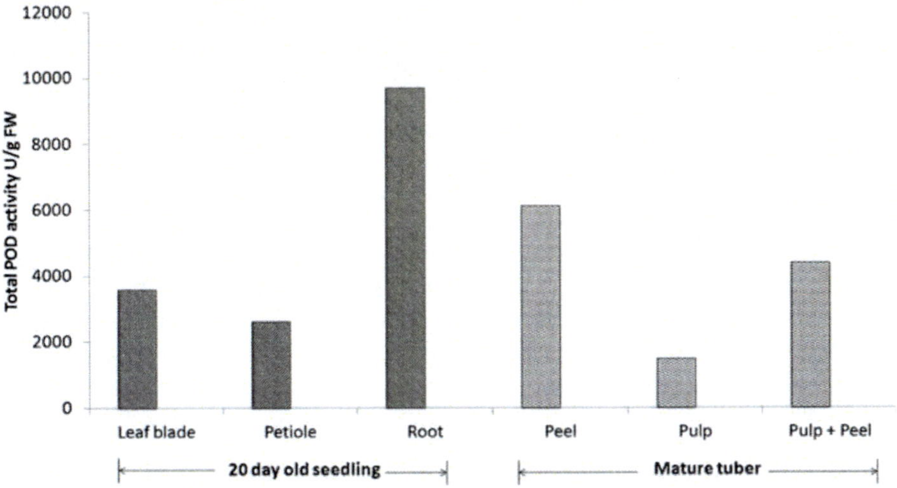

Fig. 1.3 Peroxidase activity in red beet (Figure plotted on the basis of data from Thimmaraju et al. 2005)

extraction (Delgado-Vargas et al. 2000). Aqueous extraction increases pigment stability and the pigment may be further stabilized by slight acidification of the extraction medium with ascorbic acid, which renders color constituents more stable and resistant to oxidation, both chemical and by endogenous polyphenol oxidases (PPOs) (Escribano et al. 2002; Strack et al. 2003). In the case of hairy root cultures of yellow beetroots, where tyrosinase activity was found to be high, addition of ascorbic acid was essential to avoid losses of betaxanthin and miraxanthin V, as well as to avoid the appearance of artefacts formed by degradation products (Strack et al. 2003).

Red beets have several endogenous enzymes such as β-glucosidases, PPOs and peroxidases, which if not properly inactivated by blanching may account for betalain degradation and color losses (Escribano et al. 2002; Lee and Smith 1979). The optimum pH for enzymatic degradation of both betacyanins and betaxanthins is reported as approximately 3.4 (Shih and Wiley 1981). Betacyanins are more susceptible than betaxanthins to degradation by peroxidases, whereas betaxanthins are more susceptible to oxidation by hydrogen peroxide (Wasserman et al. 1984). Generally, unpeeled whole beets are processed and more than 30% color is lost by removal of the peels. This is noteworthy because the greatest PPO activity, which is deleterious to both betacyanins and betaxanthins, is located in the peel, as is the case with peroxidases (Fig. 1.3). The presence of these enzymes at the peel part of beet tuber indicates their apparent participation in defence functions either by scavenging the peroxides or by oxidizing other molecules formed at the surface. Blanching before extraction inactivates the unfavorable enzymatic action. In theory, the oxidizing and hydrolysing activities of PPO action require monophenolic or diphenolic structures that are rarely found in betaxanthins and betacyanins and are formed only after prior hydrolysis by glucosidase activity. Hence for enzymatic betalain degradation, a concerted action of glucoside-cleaving enzymes, PPOs and peroxidases is required. Martinez-Parra and Munoz (2001) confirmed that horseradish peroxidase (HRP) catalyzes the oxidation of betanin. Moreover, characteristics of red beet peroxidases were mostly similar to that of HRP (Thimmaraju et al. 2005, 2007) in cultured hairy roots where PPOs were negligible. Since not much pigment degradation was observed in beet extracts from hairy roots, despite a high peroxidase activity, the main pigment degrading enzyme could thus be PPO.

Factors such as low water content (achieved by spray drying), lower pH (4.0–5.5), presence of antioxidants (ascorbic and iso-ascorbic acids), and absence of metal cations (Fe, Cu, Sn and Al), heat, oxygen and light increase the stability (shelf life) of the extracted beetroot pigments (von Elbe et al. 1974; Azeredo et al. 2007; Azeredo 2009; Haber et al. 1979; Havlikova et al. 1983; Pedreno and Escribano 2001; Vitti et al. 2005). Reports on the stability evaluation of red beetroot color in various pharmaceutical matrices (Pai and D'Mello 2004) and soft drinks (Havlikova et al. 1985) are available. While temperature during processing and storage is the most important factor governing the stability of the beetroot pigments (Kujala et al. 2000; Patkai et al. 1997), extraction methods employing low electrical fields appear to be superior to those utilizing cryogenic freezing (Zvitov et al. 2003). Several studies comparing various extraction methods for beetroot colorants are available (Gasztonyi et al. 2001; Kujala et al. 2001b; Vitti et al. 2005).

Since betalains are sensitive to oxidation and temperature, application of the newer technologies, such as pulsed electric field treatment (Chalermchat et al. 2004; Fincan et al. 2004; Kannan 2011; Lopez et al. 2009), gamma-irradiation (Nayak et al. 2006) and ultrasound (Sivakumar et al. 2009), in conjunction with traditional extraction methods, such as solid–liquid extraction, diffusion extraction, reverse osmosis and ultrafiltration, appear promising. Since most pigments are stored within the vacuoles of their cells and not normally excreted by cells (Cormier 1997), newer technologies are helpful for obtaining higher pigment yields. Accordingly, for small-scale extraction, methods such as diffusion, reverse osmosis and solid–liquid ultrafiltration have been applied (Wiley and Lee 1978; Wiley et al. 1979; Lee et al. 1982). Fermentation is applied to remove the carbohydrates and nitrogenous compounds in beet juice that contribute to 80% of the juice, which in turn enhances the betanin concentration. In general, small beets are favored because they accumulate higher betalain concentrations. Tissue comminution (gradual size reduction) is usually performed by milling, followed by acidification of the resulting mash through the addition of citric acid to reach pH 4. Lowering the pH will preclude the action of PPOs, while peroxidases may still be active until the filtered juice is heated to a temperature exceeding 75°C. Through acidification and a lower thermal load (such as pasteurization instead of sterilization) suffices to secure microbial stability without seriously affecting pigments. In addition to chamber filtration, ultrafiltration has been successfully applied. More recent approaches using pulsed electric fields for pigment extraction have not yet entered industrial practice. High pigment–low solid beets were suggested for food coloring purposes. Beet juice is traditionally obtained by hydraulic pressing, where betalain recovery is less than 50%; however, this process has been improved by using macerating enzymes. Several problems are associated with red beet extracts such as great color variability and their beet-like or earthy odor. Betalains have high molar absorptive and thus small quantities of pure pigments (<50 ppm calculated as betanin) are required to reach the desired hue for most of the applications. The beet extract contains large amounts of sugars, which results in low tinctorial power. Therefore, fermenting red beet juice with yeast species has been considered; this is discussed in a subsequent section.

While many pigment separation techniques use mechanical separations, an electric field can also be effectively used for separation of pigment molecules from total extract, the pulsed electric field (PEF) method. In a study by Loginova et al. (2011), PEF treatment for red beet color extract was carried out using the trains of monopolar rectangular 100-μs pulses with electric field strength $E = 375–1,500\ \mathrm{V\,cm^{-1}}$ in a time scale of 0–0.2 s, where the temperature ranged from 30°C to 80°C. PEF treatment was found to accelerate the extraction of betalains and reduce the time of extraction. Electroporation was shown to be responsible for intensification of the release of colorants through aqueous extraction. Temperature increase resulted in a significant acceleration of colorant degradation, with minor degradation at treatment for 5 h at 30°C and complete degradation at 80°C for 1 h. PEF treatment removed barriers and the extraction process was controlled by unrestricted diffusion with small activation energy. At moderate temperatures, 30–50°C, pigment degradation was lower.

1.4.6 Pigment Concentration by Drying

Convective hot air drying was optimized with the objective of maximum color retention of red beet. The process was mathematically modelled as a function of hot air temperature, batch time and moisture (Gokhale and Lele 2011). A new semi-theoretical model was tested with experimental data (50–120°C) and was found to be better than nine other reported models. The estimated effective moisture diffusivity was 3.01×10^{-9} to $7.21 \times 10^{-7}\ \mathrm{m^2/s}$ and it obeyed Arrhenius's equation. Color and rehydration ratio were used to assess the quality of beet powder. An unusual trend of color minima was also observed that was attributed to the physical phenomena of surface moisture. The final color of the beet was temperature dependent and the maximum color retention was achieved at the lowest drying temperature. The best drying condition required sequential reductions in temperature (120–50°C) resulting in good color retention. This reduced the batch time to 4 h compared with the batch time of 6 h for conventional isothermal drying at 50°C (Gokhale and Lele 2011).

A large-scale chromatographic purification was applied to obtain a betalain-rich (41%) red beet extract with significantly reduced amounts of sugars, which was achieved without using organic solvents, through a novel proprietary process by FutureCeuticals, Inc., USA (Pietrzkowski and Thresher 2008). The quality of this dry material was analyzed and found to contain mainly betanin/isobetanin and elevated levels of neobetanin pigments. In addition there were also a large number of other decarboxylated and dehydrogenated betanin derivatives in the new product (Nemzer et al. 2011).

Various commercial sources (marketed under the trade name "Betanin" and Red Food Color E162) are currently available that are produced by mixing powdered red beetroot extracts with dextrins and are stabilized with ascorbic acid (antioxidant) and, to enhance water solubility, an acidic pH of 5.4 rendered with citric acid. Such commercial extracts have been shown to possess strong bio-efficacies (Kapadia et al. 1996, 2003, 2011; Lechner et al. 2010).

1.4.7 Encapsulation of Beet Juice

The process of spray drying of betalain from red beet roots, using different carriers: maltodextrin, gum acacia and soluble starch, and the effect of these carriers, in varying percentages, with freshly extracted beet root juice was studied during spray drying, where the chamber temperature ranged from 150°C to 165°C. (Koul et al. 2002). It was observed that the percent yield of betalain was indirectly proportional to the quantum of carriers added to the extract. The shelf life studies of the spray-dried betalain dry powder over 180 days showed that the spray-dried product was stable at temperatures between –4°C and 20°C.

1.4.8 Irradiation of Fresh Cut Beet

The changes in color and the mechanical behavior of fresh-cut red beetroot were analyzed after treating with low doses (1 and 2 kGy) of gamma radiation. Irradiation seemed to contribute to higher cell–cell adhesion through increasing of calcium-cross linking at the middle lamellae regions, in addition to an increment of cross-links of polymers into the cell wall. Peroxidase (POX), polyphenol oxidase (PPO) and phenylalanine ammonia-lyase (PAL) activities, as well as the levels of compounds related to the response to the oxidative stress of plant metabolism were recorded with the purpose of understanding the influence of the processing on tissue characteristics (Latorre et al. 2010). Chemical modifications produced in the cell walls, as a response to higher levels of H_2O_2 and subsequent peroxidase-mediated effects, were visualized structurally as a more elastic behavior of irradiated tissues and rigidification of cell walls of treated roots, although puncture tests did not reveal significant differences. These observations indicated that irradiation doses of 1 or 2 kGy produced biochemical changes in cellular contents as well as in the cell wall constitutive networks, which might not be sensed by consumers, since the objective evaluation through a puncture test did not reveal significant differences. At the same time, the mentioned changes involved an increase in the antioxidant capacity of red beet root tissue, showing that the studied doses could be interesting for use in the frame of a combined technique for red beet processing. Betacyanin and betaxanthin contents decreased by 11 and 19% compared with control after γ-irradiation at 2 kGy, whereas pigment concentration was not affected at 1 kGy irradiation.

1.5 Stability of Betalains

Several aspects of betalain stability have been discussed in the context of pigment extraction. Steps involved in progressive losses of betalain and other nutraceutical biomolecules during processing of red beet, according to Patkai et al. (1997), are as follows: betanin content: (1) raw material, 100%; (2) blanching, 99.8%; (3) peeling, 99.4%; (4) crushing, homogenization, 91.6%; (5) pasteurization, 50.1%; (6) storage, 60 days at 5°C, 46.9% and (7) storage, 60 days at 20°C, 31.3%; and vulgaxanthin content: (1) 100%, (2) 84.1%, (3) 82.9%, (4) 67.5%, (5) 42.7%, (6) 36.4% and (7) 45.2%. Chapter 3 discusses the degradative losses under various processing and storage conditions.

1.6 Applications of Beet Pigments

1.6.1 Food Applications

The intense hue imparted by red beet extract has attracted its usage ever since the crop was domesticated and bred. Betalains have been used as food colorants since the beginning of twentieth century. Early food applications involved the use of pokeberry juice and in red wine to improve redness. The commercial availability of betalains around the world is currently restricted by legislation to juice or powders obtained from aqueous extracts of red beet. Beet color, in the form of beet juice concentrate or beet powder, is approved as a food additive and for use in drugs and cosmetic products in Europe and North America. Red beet is the only allowed source of betalain approved for foods in the United States (Code of Federal Regulations, 21 CFR 73, 40). Betalains have a wide range of desirable biological activities, including antioxidant, anti-inflammatory, hepato-protective and anti-cancer properties. Beet pigments are commercialized as juice concentrates (vacuum drying up to 60–65% total solids), and the FDA classifies them as vegetable juices, which are commonly spray dried with maltodextrin to obtain beet powder. The product contains 0.3–1% betalains, 75–80% sugars and 10% protein. Beet juice is traditionally obtained by hydraulic pressing, where betalain recovery is less than 50%; however, this process has been improved by using macerating enzymes. The current market for all food colorants is estimated at US $1 billion, with natural pigments comprising one-fourth of the total market. However, the market for synthetic colorants has tended to decline in favor of natural colorants (Fletcher 2006).

Typical food commodities colored with betalains include dairy products, fruit fillings for bakery products, relishes, various instant products, confectionery, meat substitutes and sausages. Beetroot color is used also by the pharmaceutical industry as a coloring agent in drug formulations, in both solid and liquid forms (Pai and D'Mello 2004).

Betalains require low thermal treatment, packaging under low water activity and oxygen free, and storage in the dark (Rayner 1993). When the effect of temperature treatment at 85°C for 8 h on the pattern of red color was evaluated in fresh juice, the initial hue angle of 358°, indicative of the typical red–purple appearance, shifted unexpectedly to 62°, resulting in a yellow–orange solution (Herbach et al. 2004). Such a change was found to occur because of the formation of yellow neobetanin and orange–red betanin degradation products that were identified by high-performance liquid chromatography (HPLC) combined with mass spectrometry (MS). Neobetanin formation resulting from heat exposure was proven for the first time by a combination of HPLC and MS techniques. These compounds were found to be responsible for the shift in color of the red beet juice to orange color during thermal treatment (Herbach et al. 2004). Betalains show high stability at low water activities since the reaction for their degradation involves water. Upon drying, betalains exhibit better stability, for which spray drying was found to be feasible. The various betalain degradation factors are extensively studied; water activity, temperature and oxygen play important roles (von Elbe 1975; von Elbe et al. 1974; von Elbe and Attoe 1985). Further details on betalain stability are available in Chap. 3.

The problems associated with beet extracts are beet odour, large color variability and low color yield. Betalains are less commonly used than anthocyanins and carotenoids, although these water-soluble pigments, stable between pH 3 and 7, are well suited for coloring low-acid food. Beet concentrates are used in dosage levels of 0.1–1.0% to obtain strawberry shades. Beet concentrates have been particularly suitable for use in foods like yoghurts, drinks, ice cream, hard candies, salad dressings, cake mixes, finished meat products and meat substitutes, gelatin desserts, powdered drink mixes, etc. Their use is limited because of the stability properties. The main problems associated with red beet extracts are their wide variations and earthy odour. Betacyanin, when applied to food, vary in color from purplish red to orange red, depending on the genotype of the plants from which the color is extracted (Cai et al. 1998). The betaxanthins, which impart a yellowish red hue to food products, are also susceptible to heat, and found to be as unstable as red betacyanins at high temperatures (>40°C), but are stable at 40°C under air- and light-excluded conditions. Dried betaxanthins showed a much better storage stability (95.6% pigment retention) than corresponding aqueous solutions (12.5%) at 22°C after 20 weeks of storage. Refrigeration (4°C) significantly increased pigment retention of aqueous betaxanthins to 76.2% after 20 weeks of storage. Purple pitaya, *Hylocereus polyrhizus*, betacyanins in a juice matrix with addition of selected additives like ascorbic acid, isoascorbic acid and citric acid showed an improvement in structural and chromatic stability (Herbach et al. 2006). The presence of matrix compounds efficiently improved the betacyanin stability effect of organic acid supplements, the best results were obtained in purple pitaya juice heated at pH 4 with addition of 1.0% ascorbic acid. The pH conditions clearly influenced the reaction paths of betacyanin degradation; pH 6 led to enhanced hydrolytic cleavage of the aldimine bond, whereas, at pH 4, other degradation mechanisms, i.e., decarboxylation and dehydrogenation, were favored. The enormous loss of betacyanin occurring at pH 6, however, regenerated in juice matrix, with lesser degree of regeneration in pigment fortified solution without juice matrix. A 24-h cool storage period following thermal treatment for maximum pigment regeneration, especially after thermal treatment at elevated pH, is crucial. In short, the heat stability of betacyanins in purple pitaya juice stabilized with ascorbic acid is suitable for food coloring. The stability of purple pitaya juice with added ascorbic acid during storage is currently being investigated (Herbach et al. 2006).

The suitability of both anthocyanins and betalains in processed cheese evaluated at $6 \pm 1\,^{\circ}C$ for 40 days in light-impermeable packaging showed that the half-life time and percentage color retention values obtained for anthocyanin and betalain extracts added to the cheeses were acceptable and adequate for the shelf life of the particular product (Prudencio et al. 2008). When the effect of beet and honey on quality attributes and carotene retention of carrot-fortified milk product during storage at 30°C was evaluated (Bandyopadhyay et al. 2008) on the basis of changes in pH, free fatty acids and sensorial properties, it was found that the addition of beet juice and honey to the product were synergistic, i.e., they reduced the acidity and free fatty acid formation and carotene degradation. Sensorial properties were enhanced because of the addition of honey, improving the overall product quality. This study showed that the addition of an equal mixture of beet and honey showed positive effects in quality improvement and carotene retention of carrot-fortified milk product.

1.6.2 Bioavailability of Betalains

A very few reports are available on the bioavailability and metabolism of betalains. The experiments have been carried out with oral administration of red beet juice to human volunteers. The intake of red beet juice or cactus pear revealed that betalains are bioavailable (Sembries et al. 2006; Netzel et al. 2005, Kanner et al. 2001). It is reported that from gut they are absorbed in their intact form into the systemic circulation, and found excreted through urine, which indicated that hydrolysis is not a condition required for absorption. The maximum concentration in plasma was observed after 3 h of the intake at a concentration of 0.2 μM and 6.9 μM of betanin and indicaxanthin, respectively (Tesoriere et al. 2004a, b). The bioavailability ranged from 0.28% to 0.9% for betanin and isobetanin after red beet juice administration (Netzel et al. 2005). When some volunteers consumed 300 mL of red beet juice (~120 mg of antioxidants), betacyanins were identified in urine after 2–4 h, where 0.5–0.9% of the ingested pigment was traced (Kanner et al. 2001). Betalains are cationized compounds, having a positive nitrogen moiety, the polyene, which may help in absorption in a similar way to anthocyanins and flavonoids with a cationic nature. In one study, it was found that betalains are able to pass through red blood cell membranes. The concentration inside the cells was 30 μM betanin and 1 μM indicaxanthin (Tesoriere et al. 2005).

A much deeper study than the above for evaluating the bioavailability of betalains was conducted using both cactus pear and red beet. In a study simulating oral, gastric and small intestinal digestion, the digestive stability of purified pigments was assessed (Tesoriere et al. 2008). There was a minor loss of indicaxanthin only in the gastric-like environment, whereas a decrease of vulgaxanthin I was evident in all digestion steps; no effects due to the food matrix were observed. Contrarily, the food matrix helped prevent the decay of betanin and isobetanin in a simulated gastric environment. Betacyanin from both purified and food samples was traceable in the small intestine phase of digestion simulation. After digestion, betalamic acid accumulated in the case of purified pigments, whereas such accumulation was absent in case of food-borne betalains. In the postintestinal (PI) digesta, betaxanthins were fully soluble in the aqueous (bioaccessible) fraction after ultracentrifugation. However, there was an incomplete release of betacyanins from the food matrix. In case of PI digesta, a dose-dependent inhibition of oxidation of methyl linoleate in methanol was observed, although such effects could not be correlated with the level of betalain. Therefore, the stability of dietary betalains in the digestive tract is directly related with its bioaccessibility, although the contribution by additional factors such as the food matrix and style of processing may also affect betalain bioaccessibility.

1.6.3 Fermentation of Red Beet Juice

Since red beet juice is rich in sugars, the presence of glucose moieties in betacyanin molecule reduces its hue strength. Removal of glucose by fermentation has been suggested for intensification of color, although such a molecule (the betanidin) may not be stable unless encapsulated or converted into other forms. Accordingly, the addition of brewer's yeast, *Lactobacillus* (*L*), (Rakin et al. 2007) and sugar fermenting *Saccharomyces* bacteria (Drdák et al. 1992) to beetroot juice was found to be helpful in enhancing pigment quality, probably because of lactic acid, which is also expected to improve the nutritive quality of the product.

In a fermentation process, the rate of pH decrease is very important, because the resultant low pH minimizes the influence of spoilage bacteria, particularly at the beginning of the fermentation, when the substrate is rich in sugars. An evaluation of lactic acid fermentation of red beet juice inoculated with *Bifidobacterium* BB12 and supplemented of commercial prebiotics containing inulin was carried out (Buruleanu et al. 2009). The acidification time required for the lactic acid fermentation of red beet juices was about 4 h, which was undesirable from the viewpoint of juice preservation, because the desirable period for this process is only 2h. Although the initial titrable acidity was the same in beet juice with and without prebiotic, at the end of the fermentation period of 48 h, it was found that supplementation with 6% inulin exerted a favorable effect on fermentation dynamics. In general, the red beet juice presented a substrate-inhibitory property, inhibiting bifidobacteria development. The inulin addition, although it did not affect pH, was used for bifidobacteria growth for progressive continuation of the fermentation process.

The conditions for lactic acid fermentation based on a mixture of beetroot juice (*Beta vulgaris* L.) and carrot juice (*Daucus carota* L.) and a different content of brewer's yeast autolysate with *Lactobacillus plantarum* A112 and with *Lactobacillus acidophilus* NCDO 1748 has been studied (Rakin et al. 2004). Both cultures showed good biochemical activity in these mixtures. The production of lactic acid has been stimulated using a higher content of brewer's yeast autolysate. In these mixtures, *L. plantarum* A112 showed better growth and lactic acid production than *L. acidophilus* NCDO 1748.

1.6.4 Snack Food Products

Dry cubes: Due to health concerns, there is a shift towards fat-free products with natural components. Many fruits and vegetable cubes, owing to their proven health benefits, have been targeted to fill this slot. Such products are formed by cutting the selected fruit and vegetable cubes/slices into any desirable shape and infiltrating them with low-calorie sweeteners, such as xylitol, sorbitol, fructo-/xylo-oligosaccharides (FOS/XOS) and other non-calorific edible fibers such as pectins, xanthan gums and other polysaccharides that are known to offer an array of health benefits. The bright red color, chewy texture and good amount of sugars, dietary fiber and nitrogenous pigments in red beet makes it an ideal material for developing healthy snacks. While vacuum infiltration of soluble carbohydrates is done by conventional methods, the most common method for removal of moisture is by convective drying, using hot air (Lewicki 2006), including beetroots (Kaminski et al. 2004; Shynkaryk et al. 2008). Such methods have several limitations as well as disadvantages, such as long processing time, high cost incurred due to processing at temperatures, degradation of texture and color, alteration in taste and degradative loss of important nutritional substances (Marfil et al. 2008). Therefore, many novel techniques have been developed, although they come with certain technical inconveniences. A combination of vacuuming with microwave drying produced red beet cubes with the least volume loss and the highest retention of quality traits such as color, texture, flavor and antioxidant properties (Figiel 2010).

1.6.5 Betalains for Photocells

Dye-sensitized solar cells (DSSCs), a new type of solar cell developed in 1991, have gained importance due to their environmental friendliness and low cost of production. A DSSC is composed of a nano-crystalline porous semiconductor electrode-absorbed dye, a counter electrode, and an electrolyte containing iodide and tri-iodide ions. In DSSCs, the dye as a sensitizer plays a key role in absorbing sunlight and transforming solar energy into electric energy. Numerous metal complexes and organic dyes have been synthesized and utilized as sensitizers. Since betalains have a high light absorption capacity in the visible wavelength range, their application for dye-sensitized solar cells was evaluated (Zhang et al. 2008).

A photo-anode was fabricated from nanocrystalline TiO_2 on a transparent conductive glass pre-treated with ethanolic HCl solution and then sensitized with raw beet extract or only betanin (betaxanthin removed). The betanin-sensitized solar cell gave a maximum photocurrent of 2.42 mA/cm^2 and open-circuit photovoltage of 0.44 V in the presence of methoxypropionitrile containing I–/I3– redox mediator. Such photocurrents and photovoltages were also observed when yellow betaxanthin and a brown betalain-derived oxidation product were used in DSSC. Photon-to-electron conversion efficiencies (IPCEs) showed maximum values of 14% and 8% for betaxanthin- and betanin-based solar cells, respectively. In this study, betaxanthin showed a higher maximum IPCE than that of a DSSC sensitized with purple betanin. Although the light-harvesting properties of the yellow pigments are less favorable for solar energy conversion compared with purple betanin, the higher photocurrents obtained with betaxanthin result in higher energy conversion efficiency. However in a recent study, removal of betaxanthin from aqueous beet pigment extract and the use of betacyanin markedly improved the efficacy of DSSC (Dumbravă et al. 2012). For both betaxanthin- and betanin-based DSSCs, less than 100% IPCE values at wavelengths well within the absorption band of the adsorbed dye reflect the undesirable consequences of recombination and possibly also the less than maximally efficient quantum yield for electron injection. The presence of carboxylic acid functions in the betalains presents an advantage for anchoring the dye to the TiO_2 surface by manipulating the surface acid–base chemistry.

1.7 Health Benefits

Among the components of red beet, the unique health benefits are linked to its nitrate content, including the nitrogen and amino acid of betalain pigments. Dietary nitrate, such as that found in the beetroot, is thought to modulate nitric oxide, the biological messenger, which is taken up by endothelial cells to trigger signals for relaxation of smooth muscles. The resulting vasodilation process increases the flow of blood in arteries (Webb et al. 2008). In another study, drinking 500 ml of beetroot juice led to a reduction in blood pressure within 1 h, which was more pronounced after 3–4 h, with measurable effect up to 24 h after drinking the juice. Such effects correlated with high nitrate concentrations in the blood and the drop in blood pressure (Wink and Paolocci 2008). A recent clinical study has shown that beetroot juice helps in quickly and significantly lowering blood pressure, particularly in hypertensive patients; the higher the blood pressure, the greater was the impact (Ahluwalia 2011). A positive effect of beetroot juice on human exercise and performance was demonstrated by a study conducted by Exeter University, UK, in which it was found that after drinking a half liter of beetroot juice several hours before setting off, cyclists were able to ride up to 20% longer than those who drank a placebo black-currant juice (Lansley et al. 2011). These beneficial effects of red beets highlight the potential of a "natural" low-cost approach for the treatment of cardiovascular disease as well as for improving athletic performances. Beet juice is also known to stimulate the function of liver cells and protect the liver and bile ducts. Since beet juice is slightly alkaline, it is suggested for the treatment of acidosis. Drinking beet juice regularly can help relieve constipation, because of its high pectin content. Some people have also experienced that the mixing of red beet juice with carrot juice excellently healed gout, kidney and gall bladder problems, although such experiences need a great depth of research for establishing such facts.

1.8 Carbohydrate Polymers from Beet and Their Applications

Beet pulp is a by-product, available after pigment extraction from red beet or after sugar extraction from sugar beet. Although slight variations may be expected among different tuberous varieties due to differences in their genetic compositions and their cultivation conditions, essentially they all contain mainly cell wall polysaccharides such as arabinans and arabinoxylans hemicelluloses (Sun and Hughes 1998), highly methylated and acetylated pectins (Oosterveld et al. 2000) and cellulose microfibrils (Dinand et al. 1996). These components find enormous applications in food products because of increased awareness of the health benefits of dietary fibers. The anti-cancer effects of red beet fibers are presented in Chap. 7. Therefore, while processing beet root fibers, research was focused first on adding value to isolated components of the pulp such as alimentary fibers (Michel et al. 1988), cellulose microfibrils (Dinand et al. 1999), pectins (Oosterveld et al. 2001; Turquois et al. 1999) or ferulic acid (Micard et al. 1994). Raw pulp was proposed as a valuable substrate for the production of xanthan gums for microbial cultivation (Yoo and Harcum 1999), for applications in complexation of divalent cations (Dronnet et al. 1998), to use as source of polyols in the process of urethanes and polyurethanes production (Pavier and Gandini 2000), to enrich fiber in biodegradable composites (Rouilly and Rigal 2002; Liu et al. 2007; Leitner et al. 2007; Mohamed et al. 2008; Baar et al. 1997) or for use in paper manufacture (Wong and Bregola 1997; Fiserova et al. 2007). Development of beet pulp-based materials has been considered through the breakdown of the cell structure by chemical or enzymatic hydrolysis (Berghofer et al. 1992) or by thermo–mechanical processing. Mechanical twin-screw extrusion processing was found to make the material thermoplastic composed of cellulose micro-fibrils embedded in a matrix constituted of hemicelluloses and pectins that was amenable for further molding by injection (Rouilly et al. 2006a, b). In view of the large demand for biofilms, scientific work has focused on composites made of mixtures of natural and synthetic biodegradable polymers (Avérous 2004), rather than pure biological materials due to their poor malleable, tensile and hydrophobic properties. Rouilly et al. (2009) screened a large number of plasticizers to develop composite material from processed beet pulp. Some of the common plasticizers like urea and xylitol were found to provide higher tensile strength than widely used glycerol; the use of xylitol further narrowed down to obtain extruded pulp biofilms with tensile strength and improved water resistance using glycidyl methacrylate as cross linker (Rouilly et al. 2009).

Pectic substances from red beet were isolated in var. *Conditiva* and structural analysis was carried out using enzymatic degradation, and after methylation, yielding three different sub-units: a homogalacturonan (75%), a highly ramified rhamnogalacturonan I (RG-I) (Strasser and Amadò 2001) and a typical rhamnogalacturonan II (RG-II (Strasser and Amadò 2002)). RG-I consisted of a highly ramified backbone composed of nearly equal amounts of rhamnose and galacturonic acid. Side chains, mainly arabinans, galactans and type-II arabinogalactans, were attached to the RG-I backbone (Strasser and Amadò 2001). Cell wall material from fully mature red beet (*Beta vulgaris* L. var. *conditiva*) was isolated as an alcohol-insoluble residue (AIR). The chelator-soluble pectin obtained by cyclohexane-*trans*-1,2-diaminotetraacetate (CDTA) extraction of the AIR was fractionated by anion exchange chromatography (AEC). The main fraction was further fractionated by gel filtration chromatography (GFC). Fractions from both chromatographic systems were stepwise degraded by endo-polygalacturonase, endo-β-(1→4)-D-galactanase, endo-α-(1→5)-L-arabinanase and α-L-arabinofuranosidase. Degradation products were further fractionated by GFC or by AEC. Polymeric fractions were investigated by methylation analysis where selected fractions were additionally methylated with trideuteromethyliodide to enable the detection of O-methyl-substituted sugars. These chemical derivitizations demonstrated that some arabinans were connected via short galactan chains directly or indirectly to this backbone. Type-II arabinogalactans were formed by "inner" chains consisting of (1→3)-linked galactans and short "outer" chains composed of an average number of one to three (1→6)-linked galactose residues. Terminal arabinofuranoses were linked via the O-3 position to galactose residues. Nearly all non-reducing ends consisted of glucuronic acid.

After acid hydrolysis of the cell wall fraction of the same variety of red beet (*B. vulgaris* var. *conditiva*) for 2 and 3 h, the pectin yields obtained, respectively, were 24 and 31 g/100 g for pH 1.5 and 11 and 17 g/100 g for pH 2.0 (Fissore et al. 2010). The rhamno-galacturonans were higher in red beet (Fissore et al. 2010) than in sugar beet (Levigne et al. 2002). The rheological properties of red beet pectin require proper characterization through technical evaluation before the same can be used for product development. Results of such rheological characteristics of pectin-enriched products of red beet by-products confirmed their suitability for use in food formulations, particularly when extracted after an alkaline pre-treatment followed by enzymatic hydrolysis with either cellulase or hemicellulase (Fissore et al. 2012). True gels of beet pectins could be developed in the presence of Ca^{2+} in water, where junction zones of homogalacturonan (HG) side chains mediated by Ca^{2+} were able to build up rigid networks. Such gelling properties are indicative of their suitability for application in food products where calcium enrichment is desirable. Beet pectin (2.00% w/v) was also used to constitute milk model systems. When skim milk was used, more transient and weaker gel networks were formed when compared with Ca^{2+}-aqueous systems, where the associated formation of a [κ-casein calcium cross-linked low methoxyl pectin] complex was dampened by milk fat globules. Relaxation spectra of pectin–milk systems were in general extended to large relaxation times (10^4 s) for most of the pectin fractions, which is a typical characteristic of structured carbohydrate polymers. Since all pectin fractions showed very similar chemical composition and molecular weight (average value and distribution), it was suggested that some differences in the rheological performance of each pectin product came from the different length of arabinans and distribution of rhamnose kinks (RG-I, random coil) as well as from the length of demethylated HG chains (semi-flexible coils).

Morris et al. (2010) carried out the characterisation of the rhamnogalacturonan-I (RG-I) and homogalacturonan (HG) fractions of beet pectin. The HG fraction, prepared by mild acid hydrolysis of acid extracted pectin, was found to have a relatively low weight average molar mass (20,000 g/mol), but a rather high intrinsic viscosity (77 ml/g), which made the material rigid in solution (Lp = 9.8 nm). Lower molar mass pectins were found to be richer in HG regions and pectins of higher molar mass were richer in RG-I regions. These characteristics show that the presence of more of HG units was important for high intrinsic viscosity, but had less effect on molar mass; the inverse is true if the pectin has more RG-I regions. Thus by selective hydrolysis of beet pectins, their functionality can be changed for various applications.

1.9 Enzymes from Beet

Polyphenol oxidase (PPO) and peroxidise (POD), the two enzymes involved in enzymatic browning reactions, have been partially purified and extracted from different fractions of beet root (Escribano et al. 2002). PPO is mainly located in the membrane fraction and is also found in the soluble fraction. In both cases, PPO was found in its latent state. However, POD activity was higher in the soluble fraction than in the membrane fraction. Kinetic characterization of the PPO for both soluble and membrane fractions was studied and it was found that The *K*m value for the soluble PPO was similar to the one obtained using beet leaves instead of beet roots, whereas for membrane-bound PPO, the *K*m value was slightly lower than the one obtained using beet leaves. Different parts of red beet showed high levels of peroxidases, with a higher level at the peel portions and seedling roots (Thimmaraju et al. 2005), indicating their probable antioxidant and defense-related functions. When this source of peroxidase, particularly the acetone powders of seedling roots, was used for curing green vanilla beans, there was a rapid (10 days) conversion of glucovanillin to vanilla flavor, reaching much higher levels than that obtained after 40 days in control beans (Sreedhar et al. 2009). Cultured beet cells and hairy roots largely produce peroxidases, some of which are released into the culture medium (Rudrappa et al. 2007, Neelwarne and Rudrappa 2009). Since peroxidases of red beet are characteristically comparable to the vastly used horseradish peroxidase, with an added advantage of higher thermo-stability, red beet peroxidases are expected to find enormous applications in developing biochemical reagents of industrial and environmental importance. Further details are available in Chap. 12 of this book.

Understanding the chemistry of cell wall biosynthesis in plant cells has remained an evergreen area of carbohydrate chemistry for which beet protoplasts (Wiśniewska and Majewska-Sawka 2008) and red beet cell wall construction enzymes have served as experimental models. Normally β-1-3-glucan polymers are the major structural components of fungal cell walls, whereas plant celluloses are normally made up of β-1-4-glucan units, although, rarely, the former type of polymers are formed during pathogen attack and wounding (Østergaard et al. 2002). Red beet microsomal β-1-3-glucan synthase has been studied extensively as a model system (Sloan et al. 1987), recently this enzyme was characterized in Arabidopsis, the other vastly studied model plant (Østergaard et al. 2002).

1.10 Other Interesting Compounds of Red Beet

1.10.1 Glycine Betain

Betaine, chemically known as *N,N,N*-trimethylglycine, is a small non-pigment molecule of *N*-trimethylated amino acid (Fig. 1.4). It exists in biological materials, in zwitterionic form at neutral pH. Although glycine betaine was chiefly detected and characterized in beets, it is now known that betaine is widely distributed in plants, animals and microorganisms.

Betaine is an osmoregulant that accumulates in plants after exposure to higher levels of sodium and other salts. Two cultivars of red beet, Scarlet Supreme and Ruby Queen, when exposed to high salts, showed linearity in the synthesis of glycine betaine with the increase in salt levels, indicative of the high salt tolerance levels of beets. Betaine, like folates, is a nutrient involved in one carbon metabolism. Betaine recently gained importance due to its cancer-preventing properties. Nevertheless, a recently concluded study of betaine usage that collected data for the last 20 years in post-menopausal women at risk of breast cancer showed neither an adverse effect nor any beneficial effects on breast cancer (Cho et al. 2010). Betaine is known to heal the liver, lower homocysteine, and improve the production of stomach acid. People who have the genetic condition of homocystinuria, in which homocysteine levels build up in the body, are at much higher risk of developing osteoporosis and heart disease as early as in their 20s. Low levels of synthetic betaine supplements are recommended for suppressing homocysteine levels when too much homocysteine builds up (Alfthan et al. 2004), and is approved by the FDA. Betaine has also been found to impart anti-atherogenic effects (Atkinson et al. 2008). However, since synthetic betaine also has a risk of increasing cholesterol levels, the natural source, the red beet, serves as an effective safer alternative. Since red beets also have vaso-dilatory functions because of nitric oxide induction, the vegetable appears to function as a wonderful remedy for circulatory problems.

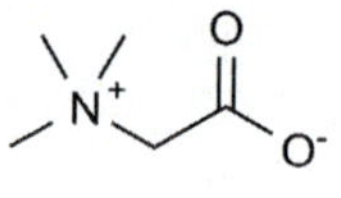

Fig. 1.4 Trimethylglycine (TMG)

1.10.2 Geosmin in Beets

Geosmin, meaning earth smell, is the typical undesirable earthy flavour of red beets. Until lately, it was not known whether geosmin is a by-product of beet metabolism or synthesized by earth-bound *Streptomycetes* (Dionigi et al. 1992; Bentley and Meganathan 1981) and then taken up by the beets during maturation. However, its endogenous biosynthesis in red beet was unambiguously demonstrated by Lu et al. (2003) using aseptically grown red beet seedlings, although the flavour had also been earlier sensed in cultured beet cells and hairy roots and remained unreported. This undesirable flavour was found to be six times higher in the peel than in the pulp. The externally supplied geosmin was not taken up by beet plants (Lu et al. 2003). Since the presence of this flavour is a major setback for the beet pigment industry, more work on the biosynthesis of geosmin (Fig. 1.5) in red beet is needed to suppress this flavour either through breeding or by metabolic engineering.

1.10.3 Azetidine

Many plants synthesize non-protein amino acids for use as defense compounds (Bell 2003). Although some of them are useful drugs for humans (e.g., serotonin and L-DOPA), others are of some concern. One such compound, L-azetidine-2-carboxylic acid (L-Aze), has been traced to red beet tubers, although the compound was not traceable to the leaves of any of the beet varieties (Rubenstein et al. 2006). L-Aze (Fig. 1.6a) structurally resembles the amino acid L-proline (Fig. 1.6b). Because of such great structural similarities, L-Aze has been found to mis-incorporate into protein in place of proline, resulting in the alteration of collagen, keratin, hemoglobin, and protein folding and hence causing teratogenic effects, as demonstrated in animal models (Rubenstein et al. 2006). Although the role of L-Aze in humans remains unexplored, a high level of precision is normally maintained in all biochemical reactions in mammals, particularly in evolutionarily advanced human beings, where the mis-incorporation of wrong molecules are unlikely and rare, because even in plants, such mis-incorporations were found to occur in susceptible plants (Fowden et al. 1979).

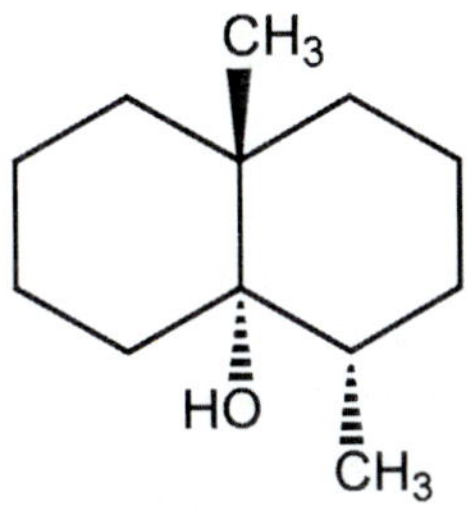

Fig. 1.5 Geosmin

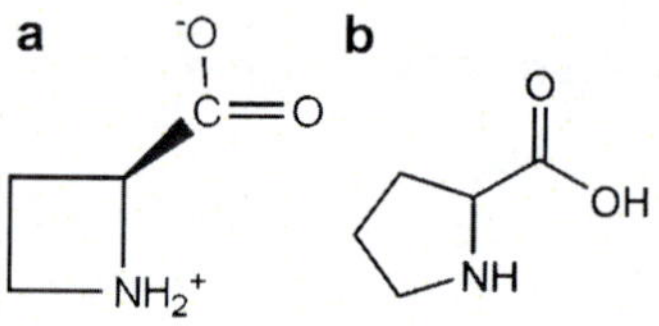

Fig. 1.6 (**a**) Azetidine-2-carboxylic acid (L-Aze). (**b**) L-proline

1.11 Red Beet: A Model System for Basic Studies

1.11.1 Vacuolar studies: See Chap. 4.

1.11.2 Mitochondrial and Respiratory Metabolism: See Chap. 5.

1.11.3 Accumulation of Sodium by Beets

The potential for red beet to accumulate high concentrations of sodium (Table 1.3) with no adverse effects on growth suggests that beet may be good crop for life support applications proposed for space (Wheeler and Sage 2006). The effect of exposure of plants to sodium (Na) and salinity on glycine betaine accumulation in tissues was studied in red beet cultivars, cv. Scarlet Supreme and cv. Ruby Queen (Subbarao et al. 2000). In their study, red beet plants grown for 42 days in a growth chamber using a technique of re-circulating nutrient film were found to accumulate higher glycine betaine levels under high salinity. There was a direct linearity between glycine betaine accumulation and higher tissue levels of sodium in both cultivars, where sodium accounted for 80–90% of the total cation uptake. Here the metabolic tolerance to high tissue levels of sodium was related to the ability of red beet to synthesize and regulate glycine betaine production. The regulatory control for partitioning sodium and glycine betaine between the vacuole and the cytoplasm was also found to be an important aspect in beets, and such sodium-accumulating plants find applications in the removal of sodium from waste water streams. Therefore, beets find enormous applications for mineral removal from waste waters, avoiding their fouling with algal and bacterial growths, allowing redistribution of biominerals through food, fodder and bio-fertilizer chains.

1.12 Beet Storage

Vegetable commodities, both foliage and tubers continue their respiration even after their harvest. The respiration is enhanced in foliage due to their large surface areas, high desiccation rates and presence of stomata, whereas, in tubers, respiration is curtailed due to their inherent storage physiologies. This being the case, foliage beet needs utmost care for their preservation from harvest time until they are consumed.

Bunched beets are best kept at >98% relative humidity (RH) at 0°C (32°F) for approximately 10–14 days, whereas topped beets are best stored at 1–2°C (32.9–35.6°F) and 98% RH. During storage, more black spot and rot occurred at 0–1°C (32–33.8°F) than at higher temperatures (Schouten and Schaik 1980).

Because of their low respiration, red beet tubers could be stored under air-ventilation for 4–6 months and in mechanical refrigerated storage for as long as 8–10 months. Expectedly, large roots could be stored longer than small ones. In cold countries with prolonged winters, red beets stored in pits and trenches with insulation maintaining 0.5°C (31.1°F) to 5°C (41°F), lasted longer than surface stored beets. More details on respiratory metabolism of beets are available in Chap. 5.

1.13 Concluding Remarks

The enormous current research interest in red beet, because of its rich nutritional and nutraceutical components, establishes that undoubtedly red beet is a marvellous vegetable, easy to grow and easy to process its products. Its strong antioxidant and vasodilation properties, imparted by pigments, flavonoids and organic nitrogen, need deeper investigations to reap the enormous array of health benefits. The richness of organic iron and folates in red beet needs extensive popularization in many developing countries where both iron-deficiency anemia and infant death rate due to neural tube defects are very high (Micronutrient initiative 2009).

Despite a growing interest in betalains as natural food colorants, very few enzymes involved in betalain synthesis have been purified and characterized until now, with very little work on genes coding such enzymes. Contrarily, enzymes involved in betalains degradation are extensively addressed. While many other experimental model plants have enjoyed the application of modern metabolomics studies, only the sugar beet has fairly been the target of such studies, with a particular emphasis on sugar enhancement or for the improvement of other agronomical traits (Gurel et al. 2008). Red beet has not been the target of such studies, despite its important nutritional composition and wide array of bioactive molecules. Along similar lines to the sugar beet, enhancement of sugar may also be of great relevance in colored beets since extracts from such beets impart natural sweetness to the product along with its rich color. Although the biotechnological production of betalain pigments cannot compete with conventional production, it offers the great advantage of obtaining characteristically consistent product, where it is possible to achieve the desired quality. Another important aspect worth addressing in red colored beets is the step forward from betalamic acid to the formation of either betacyanin or betaxanthin. Although this terminal step is a non-enzymatic condensation step, surprisingly, each variety or a cell/callus or hairy root line maintains a specific pigment ratio. How each beet genotype in nature retains this precision is not known. Although L-DOPA levels are suggested to be precisely regulated through cyclo-DOPA and DOPA quinone formation in red beet (see Chap. 2), deeper biochemical study is needed to regulate these steps so that the beet can also serve as natural rich source of L-DOPA, much needed as a precursor for dopamine, an anti-depressant and a neurotransmitter in the brain, which probably works as a therapy for people with Parkinson's disease.

References

Ahluwalia A. 2011. Health benefits of red beet. http://www.livestrong.com/article/110787-benefits-beet-juice/. Accessed 25 Nov 2011.

Akita, T., Y. Hina, and T. Nishi. 2000. Production of betacyanins by a cell suspension culture of table beet (*Beta vulgaris* L.). *Bioscience, Biotechnology, and Biochemistry* 64: 1807–1812.

Akita, T., Y. Hina, and T. Nishi. 2001. Effect of zinc deficiency on betacyanin production in a cell suspension culture of table beet (*Beta vulgaris* L.). *Bioscience, Biotechnology, and Biochemistry* 65: 962–965.

Akita, T., Y. Hina, and T. Nishi. 2002. New medium composition for high betacyanin production by a cell suspension culture of table beet (*Beta vulgaris* L.). *Bioscience, Biotechnology, and Biochemistry* 66: 902–905.

Alfthan, G., K. Tapani, K. Nissinen, et al. 2004. The effect of low doses of betaine on plasma homocysteine in healthy volunteers. *The British Journal of Nutrition* 92: 665–670.

Armuganathan, K., and E.D. Earle. 1991. Nuclear DNA content of some important plant species. *Plant Molecular Biology Reporter* 9: 208–218.

Arnaud, J.F., F. Viard, M. Delesclure, and J. Cuguen. 2003. Evidence for gene flow via seed dispersal from crop to wild relatives in *Beta vulgaris* (Chenopodiaceae): Consequences for the release of genetically modified crop species with weedy lineages. *Proceedings of the Royal Society of London Series B-Biological Science* 270: 1565–1571.

Asensi-Fabado, M.A., and S. Munne-Bosch. 2010. Vitamins in plants: Occurrence, biosynthesis and antioxidant function. *Trends in Plant Science* 15: 582–592.

Atkinson, W., J. Elmslie, M. Lever, S.T. Chambers, and P.M. George. 2008. Dietary and supplementary betaine: Acute effects on plasma betaine and homocysteine concentrations under standard and postmethionine load conditions in healthy male subjects. *American Journal of Clinical Nutrition* 87: 577.

Avérous, L. 2004. Biodegradable multiphase systems based on plasticized starch: A review. *Journal of Macromolecular Science, Part C-Polymer Reviews* 44: 231–274.

Azeredo, H.M.C. 2009. Betalains: Properties, sources, applications, and stability – A review. *International Journal of Food Science and Technology* 44: 2365–2376.

Azeredo, H.M.C., C.R.A. Santos, C.B.K. Mendes, and R.A.M. Iranilde. 2007. Betacyanin stability during processing and storage of a microencapsulated red beetroot extract. *American Journal of Food Technology* 2: 307–312.

Baar, A., W. Gebel, U. Imhof, and H. Mihalik. 1997. Process for producing mouldings with a barrier layer made of biodegradable material, and mouldings produced according to this process. WO Patent n 97/10293.

Bandyopadhyay, M., R. Chakraborty, and U. Raychaudhuri. 2008. Effect of beet and honey on quality improvement and carotene retention in a carrot fortified milk product. *Innovative Food Science and Emerging Technologies* 9: 9–17.

Bartsch, D., M. Lehnen, J. Clegg, S.I. Pohl-orf, and N.C. Eustrand. 1999. Impact of gene flow from cultivated beet on genetic diversity of wild sea beet populations. *Molecular Ecology* 8: 1733–1741.

Bell, E.A. 2003. Nonprotein amino acids of plants: Significance in medicine, nutrition, and agriculture. *Journal of Agricultural and Food Chemistry* 51: 2854–2865.

Bennett, M.D., and J.B. Smith. 1976. Nuclear DNA amounts in angiosperms. *Philosophical Transactions of the Society of London Series* 274: 227–274.

Bentley, R., and R. Meganathan. 1981. Geosmin and methylisoborneol biosynthesis in *Streptomycetes*: Evidence for an isoprenoid pathway and the absence in non-differentiating isolates. *FEBS Letters* 125: 220–222.
Berghofer, E., B. Grzeskowiak, N. Mundlinger, G. Schleining, and H. Zenz. 1992. Biologisch abbaubares Verpackungs- und/oder Hüll-Material und Verfahren zuseiner Herstellung. EP Patent n 0,506,650.
Buruleanu, L., I. Manea, M.G. Bratu, D. Avram, and C.L. Nicole. 2009. Effects of prebiotics on the quality of lactic acid fermented vegetable juices. *Ovidius University Annals of Chemistry* 20: 102–107.
Cai, Y., M. Sun, and H. Corke. 1998. Colorant properties and stability of *Amaranthus* betacyanin pigments. *Journal of Agricultural and Food Chemistry* 46: 4491–4495.
Cai, Y.Z., M. Sun, and H. Corke. 2005. Characterization and application of betalain pigments from plants of the Amaranthaceae. *Trends in Food Science and Technology* 16: 370–376.
Chalermchat, Y., P. Dijmek, and M. Fincan. 2004. Pulsed electric field treatment for solid-liquid extraction of red beetroot pigment: Mathematical modeling of mass transfer. *Journal of Food Engineering* 64: 229–236.
Cho, E., M. Holmes, S. Hankinson, and W. Willett. 2010. Choline and betaine intake and risk of breast cancer among post-menopausal women. *British Journal of Cancer* 102: 489–494.
Cormier, F. 1997. Food colorants from plant cell cultures. In *Recent advances in phytochemistry*, Functionality of phytochemicals, vol. 31, ed. J.T. Romeo and T. Johns, 201–222. New York: Plenum Press.
Delgado-Vargas, F., A. Jiménez, and O. Paredes-López. 2000. Natural pigments: Carotenoids, anthocyanins, and betalains – Characteristics, biosynthesis, processing, and stability. *Critical Reviews in Food Science and Nutrition* 40: 173–289.
Desel, C., R. Jansen, D.D. Gue, and T. Schmidt. 2002. Painting of parental chromatin in *Beta* hybrids by multicolor fluorescent in situ hybridization. *Ann Bot London* 89: 171–181.
Dinand, E., H. Chanzy, and M. Vignon. 1996. Parenchymal cell cellulose from sugar beet pulp: Preparation and properties. *Cellulose* 3: 183–188.
Dinand, E., H. Chanzy, and R. Vignon. 1999. Suspensions of cellulose microfibrils from sugar beet pulp. *Food Hydrocolloids* 13: 275–283.
Dionigi, C.P., D.F. Millie, A.M. Spanier, and P.B. Johnsen. 1992. Spore and geosmin production by *Streptomyces tendae* on several media. *Journal of Agricultural and Food Chemistry* 40: 122–125.
Drdák, M., R.C. Altamirano, A. Rajmiankova, P. Simko, J. Karovicova, and D. Benkovska. 1992. Red beet pigment composition effect of fermentation by different strains of *Saccharomyces cerevisiae*. *Journal of Food Science* 57: 935–936.
Dronnet, V., M. Axelos, C. Renard, and J.F. Thibault. 1998. Improvement of the binding capacity of metal cations by sugar-beet pulp. 2. Binding of divalent metal cations by modified sugar-beet pulp. *Carbohydrate Polymers* 35: 239–247.
Dudareva, N.A., E.V. Kiseleva, A.E. Boyarintseva, A.G. Maystrenko, N.B. Khristolyubova, and R.I. Salganik. 1988. Structure of the mitochondrial genome of *Beta vulgaris* L. *TAG Theoretical and Applied Genetics* 76: 753–759.
Dumbravă, A., I. Enachea, C.I. Opreac, A. Georgescu, and M.A. Gîrţu. 2012. Toward a more efficient utilisation of betalains as pigments for dye-sensitized solar cells. *Digest Journal of Nanomaterials and Biostructures* 7: 339–351.
Eagen, K.A., and I.L. Goldman. 1996. Assessment of RAPD marker frequencies over cycles of recurrent selection for pigment concentration and percent solids in red beet (*Beta vulgaris* L.). *Molecular Breeding* 2: 107–115.
Eastwood, M., and H. Nyhlin. 1995. Beeturia and colonic oxalic acid. *Quarterly Journal of Medicine* 88: 711.
El-Tantawy, E.M., and G.S.A. Eisa. 2009. Growth, yield, anatomical traits and betanine pigment content of table beet plants as affected by nitrogen sources and spraying of some nutrients. *Journal of Applied Sciences Research* 5: 1173–1184.
Escribano, J., F. Gandía-Herrero, N. Caballero, and M.A. Pedreño. 2002. Subcellular localization and isoenzyme pattern of peroxidase and polyphenol oxidase in beet root (*Beta vulgaris* L.). *Journal of Agricultural and Food Chemistry* 50: 6123–6129.

Figiel, A. 2010. Drying kinetics and quality of beetroots dehydrated by combination of convective and vacuum-microwave methods. *Journal of Food Engineering* 98: 461–470.

Fincan, M., F. Devito, and P. Dijmek. 2004. Pulsed electric field treatment for solid-liquid extraction of red beetroot pigment. *Journal of Food Engineering* 64: 381–388.

Fiserova, M., J. Gigac, and R. Butas. 2007. Influence of sugar beet pulp on bond strength and structure of paper. *Wood Research* 52: 59–74.

Fissore, E.N., N.M.A. Ponce, P.M. de Escalada, C.A. Stortzs, A.M. Rojas, and L.N. Gerschenson. 2010. Characterization of acid-extracted pectin-enriched products obtained from red beet (*Beta vulgaris* L. var. conditiva) and butternut (*Cucurbita moschata* Duch ex Poiret). *Journal of Agricultural and Food Chemistry* 58: 3793–3800.

Fissore, E.N., A.M. Rojas, and L.N. Gerschenson. 2012. Rheological performance of pectin-enriched products isolated from red beet (*Beta vulgaris* L. var. *conditiva*) through alkaline and enzymatic treatments. *Food Hydrocolloids* 26: 249–260.

Flavell, R.B., M.D. Bennet, and J.B. Smith. 1974. Genome size and the proportion of repeated nucleotide sequence DNA in plants. *Biochemical Genetics* 12: 257–269.

Fletcher, A. 2006. Lycopene colorant achieves regulatory approval. http://www.foodnavigator.com/Legislation/Lycopene-colorant-achieves-regulatory-approval. Accessed 25 Nov 2011.

Fowden, L., P.J. Lea, and E.A. Bell. 1979. The non-protein amino acids in plants. *Advances in Enzyme Research* 50: 117–175.

Francis, F.J. 1986. *Handbook of food colorant patents*, 1st ed. Westport: Food and Nutrition Press.

Fransz, P.F., C. Alonso-Blanco, T.B. Liharska, et al. 1996. High-resolution physical mapping in *Arabidopsis thaliana* and tomato by fluorescence in situ hybridization to extended DNA fibres. *The Plant Journal* 9: 421–430.

Gabelman, W.H., I.L. Goldman, and D.N. Breitbach. 2002. High pigment beet. Google Patents.

Gasztonyi, M.N., H. Daood, M.T. Hajos, and P. Biacs. 2001. Comparison of red beet (*Beta vulgaris* var conditiva) varieties on the basis of their pigment components. *Journal of the Science of Food and Agriculture* 81: 932–933.

Gennari, L., M. Felletti, M. Blasa, C. Celeghini, A. Corallini, and P. Ninfali. 2011. Total extract of *Beta vulgaris* var. Cicla seeds versus its purified phenolic components: Antioxidant activities and antiproliferative effects against colon cancer cells. *Phytochemical Analysis* 22: 272–279.

Georgiev, V., M. Ilieva, T. Bley, and A. Pavlov. 2008. Betalain production in plant in vitro systems. *Acta Physiologiae Plantarum* 30: 581–593.

Georgiev, V., M. Ilieva, T. Bley, and A. Pavlov. 2010a. Review: Betalain production in plant in vitro systems. *Acta Physiologiae Plantarum* 30: 581–593.

Georgiev, V.G., J. Weber, E.-M. Kneschke, P.N. Denev, T. Bley, and A.I. Pavlov. 2010b. Antioxidant activity and phenolic content of betalain extracts from intact plants and hairy root cultures of the red beetroot *Beta vulgaris* cv. Detroit Dark Red. *Plant Foods for Human Nutrition* 65: 105–111.

Gokhale, S., and S. Lele. 2011. Dehydration of red beet root (*Beta vulgaris*) by hot air drying: Process optimization and mathematical modeling. *Food Science and Biotechnology* 20: 955–964.

Grotewold, E. 2006. The genetics and biochemistry of floral pigments. *Annual Review of Plant Biology* 57: 761–780.

Gurel, E., S. Gurel, and P.G. Lemaux. 2008. Biotechnology applications for sugar beet. *Critical Reviews in Plant Sciences* 27: 108–140.

Haber, G.J., C.T. Tan, and J. Wu. 1979. Stable beet color composition. US Patent No 4132793.

Hanelt, P., R. Büttner, R. Mansfeld, and R. Kilian. 2001. *Mansfeld's encyclopedia of agricultural and horticultural crops – Institute of Plant Genetics and Crop Plant Research*, 235–241. Berlin: Springer.

Hansen, A., C. Plever, H. Pedersen, et al. 1994. Efficient in vitro chromosome doubling during *Beta vulgaris* ovule culture. *Plant Breeding* 112: 89–95.

Havlikova, L., K. Mikova, and V. Kyzlink. 1983. Heat stability of betacyanins. *Zeitschrift für Lebensmittel-Untersuchung und -Forschung* 177: 247–250.

Havlikova, L., K. Mikova, and V. Kyzlink. 1985. Red beet pigments as soft drink colorants. *Food/Nahrung* 29: 723–730.

Herbach, K., F. Stintzing, and R. Carle. 2004. Impact of thermal treatment on color and pigment pattern of red beet (*Beta vulgaris* L.) preparations. *Journal of Food Science* 69: C491–C498.

Herbach, K., F. Stintzing, and R. Carle. 2006. Betalin stability and degradation-structure and chromatic aspects. *Journal of Food Science* 71: R41–R50.

Heslop-Harrison, J.S., and T. Schwarzacher. 1993. Molecular cytogenetics – Biology and applications in plant breeding. In *Chromosomes today*, vol. XI, ed. A. Chandly and A.T. Sumner, 191–198. London: Chapmann & Hall.

Ivanov, M.K., A.S. Revenko, and G.M. Dymshits. 2004. Cytoplasmic male sterility-associated structural variation of the mitochondrial genome regions containing rps3 and orf215 in sugar beet *Beta vulgaris* L. *Molecular Biology* 38: 345–350.

Jackman, R.I., and J.L. Smith. 1996. Anthocynins and betalain. In *Natural food colorants*, ed. C.F. Hendry and J.D. Houghton, 244–309. London: Blackie Academic and Professional.

Jiratanan, T., and R.H. Liu. 2004. Antioxidant activity of processed table beets (*Beta vulgaris* var, conditiva) and green beans (*Phaseolus vulgaris* L.). *Journal of Agricultural and Food Chemistry* 52: 2659–2670.

Joshi, R., S. Adhikari, B.S. Patro, S. Chattopadhya, and T. Mukherjee. 2001. Free radical scavenging behavior of folic acid: Evidence for possible antioxidant activity. *Free Radical Biology & Medicine* 30: 1390–1399.

Kaminski, W., E. Tomczak, and E. Skorupska. 2004. Estimation of the effect of shape and temperature on drying kinetics using MLP. *Drying Technology* 22: 191–200.

Kannan, V. 2011. Extraction of bioactive compounds from whole red cabbage and beetroot using pulsed electric fields and evaluation of their functionality. Master of Science Thesis in Food Science and Technology, University of Nebraska, Lincoln. http://digitalcommons.unl.edu/foodscidiss/11. Accessed 4 Feb 2011.

Kanner, J., S. Harel, and R. Granit. 2001. Betalains a new class of dietary cationized antioxidants. *Journal of Agricultural and Food Chemistry* 49: 5178–5185.

Kapadia, G.J., H. Tokuda, T. Konoshima, and H. Nishino. 1996. Chemoprevention of lung and skin cancer by *Beta vulgaris* (beet) root extract. *Cancer Letters* 100: 211–214.

Kapadia, G.J., M.A. Azuine, R. Sridhar, Y. Okuda, A. Tsuruta, E. Ichiishi, T. Mukainake, M. Takasaki, T. Konoshima, H. Nishino, and H. Tokuda. 2003. Chemoprevention of DMBA-induced UV-B promoted, NOR-1-induced TPA promoted skin carcinogenesis, and DEN-induced phenobarbital promoted liver tumors in mice by extract of beetroot. *Pharmacological Research* 47: 141–148.

Kapadia, G.J., M.A. Azuine, G. Subba Rao, et al. 2011. Cytotoxic effect of the red beetroot (*Beta vulgaris* L.) extract compared to doxorubicin (adriamycin) in the human prostate (PC-3) and breast (MCF-7) cancer cell lines. *Anti-Cancer Agents in Medicinal Chemistry (Formerly Current Medicinal Chemistry)* 11: 280–284.

Koul, V.K., M.P. Jain, S. Koul, V.K. Sharma, C.L. Tikoo, and S.M. Jain. 2002. Spray drying of beet root juice using different carriers. *Indian Journal of Chemical Technology* 9: 442–445.

Križnik, B., and D. Pavoković. 2010. Enhancement of betanin yield in transformed cells of sugar beet (*Beta vulgaris* L.). *Acta Botanica Croatica* 69: 173–182.

Kubo, T., S. Nishizawa, and T. Mikami. 1999. Alterations in organization and transcription of the mitochondrial genome of cytoplasmic male sterile sugar beet (*Beta vulgaris* L.). *Molecular and General Genetics MGG* 262: 283–290.

Kujala, T.S., J.M. Loponen, K.D. Klika, and K. Pihlaja. 2000. Phenolic and betacyanins in red beetroot (*Beta vulgaris*) root: Distribution and effect of cold storage on the content of total phenolics and three individual compounds. *Journal of Agricultural and Food Chemistry* 48: 5338–5342.

Kujala, T., K. Klika, V. Ovcharenko, J. Laponen, M. Vienola, and K. Pihlaja. 2001a. 5,5′-6,6′-Tetrahydroxy-3,3′-biindolyl from beetroot (*Beta vulgaris*) peel extract. *Zeitschrift für Naturforschung. C* 56: 714–718.

Kujala, T., J. Laponen, and K. Pihlaja. 2001b. Betalains and phenolics in red beetroot (*Beta vulgaris*) peel extract. *Zeitschrift für Naturforschung. C* 56: 343–348.

Kujala, T.S., M.S. Vienola, K.D. Klika, J.M. Loponen, and K. Pihlaja. 2002. Betalain and phenolic composition of four beetroot (*Beta vulgaris*) cultivars. *European Food Research and Technology* 214: 505–510.

Lange, W., W.A. Brandenburg, and T.S.M.D.E. Bock. 1999. Taxonomy and cultonomy of beet (*Beta vulgaris* L.). *Botanical Journal of the Linnean Society* 130: 81–96.

Lansley, K.E., P.G. Winyard, J. Fulford, A. Vanhatalo, and S.J. Bailey. 2011. Dietary nitrate supplementation reduces the O2 cost of walking and running: A placebo-controlled study. *Journal of Applied Physiology* 10: 591–600.

Latorre, M.E., P. Narvaiz, A.M. Rojas, and L.N. Gerschenson. 2010. Effects of gamma irradiation on bio-chemical and physico-chemical parameters of fresh-cut red beet (*Beta vulgaris* L. var. *conditiva*) root. *Journal of Food Engineering* 98: 178–191.

Lechner, J.F., L.S. Wang, C.M. Rocha, B. Larue, C. Henry, M. McIntyre Colleen, K.M. Riedl, S.J. Schwartz, and G.D. Stoner. 2010. Drinking water with red beetroot color antagonizes esophageal carcinogenesis in N-nitrosomethylbenzykamine-treated rats. *Journal of Medicinal Food* 13(3): 733–739.

Lee, C.Y., and N.L. Smith. 1979. Blanching effect on polyphenoloxidase activity in table beets. *Journal of Food Science* 44: 82–86.

Lee, Y.N., R.C. Wiley, M.J. Sheu, and D.V. Schlimme. 1982. Purification and concentration of betalaines by ultrafiltration and reverse osmosis. *Journal of Food Science* 47: 465–471, 475.

Lee, J.H., C.W. Son, M.Y. Kim, M.H. Kim, H.R. Kim, E.S. Kwak, S. Kim, and M.R. Kim. 2009. Red beet (*Beta vulgaris* L.) leaf supplementation improves antioxidant status in C57BL/6 J mice fed high fat high cholesterol diet. *Nutrition Research and Practice* 3: 114–121.

Leitner, J., B. Hinterstoisser, M. Wastyn, J. Keckes, and W. Gidl. 2007. Sugar beet cellulose nanofibril-reinforced composites. *Cellulose* 14: 419–425.

Levigne, S., M.C. Ralet, and J.F. Thibault. 2002. Characterisation of pectins extracted from fresh sugar beet under different conditions using an experimental design. *Carbohydrate Polymers* 49: 145–153.

Levin, D.A., J. Francisco-Ortega, and R.K. Jansen. 1996. Hybridization and the extinction of rare plant species. *Conservation Biology* 10: 10–16.

Lewicki, P.P. 2006. Design of hot air drying for better foods. *Trends in Food Science and Technology* 17: 153–163.

Liu, L.S., V.L. Finkenstadt, C.K. Liu, T. Jin, M.L. Fishman, and K.B. Hicks. 2007. Preparation of poly (lactic acid) and pectin composite films intended for applications in antimicrobial packaging. *Journal of Applied Polymer Science* 106: 801–810.

Loginova, K., N. Lebovka, and E. Vorobiev. 2011. Pulsed electric field assisted aqueous extraction of colorants from red beet. *Journal of Food Engineering* 106: 127–133.

Lopez, N., E. Puertolas, S. Condon, J. Raso, and I. Alvarez. 2009. Enhancement of the extraction of betanine from red beetroot by pulsed electric fields. *Journal of Food Engineering* 90: 60–66.

Lu, G., C.G. Edwards, J.K. Fellman, et al. 2003. Biosynthetic origin of geosmin in red beets (*Beta vulgaris* L.). *Journal of Agricultural and Food Chemistry* 51: 1026–1029.

Mabry, T.J., and A.S. Dreiding. 1968. The betalaines. In *Recent advances in phytochemistry*, vol. 1, ed. R. Mabry, E. Alston, and V.C. Runeckles, 145–160. New York: Appleton Century Crofts.

Mabry, T.J., A. Taylor, and B.L. Turner. 1963. Betacyanins and their distribution. *Phytochemistry* 2: 61–64.

Marfil, P., E. Santos, and V. Telis. 2008. Ascorbic acid degradation kinetics in tomatoes at different drying conditions. *LWT- Food Science and Technology* 41: 1642–1647.

Marggraf, A.S. 1749. Expériences chymiques, faites dans le dessein de tirer un véritable sucre de diverse oplantes, qui croissent dans nos contrées'. Histoire de l'Académie Royale des Sciences et Belles Lettres, Berlin, pp. 79–90.

Martinez-Parra, J., and R. Munoz. 2001. Characterization of betacyanin oxidation catalyzed by a peroxidase from *Beta vulgaris* L. roots. *Journal of Agricultural and Food Chemistry* 49: 4064–4068.

McGinnis, E.E., M.H. Meyer, and A.G. Smith. 2010. Sweet and sour: A scientific and legal look at herbicide-tolerant sugar beet. *The Plant Cell Online* 22: 1653–1657.

McGrath, J.M., M. Saccomani, P. Stevanato, and E. Biancardi. 2007. Beet. In *Genome mapping and molecular breeding in plants*, Vegetables, vol. 5, ed. C. Kole, 191–207. Berlin: Springer.

Mesbah, M., T.S.M. de Bock, J.M. Sandbrink, R.M. Klein-Lankhorst, and W. Lange. 1997. Molecular and morphological characterization of monosomic additions in *Beta vulgaris*, carrying extra chromosomes of *B. procumbens* or *B. patellaris*. *Molecular Breeding* 3: 147–157.

Mesbah, M., J. Wennekes-Van Eden, J.H. De Jong, T.S. De Bock, and W. Lange. 2000. FISH to mitotic chromosomes and extended DNA fibres of *Beta procumbens* in a series of monosomic additions to beet (*B. vulgaris*). *Chromosome Research* 8: 285–293.

Micard, V., C. Renard, and J.F. Thibault. 1994. Studies on enzymic release of ferulic acid from sugar-beet pulp. *Lebensmittel-Wissenschaft und-Technologie* 27: 59–66.

Michel, F., J.F. Thibault, J.L. Barry, and R. de Baynast. 1988. Preparation and characterisation of dietary fibre from sugar beet pulp. *Journal of the Science of Food and Agriculture* 42: 77–85.

Micronutrient Initiative. 2009. Investing in the future: A united call to action on vitamin and mineral deficiencies: Global report. 978-1-894217-31-6.

Mohamed, A.A., V. Finkenstadt, and D. Palmquist. 2008. Thermal properties of extruded/injection-molded poly (lactic acid) and biobased composites. *Journal of Applied Polymer Science* 107: 898–908.

Morris, G.A., M.-C. Ralet, E. Bonnin, J.-F. Thibault, and S.E. Harding. 2010. Physical characterisation of the rhamnogalacturonan and homogalacturonan. *Carbohydrate Polymers* 82: 1161–1167.

Nayak, C.A., S. Chethana, N.K. Rastogi, and K. Raghavarao. 2006. Enhanced mass transfer during solid-liquid extraction of gamma-irradiated red beetroot. *Radiation Physics and Chemistry* 75: 173–178.

Neelwarne, B., and T. Rudrappa. 2009. Bioreactor for cultivation of red beet hairy roots and in situ recovery of primary and secondary metabolites. *Engineering Life Sciences* 9: 227–238.

Nemzer, B., Z. Pietrzkowski, A. Spórna, and P.W. Stalica. 2011. Betalainic and nutritional profiles of pigment-enriched red beet root (*Beta vulgaris* L.) dried extracts. *Food Chemistry* 1: 42–53.

Netzel, M., F. Stintzing, D. Quaas, G. Strass, R. Carle, R. Bitsch, I. Bitsch, and T. Frank. 2005. Renal excretion of antioxidative constituents from red beet in humans. *Food Research International* 38: 1051–1058.

Nilsson, T. 1970. Studies into the pigments in beetroot (*Beta vulgaris* L. ssp. *vulgaris* var. *rubra* L.). *Lantbrukhogskolans Annaler* 36: 179–219.

Nilsson, T. 1973. The pigment content in beetroot with regard to cultivar, growth, development and growing conditions. *Swedish Journal of Agricultural Research* 3: 187–200.

Ninfali, P., M. Bacchiocca, A. Antonelli, E. Biagiotti, A.M. Gioacchino, G. Piccoli, V. Stocchi, and G. Brandi. 2007. Characterization and biological activity of main flavonoids from Swiss chard (*Beta vulgaris* subspecies *cycla*). *Phytomedicine* 14: 216–221.

Nottingham. S. 2004. Beetroot e-book- 2004. *The Times*, London. http://stephennottingham.co.uk. Accessed 28 Jan 2011.

Oosterveld, A., G. Beldman, H.A. Schols, and A.G.J. Voragen. 2000. Characterization of arabinose and ferulic acid rich pectic polysaccharides and hemicelluloses from sugar beet pulp. *Carbohydrate Research* 328: 185–197.

Oosterveld, A., I. Pol, G. Beldman, and A. Voragen. 2001. Isolation of feruloylated arabinans and rhamnogalacturonans from sugar beet pulp and their gel forming ability by oxidative cross-linking. *Carbohydrate Polymers* 44: 9–17.

Østergaard, L., M. Petersen, O. Mattsson, and J. Mundy. 2002. An Arabidopsis callose synthase. *Plant Molecular Biology* 49: 559–566.

Owen, F.V. 1942. Inheritance of cross- and self-sterility and self-fertility in *Beta vulgaris* L. *Journal of Agricultural Research* 64: 679–698.

Pai, S., and P. D'Mello. 2004. Stability evaluation of beetroot colour in various pharmaceutical matrices. *Indian Journal of Pharmaceutical Sciences* 66: 696–699.

Patkai, G.Y., J. Barta, and I. Varsanyi. 1997. Decomposition of anticarcinogen factors of the beetroot during juice and nectar production. *Cancer Letters* 114: 105–106.

Pavier, C., and A. Gandini. 2000. Oxypropylation of sugar beet pulp. 1. Optimisation of the reaction. *Industrial Crops and Products* 12: 1–8.

Pavlov, A., P. Kovatcheva, V. Georgiev, I. Koleva, and M. Ilieva. 2002. Biosynthesis and radical scavenging activity of betalains during the cultivation of red beet (*Beta vulgaris*) hairy root cultures. *Zeitschrift für Naturforschung. C, Journal of Bioscience* 57: 640–644.

Pavoković, D., G. Rusak, V. Besendorfer, and M. Krsnik-Rasol. 2009. Light-dependent betanin production by transformed cells of sugar beet. *Food Technology and Biotechnology* 47: 153–158.

Pedreno, M.A., and J. Escribano. 2001. Correlation between antiradical activity and stability of betanine from *Beta vulgaris* L. roots under different pH, temperature and light conditions. *Journal of the Science of Food and Agriculture* 81: 627–631.

Pietrzkowski, Z. and W.C. Thresher. 2008. Solid betalain compositions and methods. Google Patents.

Prudencio, I.D., E.S. Prudêncio, E.F. Gris, et al. 2008. Petit suisse manufactured with cheese whey retentate and application of betalains and anthocyanins. *LWT- Food Science and Technology* 41: 905–910.

Pyo, Y.-H., T.-C. Lee, L. Logendra, and R.T. Rosen. 2004. Antioxidant activity of phenolic compounds of Swiss chard (*Beta vulgaris* subspecies cicla) extracts. *Food Chemistry* 85: 19–26.

Rakin, M., J. Baras, and M. Vukasinovic. 2004. The influence of brewer's yeast autolysate and lactic acid bacteria on the production of a functional food additive based on beetroot juice fermentation. *Food Technology and Biotechnology* 42: 109–113.

Rakin, M., M. Vukasinovic, S. Silver-Marinkovic, and M. Maksimovic. 2007. Contribution of lactic acid fermentation to improve nutritive quality of vegetable juices enriched with brewer's yeast autolysate. *Food Chemistry* 100: 599–602.

Rayner, P.B. 1993. Food and drink colors from natural sources. *Food Marketing Technology* 7: 9–10.

Rouilly, A., and L. Rigal. 2002. Agro-materials: A bibliographic review. *Journal of Macromolecular Science: Part C: Polymer Reviews* 42: 441–479.

Rouilly, A., J. Jorda, and L. Rigal. 2006a. Thermo-mechanical processing of sugar beet pulp. I. Twin-screw extrusion process. *Carbohydrate Polymers* 66: 81–87.

Rouilly, A., J. Jorda, and L. Rigal. 2006b. Thermo-mechanical processing of sugar beet pulp. II. Thermal and rheological properties of thermoplastic SBP. *Carbohydrate Polymers* 66: 117–125.

Rouilly, A., C. Geneau-Sbartaï, and L. Rigal. 2009. Thermo-mechanical processing of sugar beet pulp. III. Study of extruded films improvement with various plasticizers and cross-linkers. *Bioresource Technology* 100: 3076–3081.

Rubenstein, E., H. Zhou, K.M. Krasinska, A. Chien, and C.H. Becker. 2006. Azetidine-2-carboxylic acid in garden beets (*Beta vulgaris*). *Phytochemistry* 67: 898–903.

Rudrappa, T., V. Lakshmanan, R. Kaunain, N.M. Singara, and B. Neelwarne. 2007. Purification and characterization of an intracellular peroxidase from genetically transformed roots of red beet (*Beta vulgaris* L.). *Food Chemistry* 105: 1312–1320.

Saccomani, M., P. Stevanato, D. Trebbi, J.M. McGrath, and E. Binacardi. 2009. Molecular and morpho-physiological characterization of sea, ruderal and cultivated beets. *Euphytica* 169: 19–29.

Schmidt, T., and J. Heslop-Harrison. 1996. High-resolution mapping of repetitive DNA by in situ hybridization: Molecular and chromosomal features of prominent dispersed and discretely localized DNA families from the wild beet species *Beta procumbens*. *Plant Molecular Biology* 30: 1099–1113.

Schmidt, T., S. Kubis, and J.S. Heslop-Harrison. 1995. Analysis and chromosomal localization of retrotransposons in sugar beet (*Beta vulgaris* L.): Lines and Tyl-*copia*-like elements as major components of the genome. *Chromosome Research* 3: 335–345.

Schmidt, T., C. Jung, J. Heslop-Harrison, and M. Kleine. 1997. Detection of alien chromatin conferring resistance to the beet cyst nematode (*Heterodera schachtii* Schm.) in cultivated beet (*Beta vulgaris* L.) using in situ hybridization. *Chromosome Research* 5: 186–193.

Schouten, S., and A.C.R. Schaik. 1980. Storage of red beets. *Symposium on Postharvest Handling of Vegetables* 116: 25–30.

Scott, J., F. Rébeillé, and J. Fletcher. 2000. Folic acid and folates: The feasibility for nutritional enhancement in plant foods. *Journal of the Science of Food and Agriculture* 80: 795–824.

Sembries, S., G. Dongowski, K. Mehrlander, F. Will, and H. Dietrich. 2006. Physiological effects of extraction juices from apple, grape, and red beet pomaces in rats. *Journal of Agricultural and Food Chemistry* 54: 10269–10280.

Sepulveda-Jimenez, G., P. Rueda-Benitez, H. Porta, and M. Rocha-Sosa. 2004. Betacyanin synthesis in red beet (*Beta vulgaris*) leaves induced by wounding and bacterial infiltration is preceded by an oxidative burst. *Physiological and Molecular Plant Pathology* 64: 125–133.

Shen, D., Z. Wang, and M. Wu. 1987. Gene mapping on maize pachytene chromosomes by in situ hybridization. *Chromosoma* 95: 311–314.

Shih, C.C., and R.C. Wiley. 1981. Betacyanin and betaxanthine-decolorizing enzymes in the beet (*Beta vulgaris* L.) root. *Journal of Food Science* 47: 164–166, 172.

Shynkaryk, M., N. Lebovka, and E. Vorobiev. 2008. Pulsed electric fields and temperature effects on drying and rehydration of red beetroots. *Drying Technology* 26: 695–704.

Sivakumar, V., J.L. Anna, J. Vijayeeswarri, and G. Swaminathan. 2009. Ultrasound assisted enhancement in natural dye extraction from beetroot for industrial applications and natural dyeing of leather. *Ultrasonics Sonochemistry* 16: 782–789.

Skaracis, G.N. 1994. Optimizing control of *Cercospora* leaf spot in Greece: Prospects and limits in forecasting the disease. Commission Mediterraneenne de l'IIRB, 161–174.

Sloan, M.E., P. Rodis, and B.P. Wasserman. 1987. CHAPS solubilization and functional reconstitution of β-glucan synthase from red beet (*Beta vulgaris* L.) storage tissue. *Plant Physiology* 85: 516–522.

Sreedhar, R., K. Roohie, P. Maya, L. Vnkatachalam, and N. Bhagyalakshmi. 2009. Biotic elicitors enhance flavour compounds during accelerated curing of vanilla beans. *Food Chemistry* 112: 461–468.

Staginnus, C., B. Huettel, C. Desel, T. Schmidt, and G. Kahl. 2001. A PSR-based assay to detect En/Spm-like transposon sequences in plants. *Chromosome Research* 9: 591–605.

Strack, D., T. Vogt, and W. Schliemann. 2003. Recent advances in betalain research. *Phytochemistry* 62: 247–269.

Strasser, G., and R. Amadò. 2001. Pectic substances from red beet (*Beta vulgaris L. var. conditiva*). Part I. Structural analysis of rhamnogalacturonan I using enzymic degradation and methylation analysis. *Carbohydrate Polymers* 44: 63–70.

Strasser, G., and R. Amadò. 2002. Pectic substances from red beet (Beta vulgaris L. var. conditiva). Part II. Structural characterization of rhamnogalacturonan II. *Carbohydrate Polymers* 48: 263–269.

Subbarao, G.V., R.M. Wheeler, and G.W. Stutte. 2000. Feasibilities of substituting sodium for potassium in crop plants for advanced life support system. *Life Support and Biosphere Science* 7: 225–232.

Sun, R., and S. Hughes. 1998. Fractional extraction and physico-chemical characterization of hemicelluloses and cellulose from sugar beet pulp. *Carbohydrate Polymers* 36: 293–299.

Tesoriere, L., M. Allegra, D. Butera, and M.A. Livrea. 2004a. Absorption, excretion, and distribution of dietary antioxidant betalains in LDLs: Potential health effects of betalains in humans. *American Journal of Clinical Nutrition* 80: 941–945.

Tesoriere, L., D. Butera, A.M. Pintaudi, M. Allegra, and M.A. Livrea. 2004b. Supplementation with cactus pear (*Opuntia ficus-indica*) fruit decreases oxidative stress in healthy humans: A comparative study with vitamin C. *American Journal of Clinical Nutrition* 80: 391–395.

Tesoriere, L., D. Butera, M. Allegra, M. Fazzari, and M.A. Livrea. 2005. Distribution of betalain pigments in red blood cells after consumption of cactus pear fruits and increased resistance of the cells to ex vivo induced oxidative hemolysis in humans. *Journal of Agricultural and Food Chemistry* 53: 1266–1270.

Tesoriere, L., M. Fazzari, F. Angileri, C. Gentile, and M.A. Livrea. 2008. In vitro digestion of betalainic foods. Stability and bioaccessibility of betaxanthins and betacyanins and antioxidative potential of food digesta. *Journal of Agricultural and Food Chemistry* 56: 10487–10492.

Thimmaraju, R., N. Bhagyalakshmi, M.S. Narayan, and G.A. Ravishankar. 2003a. Kinetics of betalaine release under the influence of physical factors. *Process Biochemistry* 38: 1067–1074.

Thimmaraju, R., N. Bhagyalakshmi, M.S. Narayan, and G.A. Ravishankar. 2003b. Food grade chemical and biological agents assist the release of betalaines from hairy root cultures of *Beta vulgaris*. *Biotechnology Progress* 19: 1274–1282.

Thimmaraju, R., V. Kumar, N. Bhagyalakshmi, and G.A. Ravishankar. 2005. Peroxidase production from hairy root cultures of red beet (*Beta vulgaris*). *Electronic Journal of Biotechnology* 8: 185–196.

Thimmaraju, R., B. Neelwarne, and G.A. Ravishankar. 2004. In situ and ex situ adsorption and recovery of betalains from hairy root cultures of *Beta vulgaris*. *Biotechnology Progress* 20: 777–785.

Thimmaraju, R., B. Neelwarne, L. Venkatachalam, R.V. Sreedhar, and G.A. Ravishankar. 2006. Elicitation of peroxidase activity in genetically transformed root cultures of *Beta vulgaris* L. *Electronic Journal of Biotechnology* 9: 512–520.

Thimmaraju, R., L. Venkatachalam, M.S.K. Roohie, M.S. Narayan, and N. Bhagyalakshmi. 2007. Purification and characterization of an intracellular peroxidase from genetically transformed roots of red beet (*Beta vulgaris* L.). *Food Chemistry* 105: 1312–1320.

Turquois, T., M. Rinaudo, F. Taravel, and A. Heyraud. 1999. Extraction of highly gelling pectic substances from sugar beet pulp and potato pulp: Influence of extrinsic parameters on their gelling properties. *Food Hydrocolloids* 13: 255–262.

Vitti, M.C.D., L.K. Yamamoto, F.F. Sasaki, J.S. Aguila, R.A. Kluge, and A.P. Jacomimo. 2005. Quality of minimally processed beet roots stored in different temperatures. *Brazilian Archives of Biology and Technology* 48: 503–510.

von Elbe, J.H. 1975. Stability of betalaines as food colors. *Food Technology* 5: 42–44.

von Elbe, J.H., and E.L. Attoe. 1985. Oxygen involvement in betanine degradation – Measurement of active oxygen species and oxidation-reduction potentials. *Food Chemistry* 16: 49–67.

Von Elbe, J.H., I.Y. Maing, and C.H. Amundson. 1974. Color stability of betanin. *Journal of Food Science* 39: 334–337.

Wasserman, B.P., L.L. Eiberger, and M.P. Guilfoy. 1984. Effect of hydrogen peroxide and phenolic compounds on horseradish peroxidase-catalyzed decolorization of betalain pigments. *Journal of Food Science* 49: 536–538.

Webb, A.J., N. Patel, S. Loukogeorgakis, et al. 2008. Acute blood pressure lowering, vasoprotective, and antiplatelet properties of dietary nitrate via bioconversion to nitrite. *Hypertension* 51: 784–790.

Wheeler, R.M., and J.C. Sage. 2006. Crop production for advanced life support systems. *Technical Reports*. Accessed on 25th November 2011. Paper 1. http://docs.lib.purdue.edu/nasatr/1.

Wiley, R.C., and Y.N. Lee. 1978. Recovery of betalaines from red beets by a diffusion-extraction procedure. *Journal of Food Science* 43: 1056–1058.

Wiley, R.C., Y.N. Lee, J.J. Saladini, R.C. Wyss, and H.H. Topalian. 1979. Efficiency studies of a continuous diffusion apparatus for the recovery of betalaines from red table beet. *Journal of Food Science* 44: 208–211.

Wink, D.A., and N. Paolocci. 2008. Mother was right: Eat your vegetables and do not spit! When oral nitrate helps with high blood pressure. *Hypertension* 51: 617–619.

Wiśniewska, E., and A. Majewska-Sawka. 2008. The differences in cell wall composition in leaves and regenerating protoplasts of *Beta vulgaris* and *Nicotiana tabacum*. *Biologia Plantarum* 52: 634–641.

Wong, E., and M. Bregola. 1997. Use of sugar beet pulp for making paper or cardboard. WO 97/30215.

Yoo, S.D., and S.W. Harcum. 1999. Xanthan gum production from waste sugar beet pulp. *Bioresource Technology* 70: 105–109.

Zhang, D., S.M. Lanier, J.A. Downing, J.L. Avent, J. Lum, and J.L. McHale. 2008. Betalain pigments for dye-sensitized solar cells. *Journal of Photochemistry and Photobiology A: Chemistry* 195: 72–80.

Zhong, X., P.F. Fransz, J. Wennekes-van Eden, et al. 1996a. High-resolution mapping on pachytene chromosomes and extended DNA fibres by fluorescence in-situ hybridisation. *Plant Molecular Biology Reporter* 14: 232–242.

Zhong, X.B., J.H. de Jong, and P. Zabel. 1996b. Preparation of tomato meiotic pachytene and mitotic metaphase chromosomes suitable for fluorescence in situ hybridization (FISH). *Chromosome Research* 4: 24–28.

Zohary, D., and M. Hopf. 2000. *Domestication of plants in the old world: The origin and spread of cultivated plants in West Asia, Europe, and the Nile Valley*. Oxford: Oxford University Press.

Zvitov, R., A. Schwartz, and A. Nussinvitch. 2003. Comparison of betalain extraction from beet (*Beta vulgaris*) by low electrical field versus cryogenic freezing. *Journal of Texture Studies* 34: 83–94.

Chapter 2
Biosynthesis and Regulation of Betalains in Red Beet

Hiroshi Sekiguchi, Yoshihiro Ozeki, and Nobuhiro Sasaki

Abstract The red beet (*Beta vulgaris*) gets its distinctive color from betalains, one of the major plant pigment groups. Betalain-bearing plants appeared among plants of the families Caryophyllales, of which red beet is one of the main crops. Betalain pigments are classified into two groups—red betacyanin and yellow betaxanthin—and they all contain two atoms of nitrogen, unlike the other major plant pigments, the anthocyanins and flavonoids. Whereas the biosynthetic pathways of anthocyanins and flavonoids are reasonably well known, those of the betalains remain unclear. However, several important genes encoding the enzymes involved in betalain biosynthesis have now been identified. In this section, several distinctive enzymes of betalain biosynthesis are described and the unsolved reaction steps of the proposed pathway are discussed.

2.1 Introduction

The root of the widely cultivated red beet (*Beta vulgaris*) gets its distinctive crimson-red color from betacyanins belonging to the betalain family of pigments, which are water-soluble nitrogenous substances. Betalains are classified into two groups according to their color: red/purple betacyanins and yellow betaxanthins. Only plants in the family Caryophyllales contain betalain pigments, except for Caryophyllaceae and Molluginaceae (Clement and Mabry 1996). The mutual exclusiveness between betalains and anthocyanins, which comprise one of the major red pigment groups occurring in higher plants, is well known, although it remains uncertain why betalains and anthocyanins do not coexist in plants.

H. Sekiguchi • Y. Ozeki • N. Sasaki (✉)
Department of Biotechnology and Life Science, Tokyo University of Agriculture and Technology, 2-24-16 Naka-cho, Koganei, Tokyo, 184-8588, Japan
e-mail: fwhikky2003@yahoo.co.jp; ozeky@cc.tuat.ac.jp; nove@cc.tuat.ac.jp

B. Neelwarne (ed.), *Red Beet Biotechnology: Food and Pharmaceutical Applications*,
DOI 10.1007/978-1-4614-3458-0_2, © Springer Science+Business Media New York 2013

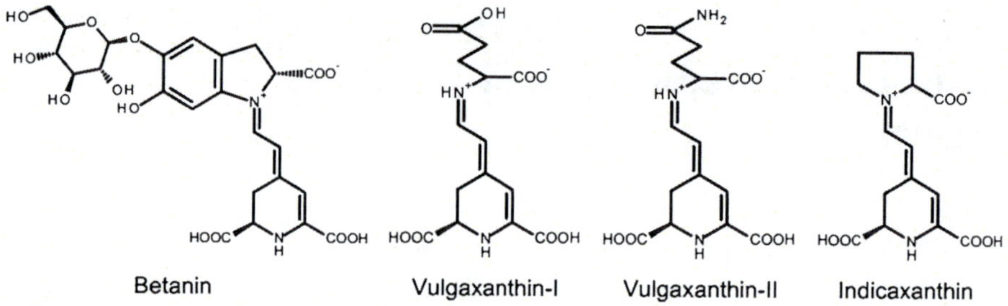

Fig. 2.1 Chemical structure of betalains

The chemical structures of representative betalains are shown in Fig. 2.1. Betalain molecules contain two atoms of nitrogen in their basic skeleton, which is not the case in the other major plant pigments of similar color, the anthocyanins and flavonoids. Betacyanin aglycones are synthesized from *cyclo*-dihydrophenylalanine (DOPA) or often from 2-descarboxy-*cyclo*-DOPA and betalamic acid; for reviews, see Strack et al. (2003) and Tanaka et al. (2008). Betacyanins usually exist as -*O*-glucosides, which are glucosylated at the 5- or 6-positions of the betanidin aglycone. Some betacyanins are modified with additional sugar moieties and often with organic acids. The main pigments in the root of the red beet are betanin (betanidin 5-*O*-glucoside), isobetanin (isobetanidin 5-*O*-glucoside), vulgaxanthin-I and -II and indicaxanthin (Fig. 2.1) (Kujala et al. 2001). Isobetanidin is a diastereoisomer of betanidin modified at the C15 position. Betaxanthins are synthesized from amino acids or amines and betalamic acid. All betalain molecules contain betalamic acid as the chromophore. It is believed that betalains are synthesized via several steps from tyrosine (Fig. 2.2). Interestingly, this pathway is very simple, in contrast to the pathways of anthocyanins, which involve at least 10 steps (Tanaka et al. 2008).

The first step of betalain biosynthesis is hydroxylation of tyrosine to form DOPA (Fig. 2.2I). In the second step, DOPA is converted to 4,5-*seco*-DOPA by DOPA 4,5-dioxygenase, followed by spontaneous cyclization to form betalamic acid (Fig. 2.2II–III). The formation of betaxanthins is completed by conjugation with amino acids or amines (except for *cyclo*-DOPA derivatives) (Fig. 2.2IV). To synthesize betacyanins, *cyclo*-DOPA formation is required in the biosynthetic pathway; *cyclo*-DOPA in turn is thought to be synthesized from tyrosine or DOPA by polyphenol oxidase (PPO). Several modification steps of betacyanin formation have been reported, with modification at the *cyclo*-DOPA step (Fig. 2.2VI–VII), at the betanidin aglycone step (Fig. 2.2VIII) and at the step after betanin formation (Fig. 2.2IX). The final formation of betalains involves the condensation of betalamic acid and amino acids, amines or *cyclo*-DOPA (derivatives), probably via spontaneous reactions (Schliemann et al. 1999). Recently, another pathway for the synthesis of 2-descarboxybetanidin from dopamine/betaxanthin via PPO has been proposed (Fig. 2.2X) (Gandia-Herrero et al. 2005a, b).

Although the pathway of anthocyanin biosynthesis has mostly been elucidated, the biosynthetic pathway of betalains remains unclear. However, since 2004, several

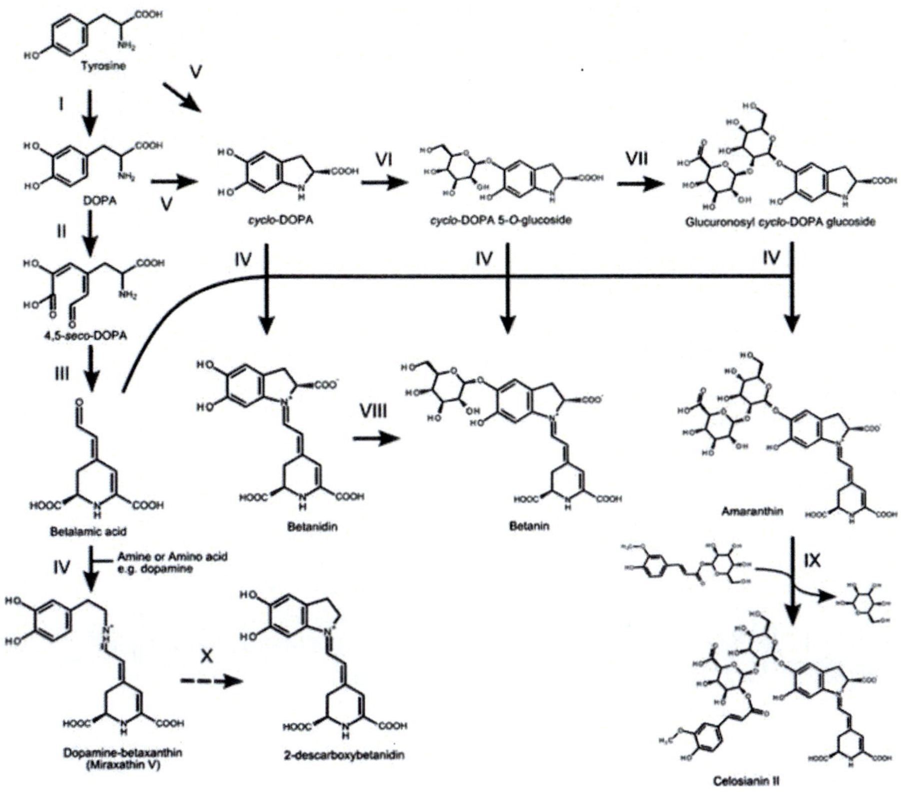

Fig. 2.2 Betalain biosynthetic pathways. *I* Tyrosine hydroxylation. *II* Extra ring cleavage of DOPA catalyzed by DOPA 4,5-dioxygenase. *III* Spontaneous cyclization leading to betalamic acid. *IV* Spontaneous condensation with betalamic acid. *V* Oxidation catalyzed by polyphenol oxidase and spontaneous cyclization. *VI cyclo*-DOPA 5-*O*-glucosyltransferase (GT) reaction. *VII cyclo*-DOPA glucoside glucuronosyltransferase reaction. *VIII* Betanidin 5-*O*-GT reaction. *IX* Acylation of betacyanin catalyzed by acyl glucose-dependent acyltransferase. *X* Oxidation of dopamine-betaxanthin catalyzed by polyphenol oxidase (PPO), producing 2-descarboxybetanidin

enzymes involved in betalain biosynthesis and the genes encoding them have been isolated and characterized.

2.2 Tyrosine Hydroxylase

It is believed that the first step in the betalain biosynthetic pathway is the hydroxylation of tyrosine (Fig. 2.2I). Several types of hydroxylases are known in the anthocyanin biosynthetic pathway. Thus, P450 monooxygenase catalyzes the addition of an oxygen atom to a substrate molecule as a hydroxyl group. Cinnamate 4-hydroxylase and flavanone 3′(5′)-hydroxylase are also classified in this group of enzymes.

The 2-oxoglutarate-dependent dioxygenases add an oxygen molecule both to a substrate molecule and to 2-oxoglutarate. PPO-type tyrosinases catalyze the addition of an oxygen atom to the *ortho* position of the monophenol and the oxidization of *ortho*-diphenol. One of the PPOs involved in aurone biosynthesis has been demonstrated in flavonoid biosynthesis and its enzymatic properties have been reported (Nakayama et al. 2000).

In relation to the biosynthesis of betalains, several tyrosine hydroxylase activities have been detected using crude protein extracts of plant materials. Steiner et al. (1999) succeeded in detecting tyrosine hydroxylase activity using crude enzyme preparations from cell cultures and plants of *Portulaca grandiflora* and from the hairy roots and cell cultures of *B. vulgaris* (Steiner et al. 1999; Steiner et al. 1996). They also characterized a tyrosine hydroxylase partially purified from callus cultures of *P. grandiflora*. Yamamoto et al. also reported the partial purification of tyrosine hydroxylase from callus cultures of *P. grandiflora* (Yamamoto et al. 2001). A cDNA coding for a PPO was isolated from *Phytolacca americana*, a betalain-producing plant, suggesting that this enzyme might be involved in the synthesis of betacyanin (Joy et al. 1995). The changes in the *PPO* mRNA levels paralleled betacyanin accumulation in the ripening stages of *P. americana* berries. However, whether the PPO catalyzes tyrosine hydroxylation or DOPA oxidization has not been determined. Despite the importance of the DOPA molecule as the precursor substance of alkaloids in general, not only of betalains, tyrosine hydroxylases have still not been isolated and the genes encoding plant tyrosine hydroxylases remain unidentified in higher plants.

2.3 DOPA Dioxygenase

DOPA dioxygenase (DOD) catalyzes the 4,5-extradiol cleavage of DOPA to form betalamic acid (Fig. 2.2II). DOD is one of the most characteristic enzymes in the biosynthetic pathway of betalains, because all betalain molecules contain betalamic acid as their chromophore (Fig. 2.1). DOD was characterized initially from fungi; it was known that some fungi produce betalain, and their DOD activities had been reported (Giord and Zryd 1991). In 1997, DOD was purified, and further characterized from *Amanita muscaria* (Hinz et al. 1997; Mueller et al. 1997). The characterization revealed two distinct enzymatic activities: DOPA 4,5-dioxygenase and DOPA 2,3-dioxygenase. The results indicate that muscaflavin, as well as betalamic acid, is generated by fungal DOD reactions. However, plant DOD clearly differs from fungal DOD, because muscaflavin has so far never been detected in plants. Indeed, for years, all attempts to isolate the plant gene for DOD using fungal DOD protein and gene information failed. In 2004, Christinet et al. isolated cDNA candidates for DOD from the petals of *P. grandiflora* using the cDNA subtraction method (Christinet et al. 2004). The *DOD* candidates were selected based on the detection of candidate gene expression in the colored petals of *P. grandiflora*. One of them clearly has DOD activity in vivo; this function was confirmed by introducing candidate cDNA into epidermal cells of *P. grandiflora* using microprojectile bombardment, followed by the confirmation of betalain production by high-performance liquid chromatography analysis of extracts of the transformed cells. In 2009, one cDNA homologous for *DOD* was isolated from the petals of *Mirabilis jalapa* and defined as *MjDOD* (Sasaki et al. 2009). *MjDOD* gene expression was detected in colored petals but not in leaves, stems or white petals. The *MjDOD* gene expression increased at later developmental stages in petals of *M. jalapa*. Furthermore, the recombinant MjDOD protein produced in *Escherichia coli* cells shows DOD enzymatic activity. MjDOD requires ascorbic acid for its enzymatic activity, which enables the iron ion harbored in the enzyme to be maintained in its proper redox state. Suitable conditions for assaying DOD activity have been determined but all attempts to detect DOD activity in intact plants have failed. Native DOD might be unstable or the tissues might contain large amounts of substances that inhibit DOD activity.

Homologous *DOD* genes in anthocyanin-producing plants have been reported (Christinet et al. 2004). However, the roles of those genes remain unclear. It will be interesting to determine how and when betalain-producing plants evolved DODs with the ability to produce betalamic acid from DOPA.

2.4 Steps in Betacyanin Biosynthesis

By 2000, more than 50 molecular species of betalains had been reported (Delgado-Vargas et al. 2000). The numbers of betalain molecules reported are increasing annually because of the development of high-precision liquid chromatography/mass spectrometry instruments (Jerz et al. 2010; Wybraniec et al. 2010, 2009). Interestingly, all betacyanins are modified with a glucose moiety (not other sugars) at the 5- or 6-position of the betanidin aglycone. By contrast, almost all anthocyanins are modified with a glucose moiety or often with other sugar moieties. The reason for why the first sugar moieties are limited to glucose is unclear. However, such selectivity is limited to the plant species included under families of Caryophyllales that produce betacyanin pigments.

Two betacyanin glucosylation steps have been proposed. One is glucosylation at a betanidin aglycone, as in anthocyanin biosynthesis (Fig. 2.2VIII). This proposed pathway, acting via a betanidin aglycone, has been supported by plant feeding experiments. Sciuto et al. supplied betanidin exogenously to the pear *Opuntia dillenii* where betanin formation was confirmed in fruit extracts (Sciuto et al. 1972). In 1992, UDP-glucose:betanidin 5-*O*- and 6-*O*-glucosyltransferase (GT) activities were detected in crude protein extracts from cell cultures of *Dorotheanthus bellidiformis* (Heuer and Strack 1992). The cDNAs encoding betanidin 5- and 6-GTs were isolated and the recombinant proteins were produced using an *E. coli* protein expression system. Both recombinant GT enzymes could provide glucose to flavonoids, and a phylogenetic tree analysis based on the amino acid sequences implied that betanidin GTs had been derived from flavonoid GTs through evolution (Vogt 2002). The other proposed step in betacyanin glucosylation is the glucosylation at *cyclo*-DOPA, the precursor of betacyanins. The formation of betacyanin is then completed by conjugation of *cyclo*-DOPA glucosides (derivatives) with betalamic acid (Fig. 2.2VI–VII). This route is supported by evidence that *cyclo*-DOPA was a more efficient precursor for producing amaranthin (glucuronosylbetanin) than for betanidin aglycone or betanin when these substrates were supplied exogenously to *Celosia plumosa* seedlings (Sciuto et al. 1974). Furthermore, *cyclo*-DOPA 5-*O*-glucoside accumulation was confirmed in young beet plants (Wyler et al. 1984) and in the peels of their roots (Kujala et al. 2001). In addition, the white petals of *M. jalapa* contain a large amount of *cyclo*-DOPA glucoside (Tanaka et al. 2008). Therefore, the possibility of a pathway via a *cyclo*-DOPA glucosylation step has long been discussed. In 2004, UDP-glucose:*cyclo*-DOPA 5-*O*-glucosyltransferase (cD5GT) activity was detected in crude protein extracts from betalain-producing plants (Sasaki et al. 2004). Furthermore, glucuronosyltransferase activity toward *cyclo*-DOPA glucosides was demonstrated using crude enzyme preparations from the inflorescences of *Celosia cristata* (Sasaki et al. 2005a). The cDNA encoding *cD5GT* was isolated from petals of *M. jalapa* and *C. cristata* (*MjcD5GT* and *CccD5GT*, respectively) and the recombinant cD5GT proteins produced using *E. coli* cells were confirmed to possess cD5GT activity (Sasaki et al. 2005b). The accumulation of *MjcD5GT* mRNA increased during the development of the petals of *M. jalapa* but the mRNAs were barely detected in stems and leaves that did not accumulate betacyanins. These results suggested that *MjcD5GT* is involved in betacyanin biosynthesis and that the biosynthetic pathway of betacyanins involves a modification at the *cyclo*-DOPA step.

Further steps in the modification of betacyanins have been reported. Enzymatic activity involved in the transfer of hydroxycinnamic acid (HCA) from HCA-glucose to betanin to produce HCA-betanins (Fig. 2.2IX) has been demonstrated (Bokern and Strack 1988; Bokern et al. 1992). Acyl glucose-dependent acyltransferase activities targeting anthocyanins were shown in crude extracts prepared from cultured carrot cells and carnation petals and purified from the petals of *Clitoria ternatea* (Noda et al. 2007). Same types of enzymes involved in other plant secondary metabolism were identified as isobutyryltransferase from *Lycopersicon pennellii* and sinapoyltransferases from Arabidopsis and *Brassica napus* (Lehfeldt et al. 2000; Li et al. 1999). Because these proteins were homologous to the serine carboxypeptidases, these enzymes were classified as members of the serine carboxypeptidase-like (SCPL) acyltransferases (Li and Steffens 2000). Unlike the enzymes catalyzing the reaction steps of betalain and anthocyanin biosynthesis, the SCPL acyltransferases are thought to be located in vacuoles (Hause et al. 2002). Therefore, the acylation of betacyanins by aromatic organic acids also seems to occur in vacuoles. Recently, acyl glucose-dependent glucosyltransferases involving anthocyanin biosynthesis were identified as the vacuolar proteins from the carnation and delphinium (Matsuba et al. 2010). Other vacuolar enzymes might also be involved in the synthesis of plant pigments (see Chap. 3).

2.5 Regulation of Betalain Biosynthesis

In the latter half of last century, the effects of light conditions, plant growth regulators and DOPA supplementation on betalain biosynthesis were examined in several betalain-bearing plants. Betacyanin synthesis in suspension cultures of *P. americana* was stimulated by the addition of 2,4-dichlorophenoxyacetic acid—a synthetic auxin—and inhibited by the addition of abscisic acid (Hirano et al. 1996; Sakuta et al. 1991). Inhibition of amaranthin synthesis by gibberellic acid was shown in *Amaranthus caudatus* (Stobart and Kinsman 1977). In *A. tricolor* seedlings, betacyanin synthesis was promoted by the exposure to cytokinins under dark conditions. Light- and temperature-sensitive control of betacyanin biosynthesis was shown in *A. tricolor* (Elliott 1979). The influence of medium and light conditions on cultured cells of *Portulaca* sp. was examined (Kishima et al. 1991; Noda and Adachi 2000), and the induction of betalain pigments in *Portulaca* sp. callus by irradiation with blue light was reported (Kishima et al. 1995). However, betalains are synthesized in beetroots continuously without light, because these usually grow underground. In the case of anthocyanin, similar regulation of the production of pigments is present in some vegetative plants, such as the black carrot, red radish and purple sweet potato.

Control of the expression of the genes involved in anthocyanin biosynthesis has been analyzed using molecular biology. Three transcriptional factors—MYB, basic helix-loop-helix (bHLH) and WD40—fundamentally regulate the expression of these genes (Hichri et al. 2011). Despite their inability to produce anthocyanins, several betalain-bearing plants harbor the genes homologous to dihydroflavonol 4-reductase and anthocyanidin synthase in their genomes (Shimada et al. 2004, 2005). Promoter analysis implied that the regulation of the expression of these genes is controlled by the same basic system that controls anthocyanin biosynthesis. Recently, two DOD-encoding genes were isolated from the cultured cells of *P. americana* and their promoter regions were isolated (Takahashi et al. 2009). Analysis of their *cis*-elements revealed sequences that can bind to MYB-type, bHLH-type and environmental stress-responsive transcription factors. Further analyses of the promoter region are required to understand the regulation of the genes involved in producing betalains.

2.6 Conclusions

For a long time, the enzymes and genes involved in betalain biosynthesis in higher plants were unknown. Since 2004, two of the important enzymes and genes encoding them have been identified. Further analysis in the fields of biochemistry and molecular biology might contribute to understanding the biosynthetic mechanism of betalains in plants. Fundamental questions that need to be answered through future research are why the betalain-bearing plants are limited to plants in the families of

Caryophyllales and why these plants never harbor the two pigments—betalains and anthocyanins—together, despite containing other flavonoids (Mabry 2001). These questions may now be answered using molecular biological analysis of the genes encoding DODs and GTs, which catalyze betanidin or *cyclo*-DOPA glucosylations. However, the two other reaction steps thought to be most important for betalain biosynthesis have not yet been identified: the synthesis of DOPA from tyrosine and the synthesis of *cyclo*-DOPA. Applying current molecular biological approaches such as transcriptome, proteome and metabolome analyses to the betalain-bearing plants might help clarify these reaction steps.

References

Bokern, M., and D. Strack. 1988. Synthesis of hydroxycinnamic acid esters of betacyanins via 1-*O*-acylglucosides of hydroxycinnamic acids by protein preparations from cell suspension cultures of *Chenopodium rubrum* and petals of *Lampranthus sociorum*. *Planta* 174: 101–105.

Bokern, M., S. Heuer, and D. Strack. 1992. Hydroxy cinnamic acid transferase in the biosynthesis of acylated betacyanins: purification and characterization from cell cultures of *Chenopodium rubrum* and occurrence in some other members of the Caryophyllales. *Botanica Acta* 105: 146–151.

Christinet, L., F.X. Burdet, M. Zaiko, U. Hinz, and J.P. Zryd. 2004. Characterization and functional identification of a novel plant 4,5-extradiol dioxygenase involved in betalain pigment biosynthesis in *Portulaca grandiflora*. *Plant Physiology* 134: 265–274.

Clement, J.S., and T.J. Mabry. 1996. Pigment evolution in the Caryophyllales: A systematic overview. *Bot. Acta* 109: 360–367.

Delgado-Vargas, F., A.R. Jimenez, and O. Paredes-Lopez. 2000. Natural pigments: Carotenoids, anthocyanins, and betalains—Characteristics, biosynthesis, processing, and stability. *Critical Reviews in Food Science and Nutrition* 40: 173–289.

Elliott, D.C. 1979. Temperature-sensitive responses of red light-dependent betacyanin synthesis. *Plant Physiology* 64: 512–524.

Gandia-Herrero, F., J. Escribano, and F. Garcia-Carmona. 2005a. Characterization of the monophenolase activity of tyrosinase on betaxanthins: The tyramine-betaxanthin/dopamine-betaxanthin pair. *Planta* 222: 307–318.

Gandia-Herrero, F., J. Escribano, and F. Garcia-Carmona. 2005b. Betaxanthins as substrates for tyrosinase. An approach to the role of tyrosinase in the biosynthetic pathway of betalains. *Plant Physiology* 138: 421–432.

Giord, P.A., and J.P. Zryd. 1991. Biogenesis of betalains: Purification and partial characterization of DOPA 4,5-dioxygenase from *Amanita muscaria*. *Phytochemistry* 30: 169–174.

Hause, B., K. Meyer, P.V. Viitanen, C. Chapple, and D. Strack. 2002. Immunolocalization of 1-*O*-sinapoylglucose: Malate sinapoyltransferase in *Arabidopsis thaliana*. *Planta* 215: 26–32.

Heuer, S., and D. Strack. 1992. Synthesis of betanin from betanidin and UDP-glucose by a protein preparation from cell suspension cultures of *Dorotheanthus bellidiformis* (Burm. f.) N. E. Br. *Planta* 186: 626–628.

Hichri, I., F. Barrieu, J. Bogs, C. Kappel, S. Delrot, and V. Lauvergeat. 2011. Recent advances in the transcriptional regulation of the flavonoid biosynthetic pathway. *Journal of Experimental Botany*. doi:10.1093/jxb/erq442 online.

Hinz, U.G., J. Fivaz, P.A. Girod, and J.P. Zyrd. 1997. The gene coding for the DOPA dioxygenase involved in betalain biosynthesis in *Amanita muscaria* and its regulation. *Molecular and General Genetics* 256: 1–6.

Hirano, H., M. Sakuta, and A. Komamine. 1996. Inhibition of betacyanin accumulation by abscisic acid in suspension cultures of *Phytolacca americana. Zeitschrift Fur Naturforschung C* 51c: 818–822.

Jerz, G., S. Wybraniec, N. Gebers, and P. Winterhalter. 2010. Target-guided separation of *Bougainvillea glabra* betacyanins by direct coupling of preparative ion-pair high-speed countercurrent chromatography and electrospray ionization mass-spectrometry. *Journal of Chromatography. A* 1217: 4544–4554.

Joy, R.W.T., M. Sugiyama, H. Fukuda, and A. Komamine. 1995. Cloning and characterization of polyphenol oxidase cDNAs of *Phytolacca americana. Plant Physiology* 107: 1083–1089.

Kishima, Y., K. Nozaki, R. Akashi, and T. Adachi. 1991. Light-inducible pigmentation in Portulaca callus; selection of a high betalain producing cell line. *Plant Cell Reports* 10: 304–307.

Kishima, Y., A. Shimaya, and T. Adachi. 1995. Evidence that blue light induces betalain pigmentation in *Portulaca callus. Plant Cell Tissue and Organ Culture* 43: 67–70.

Kujala, T., J. Loponen, and K. Pihlaja. 2001. Betalains and phenolics in red beetroot (*Beta vulgaris*) peel extracts: Extraction and characterisation. *Zeitschrift für Naturforschung C* 56: 343–348.

Lehfeldt, C., A.M. Shirley, K. Meyer, M.O. Ruegger, J.C. Cusumano, P.V. Viitanen, D. Strack, and C. Chapple. 2000. Cloning of the SNG1 gene of Arabidopsis reveals a role for a serine carboxypeptidase-like protein as an acyltransferase in secondary metabolism. *The Plant Cell* 12: 1295–1306.

Li, A.X., and J.C. Steffens. 2000. An acyltransferase catalyzing the formation of diacylglucose is a serine carboxypeptidase-like protein. *Proceedings of the National Academy of Sciences of the United States of America* 97: 6902–6907.

Li, A.X., N. Eannetta, G.S. Ghangas, and J.C. Steffens. 1999. Glucose polyester biosynthesis. Purification and characterization of a glucose acyltransferase. *Plant Physiology* 121: 453–460.

Mabry, T.J. 2001. Selected topics from forty years of natural products research: betalains to flavonoids, antiviral proteins, and neurotoxic nonprotein amino acids. *Journal of Natural Products* 64: 1596–1604.

Matsuba, Y., N. Sasaki, M. Tera, M. Okamura, Y. Abe, E. Okamoto, H. Nakamura, H. Funabashi, M. Takatsu, M. Saito, H. Matsuoka, K. Nagasawa, and Y. Ozeki. 2010. A novel glucosylation reaction on anthocyanins catalyzed by acyl-glucose-dependent glucosyltransferase in the petals of carnation and delphinium. *The Plant Cell* 22: 3374–3389.

Mueller, L.A., U. Hinz, M. Uze, C. Sautter, and J.P. Zryd. 1997. Biochemical complementation of the betalain biosynthetic pathway in *Portulaca grandiflora* by a fungal 3,4-dihydroxyphenylalanine dioxygenase. *Planta* 203: 260–263.

Nakayama, T., K. Yonekura-Sakakibara, T. Sato, S. Kikuchi, Y. Fukui, M. Fukuchi-Mizutani, T. Ueda, M. Nakao, Y. Tanaka, T. Kusumi, and T. Nishino. 2000. Aureusidin synthase: A polyphenol oxidase homolog responsible for flower coloration. *Science* 290: 1163–1166.

Noda, N., and T. Adachi. 2000. Induction of betacyanin synthesis and pigment accumulation in cell suspension cultures of Portulaca. *Plant Biotechnology* 17: 27–34.

Noda, N., T. Higaki, S. Toshio, K. Kazuma, T. Sasaki, S. Hasegawa, and M. Suzuki. 2007. The subcellular localization and substrate specificity of 1-*O*-acylglucose dependent anthocyanin acyltransferase from blue petal of butterfly pea (*Clitoria ternatea*). *Plant Cell Physiology Supplement* 48: 222.

Sakuta, M., H. Hirano, and A. Komamine. 1991. Simulation by 2,4-dichlorophenoxyacetic acid of betacyanin accumulation in suspension cultures of *Phytolacca americana. Physiologia Plantarum* 83: 154–158.

Sasaki, N., T. Adachi, T. Koda, and Y. Ozeki. 2004. Detection of UDP-glucose: *cyclo*-DOPA 5-*O*-glucosyltransferase activity in four o'clocks (*Mirabilis jalapa* L.). *FEBS Letters* 568: 159–162.

Sasaki, N., Y. Abe, K. Wada, T. Koda, Y. Goda, T. Adachi, and Y. Ozeki. 2005a. Amaranthin in feather cockscombs is synthesized via glucuronylation at the *cyclo*-DOPA glucoside step in the betacyanin biosynthetic pathway. *Journal of Plant Research* 118: 439–442.

Sasaki, N., K. Wada, T. Koda, K. Kasahara, T. Adachi, and Y. Ozeki. 2005b. Isolation and characterization of cDNAs encoding an enzyme with glucosyltransferase activity for *cyclo*-DOPA from four o'clocks and feather cockscombs. *Plant & Cell Physiology* 46: 666–670.

Sasaki, N., Y. Abe, Y. Goda, T. Adachi, K. Kasahara, and Y. Ozeki. 2009. Detection of DOPA 4,5-dioxygenase (DOD) activity using recombinant protein prepared from *Escherichia coli* cells harboring cDNA encoding DOD from *Mirabilis jalapa*. *Plant & Cell Physiology* 50: 1012–1016.

Schliemann, W., N. Kobayashi, and D. Strack. 1999. The decisive step in betaxanthin biosynthesis is a spontaneous reaction. *Plant Physiology* 119: 1217–1232.

Sciuto, S., G. Oriente, and M. Piattelli. 1972. Betanidin glucosylation in *Opuntia dillenii*. *Phytochemistry* 11: 2259–2262.

Sciuto, S., G. Oriente, M. Piattelli, G. Impelizzeri, and V. Amico. 1974. Biosynthesis of amaranthin in *Celosia plumosa*. *Phytochemistry* 13: 947–951.

Shimada, S., K. Takahashi, Y. Sato, and M. Sakuta. 2004. Dihydroflavonol 4-reductase cDNA from non-anthocyanin-producing species in the Caryophyllales. *Plant & Cell Physiology* 45: 1290–1298.

Shimada, S., Y.T. Inoue, and M. Sakuta. 2005. Anthocyanidin synthase in non-anthocyanin-producing Caryophyllales species. *The Plant Journal* 44: 950–959.

Steiner, U., W. Schliemann, and D. Strack. 1996. Assay for tyrosine hydroxylation activity of tyrosinase from betalain-forming plants and cell cultures. *Analytical Biochemistry* 238: 72–75.

Steiner, U., W. Schliemann, H. Bohm, and D. Strack. 1999. Tyrosinase involved in betalain biosynthesis of higher plants. *Planta* 208: 114–124.

Stobart, A.K., and L.T. Kinsman. 1977. The hormonal control of betacyanin synthesis in *Amaranthus cautatus*. *Phytochemistry* 16: 1137–1142.

Strack, D., T. Vogt, and W. Schliemann. 2003. Recent advances in betalain research. *Phytochemistry* 62: 247–269.

Takahashi, K., E. Takamura, and M. Sakuta. 2009. Isolation and expression analysis of two DOPA dioxygenases in *Phytolacca americana*. *Zeitschrift für Naturforschung. C* 64: 564–573.

Tanaka, Y., N. Sasaki, and A. Ohmiya. 2008. Biosynthesis of plant pigments: Anthocyanins, betalains and carotenoids. *The Plant Journal* 54: 733–749.

Vogt, T. 2002. Substrate specificity and sequence analysis define a polyphyletic origin of betanidin 5- and 6-*O*-glucosyltransferase from *Dorotheanthus bellidiformis*. *Planta* 214: 492–495.

Wybraniec, S., P. Stalica, G. Jerz, B. Klose, N. Gebers, P. Winterhalter, A. Sporna, M. Szaleniec, and Y. Mizrahi. 2009. Separation of polar betalain pigments from cacti fruits of *Hylocereus polyrhizus* by ion-pair high-speed countercurrent chromatography. *Journal of Chromatography. A* 1216: 6890–6899.

Wybraniec, S., P. Stalica, A. Sporna, and Y. Mizrahi. 2010. Profiles of betacyanins in epidermal layers of grafted and light-stressed cacti studied by LC-DAD-ESI-MS/MS. *Journal of Agricultural and Food Chemistry* 58: 5347–5354.

Wyler, H., U. Meuer, J. Bauer, and L. Stravs-Mombelli. 1984. *Cyclo*-dopa glucoside (=(2S)-5-β-d-glucopyranosyloxy)-6-hydroxyindoline-2-carboxylic acid) and its occurrence in red beet (*Beta vulgaris* var. rubra L.). *Helvetica Chimica Acta* 67: 1348–1355.

Yamamoto, K., N. Kobayashi, K. Yoshitama, S. Teramoto, and A. Komamine. 2001. Isolation and purification of tyrosine hydroxylase from callus cultures of *Portulaca grandiflora*. *Plant & Cell Physiology* 42: 969–975.

Chapter 3
Stability of Betalain Pigments of Red Beet

Shivapriya Manchali, Kotamballi N. Chidambara Murthy, Shruthi Nagaraju, and Bhagyalakshmi Neelwarne

Abstract Although pigments of biological origin are intensely researched for their market potentials and health benefits, their stability, from the maturity status of the source material to their processing and products into which the pigments are incorporated and the storage of those products, is shrouded with many problems. Both extrinsic and intrinsic factors affect the prospective arrays of applications of natural pigments such as betalains. Although the maturity of the red beet may not be the major quality trait in sourcing, the various processing steps, starting from the selection of the extraction medium, extraction conditions, product concentration, storage, transportation and incorporation into products, are huge tasks. The extrinsic factors affecting betalain stability include oxygen, temperature, pH, light and chemicals of processing and food products. On the other hand, the intrinsic factors in the red beet, apart from the quality

S. Manchali (✉)
Department of Biotechnology and Crop Improvement, University of Horticultural Sciences, Bagalkot, GKVK campus, Bengaluru 560 065, India
e-mail: priyachidambara@gmail.com

K.N.C. Murthy
Triesta Science, Health Care Global Oncology Hospital, Sampangiramnagar, Bengaluru 560 027, India

Fruit and Vegetable Improvement Centre, Texas A&M University, College Station, TX 77843-2119, USA
e-mail: kncmurthy@gmail.com

S. Nagaraju
Triesta Science, Health Care Global Oncology Hospital, Sampangiramnagar, Bengaluru 560 027, India

B. Neelwarne
Plant Cell Biotechnology Departments, Central Food Technological Research Institute (Council of Scientific and Industrial Research), Mysore 570 020, India
e-mail: bneelwarne@cftri.res.in

B. Neelwarne (ed.), *Red Beet Biotechnology: Food and Pharmaceutical Applications*, DOI 10.1007/978-1-4614-3458-0_3,

of the source material, are mainly enzymes: β-glucosidases, polyphenoloxidases (PPOs) and peroxidases (PODs). The current chapter is a compilation of research progress made in understanding the factors affecting the stability of betalains, the chemical changes taking place during degradation and their further fate. Improvements in processing/extraction techniques and storage conditions that impart stability to betalain pigments are discussed. Additionally, application of fermentation and other advanced processes used for the preparation of pigments and food formulations are discussed.

3.1 Introduction

Color is one of the important attributes of food; it is considered an indicator of quality and is an essential sensorial characteristic for consumer acceptance. Pigments from natural sources are gaining significant importance because of their enormous health benefits on one hand and because of the association of synthetic dyes with various health hazards, particularly in initiating and progression of chronic diseases, on the other hand. Among the naturally derived food colorants, carotenoids, chlorophylls, anthocyanins and betalains are commonly used (Stintzing and Carle 2007). Nature produces a variety of colorful compounds that are an integral part of the food prepared from such raw materials, and hence color appearance serves as a psychological descriptive measure for the judgement of food quality. Therefore, any deviation in the appearance of a food item from the conventional mental picture keeps the consumer from accepting such changes. Therefore, the color of natural food ingredients plays a crucial role for food. When natural pigments that can impart color even to processed foods are isolated from their source, their stability in terms of appearance and their molecular integrity become vital. The length of their stability is not just until the product reaches the consumer, but also after their consumption, so that the expected biological functions of pigments are precisely imparted.

Water-soluble pigments include anthocyanins, betalains and carminic acid, whereas carotenoids and chlorophylls constitute oil/fat-soluble pigments (Azeredo 2009). In addition to their color-imparting properties, betalains provide protection to humans against stress-related disorders by inhibiting lipid oxidation and peroxidation (Kanner et al. 2001; Kaur and Kapoor 2002; Reddy et al. 2005), impart anti-inflammatory effects (Gentile et al. 2004) and offer superior antiradical and antioxidative activities (Butera et al. 2002; Cai et al. 2003; Stintzing et al. 2005; Tsai et al. 2009). Natural pigments are known for their safety, and their stability under different processing conditions is critical. Hence, several studies were performed to understand the stability and chemical alterations of natural pigments. Stability of pigments depends on various factors, such as the maturity of the raw material, post-harvest handling and storage, extraction process, concentration, encapsulation (if applicable) and stabilization; all of these characteristics need to harmonize with the product chemistry and bio-availability when consumed.

3.2 Chemistry of Betalains and Their Structure-Stability Relationships

Betalains are aromatic indole molecules derived from tyrosine molecules (see Chap. 2) consisting of a chromophore and a sugar moiety; the molecules derive their names from "beta" and the pigments are dominant in beet root sourced for the commercial production of betalain pigments. The chief components of betalains of red beet are betacyanins and betaxanthins, which appear violet–red and yellow–orange to the human eye, respectively. The purple betacyanin is the dominant pigment in red beet, whereas yellow beets display a dominant presence of yellowish-orange betaxanthis. The other commercial variety of *Beta vulgaris*, the sugar beet, does not generally display these pigments, although sometimes pigments are expressed under stress and specific culture conditions (Pavokovic et al. 2009). In both the pigment structures, betalamic acid is a common chromophore. Apart from red beet, betalains also occur in other plants of Caryophyllales, which, during evolution, have replaced anthocyanins with betalains; although the genes for anthocyanin pathway enzymes are retained in the genomes of some of these plants (Tanaka et al. 2008; Strack et al. 2003).

Conversion of betalamic acid to betaxanthin occurs by the condensation of betalamic acid with one of the amino compounds, whereas the condensation with *cyclo*-Dopa [*cyclo*-3-(3,4-dihydroxyphenylalanine)] results in the formation of betacyanins, as shown in Chap. 2. These structural changes offer different levels of stability to each molecule. Many researchers have checked the stability of these pigments and reported a higher stability of betacyanins (betanin) than betaxanthins at room temperature (Sapers and Hornstein 1979) and also upon heating (Aronoff and Aronoff 1948; Herbach et al. 2004). Between the two pigments of red beet, vulgaxanthin I was prone to higher oxidation (Saguy 1979; Saguy et al. 1984) and hence less stable than betanin at acidic pH. A study conducted by Singer and von Elbe (1980) demonstrated that the half-life value of thermally treated betanin was 11 times higher than vulgaxanthin I. Most of the betalain stability research has been focused on betanin (betanidin 5-*O*-β-glucoside; Fig. 3.1), the most abundant betalain in red beet (*Beta vulgaris* L.). A great variety of structurally differing betacyanin formations occur as a result of glucosylation as well as acylation of the resulting 5-*O*- or 6-*O*-glucoside (Herbach et al. 2006), as shown in Fig. 3.1.

For both betaxanthins and betacyanins, the absorption maximum is influenced by the particular substitution pattern of the betanidin backbone. In the case of betacyanin, the glycosylation of betanidin (the aglycon) occurs bringing a hypsochromic shift of the resulting betacyanin, where glucose attached at C6 was observed to be less effective than C5C glycosylation (Stintzing et al. 2004). Generally, esterification with aliphatic acyl moieties was found to impart little impact on the absorption maxima of betacyanins (Wybraniec et al. 2001; Stintzing et al. 2002), whereas acylation with aromatic acids was found to bring about a bathochromic shift (Heuer et al. 1992). A later study by these authors established that C6 attachment of acylglucosides enhances the bathochromic shift, possibly resulting in a more rigid conformation (Heuer et al. 1994), which could result in copigmentation-like intra-molecular association (Schliemann and Strack 1998).

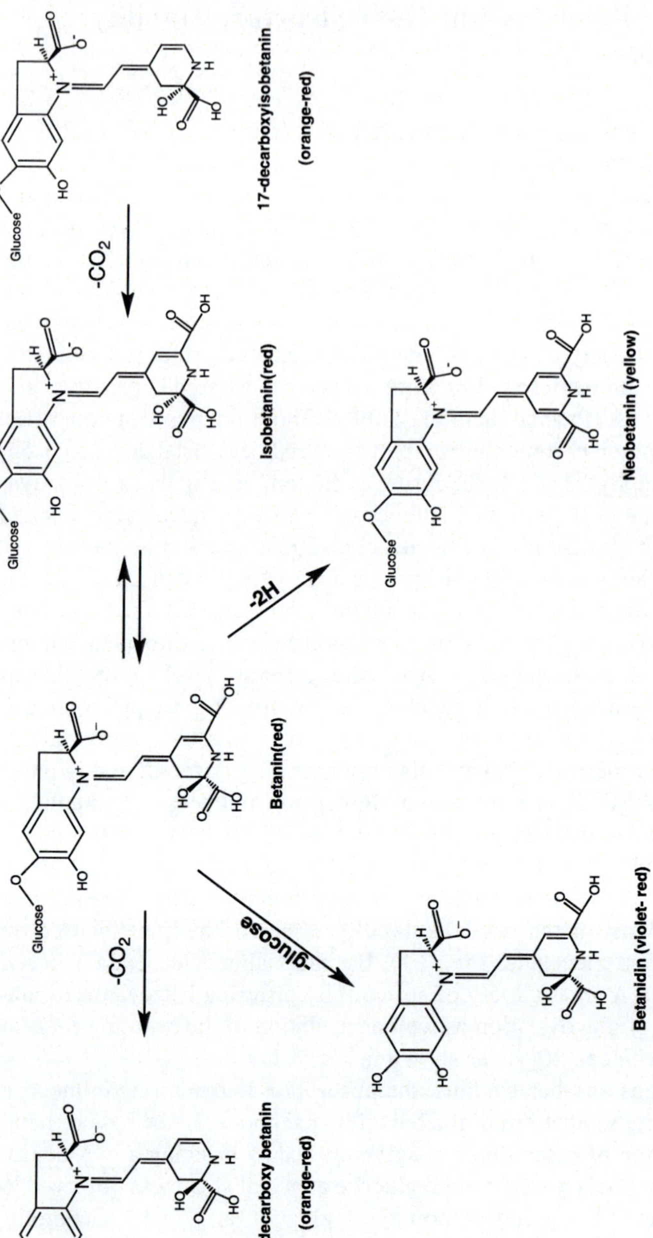

Fig. 3.1 Major degradation products of betanin (Based on Herbach et al. 2006)

3.3 Stability of Betalains

Among the factors responsible for the destabilization of betalains, both external and internal (enzymes and other components of red beet) factors play important roles. Table 3.1 lists the major factors contributing to the degradation of pigments. Most of the degradation processes are initiated by dehydrogenation, decarboxylation or oxidation of the parent molecule. The major sites prone for removal of hydrogen bonds, carboxylic acid and glucose moiety are depicted in Fig. 3.2.

Table 3.1 Influence of various indigenous and exogenous factors on the stability of red beet pigments

Type of factor	Degradation/synthesis products	Reference
pH	Acidic pH: condensation of betalemic acid with amines	Cai et al. (2001a), Herbach et al. (2006)
	Alkaline pH: hydrolysis of aldimine bond resulting in ferulic acid	
Temperature	Betalamic acid and cyclodopa-5-O-glycoside neobetanin, 17-decarboxy-betacyanins, 2-decarboxy-betacyanins, 2,17-bidecarboxybetanin and its isoform, and 14,15-dehydrogenated betacyanin (neobetanin) 2,15,17-tridecarboxy-2,3-dehydro-neobetanin	Schwartz and von Elbe (1983) Herbach et al. (2004) Wybraniec (2005) Nemzer et al. (2011)
Water activity (W_A)	Cleavage of aldimine bond	Azeredo (2009)
Hydrolysis	Alkaline: ferulic acid,	Strack et al. (2003)
	Acidic: betalemic acid	Herbach et al. (2007)
Oxidation	5,6-Dihydroxyindole, betanidin quinonoid, 2-decarboxy-2,3-dehydrobetanidin	Wybraniec and Michałowski (2011)
Nitrogen atmosphere	Prevents degradation up to 95%	Attoe and von Elbe (1982)
Metal ions	Hydrolysis and reduction of betalains	Azeredo (2009)
Dehydrogenation	Neobetanin (yellow)	Azeredo (2009)
Decarboxylation	17-decorboxybetanin, 15-decarboxybetanin	Azeredo (2009)
Isomerization	Isobetanin	Azeredo (2009)
Enzymes		
a. Peroxidase	Betalamic acid and oxidized cyclo-DOPA 5-O-beta-D-glucoside polymers	Martinez-Parra and Munoz (2001)
b. Glucosidases	Catalyzes hydrolysis of betalains	Wybraniec and Michałowski (2011)
c. Decarboxylase	Betaxanthin and isobetanin derivatives	Lee et al. (2005)
Level of glycosylation	Betanidine	Wybraniec and Michałowski (2011)
Esterification	Celosianin I and II (*p*-coumaroyl-and feruloylamaranthins), lampranthin I and II (*p*-coumaroyl, feruloylbetanins, betanidin 5-O-β-[1″→2′]-glucuronosyl-β-glucoside) and betanin (betanidin 5-O-β-glucoside)	Bokern and Strack (1988)
Acylation	Betalain acylated with ferulic, *p*-coumaric, or 3-hydroxy-3-methylglutaric acids	Cai et al. (2001a)
Epimerization	Isobetanidins	Shahidi (1999)
Encapsulation	Increases stability at <20°C	Gandía-Herrero et al. (2010)
Fermentation	Isomerization and dehydrogenation	Azeredo (2009)

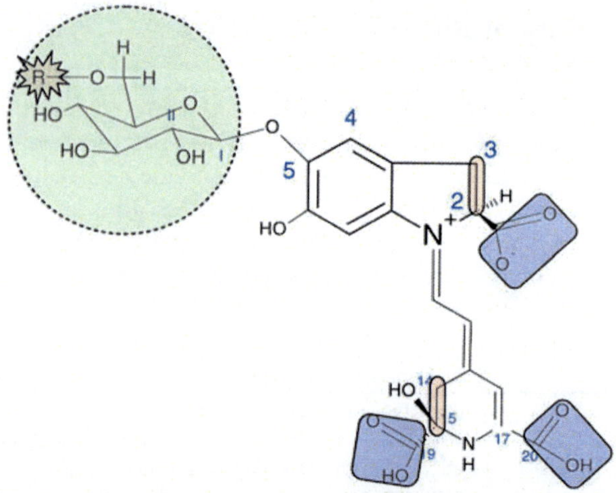

Fig. 3.2 Functional groups prone to degradation in betacyanins. The sites prone for decarboxylation are shown in *blue squares*, for dehydrogenation are in *pink* and for deglycosylation in *green circles* (Adapted with modifications from Stintzing and Carle 2007)

The earliest reports on stability suggest that betanins may undergo degradation through hydrolysis to form betalamic acid and cyclo-DOPA 5-O-glucoside (von Elbe et al. 1974). Interestingly, part of the degraded betanin was found to be reformed on cold storage of heated extracts upon condensation (Schwartz et al. 1981). Exposure of betanin to high temperature caused isomerisation and decarboxylation, resulting in the formation of C15-stereoisomer isobetanin and 15-descarboxybetanin, respectively (Schwartz and von Elbe 1983). The application of advanced analytical research tools helped in better understanding the degradation patterns; the resulting analytical data were helpful in revising old concepts. In red beet, recent studies demonstrated that the dehydrogenation of betacyanins at C14/C15 resulted in corresponding neo-compounds, causing a shift from a red to a yellowish hue. Decarboxylation at C17 and/or C2 and dehydrogenation at C14/C15 are reported to be more important in destabilizing the pigments (Herbach et al. 2006). HPLC techniques compared with spectroscopy techniques have been found to be more suitable for monitoring chemical changes during storage and processing (Stintzing and Carle 2007). In betacyanins, C19, C20 and the carboxylic acid group attached to C2 are the sites more prone for decarboxylation, whereas dehydrogenation includes the bond between C2, C3, C14 and C15 (Fig. 3.2) (Stintzing and Carle 2007).

Esterification of betacyanins with aliphatic acid is known to enhance its stability. Additionally, substitution with aromatic acid is known to increase the stability of the pigment by decreasing the susceptibility of compound for hydrolytic attachment. From a stereochemical point of view, substitution at the 6-O position more effectively increases the stability than substitution at the 5-O position (Schliemann and Strack 1998). Acetylation of pigment is known to prevent cleavage of compounds initiated by β-glucosidase, which may be present exogenously during processing or may be of endogenous origin, such as is known to occur in red beet. Another important factor responsible for enhanced stability of betalains is the concentration of the pigment molecule itself, as observed in the case of betacyanin (Moßhammer et al. 2005a).

3.3.1 Water Activity

Water can significantly influence the hydrolytic activity. The content of water/moisture in the extract can influence the degradation significantly. As in many food ingredients, a product with moisture content of less than 5% is known to be beneficial for preventing the degradation of betalain pigments; in other words pigments stored below a water activity (a_w) of 0.64 are suitable for long stability of red beet pigments (Serris and Biliaderis 2001). Plant ingredients like pectin and sugar acids have a low a_w, and that helps in the protection from degradation of the other similar pigment – the anthocyanins. The plant matrix structure also plays an important role in the stabilization of betalains and other pigments (Herbach et al. 2006). Thus, red beet and the cactus pear (pitaya), which are both major sources of betalain pigments, contain high levels of pectins, which probably render better stability to the pigment. Thus, whole extracts, rather than purified beet pigments, have been quite successful as commercial products. Storage of betalains in gelatin shells through spray drying offered higher stability compared with pectin shells, because of the higher firmness/hardness of the latter (Driver and Francis 1979). In case of hairy roots, pigment released into water was less stable than pigment released into the nutrient culture medium (Thimmaraju et al. 2003).

3.3.2 Oxygen

Both betaxanthin and betanin are prone to degradation in the presence of oxygen. Betanin showed a linear decrease in stability with increasing oxygen concentration. In addition to oxygen, hydrogen peroxide is another important atmospheric factor responsible for rapid degradation of beet pigments (Wasserman et al. 1984), probably through rapid oxidation. Degradation of betanins follows first-order kinetics in the presence of oxygen. However, in the absence of oxygen, the reaction deviates from first-order kinetics, which suggests that the reaction can be reversible under this condition. These observations are indicative that degradation caused by oxygen may be irreversible and have greater impact on the functionality of each pigment (Attoe and von Elbe 1985; Azeredo 2009).

When pigments are in solutions, removal of dissolved oxygen is known to increase the stability of betalains. This was confirmed by the ability of ascorbic acid and isoascorbic acid (strong antioxidants) to enhance the stability of betalains (Attoe and Von Elbe 1985). This study described the role of different antioxidants in oxygen-mediated degradation of betalain pigments and demonstrated that neither phenolic nor sulfur-containing antioxidants are capable of protecting betalains from degradation. Failure of free radical scavengers to protect the degradation suggests that degradation may not involve free radical-mediated degradation pathways other than direct oxidative degradation. Thus, oxygen plays a critical role in degradation, and more so in the presence of light and above ambient temperature (Attoe and Von Elbe 1985). Another study conducted on the effects of chemicals and enzymes suggests that degradation of betalains neither involves singlet oxygen nor superoxide anions and that alteration of the oxidation–reduction potential is what causes early degradation of aglycone (von Elbe and Attoe 1985).

3.3.3 Light

Light is known to influence the stability of pigments even below 40°C. Degradation due to light is caused by the absorption of light in UV and visible wavelengths leading to excitation of electrons of betalain chromophores to a more energetic state. This causes higher reactivity or lowered activation of the molecule (Jackman and Smith 1996). Light and oxygen have shown synergistic degradation of betalains. When individually tested, light and oxygen caused 15.6% and 14.6% degradation, whereas the combination of the two caused 28.6% degradation. Through a study on the stability of red beet pigments, Herbach et al. (2004) demonstrated that the degradation of pigments by light depended on the presence of oxygen, since the degradation induced by light under anaerobic conditions was negligible. Light- induced degradation was prevented using ascorbic acid at levels of 0.1–1.0% (Herbach et al. 2005), which indicates that dissolved oxygen could also pay a role. Pigments from stem tissues of the cactus *Myrtillocactus geometrizans* have shown that at pH 5.5, light of 87.09 ± 8.53 JK^{-1} mol^{-1} energy could induce the maximum degradation of pigments (Reynoso et al. 1997). This study also reported that metal ions could induce pigment degradation. Iron imparted higher degradation effects as compared with chromium. As in the case of the degradation of chlorophylls and carotenoids, the degradation of betalains by light follows first-order kinetics, and pigment degradation from dead cells was faster than from living cells. The degradation pattern in solution depends on the organic solvent used (James 1993).

Not all light sources cause degradative effects on betalains; light of different wavelengths can produce either additive or degradative effects on betalain pigments. This was studied by irradiating callus of *Portulaca* sp., and results suggest that blue light is capable of inducing higher betalain production in callus. Further, combination of UV light is known to synergize pigment production (Kishima et al. 1995). Similarly, another study reported the influence of light on cell cultures of red beet (Akita et al. 2000). When the influence on pigmentation pattern of exposing red beet hairy roots to blue, red and far red lights individually and in combination was studied, the results suggested that blue light in combination with far infra-red light induced higher accumulation of not just the betalain pigments, but the total sugar and sucrose as well (Shin et al. 2003).

3.3.4 Temperature

Temperature is another factor; although indispensable for conventional food processing, it is responsible for the degradation of betalains and many other components that support pigment stability, although it also has positive effects, such as enzyme inactivation. Studies suggest that between temperatures of 50–60°C,

70–80°C and 60–75°C, betalains are considerably degraded. Thermal degradation of beet pigment mainly depends on the temperature level, extent of heating, presence of oxygen and concentration of pigment (Herbach et al. 2006). Heating of red beet juice in the presence of acid caused degradation of betalains to form a mixture of monocarboxylated, dicarboxylated and tricarboxylated betacyanins. Decarboxylation of purified betacyanin sample occurred by the addition of acetic acid, followed by heating at 75–80°C for longer than an hour. The resulting products were analyzed using high-performance liquid chromatography with tandem mass spectrometry (LC-MS/MS) and diode-array (LC-DAD) detection. The results showed the formation of many derivatives, which included 17-decarboxy-betacyanins, 2-decarboxy-betacyanins, 2,17-bidecarboxybetanin and its isoform and 14,15-dehydrogenated betacyanin, also known as neobetacyanin (Wybraniec 2005). In addition to decarboxylation, heating is also known to produce dehydrogenated derivatives of betalains. Some of the major dehydrogenated products include 2,17-bidecarboxy-2,3-dehydro-neobetanin, 2-decarboxy-2,3-dehydro-neobetanin, 2,15,17-tridecarboxy-2,3-dehydro-neobetanin and 14,15-dehydrogenated betacyanins (Nemzer et al. 2011). After heat treatment of beet pigment, mixtures of mono-decarboxylated, bidecarboxylated, and tridecarboxylated betacyanins were identified along with their corresponding neobetacyanins, where epimerization as well as pairing of two decarboxy betacyanin were observed after mass spectrometry (LC/MS-MS) and diode array (LC-DAD) analyses (Wybraniec 2005). Another study reported that heating of betanin at 85°C resulted in formation of major products of hydrolytic cleavage and that few other products were formed from decarboxylation and dehydrogenation (Herbach et al. 2005). Betaxanthins isolated from *Celosia argentea* varieties were found to be stable at 40°C when stored in buffer of pH 5.5 in the absence of light and air. Betaxanthin was found to be stable up to 20 weeks when stored as lyophilized powder, compared with the solution form stored at 22°C (Cai et al. 2001b), which is indicative of the negative effects imparted by water activity. These results suggest that red beet pigments can be used in products such as dairy products, ice cream and confectionaries, that are prepared at a pH of 5.5 and stored at or below 22°C. The stability of betacyanins in powdered extract of *Amaranthus* sp. was higher compared with its solution form, suggesting the suitability of dried powder as a source for commercial food colorant (Cai et al. 1998).

3.3.5 Effect of pH

The major factor that impacts the stability of beet color is pH. Betalains are known to undergo hydrolytic cleavage that is facilitated by pH. Betalains are relatively stable at a wide range of pHs between 3 and 7, which offers a wide band for their applications in a large number of food formulations (Herbach et al. 2006). However, a significant bathochromic shift was observed (Fig. 3.3) when a particular quantity of pigment was added to solutions of different pH from 2 to 9 (taken in cuvette and pigment mixed quickly before spectral reading) where the spectral shift was higher in pigment obtained from hairy roots than from normal roots. However, the visible

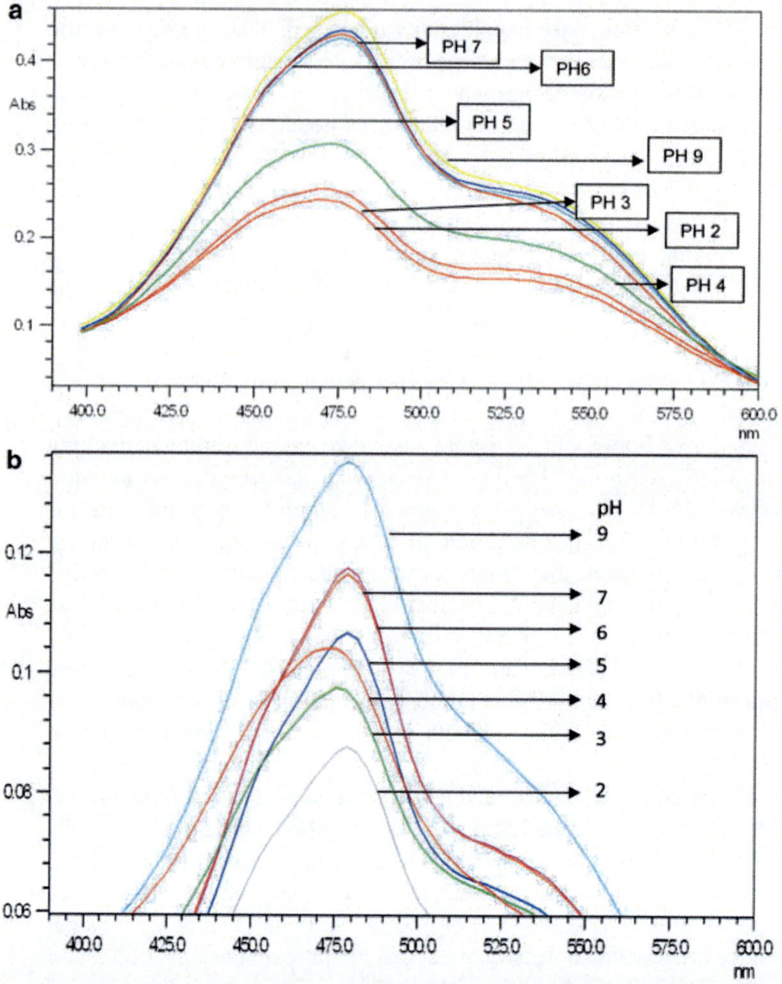

Fig. 3.3 Changes in spectral property of betalains at different pH from normal beet root (**a**) and hairy roots (**b**)

color tint in terms of Hunter value (Fig. 3.4) did not show significant color shifts between pH 3 and 9 (Thimmaraju et al. 2003).

3.3.6 *Other Factors*

Among the beet pigments, betacyanins are known to be more stable compared with betaxanthin, both when at room temperature and during thermal processing (Azeredo 2009). Hydrolytic cleavage was said to be responsible for the degradation of betacyanins (Stintzing and Carle 2007), however, later research has shown that decarboxylation and dehydrogenation are the major events responsible for the

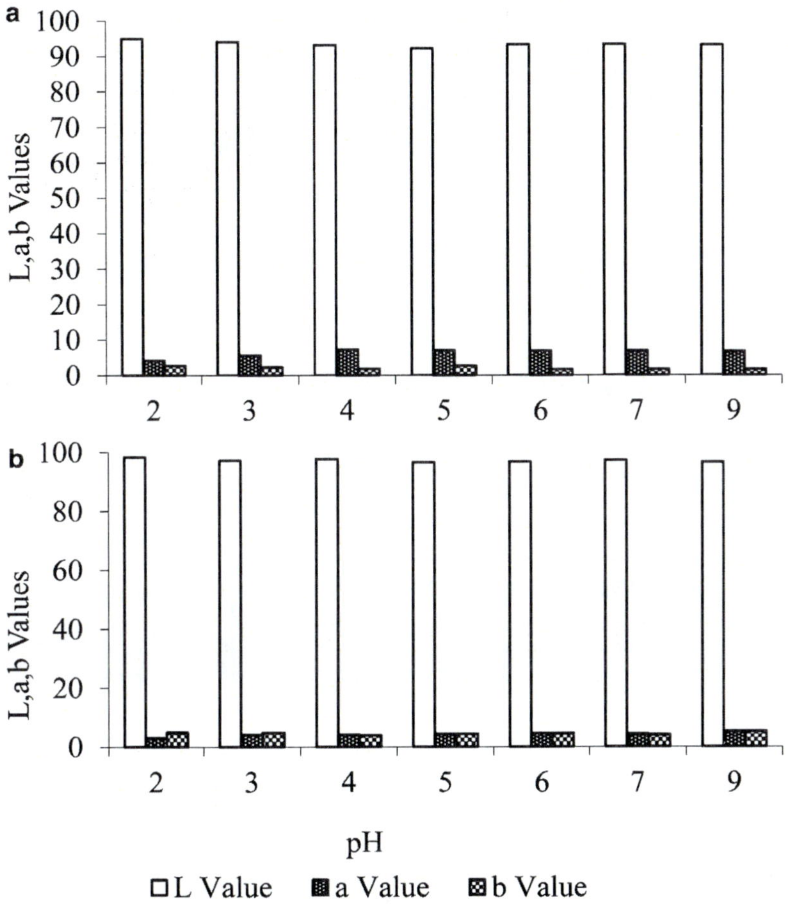

Fig. 3.4 Changes in Hunter's color property in terms of L, a and b values of betalains at different pH from normal beet root (**a**) and hairy roots (**b**). L: lightness, a: *redness*, b: *yellow*. The negative value for 'L', 'a' and 'b', represents darkness, *green* and *blue* respectively

degradation of betacyanin. Dehydrogenation of betacyanin at C14/C15 yielding a yellowish colored neo-compound is demonstrated in red beet and purple pitaya (Herbach et al. 2006). Among different betacyanins, glycosides are more stable than aglycons, which may be due to higher oxidation–reduction potentials of aglycons. Other causes of betalain degradation are improper inactivation of enzymes such as β-glucosidases, PPOs and PODs. These enzymes are known for inactivation of betalains. The optimum pH for the degradation of both betacyanins and betaxanthins is 3.4 (Shih and Wiley 1981).

Although pigment degradation due to peroxidation and the action of peroxidase (POD) are indicated for pigment degradation, betacyanin was found to be more susceptible for degradation by POD compared with betaxanthin. Since red beet is rich in PODs (Thimmaraju et al. 2005), quick inactivation of this enzyme is important during processing, for which temperature treatment is suggested. In red beet hairy

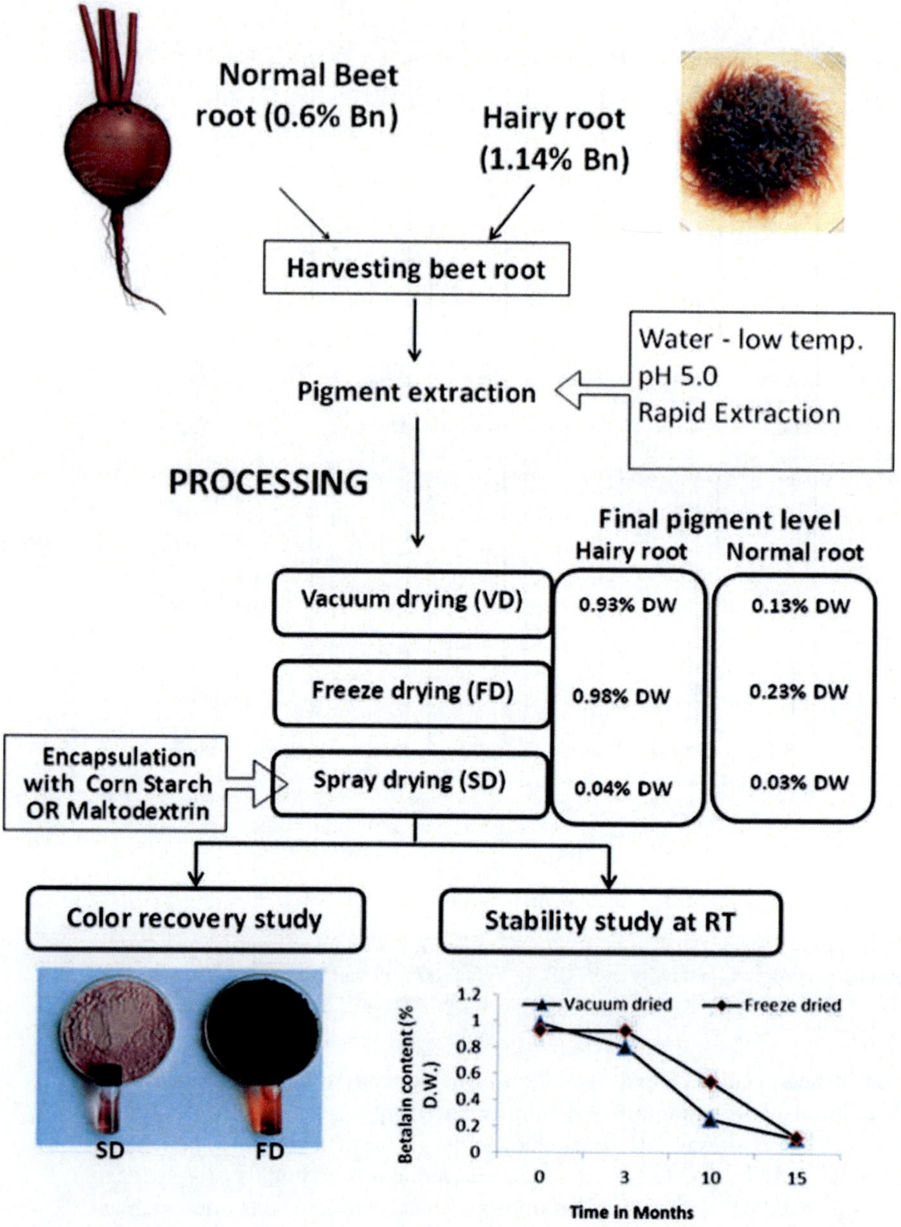

Fig. 3.5 Stability of betalains during processing (Neelwarne and Rudrappa –unpublished)

roots, acidic POD was found to be stable for longer than an hour at 50°C, whereas this enzyme extracted at pH 6 and pH 9 was very sensitive to temperature (Thimmaraju et al. 2005). Probably for these reasons, higher stability was observed in hairy root pigment extract than of normal beets (Fig. 3.5). In red beet extract, with continuous high-pressure carbon dioxide treatment at 7.5 MPa at 55°C for a short period of 30 min was found to inactivate 73% of POD and 93% of PPOs, which, in traditional thermal processing, could also be achieved at 95°C for 5 min, the latter invariably reduced betanin stability (Liu et al. 2010). Acidification of final product and isomerization of betanin are some of the major effects of this process.

The presence of catalase is known to completely degrade betaxanthins, suggesting a sensitivity of the pigment to hydrogen peroxide (Wasserman et al. 1984), while glucosidases cause aglycones. Such sensitivities to enzymes have been used advantageously to produce betalains with different levels of degradative products, each with a bathochromic shift resulting in products of different shades. Although there are a number of research publications describing the products obtained after heating, the correlation between the health benefits of these degraded products in comparison with the parent compound is elusive. In one study, it was demonstrated that the thermal degradation of beet pigments did not result in any loss of its antioxidant activity (Jiratanan and Liu 2004), which shows the utility of betalains in meat products that are cooked before consumption as well as the benefits of betalains beyond coloring.

3.4 Stability During Extraction

While betalains are wwell-known water soluble compounds, often methanol and ethanol are used up to 20–30% for their complete extraction, depending on the source of color. Slight acidification of water (by ascorbic or acetic acid) used for extraction of betalains is known to increase the stability, which is due to inactivation of PPO (Strack et al. 2003). Preheating of extract/plant material for a short period is helpful in preventing enzymatic degradation of the pigment. However, enzymatic treatment for degradation of hydrocolloids is helpful in the maximum extraction of pigments (Moßhammer et al. 2005b). The fermentation of red beet extract is known to utilize sugar, leading to an increased betacyanin content.

The extraction efficiency of betalains depends on the degree of cell permeabilization achieved in the method used for the extraction process. Among the different techniques, pulsed electric field treatment, ultrasound (Sivakumar et al. 2009), and irradiation with γ rays have shown to increase the efficiency of extraction (López et al. 2009). Pilot scale extraction techniques such as diffusion extraction, solid–liquid extraction and ultrafiltration are found to be more efficient than conventional hydraulic techniques (Azeredo 2009). Addition of antioxidants such as ascorbic acid (up to 50 mM) and vitamin E has shown to provide higher stability of pigments (Reynoso et al. 1997). Another method to increase the stability of betalain is micro-encapsulation using maltodextrin. This technique is good for obtaining an optimum health benefit, however, for colorant use, encapsulation may not be useful (Azeredo et al. 2007). In addition to use of suitable solvent, pH and extraction methodology, other factors that may help in retention of pigment stability are darkness during extraction, replacing nitrogen with oxygen, use of chelating agents and utilization of low temperature (Herbach et al. 2006).

Based on the available information, use of low light, temperature and antioxidant in a suitable cell destruction method would provide the optimum amount of betalains and help to prevent degradation.

3.5 Stability During Processing and Storage

Stability of betalains should be understood in the plant source and in the food products/formulation in which they are being used. Food processing steps may have negative effects on the stability of pigments, on the other hand, food matrix and other ingredients used in food may have positive effects. Addition of antioxidants like ascorbic acid and isoascorbic acid to food formulation is known to enhance the stability of pigments by the removal of oxygen (Attoe and von Elbe 1982). In addition to antioxidants, chelating agents such as citric acid and EDTA are also known to increase the stability of pigments in various formulations and direct storage of pigments. Glucose oxidase and β-cyclodextrin are known to provide the best stability of pigments through absorption of water and removal of dissolved oxygen, respectively (Hamburg and Hamburg 1991). Figure 3.5 depicts the major steps involved in the processing of betalains at the laboratory scale from either conventionally grown normal beet or from cultured hairy roots. Cultured hairy roots generally yield high pigment contents of uniform quality, which probably is the reason for the better stability of the final product. The lower degradation may also be due to a lower activity of polyphenol oxidase in hairy roots than in normal roots.

3.5.1 Encapsulation

Microencapsulation is often used to increase the stability and bio-efficacy of betalains. One of the studies suggests that microencapsulation of betalain with gum acacia and storing under low moisture conditions, i.e., less than 0.5, has restored pigment content up to 45 days. However, encapsulated samples stored at a_ws of 0.7 and 0.8 have shown degradation with restoration of antioxidant activity (Pitalua et al. 2010). In another study, microencapsulation of betalain-rich cactus pear juice and ethanolic extract with maltodextrin and inulin showed significant retention of pigments stored at 60°C for up to 44 days (Saénz et al. 2009). The results of these studies suggest that encapsulation is a promising method to restore the stability of red beet pigments.

3.6 Stability During Fermentation

Fermentation is generally applied to products that serve as good substrates for fermentation bacteria, where the process removes the carbohydrates and nitrogenous compounds. In beet juice, 80% of the juice solids constitute carbohydrates and nitrogenous compounds (von Elbe et al. 1974, Baráth et al. 2004), and their removal by fermentation enhances the remaining betalain concentration. Fermentation has been shown to decrease/increase the content and stability of betalain and its derivative pigments depending on the pH, the organism used and the fermentation conditions. One study reported that lactic acid-mediated fermentation resulted in a 60% loss in pigment content, whereas an un-inoculated control sample showed an 80–90% loss of betalain (Buruleanu et al. 2009). Among the different *Lactobacillus* strains used in the study, *L. delbrueckii* 0854 and *L. paracasei/casei* 0923 resulted in the maximum loss of pigment in Chrobry variety. On the other hand, *L. brevis/L. paracasei/casei* 0944/0916 resulted in the maximum loss of pigments in Czerwona Kula (Czyżowska et al. 2006). The study showed that the Czerwona Kula variety is more stable, and that fermentation can induce formation of betanidin and isobetanidin and that the optimum effect was seen with *L. paracasei/casei* 0923. Another study reported on the fermentation of cactus pear (*Opuntia ficus-indica*) juice using the wine yeast *Saccharomyces cerevisiae*. The kinetics study conducted at 50°C, 70°C and 90°C showed pigment degradation as a function of temperature. The reaction rate constants were found to be 0.0066, 0.0206 and 0.1244 min^{-1} for temperatures of 50°C, 75°C and 90°C, respectively. It was also found that fermentation did not influence the thermostability of the pigments. The fermentation process has shown a reduction in the content of total soluble solids by 44% and this may help in increasing the stability of pigments (Turker et al. 2001). Another study suggests that stability of betalain will be increased by 20–40% with fermentation process (Socaciu 2007). In a study conducted by Baráth et al. (2004), it was found that *Lactobacillus curvatus* 2770 provided the best fermentation with the fastest acidification in 24 h, although such an effect was influenced by the beet variety from which the juice was obtained. However, pectolytic enzyme treatment for juicing was found to be more useful without drastic acidification, allowing fermentation to continue longer, which the authors reasoned was due to a higher availability of dissolved oxygen (Baráth et al. 2004).

3.7 Stability in Products

The history of using betalains as food colorants goes back to the eighteenth century, when betalain was used to color pancakes. Later, traditionally, red beet was a component of red wines and pickled red beets in hamburgers. As a colorant, betalains are used in a number of food products, such as juice, sausage, cake and other confectionaries (von Elbe et al. 1974; Herbach et al. 2004; Stintzing and Carle 2007), to compensate for pigment losses during processing and to impart a different taste. A carbonated beverage has been developed and patented (Indian Patent No. 244380). The last author's research group has developed a process for the preparation of stable red beet concentrate (Indian Patent No. 237374), where both normal roots and hairy roots are extracted using a special process and the color in syrup form could be used in a wide array of products. However, the stability in the recipient product was subject to the exposure to pigment-deteriorating conditions such as light, water activity, air and temperature. Juice products of cactus pear have been researched extensively for betalain stability (Herbach et al. 2007).

Regarding the solid version of red beet pigment, there was initial success in obtaining both encapsulated and spray-dried betalains for coloring food products. As explained above, gelatin-encapsulated products display a high stability (Driver and Francis 1979). Commercially, red beet betacyanins are approved for use as food additive in the USA (No. 1600) and in the European Union (E-162), where they are considered as natural and exempt from batch certification. These commercial versions are widely used in the world (Castellar et al. 2003). When red rice color and beet color were tested in packaged pork sausages, products with red yeast (*Monascus purpureus*) rice, red beet root juice and betanin (E162), the color properties of sausages with red beet root were the nearest to fresh (control) sausages, whereas those with red yeast rice had significantly lower color values. For this product, beet color not only protected the sausages from discoloration but also extended acceptability by approximately 4 days and the willingness of habitual consumers to purchase, according to evaluation with two different types of lighting display. Therefore, red beet root juice may be envisioned as one of the customer-preferred natural colorant for use in fresh pork sausages (Martinez et al. 2006) and perhaps for other meat products. These commercial red beet color powders have also been useful in dairy products and juice centers and pastry shops where fresh products are continuously served.

3.8 Concluding Remarks

The fact that more studies on the enzymes involved in betalain degradation are addressed in red beets rather than on the enzymes involved in its biosynthesis speaks to the concern of food industries in making betalain molecules more suitable for food applications than their counterpart anthocyanins, which have only one third of the hue strength and lose their performance by shifting their tint (Stintzing and Carle 2004). In summary, betalains are excellent natural colorants with reasonably good stability and a broad range of coloring properties. With a wide range of biological and beneficial health properties, betalains are the ideal source of natural food colorant, which is; as evident from the vast utility and market of betalains. The market value of food color in 2009 was $1.45 billion, with a demand of 40,000–50,000 t, of which, betalains have a major share. The main concern in beet pigment stability is its fast browning, which may be desirable for meat products, but may not be suitable for products such as juice. Although research has shown beneficial effects of even degraded pigments (Jiratanan and Liu 2004), research has also shown that such browning can be controlled by oxygen removal and inhibition of auto-oxidative enzymes. The high content of nitrogen in beet extract can be tackled by the fermentation process – which offers additional health benefits with better pigment stability. In addition to red beet roots, cell cultures of red beet have been developed as a perpetual source of pigments (Akita et al. 2002). Red beet hairy roots offer a consortium of better opportunities packed with high pigment quantity, uniform quality and low geosmin (earthy odor) content coupled with easy processing. However, their genetically transformed nature is currently a barrier for the commercial application of such processes. Considering their high tinctorial value, more innovative approaches for extraction (see Chap. 14) as well as molecular modifications (probably with nanotechnology) may indeed improve the stability of this marvellous class of pigments.

References

Akita, T., Y. Hina, and T. Nishi. 2000. Production of betacyanins by a cell suspension culture of table beet (*Beta vulgaris* L.). *Bioscience, Biotechnology, and Biochemistry* 64: 1807–1812.

Akita, T., Y. Hina, and T. Nishi. 2002. New medium composition for high betacyanin production by a cell suspension culture of table beet (*Beta vulgaris* L.). *Bioscience, Biotechnology, and Biochemistry* 66: 902–905.

Aronoff, S., and E.M. Aronoff. 1948. Thermal degradation of dehydrated beets. 2. Chromatographic separation of red beet-root pigments. *Food Research* 13: 59.

Attoe, E.L., and J.H. von Elbe. 1982. Degradation kineticsofbetanine in solutions as influenced by oxygen. *Journal of Agricultural and Food Chemistry* 30: 708–712.

Attoe, E., and J. Von Elbe. 1985. Oxygen involvement in betanine degradation: Effect of antioxidants. *Journal of Food Science* 50(1): 106–110.

Azeredo, H. 2009. Betalains: Properties, sources, applications, and stability–a review. *International Journal of Food Science and Technology* 44(12): 2365–2376.

Azeredo, H.M.C., A.N. Santos, et al. 2007. Betacyanin stability during processing and storage of a microencapsulated red beetroot extract. *American Journal of Food Technology* 2(4): 307–312.

Baráth, A., A. Halász, E. Neméth, and Z. Zalán. 2004. Selection of LAB strains for fermented red beet juice production. *European Food Research and Technology* 218: 184–187.

Bokern, M., and D. Strack. 1988. Synthesis of hydroxycinnamic acid esters of betacyanins via 1-O-acylglucosides of hydroxycinnamic acids by protein preparations from cell suspension cultures of *Chenopodium rubrum* and petals of *Lampranthus sociorum*. *Planta* 174(1): 101–105.

Buruleanu, L., I. Manea, M.G. Bratu, D. Avram, and C.L. Nicole. 2009. Effects of prebiotics on the quality of lactic acid fermented vegetable juices. *Ovidius University Annals of Chemistry* 20: 102–107.

Butera, D., L. Tesoriere, et al. 2002. Antioxidant activities of Sicilian prickly pear (*Opuntia ficus indica*) fruit extracts and reducing properties of its betalains: Betanin and indicaxanthin. *Journal of Agricultural and Food Chemistry* 50(23): 6895–6901.

Cai, Y., M. Sun, et al. 1998. Colorant properties and stability of *Amaranthus* betacyanin pigments. *Journal of Agricultural and Food Chemistry* 46(11): 4491–4495.

Cai, M.S., et al. 2001a. Identification and distribution of simple and acylated betacyanins in the Amaranthaceae. *Journal of Agricultural and Food Chemistry* 49(4): 1971–1978.

Cai, Y., M. Sun, et al. 2001b. Chemical stability and colorant properties of betaxanthin pigments from *Celosia argentea*. *Journal of Agricultural and Food Chemistry* 49(9): 4429–4435.

Cai, Y., M. Sun, et al. 2003. Antioxidant activity of betalains from plants of the Amaranthaceae. *Journal of Agricultural and Food Chemistry* 51(8): 2288–2294.

Castellar R., J.M. Obón, M. Alacid, and J.A. Fernández-López. 2003. Color properties and stability of betacyanins from Opuntia fruits. *Journal of Agricultural and Food Chemistry* 51: 2772–2776.

Czyżowska, A., E. Klewicka, et al. 2006. The influence of lactic acid fermentation process of red beet juice on the stability of biologically active colorants. *European Food Research and Technology* 223(1): 110–116.

Driver, M.G., and F.J. Francis. 1979. Stability of phytolaccanin, betanin and FD&C Red #2 in dessert gels. *Journal of Food Science* 44: 518–520.

Gandía-Herrero, F., M. Jiménez-Atiénzar, et al. 2010. Stabilization of the bioactive pigment of *Opuntia* fruits through maltodextrin encapsulation. *Journal of Agricultural and Food Chemistry* 58(19): 10646–10652.

Gentile, C., L. Tesoriere, et al. 2004. Antioxidant betalains from cactus pear (*Opuntia ficus-indica*) inhibit endothelial ICAM-1 expression. *Annals of the New York Academy of Sciences* 1028(1): 481–486.

Hamburg, A., and A. Hamburg. 1991. The stability properties of red beet pigments: Influence of pH, temperature and some stabilizers. *European Food Research and Technology* 56: 1693–1695.

Herbach, K.M., F.C. Stintzing, et al. 2004. Impact of thermal treatment on color and pigment pattern of red beet (*Beta vulgaris* L.) preparations. *Journal of Food Science* 69(6): C491–C498.

Herbach, K., F. Stintzing, et al. 2005. Identification of heat-induced degradation products frompurified betanin, phyllocactin and hylocerenin by high-performance liquid chromatography/electrospray ionization mass spectrometry. *Rapid Communications in Mass Spectrometry* 19: 2603–2616.

Herbach, K.M., F.C. Stintzing, et al. 2006. Betalain stability and degradation – structural and chromatic aspects. *Journal of Food Science* 71(4): R41–R50.

Herbach, K.M., C. Maier, et al. 2007. Effects of processing and storage on juice colour and betacyanin stability of purple pitaya (*Hylocereus polyrhizus*) juice. *European Food Research and Technology* 224(5): 649–658.

Heuer S., S. Richter, J.W. Metzger, V. Wray, M. Nimtzt, and D. Strack. 1994. Betacyanins from bracts of *Bougainvillea glabra*. Phytochemistry 37: 761–767.

Heuer, S., V. Wray, J.W. Metzger, and D. Strack. 1992. Betacyanins from flowers of Gomphrena globosa. *Phytochemistry* 31: 1801–1807.

Jackman, R and J. Smith. 1996. Anthocyanins and betalains. In *Natural food colorants*, eds. G. Hendry, J. Houghton, 245–309. *Natural food colorants.* Glaslow, UK: Blackie Academic & Professional.

James, R.N. 1993. Rates and possible mechanism of light-dependent degradation of pigments in detritus derived from phytoplankton. *Journal of Marine Research* 51(1): 155–179.

Jiratanan, T., and R.H. Liu. 2004. Antioxidant activity of processed table beets (*Beta vulgaris* var, conditiva) and green beans (*Phaseolus vulgaris* L.). *Journal of Agricultural and Food Chemistry* 52: 2659–2670.

Kanner, J., S. Harel, et al. 2001. Betalains a new class of dietary cationized antioxidants. *Journal of Agricultural and Food Chemistry* 49(11): 5178–5185.

Kaur, C., and H.C. Kapoor. 2002. Antioxidant activity and total phenolic content of some Asian vegetables. *International Journal of Food Science and Technology* 37(2): 153–161.

Kishima, Y., A. Shimaya, et al. 1995. Evidence that blue light induces betalain pigmentation in *Portulaca callus. Plant Cell, Tissue and Organ Culture* 43(1): 67–70.

Lee, C.H., M. Wettasinghe, et al. 2005. Betalains, phase II enzyme-inducing components from red beetroot (*Beta vulgaris* L.) extracts. *Nutrition and Cancer* 53(1): 91–103.

Liu, X., Y. Gao, H. Xu, Q. Hao, G. Liu, and Q. Wang. 2010. Inactivation of peroxidase and olyphenol oxidase in red beet (*Beta vulgaris* L.) extract with continuous high pressure carbon dioxide. *Food Chemistry* 119: 108–113.

López, N., E. Puértolas, et al. 2009. Enhancement of the extraction of betanine from red beetroot by pulsed electric fields. *Journal of Food Engineering* 90(1): 60–66.

Martinez, L., I. Cilla, J.A. Beltran, and P. Roncales. 2006. Comparative effect of red yeast rice (*Monascus purpureus*), red beet root (*Beta vulgaris*) and betanin (E-162) on colour and consumer acceptability of fresh pork sausages packaged in a modified atmosphere. *Journal of the Science of Food and Agriculture* 86: 500–508.

Martinez-Parra, J., and R. Munoz. 2001. Characterization of betacyanin oxidation catalyzed by a peroxidase from *Beta vulgaris* L. roots. *Journal of Agricultural and Food Chemistry* 49(8): 4064–4068.

Moßhammer, M.R., F.C. Stintzing, et al. 2005a. Development of a process for the production of a betalain-based colouring foodstuff from cactus pear. *Innovative Food Science and Emerging Technologies* 6(2): 221–231.

Moßhammer, M., F. Stintzing, et al. 2005b. Colour studies on fruit juice blends from *Opuntia* and *Hylocereus cacti* and betalain-containing model solutions derived therefrom. *Food Research International* 38(8–9): 975–981.

Nemzer, B., Z. Pietrzkowski, A. Spórna, P.W. Stalica, W. Thresher, T. Michałowski, and S. Wybraniec. 2011. Betalainic and nutritional profiles of pigment-enriched red beet root (*Beta vulgaris* L.) dried extracts. *Food Chemistry* 127: 42–53.

Pavokovic, D., G. Rusak, V. Besendorfer, and M. Krsnik-Rasol. 2009. Light-dependent betanin production by transformed cells of sugar beet. *Food Technology and Biotechnology* 47: 153–158.

Pitalua, E., M. Jimenez, et al. 2010. Antioxidative activity of microcapsules with beetroot juice using gum Arabic as wall material. *Food and Bioproducts Processing* 88(2–3): 253–258.

Reddy, M.K., R.L. Alexander-Lindo, et al. 2005. Relative inhibition of lipid peroxidation, cyclooxygenase enzymes, and human tumor cell proliferation by natural food colors. *Journal of Agricultural and Food Chemistry* 53: 9268–9273.

Reynoso, R., F. Garcia, et al. 1997. Stability of betalain pigments from a cactacea fruit. *Journal of Agricultural and Food Chemistry* 45: 2884–2889.

Saénz, C., S. Tapia, et al. 2009. Microencapsulation by spray drying of bioactive compounds from cactus pear (*Opuntia ficus-indica*). *Food Chemistry* 114(2): 616–622.

Saguy, I. 1979. Thermostability of red beet pigments (Betanine and Vulgaxanthin–I): influence of pH and temperature. *Journal of Food Science* 44:1554–1555.

Sapers, G.M., and J.S. Hornstein 1979. Varietal differences in colorant properties and stability of red beet pigments. *Journal of Food Science* 44: 1245–1248.

Schliemann, W., and D. Strack. 1998. Intramolecular stabilization of acylated betacyanins. *Phytochemistry* 49(2): 585–588.

Schwartz, S.J., and J.H. von Elbe. 1983. Identification of betanin degradation products. *Zeitschrift für Lebensmitteluntersuchung und-Forschung A* 176(6): 448–453.

Schwartz, S., B. Hildenbrand, et al. 1981. Comparison of spectrophotometric and HPLC methods to quantify betacyanins. *Journal of Food Science* 46(1): 296–297.

Serris, G.S., and C.G. Biliaderis. 2001. Degradation kinetics of beetroot pigment encapsulated in polymeric matrices. *Journal of the Science of Food and Agriculture* 81(8): 691–700.

Shahidi, F. 1999. Chemicals via higher plant bioengineering. In *Advances in Experimental Medicine and Biology*, ed. F. Shahidi, et al. Vol. 464. New York: Kluwer Academic.

Shih, C.C., and R.C. Wiley. 1981. Betacyanine and betaxanthine decolorizing enzymes in the beet (*Beta vulgaris* L.) root. *Journal of Food Science* 47: 164–166.

Shin, K.S., H.N. Murthy, et al. 2003. Induction of betalain pigmentation in hairy roots of red beet under different radiation sources. *Biologia Plantarum* 47(1): 149–152.

Singer, J.W., and J.H. von Elbe. 1980. Degradation rates of Vulgaxanthine I. *Journal of Food Science* 45: 489–491.

Sivakumar, V., J.L. Anna, et al. 2009. Ultrasound assisted enhancement in natural dye extraction from beetroot for industrial applications and natural dyeing of leather. *Ultrasonics Sonochemistry* 16(6): 782–789.

Socaciu, C. 2007. *Food colorants: Chemical and functional properties*. Florida, USA: CRC.

Stintzing, F.C., and R. Carle. 2004. Functional properties of anthocyanins and betalains in plants, food, and in human nutrition. *Trends in Food Science & Technology* 15: 19–38.

Stintzing, F.C., and R. Carle. 2007. Betalains-emerging prospects for food scientists. *Trends in Food Science and Technology* 18(10): 514–525.

Stintzing, F.C., J. Conrad, I. Klaiber, U. Beifuss, and R. Carle. 2004. Structural investigations on betacyanin pigments by LC NMR and 2D NMR spectroscopy. *Phytochemistry* 65: 415–422.

Stintzing, F.C., A. Schieber, and R. Carle. 2002. Betacyanins in fruits from red-purple pitaya, Hylocereus polyrhizus (Weber) Britton & Rose. *Food Chemistry* 77: 101–106.

Stintzing, F.C., K.M. Herbach, et al. 2005. Color, betalain pattern, and antioxidant properties of cactus pear (*Opuntia* spp.) clones. *Journal of Agricultural and Food Chemistry* 53(2): 442–451.

Strack, D., T. Vogt, et al. 2003. Recent advances in betalain research. *Phytochemistry* 62(3): 247–269.

Tanaka, Y., N. Sasaki, and A. Ohmiya. 2008. Biosynthesis of plant pigments: Anthocyanins, betalains and carotenoids. *The Plant Journal* J54: 733–749.

Thimmaraju, R., N. Bhagyalakshmi, et al. 2003. Kinetics of pigment release from hairy root cultures of *Beta vulgaris* under the influence of pH, sonication, temperature and oxygen stress. *Process Biochemistry* 38(7): 1069–1076.

Thimmaraju, R., N. Bhagyalakshmi, V. Kumar, and G.A. Ravishankar. 2005. Peroxidase production from hairy root cultures of red beet (*Beta vulgaris*). *Electronic Journal of Biotechnology* 8: 185–196.

Tsai, P.J., C.H. Sheu, et al. 2009. Thermal and pH stability of betacyanin pigment of Djulis (*Chenopodium formosanum*) in Taiwan and their relation to antioxidant activity. *Journal of Agricultural and Food Chemistry* 58(2): 1020–1025.

Turker, N., Y. Coşkuner, et al. 2001. The effects of fermentation on the thermostability of the yellow-orange pigments extracted from cactus pear (*Opuntia ficus-indica*). *European Food Research and Technology* 212(2): 213–216.

von Elbe, J.H., and E.L. Attoe. 1985. Oxygen involvement in betanine degradation – Measurement of active oxygen species and oxidation reduction potentials. *Food Chemistry* 16(1): 49–67.

von Elbe, J., J. Klement, et al. 1974. Evaluation of betalain pigments as sausage colorants. *Journal of Food Science* 39(1): 128–132.

Wasserman, B., L. Eiberger, et al. 1984. Effect of hydrogen peroxide and phenolic compounds on horseradish peroxidase-catalysed decolorization of betalain pigments. *Journal of Food Science* 49: 536–538.

Wybraniec, S. 2005. Formation of decarboxylated betacyanins in heated purified betacyanin fractions from red beet root (*Beta vulgaris* L.) monitored by LC-MS/MS. *Journal of Agricultural and Food Chemistry* 53(9): 3483–3487.

Wybraniec, S., and T. Michałowski. 2011. New pathways of betanidin and betanin enzymatic oxidation. *Journal of Agricultural and Food Chemistry* 59(17): 9612–9622.

Wybraniec, S., I. Platzner, S. Geresh, H.E. Gottlieb, M. Haimberg, M. Mogilnitzki, et al. 2001. Betacyanins from vine cactus Hylocereus polyrhizus. *Phytochemistry* 58: 1209–1212.

Chapter 4
Red Beet as a Model System for Studying Vacuolar Transport of Primary and Secondary Metabolites

Nandini P. Shetty, Kirsten Jørgensen, and Hans J. Lyngs Jørgensen

Abstract The vacuole in a plant cell is a highly dynamic structure with important functions in maintaining the balance between synthesis, sequestration, storage and distribution of minerals/metabolites, precisely monitoring the flow in and out as well as biochemical conversions occurring in the cell. Red beet has served as a model system for studies of several vacuolar functions and made important contributions to our understanding of metabolite transporters located in the vacuolar membrane. Thus, here, we give an overview of the current progress made on unravelling transport of secondary metabolites, solutes and ions with special emphasis on contributions from red beets. Red beet is an important source of the two metabolites betalains and folates. Whereas the biosynthesis and accumulation of these metabolites have been studied in different systems, the transport processes have not been fully understood. Thus, red beet serves as a model to study not only important biosynthetic steps and the enzymes involved in betalain and folate synthesis, but also for unravelling transport and cellular signalling processes. Knowledge about red beet primary and secondary metabolism can thus provide further insight into the possible roles of various transporters, which may then be applied for regulation of the biosynthesis and recovery of these metabolites.

N.P. Shetty (✉)
Plant Cell Biotechnology Department, Central Food Technological Research Institute (Council of Scientific and Industrial Research), Mysore 570 020, India
e-mail: pcbt@cftri.res.in

K. Jørgensen • H.J.L. Jørgensen
Department of Plant Biology and Biotechnology, University of Copenhagen, Thorvaldsensvej 40, Frederiksberg C DK-1871, Denmark
e-mail: kij@life.ku.dk; hjo@life.ku.dk

B. Neelwarne (ed.), *Red Beet Biotechnology: Food and Pharmaceutical Applications*, DOI 10.1007/978-1-4614-3458-0_4,

4.1 Introduction

For a long time, red beet has been a valued food item enjoyed by many people around the world. In addition, red beet has attracted considerable attention in research during the past years due to its content of natural pigments which can be used by the food industry (Azeredo 2009; Nemzer et al. 2011; Loginova et al. 2011; Pavoković and Krsnik-Rasol 2011) as well as for textile and other industries (Sivakumar et al. 2009; Pavoković and Krsnik-Rasol 2011). Furthermore, several of the compounds found in red beets, such as the betalain pigments, have been found to have health-promoting effects like antioxidant activity (Azeredo 2009; Lee et al. 2009; Sugihara et al. 2009; Gandía-Herrero et al. 2010; Pavoković and Krsnik-Rasol 2011), making them highly interesting from a dietary point of view.

Red beet has served as a valuable model for understanding vacuole-related functions within the cell. Many important phytochemicals, such as the colored betalain pigments, are sequestered into the vacuole, making this important for studying vacuole-related structure–function relationships. This knowledge will be important for developing appropriate processes for industrial applications. At the same time, studies of vacuoles will permit utilisation of red beet as a model for further basic physiological studies. Extensive progress has been made in understanding a number of physiological processes, such as sugar transport across plant cell membranes (Doll et al. 1979), isolation techniques for vacuoles and protoplasts (Schmidt and Poole 1980), studies of the biological effects of phytoalexins on membranes (Spessard et al. 1994) and investigations of membrane potential and side effects of microelectrode measurements of cell intracellular potentials on host cell vacuoles (Doll and Hauer 1981; Lassalles et al. 1987). This review provides an update of the current research progress on plant vacuoles as well as on processes involved in transportation of compounds into and out of the vacuole, mainly emphasising studies conducted in red beet.

4.2 Vacuoles

4.2.1 General Features of the Vacuole

The vacuole has numerous functions, including maintenance of turgor pressure, protoplasmic homeostasis (Roytrakul and Verpoorte 2007), storage of metabolic products and secondary metabolites, such as sugars (Niemietz and Hawker 1988; Doll et al. 1979), production of defence-related compounds (Osbourn 1996), accumulation of nitrate and other ions (Dechorgnat et al. 2011), sequestration of xenobiotics and heavy metals (Rascio and Navari-Izzo 2011) as well as digestion of cytoplasmic constituents (Johnson et al. 1990). The major function of the vacuolar membrane (tonoplast) is the transport of ions and metabolites between cytoplasm and vacuole (Taiz 1992).

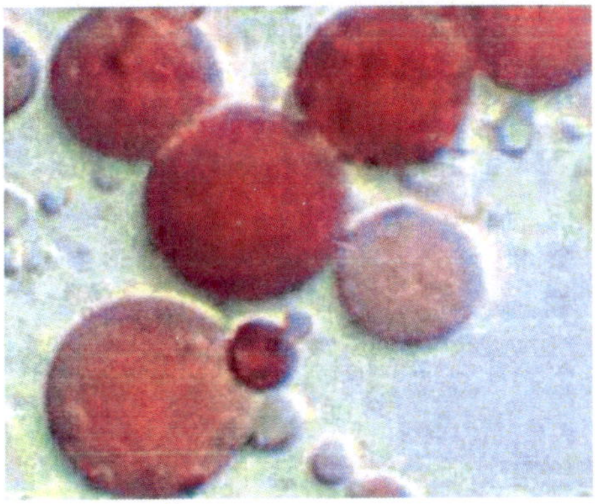

Fig. 4.1 Pigment-containing vacuoles isolated from red beet (The photograph was kindly provided by Dr. E. V. Pradedova, *Siberian Institute of Plant Physiology and Biochemistry, Russia. For more details about the author and vacuole extraction, consult the reference*: Pradedova et al. 2011)

The central vacuole in a plant cell is the most important storage compartment for secondary metabolites, proteins and other compounds, often taking up a very large proportion of the volume of the cell and indeed, it has many important roles, which result in a significant impact on the physiology of the cell and therefore on the whole plant. Like other cellular organelles, vacuoles can be studied when isolated (Fig. 4.1). Based on their mode of generation, vacuoles may be divided in two main groups, i.e., lytic and protein storage vacuoles (Zouhar and Rojo 2009; Hara-Nishimura and Hatsugai 2011). Whereas the first type primarily contains hydrolytic enzymes for degradation of various compounds in the cell as well as ions and metabolites (including pigments), the second type is used for the accumulation of different types of proteins. The formation and distribution of these different types of vacuoles have been reviewed by Zouhar and Rojo (2009). The soluble proteins in both types of vacuoles are either produced in the endoplasmic reticulum or the Golgi apparatus and moved into the vacuoles in the form of precursors, where they are subsequently assembled enzymatically (reviewed by Zouhar and Rojo 2009; Hara-Nishimura and Hatsugai 2011). The vacuolar content is acidic; this is caused by the activity of two proton pumps, one driven by adenosine triphosphate (ATP) and the other by inorganic pyrophosphate (PP_i) (Hedrich and Marten 2011).

4.2.2 *Sequestration of Different Classes of Compounds in Vacuoles*

Many plant secondary metabolites and other compounds are stored in vacuoles and vesicles, which serve as reservoirs for the future use of these molecules. Another major reason for their transport into the vacuoles is the need to remove them from the cytoplasm where they are synthesized so that further product formation is not repressed. Many of these compounds are needed for the cell to maintain normal

Table 4.1 Overview of transporters involved in transport of different metabolites and other molecules into and out of vacuoles/tonoplasts

Molecule	Transporter	Driving force	Selected references
Water	Aquaporins	Diffusion gradient	Gilliham et al. (2011)
Sucrose	Symport with protons	ATP	Ayre (2011), Doll et al. (1979), Echeverría and Gonzalez (2000), Etxeberria and Gonzales (2003)
Ions (e.g., Ca^{2+}, Na^{+}, K^{+})	H^{+}-pumps Ca^{2+}/H^{+} antiporter K^{+}/H^{+} antiporter FV channels SV channels VK channels Cation/H^{+} exchangers	ATP Passive	Gilliham et al. (2011), Hedrich and Neher (1987), Zhao et al. (2009), Pottosin et al. (2003)
Alkaloids/phenols/terpenoids	ABC transporters H^{+}-antiporter Ion and conformational trapping	ATP	Bartholomew et al. (2002), Dean and Mills (2004), Yazaki (2005)
Folate	ABC transporters	ATP	Raichaudhuri et al. (2009)
Xenobiotics (heavy metals/herbicides)	Transporter proteins H^{+}-antiporter ABC transporter	ATP	Rascio and Navari-Izzo (2011)

functions, but at the same time, they may be toxic to the cells themselves or unstable in the cytoplasm and therefore necessitate effective mechanisms for detoxification and/or compartmentalization like storage in vacuoles (Gunawardena et al. 2004; Rascio and Navari-Izzo 2011). Once transported to their storage sites, the secondary metabolites may interact with other compounds and/or proteins to form stable structures, or may be degraded by catabolic enzymes and/or chemical reactions for recycling them through other metabolic pathways (Roytrakul and Verpoorte 2007). The exact role of each cellular compartment in secondary metabolism is still not clear for most metabolites, because these cellular organelles/structures are often involved in synthesis, storage, protein turnover, transport or export of the end product as well as the biosynthetic enzymes.

Calcium and other inorganic cations participate in many vital processes in the plant cell, including maintenance of stability and strength of membranes and cell walls, maintenance of osmotic balance in the vacuole and diverse signalling events (reviewed by Gilliham et al. 2011; Isayenkov et al. 2010; Peiter 2011). The vacuole serves as a primary pool of free calcium ions in plant cells and the vacuole is a major source of Ca^{2+} for intracellular calcium signalling (Alexandre et al. 1990; Maeshima 2001; Gilliham et al. 2011; Peiter 2011). An overview of the transport mechanisms of different metabolites and other molecules are shown in Table 4.1.

Sucrose is one of the major solutes accumulating in the vacuoles, both for long-term storage purposes and for short-term storage in order to avoid repressing photosynthetic activity. Red beets have been extensively studied with respect to sucrose storage (e.g., Doll et al. 1979; Niemietz and Hawker 1988; Echeverría and Gonzalez 2000) as well as for influx and efflux of the vacuoles.

Another important type of metabolite stored in the vacuole is chlorophyll degradation products (Hörtensteiner and Kräutler 2011). Thus, chlorophyll is potentially dangerous for the cell because the molecules can absorb light energy. Therefore, if chlorophyll and its degradation products are not tightly regulated, the energy trapped may lead to uncontrolled production of reactive oxygen species, causing cell damage. The end product of chlorophyll degradation, i.e., blue-fluorescing, colorless breakdown products and subsequently non-fluorescent chlorophyll catabolites, are stored inside the vacuole (reviewed by Hörtensteiner and Kräutler 2011). Movement of these catabolites to the vacuole from the senescing chloroplasts takes place by active transport processes and involves ATP-binding cassette (ABC) transporters, but the exact nature of these is currently not known.

Yet another group of compounds stored in the vacuole is aimed at defence against biotic and abiotic stress. An example is the preformed antifungal compounds, phytoanticipins (such as cyanogenic glycosides, glucosinolates, phenols and phenolic glycosides), saponins, sulphur compounds and unsaturated lactones, which are sequestered in vacuoles or other organelles in healthy, non-infected plants (Osbourn 1996). In the vacuole, the compounds may either be present in an active form in healthy plants from where they can directly be released once the cell is damaged. Other types of compounds, such as cyanogenic glycosides and glucosinolates, occur as inactive precursors, which are activated when the plant is attacked by a pest or a pathogen or the tissue is damaged. This activation involves plant apoplastic enzymes, released from the vacuole due to tissue damage (Osbourn 1996). Other defence-related enzymes found in vacuoles include the PR-protein β-1,3-glucanase (Van den Bulcke et al. 1989).

Two types of metabolites have attracted special attention in red beet, i.e., betalains (pigments of secondary metabolism) and folates (vitamin B9 compounds). Betalains are nitrogen-containing, water-soluble pigments, synthesised from the amino acid tyrosine and primarily responsible for the color in red beets. There are two types—the red–violet betacyanins and the yellow betaxanthins, which are formed as conjugates of betalamic acid and cyclo-3-(3,4-dihydroxyphenyl)-L-alanine (cyclo-DOPA) with amino acids or amines, respectively (Strack et al. 2003, Azeredo 2009, Pavoković and Krsnik-Rasol 2011). In red beet, betanin and betanidin are the most common betacyanins and vulgaxanthin I and II are the most abundant betaxanthins (Escribano et al. 1998) (see Chap. 2). In plants, betalains play a variety of physiological roles and, owing to their bioactive properties, are sequestered in the vacuoles (Han et al. 2009). In plants, their display in flowers has been found to attract animals for pollination; in fruits, they aid in seed dispersal (Stintzing and Carle 2004); whereas in vegetative organs such as leaves and roots, they may serve as defence compounds. Furthermore, they have also been shown to function as scavengers of reactive oxygen species (Escribano et al. 1998) and function in UV protection as exemplified by the ice plant (*Mesembryanthemum crystallinum*) (Stintzing and Carle 2004). Betalains have also been suggested to possess antiviral properties in red beet (Pavoković and Krsnik-Rasol 2011), and Sepúlveda-Jiménez et al. (2004, 2005) found that betacyanin accumulated in response to wounding and infiltration with avirulent *Pseudomonas syringae* pv. *tabaci* or *Agrobacterium tumefaciens* in leaves of red beet, thus demonstrating a putative role of these compounds in defence.

Tetrahydrofolate and derivatives thereof (folates) are also stored in vacuoles in red beet roots. They are essential for nucleotide biosynthesis, gene methylation and amino acid metabolism (Orsomando et al. 2005; Raichaudhuri et al. 2009; Hanson and Gregory 2011). Orsomando et al. (2005) reported that in red beet, almost 80% of the folate in roots is polyglutamylated and up to 60% of this was found in the vacuoles (mainly 5-methyltetrahydrofolate).

The vacuole also stores enzymes that are important in reducing the damage resulting from oxidative stress on vacuolar molecular structures (Isheeva et al. 2009; Pradedova et al. 2011). Thus, in red beet taproots, Pradedova et al. (2011) found three isoforms of cyanide-sensitive Cu,Zn-activated superoxide dismutases located in the vacuolar sap, without any association with the membrane. Furthermore, they discovered several acidic and basic isoforms of phenol-dependent peroxidases in the vacuolar sap and in the membrane, being weakly associated with the tonoplast.

4.3 Biosynthesis of Secondary Metabolites and Other Compounds

The enzymes and genes involved in betalain biosynthesis are less well characterized than those of flavonoids and carotenoids. However, betalains are synthesized in the cytoplasm, but stored in vacuoles as glycosides, mainly in flowers and fruits, but also in vegetative tissues of plants such as the red beet root. The current knowledge on biosynthesis, properties and distribution of betalains has been the subject of several reviews (Strack et al. 2003; Azeredo 2009; Han et al. 2009; Gandía-Herrero et al. 2010; Pavoković and Krsnik-Rasol 2011). The biosynthesis starts in the cytoplasm, where conjugates are formed between betalamic acid and cyclo-3-(3,4-dihydroxyphenyl)-L-alanine (cyclo-DOPA), but later, the products are transported into and stored in the vacuole (Strack et al. 2003, Azeredo 2009, Pavoković and Krsnik-Rasol 2011).

Folate biosynthesis takes place in the mitochondria, but folates are present in all parts of the cells, i.e., in mitochondria, plastids, cytosol and vacuoles (Hanson and Gregory 2011), with a major proportion accumulating in the vacuole. Folates have γ-linked polyglutamyl tails that affect their activity and transport. Folates consist of pteridine and *p*-aminobenzoate, which together constitute the pteroyl moiety, and one or more glutamate residues (Raichaudhuri et al. 2009). In plants and most other organisms, the parent folate molecule, pteroyl-monoglutamate, is polyglutamylated to yield pteroyl-polyglutamates containing one to seven additional glutamate residues (Raichaudhuri et al. 2009).

Alkaloids constitute another class of compounds that accumulate in the vacuole. Several studies have been carried out to characterize the mechanism of vacuolar accumulation of alkaloids using cell cultures, protoplasts and isolated native vacuoles. In the cell, anthocyanins are believed to be synthesized at the cytosolic surface of the endoplasmic reticulum by a multi-enzyme complex (Saslowsky and Winkel-Shirley 2001), although the association of anthocyanin enzymes with the endoplasmic reticulum has not been fully established. Grotewold and Davies (2008) suggested two models concerning transport from the cytoplasmic surface of the endoplasmic reticulum to the vacuole. The first is based on direct transport of vesicles filled with anthocyanins to the vacuole whereas the second is a ligandin transporter model.

4.4 Vacuolar Transport

The tonoplast, which surrounds the vacuole and separates the vacuolar contents from the cell cytoplasm, has been investigated with respect to transport systems of secondary metabolites such as anthocyanins and indole alkaloids (see reviews by Taiz 1992; Yazaki 2005; Roytrakul and Verpoorte 2007; Martinoia et al. 2007). Since vacuoles accumulate a range of different substances, different transport systems have been developed to move these substances in and out of the vacuole.

4.4.1 Pumps and Energization of Transport

A characteristic feature of the tonoplast is the presence of a vacuolar H^+-translocating inorganic pyrophosphatase (V-PPase) as well as a more conventional H^+-ATPase (V-ATPase), which were found to be important for vacuolar uptake of most solutes (Rea and Poole 1993, Martinoia et al. 2007; Roytrakul and Verpoorte 2007). Both these enzymes are primary H^+ pumps and catalyze electrogenic H^+ translocation from the cytosol into the vacuole. Thus, both of these enzymes are involved in the creation of proton and electrochemical gradients required for secondary-active transport across the tonoplast by secondary transporters. The H^+-PPase uses inorganic pyrophosphate (PPi) instead of ATP as an energy source (Roytrakul and Verpoorte 2007). In different plants and tissues, these proton pumps are likely to be expressed differently (Maeshima 2001, Rea and Poole 1993).

Studies using red beet have made considerable contributions to our understanding of the energy requirements for sucrose transport (Willenbrink and Doll 1979; Thom et al. 1986; Getz and Klein 1995; Echeverría, and Gonzalez 2000). Three principal mechanisms for vacuolar accumulation of secondary metabolites have been described, i.e., H^+-antiport, ion and conformational trapping and ATP-binding cassette (ABC) transporters (Roytrakul and Verpoorte 2007). Antiporters use H^+ as counter ions to provide energy for transport of another molecule or ion into the vacuole. Carrier proteins and channels transport their substrates along the electric potential gradient (see, e.g., Forster et al. 2002). The ABC transporters are energised by the hydrolysis of ATP and can therefore transport a wide variety of substances over membranes against a concentration gradient. In addition to these principal ways for transportation, secondary metabolites may passively equilibrate across tonoplasts, become trapped in the vacuole by protonation, obtain altered configuration due to the acidic environment, undergo isomerization or complexation with ions, bind to phenolics or the tonoplast, or interact with other vacuolar constituents or even crystallise (reviewed by Roytrakul and Verpoorte 2007). The vacuolar uptake mechanism not only depends on the source of secondary metabolite, but also the plant species (Dean and Mills 2004; Yazaki 2005). Thus, Dean and Mills (2004) found that salicylic acid 2-*O*-β-D-glucose was transported into soybean tonoplast vesicles via an ABC transporter, whereas, in red beet, uptake appeared to involve an H^+-antiporter.

4.4.2 *Transport of Water*

A very important feature of water transport in and out of vacuoles is water channels or aquaporins, which are water-conducting Major Intrinsic Proteins (MIPs) facilitating the flow of water across membranes (Gilliham et al. 2011). Whereas water is freely transported through the aquaporins, the passage of ions and other solutes is generally prevented, especially for charged compounds. A general inhibition of aquaporins by low pH has been reported in most studies, but Sutka et al. (2005) found that only a fraction of the active tonoplast aquaporins were blocked by acidification and suggested that it was probably because only certain isoforms were blocked by protons.

4.4.3 *Transport of Sucrose*

As outlined in Sect. 4.2.2, sucrose is a major storage solute in vacuoles and therefore the transport processes of sucrose has been intensely studied, especially regarding energy requirements. Sucrose is a rather large, polar compound and therefore needs various transport systems to cross membranes (Ayre 2011). Thus, for transport across the tonoplast, different mechanisms have been suggested, i.e., facilitated diffusion, antiporters with protons and symporters with protons, but there is only convincing evidence for the symport model (Ayre 2011). Briskin et al. (1985a, b) showed that vesicle systems isolated from sugar beet taproots were highly useful for studying the sucrose transport across the tonoplast since large quantities of highly stable vesicles could easily be produced.

The enzyme sucrose synthase is very important for the cell metabolism and is involved in the transfer of sucrose into metabolic, structural and storage routes in the plant cell. The enzyme, which has been intensely studied in beets, catalyses the reversible conversion of sucrose and UDP to UDP-glucose and fructose and it has been implicated in sugar import and export, synthesis of starch and cell walls, metabolic changes related to cold stress as well as sink strength (Hesse and Willmitzer 1996; Etxeberria and Gonzalez 2003). The export of sucrose from the vacuole is an active, ATP-requiring process (Doll et al. 1979; Echeverría and Gonzalez 2000; Etxeberria and Gonzalez 2003) and from studies of red beet, Etxeberria and Gonzalez (2003) found that the tonoplast contained sucrose synthase activity, which was involved in sucrose mobilisation from the vacuole. Furthermore, kinetic analyses under conditions of low sucrose availability indicated that sucrose channelling occurred between the ATP-dependent sucrose transporter and sucrose synthase. A general update on sucrose transport and transporters was given by Kühn and Grof (2010).

4.4.4 Transport of Ions

In plants, cations are transported by cation/H^+ exchangers (CAXs), which regulate the redistribution of different cations including calcium (Ca^{2+}) in exchange for the protons generated by H^+-pumps (reviewed by Manohar et al. 2011; Isayenkov et al. 2010). In *Arabidopsis*, it has been demonstrated that a Ca^{2+}/H^+ antiporter (CAX1) together with Ca^{2+}-ATPase plays a key role in vacuolar Ca^{2+} accumulation (Zhao et al. 2009; Gilliham et al. 2011; Manohar et al. 2011).

Elevation of cytosolic free Ca^{2+} is important in many signalling pathways in plants (Sanders et al. 1999; Kudla et al. 2010; Peiter 2011). In accordance with the observation that the vacuole represents a source of Ca^{2+} for signalling, a number of Ca^{2+}-release channels has been identified at the tonoplast (Sanders et al. 1999; Kudla et al. 2010). These channels are gated by voltage, i.e., by membrane hyperpolarization and depolarization (Sanders et al. 1999; Isayenkov et al. 2010; Peiter 2011). However, the most conspicuous type of vacuolar channels is the so-called slow vacuolar (SV) channels, which are known from all plant tissues (Hedrich and Marten 2011; Isayenkov et al. 2010; Peiter 2011). These channels are non-selective cation channels, which are used to accumulate, e.g., Ca^{2+}, Na^+ and K^+, in the vacuole. They are driven by ATPase and PP_i-ase, activated by calcium or voltage and their activity is influenced by the concentration of calcium, pH, redox state and regulatory proteins (see review by Hedrich and Marten 2011). On the other hand, during resting cytosolic Ca^{2+} conditions, so-called fast vacuolar (FV) channels represent the principal passive pathway for cation uptake and release across the tonoplast (Pottosin and Martínez-Estévez 2003; Isayenkov et al. 2010; Peiter 2011). Studies from red beet have contributed substantially to our knowledge about the different types of ion channels (Hedrich and Neher 1987; Bewell et al. 1999; Pottosin and Martínez-Estévez 2003; Pottosin et al. 2003, 2005; Brito-Argáez et al. 2008).

D-*myo*-inositol 1,4,5-trisphosphate ($InsP_3$) is a second messenger involved in signal transduction (calcium homeostasis) in the cytoplasm, where it binds to its receptor and mobilizes internal calcium (Communi et al. 1995). Thus, opening of the Ca^{2+}-permeable ion channel increases the Ca^{2+} concentration in the cytosol and hence may trigger the opening of other ion channels (Krinke et al. 2007). The level of intracellular $InsP_3$ may vary in response to environmental factors like plant hormones, light, temperature, gravity, hyperosmotic stress, oxidative stress, G-proteins and pathogens (Krinke et al. 2007). Studies from sugar and red beets have also contributed to our understanding of SV channels, especially on how Ca^{2+} dynamics is influenced by $InsP_3$ (Hedrich and Neher 1987; Alexandre et al. 1990; Brosnan and Sanders 1993; Allen and Sanders 1994; Allen et al. 1995; Muir and Sanders 1996; Navazio et al. 2000). Brosnan and Sanders (1993) furthermore identified and characterized the highly specific $InsP_3$ binding site as well as possible $InsP_3$ receptors on red beet microsomes.

Salinity generally has an adverse effect on the water household of plants and plants have developed various methods to cope with this stress (see reviews by Mahajan et al. 2008; Munns and Tester 2008; Kronzucker and Britto 2011). However, sugar and red beet have a tremendous capacity to take up sodium and utilize it in non-specific functions as a substitute for potassium (Subbarao et al. 2001) and NaCl fertilisation increases the yield of beets (Wybenga and Lehr 1958). One of the reasons for this preference for sodium is the accumulation of glycine betaine, which occurs when the turgor pressure becomes too low or plants are experiencing too high levels of sodium, resulting in protection of the cytoplasm from dehydration, protection of chloroplasts from damage by sodium and prevention of sodium interference with essential potassium functions (Subbarao et al. 2001).

The vacuole also stores anions like nitrate for later use, but also for maintenance of turgor pressure, to serve as osmotica and/or as a stock, which can be utilised for growth under unfavourable conditions (Dechorgnat et al. 2011). Nitrate is moved between the two compartments, the cytoplasm and the vacuole, and the dynamics of these processes were studied by Izmailov (2004) in leaves of sugar beet and pea under different environmental conditions. Dechorgnat et al. (2011) further discussed that the transport of nitrate into the vacuole most probably involve a chloride channel/transporter, of which one, the $NO_3^-/2\ H^+$ exchanger AtCLCa, has been found to load nitrate into *Arabidopsis* vacuoles.

4.4.5 Transport of Secondary Metabolites

Plants produce a high number of secondary metabolites for a range of different purposes, like alkaloids, phenols, terpenoids etc. and these are often stored in the vacuole (Yazaki 2005). Two principal mechanisms have been suggested for vacuolar transport of secondary metabolites, i.e., H^+ antiporters and ABC transporters. Which one is used for a particular metabolite depends on the plant species as well as the source of the metabolite (Dean and Mills 2004; Yazaki 2005). Often, the transport depends on conjugation of the metabolite to glucose or glutathione (Yazaki 2005; Bartholomew et al. 2002; see Sect. 4.4.5). Although the major steps in the betalain biosynthetic pathway are known (see Sect. 4.3), we have very limited knowledge concerning how the end products are transported to the different places for storage and use. For example, it is not known whether a single pool or multiple pools of betalamic acid are responsible for the formation of both betacyanins and betaxanthins, or how these compounds are transported to the vacuole, their ultimate site of accumulation (Grotewold 2006).

Limited data are also available for folate vacuolar transport. Raichaudhuri et al. (2009) found that the multidrug resistance-associated protein (MRP)-subfamily of ABC transporters participated in vacuolar folate uptake. On the other hand, Akhtar et al. (2010) suggested that folates continuously enter and accumulate in the vacuole as polyglutamates. Here, they are hydrolyzed by γ-glutamyl hydrolase and exit as monoglutamates. Furthermore, it appeared that γ-glutamyl hydrolase has an important influence on the polyglutamyl tail length and hence folate stability and cellular content. Akhtar et al. (2010) themselves raised some concerns regarding this model. Thus, among others, it implies that there is a tonoplast transport system for polyglutamates and this is unlike all previously described folate transporters in plants, which accept only monoglutamates (Raichaudhuri et al. 2009).

4.4.6 Transport of Xenobiotics

In addition to storage of plant beneficial compounds, the vacuole may also store undesirable compounds such as heavy metals like Ni, Cd and As. Such metals are generally harmful to the cell and, in high doses, lead to accumulation of reactive oxygen species, which in turn can destroy cell structure (Rascio and Navari-Izzo 2011). Therefore, heavy metals taken up by the plant are either detoxified by merging with organic molecules, sequestered into the vacuoles or incorporated in cell walls. These processes mainly occur in the roots to protect the aerial parts of the plant (Rascio and Navari-Izzo 2011). Uptake of heavy metals into the vacuole involve special transporter proteins, i.e., cation diffusion facilitator proteins (also termed metal transporter proteins), which mediate bivalent cation efflux from the cytosol (Rascio and Navari-Izzo 2011).

Substances toxic to plants, such as herbicides, may also accumulate in vacuoles to reduce their toxicity (Bartholomew et al. 2002). Thus, among the processes employed is attachment of the undesired compound to large hydrophilic molecules, increasing water solubility and promoting recognition by relevant transporters. This may then be followed by removal of the conjugates from the cytosol into the extracellular space and/or intracellular compartments, like the vacuole. Bartholomew et al. (2002) used tonoplast vesicles isolated from red beet roots as a model system to study the transport of two sulfonylurea herbicides, chlorsulfuron and chlorimuron-ethyl before and after conjugation with glucose and glutathione. Contrary to the previous belief that only naturally occurring glucosylated plant secondary metabolites entered the vacuole by H^+-antiporters, whereas glucosylated xenobiotics entered via ABC transporters, it was found that transport of the glucosylated herbicide depended on the transmembrane H^+-gradient, whereas the glutathionated herbicide was transported by an ABC transporter-like carrier. This demonstrated that not only glucosides of endogenous metabolites, but also of xenobiotics, may be transported by H^+-antiporters.

4.5 Concluding Remarks

The knowledge emerging from studies on basic functions of cellular constituents in red beet and other crops offer a wide range of potential applications. Beets have been used as important models for studies of cellular functions like sugar transport, sequestration of phytoalexins, protoplast function as well as intracellular electric potentials (Doll et al. 1979; Doll and Hauer 1981; Schmidt and Poole 1980; Spessard et al. 1994; Lassalles et al. 1987). However, with new insights into the biological functions of cells, new possibilities arise. Thus, Manohar et al. (2011) suggested that knowledge about and manipulation of vacuolar cation/H^+ exchangers could be used for biofortification, by increasing bioavailable mineral content in agriculturally important crops such as Ca^{2+} content in edible roots. Likewise, there might be a potential in phytoremediation to remove environmental contaminations like Cd using CAX-expressing plants or even in phytomining to extract precious metals from soils. In these examples, it could be advantageous to utilise a root crop where the accumulated compounds are easily removed from the soil for further processing.

Another area of future interest is the study of membrane transport of plant secondary metabolites (Yazaki 2005; Zouhar and Rojo 2009; Isayenkov et al. 2010; Peiter 2011). Thus, discovery and characterization of the many transporters and channels in plant genomes have been used to characterize membrane transport of plant secondary metabolites. These have revealed that membrane transport is quite specific and highly regulated for each secondary metabolite. Therefore, the genes involved in biosynthesis of secondary metabolites as well as the genes involved in their transport will be important for metabolic engineering aimed at increasing the productivity of valuable secondary metabolites *in planta*.

A key reason for the interest in red beets is their content of betalains and folates. Production of these metabolites in, e.g., tissue culture or by hairy root cultures, will be advantageous for large-scale production and, at least for betalains, such production is feasible (Thimmaraju et al. 2003b; Pavoković and Krsnik-Rasol 2011). Even though betalains are stored in the vacuoles, they can be forced into the extracellular medium from the vacuole by application of transient stress to the culture (Pavoković and Krsnik-Rasol 2011) or membrane-permeabilizing agents (Thimmaraju et al. 2003a, b). Here, a thorough knowledge of transport processes in plants will play a pivotal role. By enhancing quick sequestration of secondary compounds into the vacuoles, cellular toxicity can be minimised, as has been demonstrated in tobacco. Thus, when the concentration of the highly toxic secondary metabolite nicotine was increased by genetic engineering in *Nicotiana tabacum*, it caused intrinsic toxicity to the cells, but engineering tobacco cells for over-expression of ABC transporters resulted in a decrease in the cellular toxicity (Goossens et al. 2003). Product recovery is also a feasible strategy for reaching enhanced productivities. Thus, product enhancement depends on all steps from synthesis and cellular storage to efficient and timely recovery. Therefore, for efficient production of various metabolites, vacuolar transportation is a key function that needs to be thoroughly understood and for this, red beet serves as an excellent model.

References

Akhtar, T.A., G. Orsomando, P. Mehrshahi, A. Lara-Núñez, M.J. Bennett, J.F. Gregory III, and A.D. Hanson. 2010. A central role for gamma-glutamyl hydrolases in plant folate homeostasis. *The Plant Journal* 64: 256–266.

Alexandre, J., J.P. Lassalles, and R.T. Kado. 1990. Opening of Ca^{2+} channels in isolated red beet root vacuole membrane by inositol 1,4,5-trisphosphate. *Nature* 343: 567–570.

Allen, G.J., and D. Sanders. 1994. Osmotic stress enhances the competence of *Beta vulgaris* vacuoles to respond to inositol 1,4,5-trisphosphate. *The Plant Journal* 6: 687–695.

Allen, G.J., S.R. Muir, and D. Sanders. 1995. Release of Ca^{2+} from individual plant vacuoles by both $InsP_3$ and cyclic ADP-ribose. *Science* 268: 735–737.

Ayre, B.G. 2011. Membrane-transport systems for sucrose in relation to whole-plant carbon partitioning. *Molecular Plant* 4: 377–394.

Azeredo, H.M.C. 2009. Betalains: Properties, sources, applications, and stability—A review. *International Journal of Food Science and Technology* 44: 2365–2376.

Bartholomew, D.M., D.E. Van Dyk, S.-M.C. Lau, D.P. O'Keefe, P.A. Rea, and P.V. Viitanen. 2002. Alternate energy-dependent pathways for the vacuolar uptake of glucose and glutathione conjugates. *Plant Physiology* 130: 1562–1572.

Bewell, M.A., F.J.M. Maathuis, G.J. Allen, and D. Sanders. 1999. Calcium-induced calcium release mediated by a voltage-activated cation channel in vacuolar vesicles from red beet. *FEBS Letters* 458: 41–44.

Briskin, D.P., W.R. Thornley, and R.E. Wyse. 1985a. Membrane transport in isolated vesicles from sugarbeet taproot. I. Isolation and characterization of energy-dependent, H^+-transporting vesicles. *Plant Physiology* 78: 865–870.

Briskin, D.P., W.R. Thornley, and R.E. Wyse. 1985b. Membrane transport in isolated vesicles from sugarbeet taproot. II. Evidence for a sucrose/H^+-antiport. *Plant Physiology* 78: 871–875.

Brito-Argáez, L., B. Canto-Canché, S.M.T. Hernández-Sotomayor, M. Martínez-Estevez, and I.I. Pottosin. 2008. Patch clamp characterization of a non-selective cation channel of ER membranes purified from *Beta vulgaris* taproots. *Physiologia Plantarum* 132: 399–406.

Brosnan, J.M., and D. Sanders. 1993. Identification and characterization of high-affinity binding sites for inositol trisphosphate in red beet. *The Plant Cell* 5: 931–940.

Communi, D., V. Vanweyenberg, and C. Erneux. 1995. Molecular study and regulation of D-*myo*-inositol 1,4,5-*tris*phosphate 3-kinase. *Cellular Signalling* 7: 643–650.

Dean, J.V., and J.D. Mills. 2004. Uptake of salicylic acid 2-*O*-b-D-glucose into soybean tonoplast vesicles by an ATP-binding cassette transporter-type mechanism. *Physiologia Plantarum* 120: 603–612.

Dechorgnat, J., C.T. Nguyen, P. Armengaud, M. Jossier, E. Diatloff, S. Filleur, and F. Daniel-Vedele. 2011. From the soil to the seeds: The long journey of nitrate in plants. *Journal of Experimental Botany* 62: 1349–1359.

Doll, S., and R. Hauer. 1981. Determination of the membrane potential of vacuoles isolated from red-beet storage tissue. Evidence for an electrogenic ATPase. *Planta* 152: 153–158.

Doll, S., F. Rodier, and J. Willenbrink. 1979. Accumulation of sucrose in vacuoles isolated from red beet tissue. *Planta* 144: 407–411.

Echeverría, E., and P.C. Gonzalez. 2000. ATP-induced sucrose efflux from red-beet tonoplast vesicles. *Planta* 211: 77–84.

Escribano, J., M.A. Pedreño, F. García-Carmona, and R. Muñoz. 1998. Characterization of the antiradical activity of betalains from *Beta vulgaris* L. roots. *Phytochemical Analysis* 9: 124–127.

Etxeberria, E., and P. Gonzalez. 2003. Evidence for a tonoplast-associated form of sucrose synthase and its potential involvement in sucrose mobilization from the vacuole. *Journal of Experimental Botany* 54: 1407–1414.

Forster, I.C., K. Köhler, J. Biber, and H. Murer. 2002. Forging the link between structure and function of electrogenic cotransporters: The renal type IIa Na^+/P_i cotransporter as a case study. *Progress in Biophysics and Molecular Biology* 80: 69–108.

Gandía-Herrero, F., J. Escribano, and F. García-Carmona. 2010. Structural implications on color, fluorescence, and antiradical activity in betalains. *Planta* 232: 449–460.

Getz, H.P., and M. Klein. 1995. Characteristics of sucrose transport and sucrose-induced H^+ transport on the tonoplast of red beet (*Beta vulgaris* L.) storage tissue. *Plant Physiology* 107: 459–467.

Gilliham, M., M. Dayod, B.J. Hocking, B. Xu, S.J. Conn, B.N. Kaiser, R.A. Leigh, and S.D. Tyerman. 2011. Calcium delivery and storage in plant leaves: Exploring the link with water flow. *Journal of Experimental Botany* 62: 2233–2250.

Goossens, A., S.T. Häkkinen, I. Laakso, K.-M. Oksman-Caldentey, and D. Inzé. 2003. Secretion of secondary metabolites by ATP-binding cassette transporters in plant cell suspension cultures. *Plant Physiology* 131: 1161–1164.

Grotewold, E. 2006. The genetics and biochemistry of floral pigments. *Annual Review of Plant Biology* 57: 761–780.

Grotewold, E., and K. Davies. 2008. Trafficking and sequestration of anthocyanins. *Natural Product Communications* 3: 1251–1258.

Gunawardena, A.H.L.A., J.S. Greenwood, and N.G. Dengler. 2004. Programmed cell death remodels lace plant leaf shape during development. *The Plant Cell* 16: 60–73.

Han, X.-H., Z.-J. Gao, and X.-G. Xiao. 2009. Enzymes and genes involved in the betalain biosynthesis in higher plants. *African Journal of Biotechnology* 8: 6735–6744.

Hanson, A.D., and J.F. Gregory III. 2011. Folate biosynthesis, turnover, and transport in plants. *Annual Review of Plant Biology* 62: 105–125.

Hara-Nishimura, I., and N. Hatsugai. 2011. The role of vacuole in plant cell death. *Cell Death and Differentiation* 18: 1298–1304.

Hedrich, R., and I. Marten. 2011. TPC1—SV channels gain shape. *Molecular Plant* 4: 428–441.

Hedrich, R., and E. Neher. 1987. Cytoplasmic calcium regulates voltage-dependent ion channels in plant vacuoles. *Nature* 329: 833–836.

Hesse, H., and L. Willmitzer. 1996. Expression analysis of a sucrose synthase gene from sugar beet (*Beta vulgaris* L.). *Plant Molecular Biology* 30: 863–872.

Hörtensteiner, S., and B. Kräutler. 2011. Chlorophyll breakdown in higher plants. *Biochimica et Biophysica Acta* 1807: 977–988.

Isayenkov, S., J.C. Isner, and F.J.M. Maathuis. 2010. Vacuolar ion channels: Roles in plant nutrition and signalling. *FEBS Letters* 584: 1982–1988.

Isheeva, O.D., E.V. Pradedova, and R.K. Salyaev. 2009. Multiplicity of vacuolar peroxidase isozymes in higher plants by the example of red beet (*Beta vulgaris* L.) roots. *Doklady. Biochemistry and Biophysics* 424: 53–55.

Izmailov, S.F. 2004. Saturation and utilization of nitrate pools in pea and sugar beet leaves. *Russian Journal of Plant Physiology* 51: 189–193.

Johnson, K.D., H. Höfte, and M.J. Chrispeels. 1990. An intrinsic tonoplast protein of protein storage vacuoles in seeds is structurally related to a bacterial solute transporter (GlpF). *The Plant Cell* 2: 525–532.

Krinke, O., Z. Novotná, O. Valentová, and J. Martinec. 2007. Inositol trisphosphate receptor in higher plants: Is it real? *Journal of Experimental Botany* 58: 361–376.

Kronzucker, H.J., and D.T. Britto. 2011. Sodium transport in plants: A critical review. *New Phytologist* 189: 54–81.

Kudla, J., O. Batistic, and K. Hashimoto. 2010. Calcium signals: The lead currency of plant information processing. *The Plant Cell* 22: 541–563.

Kühn, C., and C.P.L. Grof. 2010. Sucrose transporters of higher plants. *Current Opinion in Plant Biology* 13: 288–298.

Lassalles, J.-P., J. Alexandre, and M. Thellier. 1987. Microelectrode measurements on red beet vacuole. Biological effect of Na^+ or NO_3^- ions, diffusing from the microelectrode. *Plant Physiology* 85: 608–610.

Lee, J.H., C.W. Son, M.Y. Kim, M.H. Kim, H.R. Kim, E.S. Kwak, S. Kim, and M.R. Kim. 2009. Red beet (*Beta vulgaris* L.) leaf supplementation improves antioxidant status in C57BL/6J mice fed high fat high cholesterol diet. *Nutrition Research and Practice* 3: 114–121.

Loginova, K.V., N.I. Lebovka, and E. Vorobiev. 2011. Pulsed electric field assisted aqueous extraction of colorants from red beet. *Journal of Food Engineering* 106: 127–133.

Maeshima, M. 2001. Tonoplast transporters: Organization and function. *Annual Review of Plant Physiology and Plant Molecular Biology* 52: 469–497.

Mahajan, S., G.K. Pandey, and N. Tuteja. 2008. Calcium- and salt-stress signaling in plants: Shedding light on SOS pathway. *Archives of Biochemistry and Biophysics* 471: 146–158.

Manohar, M., T. Shigaki, and K.D. Hirschi. 2011. Plant cation/H^+ exchangers (CAXs): Biological functions and genetic manipulations. *Plant Biology* 13: 561–569.

Martinoia, E., M. Maeshima, and H.E. Neuhaus. 2007. Vacuolar transporters and their essential role in plant metabolism. *Journal of Experimental Botany* 58: 83–102.

Muir, S.R., and D. Sanders. 1996. Pharmacology of Ca^{2+} release from red beet microsomes suggests the presence of ryanodine receptor homologs in higher plants. *FEBS Letters* 395: 39–42.

Munns, R., and M. Tester. 2008. Mechanisms of salinity tolerance. *Annual Review of Plant Biology* 59: 651–681.

Navazio, L., M.A. Bewell, A. Siddiqua, G.D. Dickinson, A. Galione, and D. Sanders. 2000. Calcium release from the endoplasmic reticulum of higher plants elicited by the NADP metabolite nicotinic acid adenine dinucleotide phosphate. *Proceedings of the National Academy of Sciences of the United States of America* 97: 8693–8698.

Nemzer, B., Z. Pietrzkowski, A. Spórna, P. Stalica, W. Thresher, T. Michalowski, and S. Wybraniec. 2011. Betalainic and nutritional profiles of pigment-enriched red beet root (*Beta vulgaris* L.) dried extracts. *Food Chemistry* 127: 42–53.

Niemietz, C., and J.S. Hawker. 1988. Sucrose accumulation in red beet vacuoles by UDPglucose-dependent group translocation—Fact or artifact? *Australian Journal of Plant Physiology* 15: 359–366.

Orsomando, G., R.D. de la Garza, B.J. Green, M. Peng, P.A. Rea, T.J. Ryan, J.F. Gregory III, and A.D. Hanson. 2005. Plant g-glutamyl hydrolases and folate polyglutamates. *The Journal of Biological Chemistry* 280: 28877–28884.

Osbourn, A.E. 1996. Preformed antimicrobial compounds and plant defense against fungal attack. *The Plant Cell* 8: 1821–1831.

Pavoković, D., and M. Krsnik-Rasol. 2011. Complex biochemistry and biotechnological production of betalains. *Food Technology and Biotechnology* 49: 145–155.

Peiter, E. 2011. The plant vacuole: Emitter and receiver of calcium signals. *Cell Calcium* 50: 120–128.

Pottosin, I.I., and M. Martínez-Estévez. 2003. Regulation of the fast vacuolar channel by cytosolic and vacuolar potassium. *Biophysical Journal* 84: 977–986.

Pottosin, I.I., M. Martínez-Estévez, O.R. Dobrovinskaya, and J. Muñiz. 2003. Potassium-selective channel in the red beet vacuolar membrane. *Journal of Experimental Botany* 54: 663–667.

Pottosin, I.I., M. Martínez-Estévez, O.R. Dobrovinskaya, and J. Muñiz. 2005. Regulation of the slow vacuolar channel by luminal potassium: Role of surface charge. *Journal of Membrane Biology* 205: 103–111.

Pradedova, E.V., O.D. Isheeva, and R.K. Salyaev. 2011. Antioxidant defense enzymes in cell vacuoles of red beet roots. *Russian Journal of Plant Physiology* 58: 36–44.

Raichaudhuri, A., M. Peng, V. Naponelli, S. Chen, R. Sánchez-Fernández, H. Gu, J.F. Gregory III, A.D. Hanson, and P.A. Rea. 2009. Plant vacuolar ATP-binding cassette transporters that translocate folates and antifolates in vitro and contribute to antifolate tolerance in vivo. *The Journal of Biological Chemistry* 284: 8449–8460.

Rascio, N., and F. Navari-Izzo. 2011. Heavy metal hyperaccumulating plants: How and why do they do it? And what makes them so interesting? *Plant Science* 180: 169–181.

Rea, P.A., and R.J. Poole. 1993. Vacuolar H^+-translocating pyrophosphatase. *Annual Review of Plant Physiology and Plant Molecular Biology* 44: 157–180.

Roytrakul, S., and R. Verpoorte. 2007. Role of vacuolar transporter proteins in plant secondary metabolism: *Catharanthus roseus* cell culture. *Phytochemistry Reviews* 6: 383–396.

Sanders, D., C. Brownlee, and J.F. Harper. 1999. Communicating with calcium. *The Plant Cell* 11: 691–706.

Saslowsky, D., and B. Winkel-Shirley. 2001. Localization of flavonoid enzymes in *Arabidopsis* roots. *The Plant Journal* 27: 37–48.

Schmidt, R., and R.J. Poole. 1980. Isolation of protoplasts and vacuoles from storage tissue of red beet. *Plant Physiology* 66: 25–28.

Sepúlveda-Jiménez, G., P. Rueda-Benítez, H. Porta, and M. Rocha-Sosa. 2004. Betacyanin synthesis in red beet (*Beta vulgaris*) leaves induced by wounding and bacterial infiltration is preceded by an oxidative burst. *Physiological and Molecular Plant Pathology* 64: 125–133.

Sepúlveda-Jiménez, G., P. Rueda-Benítez, H. Porta, and M. Rocha-Sosa. 2005. Erratum to: "Betacyanin synthesis in red beet (*Beta vulgaris*) leaves induced by wounding and bacterial infiltration is preceded by an oxidative burst" (Physiological and Molecular Plant Pathology 64(3): 125–133). *Physiological and Molecular Plant Pathology* 66: 75.

Sivakumar, V., J.L. Anna, J. Vijayeeswarri, and G. Swaminathan. 2009. Ultrasound assisted enhancement in natural dye extraction from beetroot for industrial applications and natural dyeing of leather. *Ultrasonics Sonochemistry* 16: 782–789.

Spessard, G.O., C. Hanson, J.S. Halvorson, and J.L. Giannini. 1994. Effects of phaseollin on membrane leakage in red beet vacuoles and tonoplast vesicles. *Phytochemistry* 35: 43–47.

Stintzing, F.C., and R. Carle. 2004. Functional properties of anthocyanins and betalains in plants, food, and in human nutrition. *Trends in Food Science & Technology* 15: 19–38.

Strack, D., T. Vogt, and W. Schliemann. 2003. Recent advances in betalain research. *Phytochemistry* 62: 247–269.

Subbarao, G.V., R.M. Wheeler, L.H. Levine, and G.W. Stutte. 2001. Glycine betaine accumulation, ionic and water relations of red-beet at contrasting levels of sodium supply. *Journal of Plant Physiology* 158: 767–776.

Sugihara, N., Y. Kishimoto, M. Iizuka, E. Saita, K. Suzuki, M. Toyozaki, M. Nagai, K. Murakami, K. Takenaka, M. Kamiyama, M. Tani, and K. Kondo. 2009. Effects of red beet on LDL oxidation and reactive oxygen species (ROS) production. Atherosclerosis Supplements 10.

Sutka, M., K. Alleva, M. Parisi, and G. Amodeo. 2005. Tonoplast vesicles of *Beta vulgaris* storage root show functional aquaporins regulated by protons. *Biology of the Cell* 97: 837–846.

Taiz, L. 1992. The plant vacuole. *Journal of Experimental Biology* 172: 113–122.

Thimmaraju, R., N. Bhagyalakshmi, M.S. Narayan, and G.A. Ravishankar. 2003a. Food-grade chemical and biological agents permeabilize red beet hairy roots, assisting the release of betalaines. *Biotechnology Progress* 19: 1274–1282.

Thimmaraju, R., N. Bhagyalakshmi, M.S. Narayan, and G.A. Ravishankar. 2003b. Kinetics of pigment release from hairy root cultures of *Beta vulgaris* under the influence of pH, sonication, temperature and oxygen stress. *Process Biochemistry* 38: 1069–1076.

Thom, M., R.A. Leigh, and A. Maretzki. 1986. Evidence for the involvement of a UDP-glucose-dependent group translocator in sucrose uptake into vacuoles of storage roots of red beet. *Planta* 167: 410–413.

Van den Bulcke, M., G. Bauw, C. Castresana, M. Van Montagu, and J. Vandekerckhove. 1989. Characterization of vacuolar and extracellular b(1,3)-glucanases of tobacco: Evidence for a strictly compartmentalized plant defense system. *Proceedings of the National Academy of Sciences of the United States of America* 86: 2673–2677.

Willenbrink, J., and S. Doll. 1979. Characteristics of the sucrose uptake system of vacuoles isolated from red beet tissue. Kinetics and specificity of the sucrose uptake system. *Planta* 147: 159–162.

Wybenga, J.M., and J.J. Lehr. 1958. Exploratory pot experiments on sensitiveness of different crops to sodium: E. Red table beet. *Plant and Soil* 9: 385–394.

Yazaki, K. 2005. Transporters of secondary metabolites. *Current Opinion in Plant Biology* 8: 301–307.

Zhao, J., T. Shigaki, H. Mei, Y.-Q. Guo, N.-H. Cheng, and K.D. Hirschi. 2009. Interaction between *Arabidopsis* Ca^{2+}/H^{+} exchangers CAX1 and CAX3. *The Journal of Biological Chemistry* 284: 4605–4615.

Zouhar, J., and E. Rojo. 2009. Plant vacuoles: Where did they come from and where are they heading? *Current Opinion in Plant Biology* 12: 677–684.

Chapter 5
Plant Respiratory Metabolism: A Special Focus on the Physiology of Beetroot (*Beta Vulgaris* L.) Mitochondria

Kapuganti J. Gupta and Hardy Rolletschek

Abstract Mitochondria are important organelles for cellular energy. Apart from orchestrating cellular energy balance, mitochondria involve themselves in various processes such as synthesis of vitamins and lipids and production of reactive oxygen and nitrogen species. Understanding mitochondrial metabolism is crucial for understanding complete cellular metabolism, which is directly or indirectly involved in the biosynthesis of food and pharmaceutical products. Respiration, the chief function occurring in mitochondria, is divided into glycolysis, tricarboxylic acid (TCA) cycle, and electron transport chain. In addition to the classic operation of the pathways, mitochondria have various alternative pathways that are involved in plant responses to various stresses. Respiratory metabolism in green tissues is different from that occurring in roots, and more so in bulky tissues such as beetroots, because such storage organs have to cope with very low cellular oxygen concentrations. Bulky tissue organs like beetroots store sucrose as their storage carbohydrate. Therefore the cellular respiration in beetroots is different, especially during the storage period and sprouting period in comparison with the normal roots. Here respiratory metabolism and its association with various other metabolic pathways are described, with a special focus on beetroot, in which various alternative pathways such as NAD(P)H dehydrogenases or alternative oxidase pathways are discussed. Understanding respiratory metabolism has bearing on engineering beetroot for higher yields as well as prevention of storage losses.

K.J. Gupta (✉)
Department of Plant Physiology, University of Rostock, Albert Einstein Str 3, Rostock D-18059, Germany
e-mail: jagadis.kapuganti@uni-rostock.de

H. Rolletschek
Leibniz-Institut für Pflanzengenetik und Kulturpflanzenforschung (IPK), Corrensstr. 3, Gatersleben 06466, Germany
e-mail: rollet@ipk-gatersleben.de

B. Neelwarne (ed.), *Red Beet Biotechnology: Food and Pharmaceutical Applications*, DOI 10.1007/978-1-4614-3458-0_5, © Springer Science+Business Media New York 2013

5.1 Introduction

5.1.1 General Features of Respiratory Electron Transport Chain

Mitochondria are the power generators of the cells. They generate ATP via the respiratory electron transport chain (ETC) for cellular energy needs. Mitochondrial respiration is a complex process involving various steps: the tricarboxylic acid (TCA) cycle, electron transport and oxidative phosphorylation, as well as transport of various intermediates between the cellular compartments. Respiratory metabolism is also linked to other cellular processes, such as photosynthesis, photorespiration, and protein and lipid metabolism. Briefly, the glycolysis pathway is present in cytosol, partly in plastids, and some components of glycolysis are also associated with the mitochondrial electron transport chain (Giege et al. 2003). Glycolysis starts with the oxidation of glucose and finally delivers pyruvate. The enzyme pyruvate kinase (PK) catalyzes the terminal reaction of glycolysis by converting phosphoenol pyruvate (PEP) and ADP to ATP and pyruvate, and plays a role in the regulation of glycolysis (Podesta and Plaxton 1991). Pyruvate also plays a role in regulating cellular respiration, particularly at hypoxic conditions (Gupta et al. 2009; Zabalza et al. 2009), to which storage organs such as beet root are exposed. The TCA cycle is present in the mitochondrial matrix and it is the second major component of plant aerobic metabolism. It is also called the Krebs cycle or citric acid cycle, and the various steps involved in this process are well studied and reviewed (Millar et al. 2011; van Dongen et al. 2011). Both glycolysis and the TCA cycle are regulated by the availability of particular substrates, such as sucrose (Fernie et al. 2002; Urbanczyk-Wochniak et al. 2006), or adenylate levels in plants (Geigenberger et al. 2010). Various transgenic approaches such as gene knockouts, antisense, and studies with micro RNA have all provided insights into our understanding of the association of respiration with other important metabolic and physiologic responses, such as the functioning of stomata, biomass accumulation, and net photosynthetic efficiency. In some cases, the TCA cycle operates in a non-cyclic manner, in particular, under conditions of oxygen limitation (Sweetlove et al. 2010; Rocha et al. 2010).

The operation of the mitochondrial electron transport chain (MtETC) involves transfer of electrons from reducing equivalents such as NADH, NADPH, or succinate to molecular oxygen via four integral membrane complexes located in the inner mitochondrial membrane. These are denoted as complex I (NADH dehydrogenase), II (succinate dehydrogenase), III (cytochrome bc1 complex), and IV (cytochrome c oxidase) (details of these complexes are explained in Fig. 5.1). In addition, there is also a mobile electron transfer protein called cytochrome c. Apart from the classic pathways, several alternative pathways are present in mitochondria (Millar et al. 2011). Under various physiological and developmental stages, electrons are channeled through these alternative pathways. These are alternative oxidase (AOX), uncoupling protein (UCP), and type II NADPH dehydrogenases. Transfer of electrons from NAD(P)H to oxygen is coupled to the proton translocation from the matrix to the inner membrane space and this leads to ATP production. However, electron transfer through these alternative pathways does not lead to proton translocation, and thus no

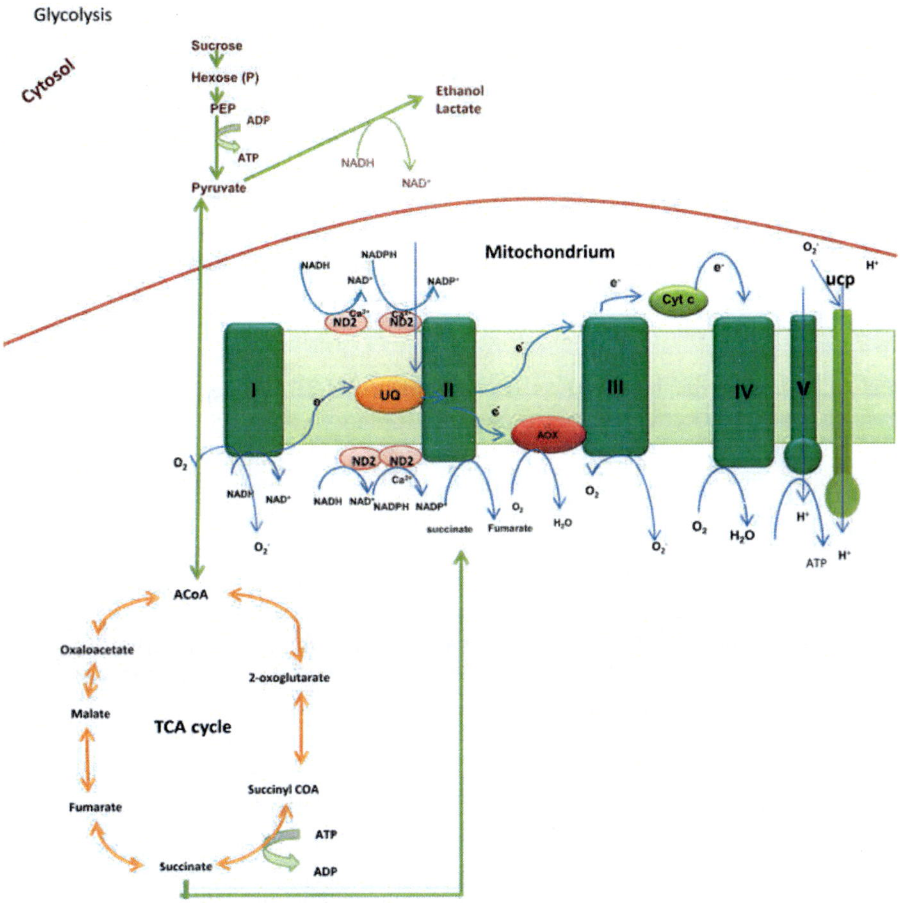

Fig. 5.1 Schematic view of the plant mitochondrial respiratory metabolism. Plant mitochondrial electron transport complexes I to V, described in the text as NADH dehydrogenase, succinate dehydrogenase, cytochrome c reductase, cytochrome c oxidase, ATP synthase respectively. *AOX* alternative oxidase, *cyt-c* cytochrome c, electron transfer flavoprotein quinine oxidoreductase; ND2 (*located inside*), type-2 NAD(P)H dehydrogenase (situated at the matrix side of the mitochondrial inner membrane); ND2 (*located outside*), type-2 NAD(P)H dehydrogenase (situated at the outside of the mitochondrial inner membrane), *UCP* plant uncoupling protein, *UQ* ubiquinone pool, *PEP* phoshoenolpyruvate, *ME* malic enzyme. To simplify understanding, various intermediate reactions are not shown

ATP production (Rasmusson et al. 2009). Here we describe briefly these complexes in the plant mitochondrial electron transport chain.

Complex I is associated with both the TCA cycle and the electron transport chain. It catalyzes the oxidation of NADH formed in the matrix to reduce ubiquinone (UQ). It contains up to 49 subunits. Complex I mutants are not lethal. Studies on knockout mutants suggest that complex I is not required for electron transport chain functionality due to the presence of alternative NAD(P)H dehydrogenases that transfer electrons directly to the ubiquinone pool (Millar et al. 2011). Complex II is called succinate dehydrogenase. It contains four subunits: one flavoprotein, one iron sulfur protein, and two membrane-anchoring subunits. Complex III is cytochrome c reductase, which donates electrons to cytochrome oxidase. Cytochrome c reductase contains 10 subunits. It is also called *bc I* complex. This is one of the sites where reactive oxygen species (ROS) production takes place. Complex III is sensitive to antimycin-A and myxothiazol. Complex IV is cytochrome c oxidase (COX). This is the terminal complex in the electron transport chain, and contains up to 14 subunits. This complex is sensitive to cyanide. Complex V is the fifth respiratory complex, is also called ATPase or F_1F_0type H^+-ATP synthase, and it catalyzes the final step in oxidative phosphorylation within the electron transport chain by converting the electrochemical gradient into ATP.

Most of the current research on alternative electron transfer is focused on non-phosphorylating bypass mechanisms such as external NAD(P)H dehydrogenases and alternative oxidase (AOX). AOX branches from the respiratory chain at the step of UQ, reducing oxygen to water without translocation of the protons. Four forms of AOX are present in dicotyledons, whereas monocotyledons have only one AOX (AOX1) (Considine et al. 2002; Karpova et al. 2002). Based on the substrate utilization for the respiration, AOX and type II NADH dehydrogenases can diminish respiratory ATP production by up to 60% and 30%, respectively (Rasmusson et al. 2008), but are important valves during stress and development (Rasmusson et al. 2009). There are at least four types of NADH dehydrogenase proteins, two on the external side of the inner mitochondrial membrane (one oxidizing NADH and one NADPH) and, similarly, two at the inner face of the inner membrane (one using NADH and the other using NADPH) (Rasmusson and Møller 1991). The roles of these different dehydrogenases have not been discovered, but studies on their properties and expression have led to several hypotheses (Rasmusson et al. 2008). Downstream to complex IV, the uncoupling protein (UCP) acts as a proton channel. The UCP passes electrons from the inner membrane space to the matrix, bypassing the ATP synthase (opposite of F_1F_0type H^+-ATP synthase). Therefore, in context with plant energy balance rearrangements, UCP may have overlapping functions with other alternative pathway proteins in the ETC, such as AOX and NAD(P)H dehydrogenases (Fig. 5.2).

5.2 Low Mitochondrial Respiration During the Storage of Beetroots

Plant respiration function differs in roots and leaves. In roots it is associated with membrane transport and control of oxygen homeostasis. In leaves it is mainly associated with redox regulation and net photosynthetic rate. In storage tubers the respiration is different between leaves and roots. During the conditions of active photosynthesis, leaves generate more carbohydrates than required for growth and metabolism. The excess quantum of these carbohydrates is transported to the storage organs and converted to starch, but in some species, such as sugarcane and sugar beet, the primary storage form is sucrose. The underground organs in plants are made to store these carbohydrates for future utilization when needed, probably

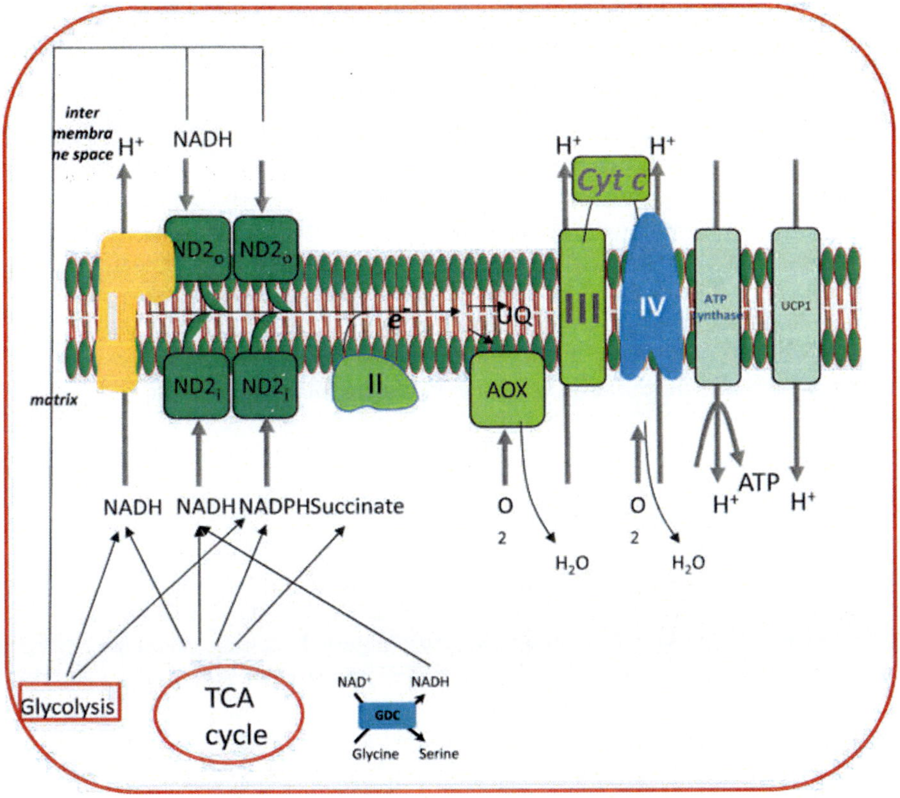

Fig. 5.2 Detailed organization of the plant mitochondrial electron transport chain and various sources of substrates being oxidized by electron transport. *I, II, III, IV* and *V* are complexes, *UCP* uncoupling protein, *AOX* alternative oxidase, *ND2* NAD(P)H dehydrogenases, *GDC* glycine decarboxylase complex, *UQ* ubiquinone pool

during the subsequent season. During the completion of the storage process, the storage tissue shifts to a phase with low respiratory rates as well as low levels of NADP, NADPH, and ADP; ATP levels have been shown to be low during the storage period but to increase during the maturation period (Shugaev et al. 2011). This makes lot of sense because both maturation and development need more energy.

5.3 Intra Mitochondrial NAD Determines the Respiratory Rate and Flux Through the TCA Cycle in Beetroots

Shugaev and Vyskrebentseva (1985) concluded that in storage organs the low abundance of mitochondrial proteins and limited operation of TCA cycle reactions (e.g., succinate dehydrogenase activity) are likely to be responsible for lower respiratory rates. Low abundance of NADH supply to the mitochondria in storage organs also leads to limited activity of NAD(P)H dehydrogenases. For instance, the

NAD content in mung bean mitochondria was 5–6 nmoles/mg protein; whereas, in potato, it was 0.8–1.7 nmoles/mg protein (Tobin et al. 1980). In mature beetroots, the intra-mitochondrial NAD content was 3.2 nmoles/mg protein, which is almost twofold higher than in turnip mitochondria (1.7 nmoles/mg protein). But external supply of NAD enhanced the oxidation of malate by 10–15% in beetroot mitochondria; whereas, in turnip, malate oxidation was enhanced by 25–45% (Soole et al. 1986). In isolated mitochondria, the inhibition of complex I by rotenone enhanced the oxidation of NAD via activation of externally facing NAD(P)H dehydrogenases. Taken together these data suggest that intra-mitochondrial NAD content may in part regulate the TCA cycle operation and thus the respiratory rate in storage organs. This mechanism could help in transition of storage organs during germination and sprouting. Because of these important functionalities, the NAD(P)H dehydrogenases and its features in beetroot mitochondria are discussed in the following section.

5.4 NADH Dehydrogenases in Beetroot Mitochondria

In contrast to the animal system, mitochondria from plants, fungi, protists, and some bacteria contain non-proton pumping external and internal NAD(P)H dehydrogenases. At least four other $NAD(P)^+$ dehydrogenases, two "internal" (facing the intermembrane space) and two "external" (facing the matrix) have been proposed in plants (Rasmusson et al. 2008) (Figs. 5.1, 5.3).

The dehydrogenases are able to oxidize the NAD(P)H and are insensitive to the complex I inhibitor, rotenone. Therefore type II dehydrogenases bypass complex I. The type II dehydrogenases transfer the electrons without pumping protons. Therefore they do not contribute to ATP production and also are not controlled by the cellular adenylate status. The physiological relevance of NADPH dehydrogenases is not well understood. One of the relevant features may be that they regulate the redox state of the cytoplasmic pools of NADH and NAPPH (Rasmusson et al. 2008; Moller and Rasmusson 1998). This would indirectly influence the flux via glycolysis and the TCA cycle (Krömer 1995).

However freshly isolated mitochondria from red beetroot did not show any NADH oxidation (Rayner and Wiskich 1983; Soole et al. 1990). In contrast to this study, Fredlund et al. (1991) showed that mitochondria isolated from fresh beetroots of two cultivars (Nina and Rubria) have the capability to oxidize externally supplied NADH but the rates of state 3 NADH oxidation was very low (25 nmoles min^{-1} mg^{-1} protein). This rate of NADH oxidation was enhanced up to twofold during cold storage (Fredlund et al. 1991). In contrast, cold-stored beetroot tubers were unable to oxidize NAD(P)H (Fredlund et al. 1991). Zottini et al. (1993) showed that mitochondria from tap roots and cold-stored tap roots contain functional NADH dehydrogenase. These authors also showed that mitochondria from sugar beet tissue culture are capable of oxidizing the NAD(P)H in a calcium-dependent manner.

A 43-kDa protein that contains NADH and NADPH dehydrogenase activity was purified (Luethy et al. 1995; Menz and Day 1996). This protein faces the inner side

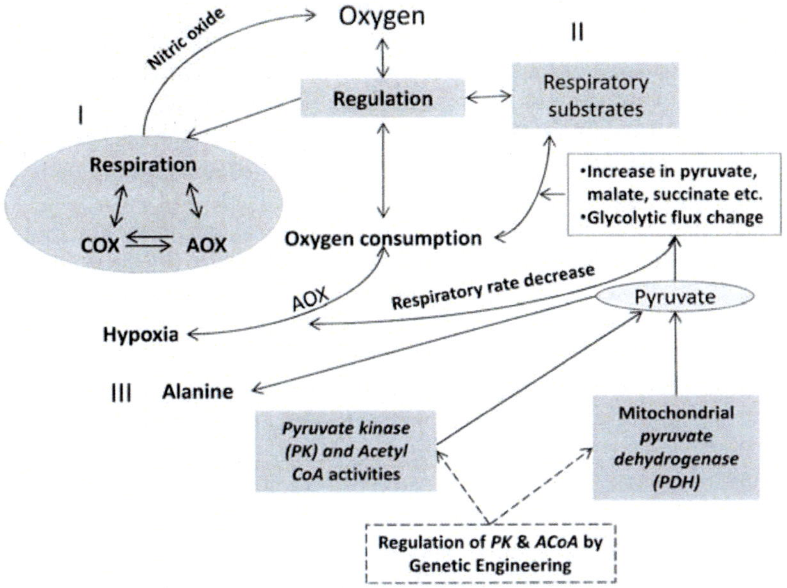

Fig. 5.3 Regulation of respiration and internal oxygen concentrations in storage organs. *I* Respiratory rate is determined by partitioning of cytochrome c oxidase (*COX*) and alternative oxidase (*AOX*). At low oxygen conditions, nitric oxide (*NO*) production takes place. By inhibiting COX, nitric oxide increases the internal oxygen concentrations. *II* Respiratory substrates play a role in oxygen consumption thereby they indirectly influence the internal oxygen concentrations in tuber tissues. *III* Pyruvate kinase, acetyl CoA, and mitochondrial pyruvate dehydrogenase determine the pyruvate and alanine production. This can be also achieved by genetic engineering by expressing acetyl CoA synthetase

of the inner mitochondrial membrane, reduces quinones in high rates, and is sensitive to dicumoral.

A fourth member of NAD(P)H DH activity was purified from red beetroot mitochondria. This is a 26-kDa protein (Rasmusson et al. 1993). Trost et al. (1995) suggested that this protein is present in cytosol, suggesting there is also a soluble form of this NADPH dehydrogenase.

5.5 Alternative Oxidase and Its Induction in Beetroot Mitochondria

Alternative oxidase (AOX) is located in the inner mitochondrial membrane (Fig. 5.2) of all plants and fungi and in a limited number of protists (Millar et al. 2011). Recent evidence suggests that AOX is also present in several prokaryotes and animal systems (Chaudhuri and Hill 1996; McDonald 2008; McDonald and Vanlerberghe 2006) and that AOX is a homodimeric protein presented towards the matrix side. AOX branches from the main respiratory chain at the level of the ubiquinone pool and mediates the four-electron reduction of oxygen to water (Fig. 5.2). When compared with electron transfer by the cytochrome chain (complex III and IV), AOX does not pump protons, therefore does not create a transmembrane potential; the decline in free energy between ubiquinol and oxygen is dissipated and mostly released as heat (Vanlerberghe et al. 2009). However it plays important roles when the cytochrome pathway is inhibited under various stress conditions. The mechanism of AOX activation could be via shifts in the level of nucleotides or reactive oxygen species (ROS) (Møller 2001). Increases in ROS in turn might act as signals (see Fig. 5.2) to induce transcription of AOX (Mackenzie and McIntosh 1999; Vanlerberghe and McIntosh 1996). This is supported by an experiment in which addition of antioxidants leads to the suppression of AOX (Maxwell et al. 1999). Another function of AOX might be attributed to the maintenance of oxygen homeostasis (Gupta et al. 2009). This hypothesis is supported by large differences in oxygen affinities of either COX or AOX pathways. The *Km* of cytochrome c oxidase (COX) is approximately 0.1 μM and for AOX is 10–20 μM, which means that AOX has 100–200 times less affinity for oxygen (with the exception shown by Millar et al. (1994) where AOX affinity was only 10 times higher). The lower affinity of AOX for oxygen could mean that, when oxygen concentrations are very high, AOX reduces excess oxygen, thereby participating in controlling the levels of ROS (Skutnik and Rychter 2009; Gupta et al. 2009). This hypothesis is one of the exceptions to the energy over-flow hypothesis proposed by Lambers (1982) and supports the model of Ribas-Carbo et al. (1995) on electron partitioning. They investigated the flow of electrons using the oxygen isotope discrimination technique, and showed that the inhibition of AOX by its inhibitor salicylhydroxamic acid (SHAM) did not lead to a decrease in total respiratory rates. These authors further concluded that COX can fully compensate oxygen consumption when AOX activity is absent. This study also provided evidence that both pathways compete for electrons, and that electron flow to each pathway is determined by its capacity (which in turn is greatly dependent on environmental and developmental conditions).

Klotz et al. (2008) measured the COX and AOX capacity in beetroots in response to cold treatment. In roots stored at low temperature (10°C and 15°C), total COX and AOX capacity were found to be up to fourfold elevated in the first 24 h after harvest, and remained elevated above initial values during the next 12 day in storage. Such an increase in the cold-induced AOX capacity in beetroots was comparable to the induction of AOX in tobacco (Vanlerberghe and McIntosh 1992). In studies with *Arabidopsis*, cold stress was found to induce the AOX capacity as a response to ROS (Sugie et al. 2006). This mechanism could also explain the AOX capacity in beetroot mitochondria (Klotz et al. 2008). These authors also found that, in beetroot tissue, the total respiration was enhanced by the addition of NADH, suggesting that mitochondria are capable of oxidizing additional substrates within the electron transport chain through external NADH dehydrogenase. However, respiration was not induced by an ADP supplement or by an uncoupler of the electron transport chain. Taken together, these findings suggest that respiration in stored beetroot is not limited by respiratory capacity, ADP availability, or cellular energy status, and other factors such as low internal oxygen concentrations should also be taken into consideration (Geigenberger 2003).

5.6 Internal Oxygen Control by Respiration and Fermentation Pathways

The internal oxygen concentrations of various plant species and organs vary depending on their developmental and metabolic state (Rolletschek et al. 2002, 2007). For instance internal oxygen concentrations are about 50 μM within phloem tissue of stems (van Dongen et al. 2003), about 100 μM within potato tubers (Geigenberger et al. 2010), and typically range between 1 and 25 μM in seeds (Borisjuk and Rolletschek 2009; Rolletschek et al. 2005a). The barriers for oxygen diffusion and active cellular metabolism/respiration, especially during the sprouting periods in bulky tissues such as beetroot, may encounter severe oxygen limitation (hypoxia), and more so during the storage period. Although the internal oxygen concentrations of beetroot have not yet been measured, the levels are expected to be in the same range as in potato tubers (~100 μM). Notably, actual internal oxygen concentrations largely depend on two factors (1) diffusion resistance, in which the diameter of the tissue prevents the oxygen diffusion to inner cells, and (2) respiratory activity of the tissue, which varies in response to environmental and developmental conditions.

Internal oxygen concentrations also play a role in the regulation of cellular metabolism. Several studies have demonstrated that assimilate uptake, metabolite pattern, and metabolic pathway activity and eventually fluxes towards storage products are affected by the in vivo oxygen supply (Geigenberger 2003; Borisjuk and Rolletschek 2009; Rolletschek et al. 2005b). The molecular mechanisms underlying the metabolic adjustments towards hypoxia are still not fully understood, but seem to involve down-regulation of genes associated with the maintenance of cell wall metabolism and ion/metabolite transport, along with up-regulation of those involved in transcriptional regulation, protein kinase activity, glycolysis, and fermentation (Klok et al. 2002: Liu et al. 2005: Thiel et al. 2011). Several genes related to the signal transduction network (calcium- or phytohormone-dependent) and specific sequence motifs in co-expressed genes have also been found. All these together demonstrate the significance of oxygen availability for mitochondrial and overall cellular metabolism.

Control of internal oxygen is essential to cope with fluctuating oxygen supplies. Regulation of internal oxygen concentrations (see Fig. 5.3) can be achieved in various ways. First, by regulation of respiration via COX and AOX. The down-regulation of AOX at low oxygen concentrations can help in reducing the overall oxygen consumption, thus maintaining an internal oxygen concentration above zero (Gupta et al. 2009). Second, respiratory substrates play a role in the regulation of internal oxygen concentration via stimulation of respiratory rates. For example, feeding pyruvate and succinate to pea roots caused an increase in respiratory rates, which correlated with a decrease in internal oxygen concentrations (Zabalza et al. 2009). Shifts in the glycolytic flux can affect the concentration of pyruvate, which becomes available for respiration, thereby regulating the internal oxygen concentration of the tissue. Important regulatory enzymes in the glycolytic pathway are PK (converting PEP into pyruvate) and PEP carboxylase. The activity of PK plays a role in determining the amount of pyruvate available for mitochondria. It has been shown that only the stimulation of glycolysis by PK resulted in an increase of oxygen consumption. Therefore, one possible mechanism for the internal oxygen concentration is via the control of PK activity. Another enzyme that controls pyruvate levels is the mitochondrial pyruvate dehydrogenase (PDH) complex that catalyzes the conversion of pyruvate into acetyl CoA. A third mechanism is the formation of alanine via alanine aminotransferase (utilizing pyruvate) (Fig. 5.3).

Apart from such metabolic intermediates/enzymes, various authors postulated a role for nitric oxide (NO) in regulating the oxygen consumption via COX (Borisjuk et al. 2007; Benamar et al. 2008). NO is a small gaseous, oxygen-sensitive molecule, able to mediate rapid and reversible oxygen balancing in plant as well as animal tissues, including adjustments to the global metabolism in response to changing oxygen. Its dynamic nature allows for the maintenance of a steady-state oxygen level and avoids the risk of the onset of anoxia (Fig. 5.3).

When oxygen concentration falls below the saturation of terminal oxidases, the respiration is inhibited and adenylate levels drop as an adaptive strategy (Geigenberger 2003). Addition of substrates such as pyruvate can stimulate respiration even at low oxygen levels. It remains to be tested if this kind of adaptive response is also found in beetroots.

5.7 Factors Affecting Respiratory Rates

In sugar beet (a variety of *Beta vulgaris*), respiration is strongly influenced by genetic factors, storage conditions, maintenance of sterile conditions preventing pathogen attacks, etc. (Shugaev et al. 2011). During the post-harvest periods, the loss of sugar in beetroot occurs due to the rapid fluctuations in respiration and the rapid inter-conversions of sugars into other storage carbohydrates. In this context, pathogen attack may play a role in changing enzyme activities (Shugaev et al. 2011). It has been shown that respiratory changes are the major causes for the variations in levels of sucrose in beetroot. As outlined above, respiratory metabolism is an oxidative process in which carbohydrates are converted into carbon dioxide/water and providing energy. Storage of beetroot tubers at low temperature maintains low respiration. During storage at room temperature, they maintain high respiratory rates and mitochondria provide constant energy for the maintenance of the metabolism. Although proteins, organic acids, and lipids act as substrates for the respiration, carbohydrates are the chief source for respiratory substrates. Another important factor that determines the respiratory rate is acetyl CoA. It plays a role in determining the flux of the TCA cycle, thereby controlling respiration. Acetyl CoA formation is catalyzed by acetyl CoA synthetase, which catalyzes the condensation of acetate and coenzyme A (coA). Engineering root-specific acetyl coA, particularly acetyl coA synthase, which specifically helps in channeling break-down of sugars as well as in binding acetate with CoA, is relevant in the case of beetroot.

5.8 Concluding Remarks

Extensive research on the respiratory metabolisms of beetroot is required to improve its quality in terms of maintenance of cell turgidity and prevention of post-harvest losses due to rapid respiration. In this context, energy-dissipating alternative pathways need to be explored to a greater extent. Some basic research needs to be undertaken to answer questions related to respiratory metabolism.

- If the mitochondrial content is regulated by its demand in the development, how do the number of mitochondria differ in leaves, roots, and storage organs, such as beetroot. This knowledge is important in the context of tubers because, in storage organs like beetroots, tissue-specific respiratory metabolism plays a role in the carbohydrate source sink relationship.
- The cell-specific functions of mitochondria need to be explored in normal and storage organs.
- The roles of non-phosphorylation pathways in storage tissues need to be explored in more detail.
- Systems biology approaches (transcriptomics, metabolomics, and proteomics) will be useful for the improvement of beetroot quality.

Acknowledgements This work was supported by Deutsche Forschungsgemeinschaft (BA 1177/8-1) (KJG, HB). I thank Ian Max Møller, Aarhus University, Denmark, for valuable suggestions on this book chapter. I thank Shruthi Segu for editorial help.

References

Benamar, A., H. Rolletschek, L. Borisjuk, M.H. Avelange-Macherel, G. Curien, A. Mostefai, R. Andriantsitohaina, and D. Macherel. 2008. Nitrite-nitric oxide control of mitochondrial respiration at the frontier of anoxia. *Biochimica et Biophysica Acta* 1777: 1268–1275.

Borisjuk, L., D. Macherel, A. Benamar, U. Wobus, and H. Rolletschek. 2007. Low oxygen sensing and balancing in plant seeds – A role for nitric oxide. *The New Phytologist* 176: 813–823.

Borisjuk L, Rolletschek H 2009.The oxygen status of the developing seed. New Phytol. 182(1): 17–30

Chaudhuri, M., and G.C. Hill. 1996. Cloning, sequencing and functional activity of the *Trypanosoma brucei brucei* alternative oxidase. *Molecular and Biochemical Parasitology* 83: 125–129.

Considine, M.J., R.C. Holtzapffel, D.A. Day, J. Whelan, and A.H. Millar. 2002. Molecular distinction between alternative oxidase from monocots and dicots. *Plant Physiology* 129: 949–953.

Fernie, A.R., A. Tiessen, M. Stitt, L. Willmitzer, and P. Geigenberger. 2002. Altered metabolic fluxes result from shifts in metabolite levels in sucrose phosphorylase-expressing potato tubers. *Plant, Cell & Environment* 25: 1219–32.

Fredlund, K.M., A.G. Rasmusson, and I.M. Møller. 1991. The effects of aging on the oxidation of external NAD(P)H in purified red beetroot (*Beta vulgaris* L.) mitochondria. *Plant Physiology* 97: 99–103.

Geigenberger, P. 2003. Response of plant metabolism to too little oxygen. *Current Opinion in Plant Biology* 6: 247–56.

Geigenberger, P., D. Riewe, and A.R. Fernie. 2010. The central regulation of plant physiology by adenylates. *Trends in Plant Science* 15: 98–105.

Giege, P., J.L. Heazlewood, U. Roessner-Tunali, A.H. Millar, A.R. Fernie, C.J. Leaver, and L.J. Sweetlove. 2003. Enzymes of glycolysis are functionally associated with the mitochondrion in *Arabidopsis* cells. *The Plant Cell* 15(9): 2140–2151.

Gupta, K.J., A. Zabalza, and J.T. van Dongen. 2009. Regulation of respiration when the oxygen availability changes. *Physiologia Plantarum* 137: 383–391.

Karpova, O.V., E.V. Kuzmin, T.E. Elthon, and K.J. Newton. 2002. Differential expression of alternative oxidase genes in maize mitochondrial mutants. *The Plant Cell* 14: 3271–3284.

Klok, E.J., I.W. Wilson, D. Wilson, S.C. Chapman, R.M. Ewing, S.C. Somerville, W.J. Peacock, R. Dolferus, and E.S. Dennis. 2002. Expression profile analysis of the low-oxygen response in *Arabidopsis* root cultures. *The Plant Cell* 14: 2481–2494.

Klotz, K.L., F.L. Finger, and M.D. Anderson. 2008. Respiration in postharvest sugar beet roots is not limited by respiratory capacity or adenylates. *Journal of Plant Physiology* 165: 1500–1510.

Krömer, S. 1995. Respiration during photosynthesis. *Annual Review of Plant Physiology and Plant Molecular Biology* 46: 45–70.

Lambers, H. 1982. Cyanide resistant respiration: A nonphosphorylating electron transport pathway acting as an energy overflow. *Plant Physiology* 55: 478–485.

Liu, F., T. VanToai, L.P. Moy, G. Bock, L. Linford, and J. Quackenbush. 2005. Global transcription profiling reveals comprehensive insights into hypoxic response in *Arabidopsis*. *Plant Physiology* 137: 1115–1129.

Luethy, M.H., J.J. Thelen, A.F. Knudten, and T.E. Elthon. 1995. Purification, characterization, and submitochondrial localization of a 58-kilodalton NAD(P)H dehydrogenase. *Plant Physiology* 107: 443–50.

Mackenzie, S., and L. McIntosh. 1999. Higher plant mitochondria. *The Plant Cell* 11: 571–585.

Maxwell, D.P., Y. Wang, and L. McIntosh. 1999. The alternative oxidase lowers mitochondrial reactive oxygen production in plant cells. *Proceedings of the National Academy of Sciences of the United States of America* 96: 8271–76.

McDonald, A.E. 2008. Alternative oxidase: An inter-kingdom perspective on the function and regulation of this broadly distributed 'cyanide resistant' terminal oxidase. *Functional Plant Biology* 35: 535–552.

McDonald, A.E., and G.C. Vanlerberghe. 2006. Origins, evolutionary history, and taxonomic distribution of alternative oxidase and plastoquinol terminal oxidase. *Comparative Biochemistry and Physiology* 1: 357–364.

Menz, R.I., and D.A. Day. 1996. Purification and characterisation of a 43-kDa rotenone-insensitive NADH dehydrogenase from plant mitochondria. *Journal of Biological Chemistry* 271: 23117–20.

Millar, A.H., F.J. Bergersen, and D.A. Day. 1994. Oxygen affinity of terminal oxidases in soybean mitochondria. *Plant Physiology and Biochemistry* 32: 847–852.

Millar, A.H., J. Whelan, K.L. Soole, and D.A. Day. 2011. Organization and regulation mitochondrial respiration in plants. *Annual Review of Plant Biology* 62: 79–104.

Møller, I.M. 2001. Plant mitochondria and oxidative stress: Electron transport, NADPH turnover, and metabolism of reactive oxygen species. *Annual Review of Plant Biology and Plant Molecular Biology* 52: 561–91.

Moller, I.M., and A.G. Rasmusson. 1998. The role of NADP in the mitochondrial matrix. *Trends in Plant Science* 3: 21–27.

Podestá, F.E., and W.C. Plaxton. 1991. Association of phosphoenolpyruvate phosphatase activity with the cytosolic pyruvate kinase of germinating mung beans. *Plant Physiology* 97: 1329–1333.

Rasmusson, A.G., and I.M. Møller. 1991. NAD(P)H dehydrogenases on the inner surface of the inner mitochondrial membrane studied using inside-out submitochondrial particles. *Physiologia Plantarum* 83: 357–365.

Rasmusson, A.G., K.M. Fredlund, and I.M. Møller. 1993. Purification of a rotenone-insensitive NAD(P)H dehydrogenase from the inner surface of the inner mitochondrial membrane of red beetroot mitochondria. *Biochimica et Biophysica Acta* 1141: 107–10.

Rasmusson, A.G., D.A. Geisler, and I.M. Møller. 2008. The multiplicity of dehydrogenases in the electron transport chain of plant mitochondria. *Mitochondrion* 8: 47–60.

Rasmusson, A.G., A.R. Fernie, and J.T. van Dongen. 2009. Alternative oxidase: A defence against metabolic fluctuations? *Physiologia Plantarum* 137: 371–82.

Rayner, J.R., and J.T. Wiskich. 1983. Development of NADH oxidation by red beet mitochondria on slicing and aging of the tissues. *Australian Journal of Plant Physiology* 10: 55–63.

Ribas-Carbo, M., J.A. Berry, D. Yakir, L. Giles, S.A. Robinson, A.M. Lennon, and J.N. Siedow. 1995. Electron partitioning between the cytochrome and alternative pathways in plant mitochondria. *Plant Physiology* 109: 829–837.

Rocha, M., F. Licausi, W.L. Araújo, A. Nunes-Nesi, L. Sodek, A.R. Fernie, and J.T. van Dongen. 2010. Glycolysis and the tricarboxylic acid cycle are linked by alanine aminotransferase during hypoxia induced by waterlogging of *Lotus japonicus*. *Plant Physiology* 152: 1501–1513.

Rolletschek, H., L. Borisjuk, M. Koschorreck, U. Wobus, and H. Weber. 2002. Legume embryos develop in a hypoxic environment. *Journal of Experimental Botany* 53: 1099–110.

Rolletschek, H., K. Koch, U. Wobus, and L. Borisjuk. 2005a. Positional cues for the starch/lipid balance in maize kernels and resource partitioning to the embryo. *The Plant Journal* 42: 69–83.

Rolletschek, H., R. Radchuk, C. Klukas, F. Schreiber, U. Wobus, and L. Borisjuk. 2005b. Evidence of a key role for photosynthetic oxygen release in oil storage in developing soybean seeds. *The New Phytologist* 167: 777–786.

Rolletschek, H., L. Borisjuk, A. Sánchez-García, C. Romero, L. Gotor, J. Rivas, and M. Mancha. 2007. Temperature-dependent endogenous oxygen concentration regulates microsomal oleate desaturase in developing sunflower seeds. *Journal of Experimental Botany* 58: 3171–3181.

Shugaev, A.G., and E.I. Vyskrebentseva. 1985. Developmental changes in the functional activity of mitochondria in sugar beet roots. *Fiziologiya Rastenii (Moscow)* 32: 259–267 (Sov. Plant Physiology, Engl. Transl.).

Shugaev A.G., D.A. Lashtabega, N.A. Shugaeva, and E.I. Vyskrebentseva. 2011. Activities of antioxidant enzymes in mitochondria of growing and dormant sugar beet roots. *Russian Journal of Plant Physiology* 58: 387–393.

Skutnik, M., and A.M. Rychter. 2009. Differential response of antioxidant systems in leaves and roots of barley subjected to anoxia and post-anoxia. *Journal of Plant Physiology* 166: 926–937.

Soole, K.L., I.B. Dry, and J.T. Wiskich. 1986. The responses of isolated plant mitochondria to external nicotinamide adenine dinucleotide. *Plant Physiology* 81: 587–592.

Soole, K.L., I.B. Dry, A.T. James, and J.T. Wiskich. 1990. The kinetics of NADH oxidation by complex I of isolated plant mitochondria. *Physiologia Plantarum* 80: 75–82.

Sugie, A., N. Naydenov, N. Mizuno, C. Nakamura, and S. Takumi. 2006. Over expression of wheat alternative oxidase gene Waox1a alters respiration capacity and response to reactive oxygen species under low temperature in transgenic *Arabidopsis*. *Genes & Genetic Systems* 81: 349–54.

Sweetlove, L.J., K.F.M. Beard, A. Nunes-Nesi, A.R. Fernie, and R.G. Ratcliffe. 2010. Not just a circle: Flux modes in the plant TCA cycle. *Trends in Plant Science* 15: 462–470.

Thiel, J., H. Rolletschek, S. Friedel, J.E. Lunn, T.H. Nguyen, R. Feil, H. Tschiersch, M. Müller, and L. Borisjuk. 2011. Seed-specific elevation of non-symbiotic hemoglobin AtHb1: Beneficial effects and underlying molecular networks in *Arabidopsis thaliana*. *BMC Plant Biology* 11: 48.

Tobin, A., B. Djerdjour, E. Journet, M. Neuburger, and R. Douce. 1980. Effect of NAD on the malate oxidation in intact plant mitochondria. *Plant Physiology* 66: 225–229.

Trost, P., P. Bonora, S. Scagliarini, and P. Pupillo. 1995. Purification and properties of NAD(P) H:(quinone-acceptor) oxidoreductase of sugar beet cells. *European Journal of Biochemistry* 234: 452–58.

Urbanczyk-Wochniak, E., B. Usadel, O. Thimm, A. Nunes-Nesi, F. Carrari, M. Davy, O. Blasing, M. Kowalczyk, D. Weicht, A. Polinceusz, S. Meyer, M. Stitt, and A.R. Fernie. 2006. Conversion of MapMan to allow the analysis of transcript data from *Solanaceous* species: Effects of genetic and environmental alterations in energy metabolism in the leaf. *Plant Molecular Biology* 60: 773–92.

van Dongen, J.T., U. Schurr, M. Pfister, and P. Geigenberger. 2003. Phloem metabolism and function have to cope with low internal oxygen. *Plant Physiology* 131: 1529–1543.

van Dongen, J.T., K.J. Gupta, S.J. Ramírez-Aguilar, W.J. Araujo, A. Nunes-Nesi, A.R. Fernie. 2011. Regulation of respiration in plants: A role for alternative metabolic pathways. *Journal of Plant Physiology* 168: 1434–1443.

Vanlerberghe, G.C., and L. McIntosh. 1992. Lower growth temperature increases alternative pathway capacity and alternative oxidase protein in tobacco. *Plant Physiology* 100: 115–119.

Vanlerberghe, G.C., L. McLntosh. 1996. Signals regulating the expression of the nuclear gene encoding alternative oxidase of plant mitochondria. *Plant Physiology* 111(2): 589–595.

Vanlerberghe, G.C., M. Cvetkovska, and J. Wang. 2009. Is the maintenance of homeostatic mitochondrial signaling during stress a physiological role for alternative oxidase? *Physiologia Plantarum* 137: 392–406.

Zabalza, A., J.T. van Dongen, A. Froehlich, S.N. Oliver, B. Faix, K.J. Gupta, E. Schma lzlin, M. Igal, L. Orcaray, M. Royuela, et al. 2009. Regulation of respiration and fermentation to control the plant internal oxygen concentration. *Plant Physiology* 149: 1087–1098.

Zottini, M., G. Mandolino, and D. Zannoni. 1993. Oxidation of external NAD(P)H by mitochondria from taproots and tissue cultures of sugar beet (*Beta vulgaris*). *Plant Physiology* 102: 579–85.

Chapter 6
Lipoperoxyl Radical Scavenging and Antioxidative Effects of Red Beet Pigments

Maria A. Livrea and Luisa Tesoriere

Abstract Aerobic life is characterized by a steady formation of reactive oxygen species and free radicals, which is almost entirely counteracted by endogenous primary and secondary antioxidant systems. Maintenance of these systems is then imperative to ensure a continuous defense to cells and to avoid conditions known as oxidative stress. Apart from antioxidant vitamins, many compounds from the plant kingdom are now considered very helpful to maintain a proper cell redox balance. Among them, betalain pigments have received recent attention. Betanin (betanidin-5-*O*-β glucoside) is the main betacyanin from red beet. Redox potential, ability to interact with lipid structures and bioavailability in humans make this molecule a potential natural antioxidant with protective effects in vivo. This review summarizes the peroxyl radical-scavenging activity of the molecule and of its aglycone betanidin, as observed in a few chemical or biological models.

6.1 Introduction

It is now acknowledged that cell and tissue wellbeing relies on an appropriate cell redox status. Indeed, a million years of evolution led aerobic organisms to produce free radicals and oxidants (reactive oxygen species [ROS]), as well as to exploit an effective antioxidant machinery to control redox-sensitive signaling pathways responsible for a variety of processes including, among others, cell differentiation and proliferation, inflammation, apoptosis and aging itself (Hancock 2009; Dröge 2002; Matsuzawa and Ichijo 2008; Giles 2006; Wu et al. 2006; Giorgio et al. 2007; Valko et al. 2007; Lee and Griendling 2008; Pan et al. 2009).

M.A. Livrea (✉) • L. Tesoriere
Dipartimento STEMBIO, Università di Palermo, Via M. Cipolla, 74, Palermo 90123, Italy
e-mail: maria.livrea@unipa.it

B. Neelwarne (ed.), *Red Beet Biotechnology: Food and Pharmaceutical Applications*,
DOI 10.1007/978-1-4614-3458-0_6, © Springer Science+Business Media New York 2013

While ROS production in a finely controlled fashion is required to maintain the natural oxidative homeostasis, uncontrolled generation and/or aggression by environmental oxidants, toxicants and heavy metals can modify the balance between pro- and antioxidative processes, resulting in the condition known as "oxidative stress", initiating biochemical events resulting in pathological conditions (Ma 2010; Martin and Barrett 2002).

Cells are endowed with primary antioxidant defenses, i.e. enzymes such as superoxide dismutase, catalase and glutathione peroxidase, that remove ROS before they may attack cell components, and various repair systems needed to cope with damaged molecules, including low molecular weight antioxidants such as glutathione and vitamins E, C, A and carotenoids. By these means, cells protect all compartments, thus preventing damage to nucleic acids, proteins and membrane lipids.

Because endogenous antioxidants are continuously consumed, the organism should be helped to keep their optimal level to avoid oxidative damage. This can be accomplished by introducing new reducing molecules to replace the consumed ones. Numerous epidemiological studies (Willett et al. 1995; Kushi et al. 1995) point out the importance of diets based on herbs, fruits, grains, and vegetables in reducing the incidence of chronic and degenerative diseases such as cancer and cardiovascular disease, the etio-pathogenesis of which is strongly supported by oxidative stress (Lin 1995; Cao et al. 1997). Indeed plants are the main source of dietary antioxidants. Apart from the antioxidant vitamins, a vast array of phytochemicals, from bioflavonoids to phytosterols and terpenoids, with potential antioxidative activity and/or ability to modulate redox-sensitive signaling pathways, have been isolated. Recently, the radical-scavenging activity and antioxidant capacity of betalains have been the object of research in our as well as in other laboratories (Kanner et al. 2001; Escribano et al. 1998; Butera et al. 2002; Livrea and Tesoriere 2004; Gliszczynska-Swiglo et al. 2006; Czapski et al. 2009).

Betalain pigments, secondary metabolites of plants of the *Caryophyllales* order, share the chemical structure of betalamic acid and include two classes of compounds, i.e. the yellow betaxanthins and red betacyanins, according to the structure bound to betalamic acid. When the latter is conjugated with amino acids or corresponding amines (including dopamine), betaxanthins arise. Betacyanins are derivatives of betanidin, the conjugate of betalamic acid with *cyclo*-DOPA, with additional substitutions through varying glycosylation and acylation patterns at C5 or C6 positions. Betanin (5-*O*-glucose betanidin) and vulgaxanthin I (glutamine–betaxanthin) are the main pigments found in raw red beet (Fig. 6.1). On the other hand, in accordance with studies showing that vulgaxanthin I is poorly stable under a number of physical and chemical conditions (Herbach et al. 2006), vulgaxanthin did not appear detectable in the steamed red beet, nor in other beet preparations such as juice and jam (Tesoriere et al. 2008).

When treating with the potential health-promoting effects of dietary compounds, it is important to consider their bioavailability, i.e. how much of the active molecule is absorbed, its eventual transformation at the level of the digestive tract and, finally, the distribution to tissues and cells. Factors such as the chemistry of the molecule, the nature of co-ingested compounds as well as the complexity of the food matrix

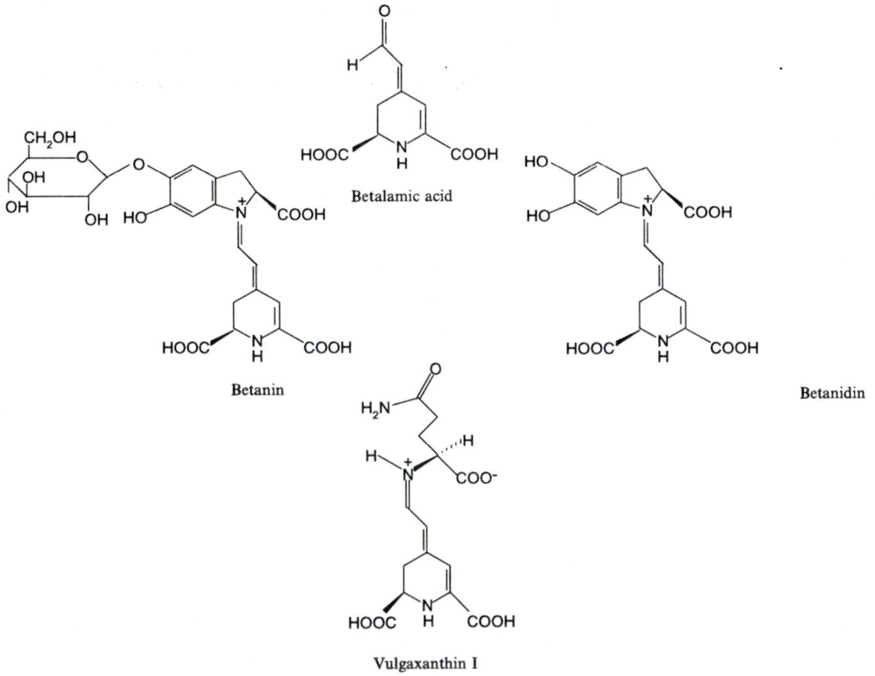

Fig. 6.1 Chemical structure of betalamic acid, vulgaxanthin I and main betacyanin derivatives

may largely affect bioavailability. Studies in humans reporting kinetics of absorption and extent of plasma concentration and urinary excretion (Kanner et al. 2001; Tesoriere et al. 2004a; Frank et al. 2005) provided evidence that discrete amounts of betanin can reach the circulation and distribute in low-density lipoproteins (LDL) (Tesoriere et al. 2004a) and red blood cells (Tesoriere et al. 2005), where the molecule presumably was involved in antioxidant protection. On this basis, investigating the activity of betanin as a lipid antioxidant and providing kinetic parameters of the activity has been a stimulating challenge for our group (Tesoriere et al. 2009). To this purpose chemical lipid systems such as methanolic solutions of methyl linoleate and soybean phosphatidylcholine liposomes have been used. In other studies, the antioxidant activity of betanin has been evaluated in more complex biological lipid matrixes such as LDL (Tesoriere et al. 2003; Allegra et al. 2007).

6.2 Oxidation of Lipids

Oxidation of membrane unsaturated lipids is believed to contribute to human ageing and disease by disrupting the structure and the packaging of the lipid components and, ultimately, by preventing membrane function. Beside causing local disruption, this process may also affect intracellular signaling, since reactive end-products of lipid peroxidation such as unsaturated aldehydes may easily migrate from membranes, causing intracellular injury and remarkable modifications of the oxidative homeostatic signaling (Uchida 2007; Echtay et al. 2003; Leitinger 2005; Leonarduzzi et al. 1997). Due to the importance of maintaining membrane integrity, numerous bioactive substances present in foods have been explored as potential lipid antioxidants.

Peroxidation of polyunsaturated lipids (PUFA) is characterized by radical chain reactions, where a single initiating free radical (R·) may cause the peroxidation of a large number of lipids (LH). In the presence of appropriate initiators, the process takes place according to a mechanism exemplified in

Initiation

$$R^{\bullet} + O_2 \rightarrow ROO^{\bullet}$$
$$ROO^{\bullet} + LH \rightarrow ROO^{\bullet} + L^{\bullet}$$

Propagation

$$L^{\bullet} + O_2 \rightarrow LOO^{\bullet}$$
$$LOO^{\bullet} + LH \xrightarrow{kp} LOOH + L^{\bullet}$$

Termination

$$2\,LOO^{\bullet} \xrightarrow{kt} \text{non - radical products}$$

where L·, LOO·, and LOOH are the alkyl and alkylperoxyl radicals and hydroperoxide generated, and kp and kt are the rate constants for propagation and termination of the radical chain, respectively. Classical chain-breaking antioxidants, such as vitamin E, inhibit the peroxidation process by scavenging the chain-carrying lipoperoxyl radicals, thus preventing the radical attack of other lipids and production of hydroperoxides. The effectiveness of these antioxidants is determined by the rate at which they actually scavenge lipoperoxyl radicals, comparable with the rate at which the radicals are produced, as well as by the number of radicals scavenged per mole of antioxidant. In the presence of a chain-breaking antioxidant, lipid peroxidation is stopped as long as the antioxidant is totally consumed, a time interval known as the inhibition period or lag time. Due to the primary importance of vitamin E (α-tocopherol) in protecting membrane lipids (Fukuzawa 2008), the comparison between kinetic parameters measured for natural antioxidants and those of vitamin E may provide an indication of the compound's effectiveness.

The oxidation of methyl linoleate (LAME) under controlled conditions is the simplest way to study the oxidation of polyunsaturated lipids, and it has widely been adopted to carry out kinetic studies with antioxidants. Since the linoleic acid has two double bonds, peroxidation occurs at the bis-allylic hydrogens and generates stoichiometric amounts of conjugated dienes (CD) lipid hydroperoxides that can be measured spectrophotometrically (Pryor and Castle 1984). Methanolic LAME solutions are oxidized by radicals thermally generated from a lipophilic azo-initiator such as AMVN (2,2′-azobis (2,4-dimethylvaleronitrile)) (Niki 1990) to ensure a linear production of lipoperoxides propagating chain reactions. Analysis of the peroxidation curve generated by monitoring the formation of CD hydroperoxides at time intervals permits the calculation of kinetic parameters for the reaction of lipoperoxyl radicals with antioxidants. The propagation rate, R_p, is measured as the amount of CD lipid hydroperoxides formed per second, either in the absence (control) or in the presence of antioxidant. The rate of chain initiation, R_i, is measured by the inhibition period (t_{inh}) produced by a known amount of α-tocopherol, following the equation

$$R_i = n[IH] / t_{inh} \tag{6.1}$$

where IH is the concentration of α-tocopherol, and n, the stoichiometric factor that represents the peroxyl radicals scavenged by each molecule of antioxidant, is assumed to be 2 (Burton and Ingold 1981).

In the curve of peroxidation in the presence of antioxidant, the inhibition period, t_{inh}, is measured as the time interval between the addition of free radical initiator and the point of intersection of the tangents to the tracts of the curve representing the inhibition and propagation phases. When inhibition periods are measured, the inhibition rate constant, K_{inh}, in solution of peroxidizing LAME is calculated as

$$K_{inh} = K_p[LH] / R_{inh} t_{inh}, \tag{6.2}$$

where [LH] is the concentration of the lipid; and k_p, the absolute rate constant for the oxidation of LAME at 50°C, is to be assumed 230 M^{-1} s^{-1} (Yamamoto et al. 1982). The inhibition rate, R_{inh}, that is the rate of production of lipid hydroperoxides during the inhibition period, is calculated by the coordinates of the intercept of the extrapolations of the parts of the curve representing the inhibition and propagation phases.

Soybean phosphatidylcholine (PC) unilamellar liposomes are a suitable membrane-mimetic system to obtain quantitative data of the peroxyl radical-scavenging activity of antioxidants, due to the peculiar composition in unsaturated fatty acids, 95% of which consist of linoleic acid. The use of a hydrophilic azo-initiator such as AAPH (2,2′-azobis(2-amidinopropane) dihydrochloride) (Niki 1990) causes a linear hydroperoxide formation, thereby R_i can be evaluated by the classic inhibitor method according to Eq. 6.1.

6.3 Antioxidant Activity of Betacyanins

Betacyanins are heterocyclic tyrosine-derived pigments. The phenol moiety and/or the cyclic amine group have been considered to confer reducing properties to this class of compounds (Kanner et al. 2001; Gliszczynska-Swiglo et al. 2006; Gandia-Herrero et al. 2010). In addition, because of their chemistry, including

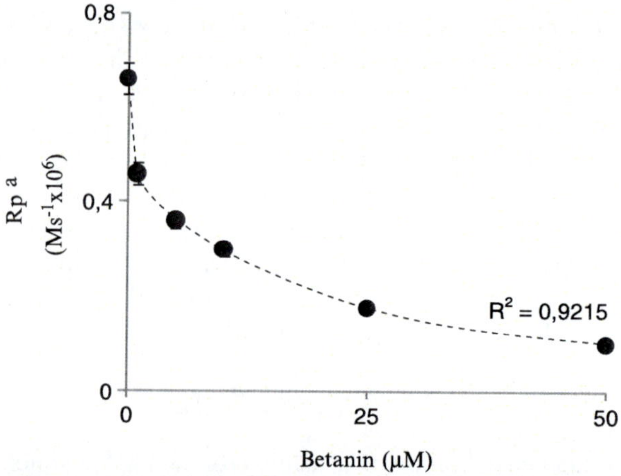

Fig. 6.2 Relationships between the propagation rate (Rp) and betanin concentrations in AMVN-induced oxidation of methyl linoleate in methanol. [a]CD-hydroperoxide formation per second

charged portions and ionizable groups as well as lipophilic moieties, these molecules may behave as amphiphilic-like compounds at physiological pH. Kinetic measurements of the peroxyl radical-scavenging activity of betanin and of its aglycone, betanidin, in organic solution and liposomes, and the identification of oxidized products, have recently provided mechanistic insights on the antioxidant properties of these compounds, consistent with the activity of the glucose-substituted monophenol and *ortho*-diphenol moieties, respectively (Tesoriere et al. 2009). Though both pigments appear to be peroxyl radical scavengers, betanidin exhibits an effectiveness higher than betanin.

(a) *Peroxyl radical-scavenging activity of betanin and betanidin in methanol*
Betanin does not cause any delay of the oxidation of LAME in methanol solution, but only a decrease of the peroxidation rate that depends exponentially on the betanin amount (Fig. 6.2). This is typical of antioxidants known as retarders. These may react so slowly with chain-carrying lipoperoxyl radicals that termination also occurs by the bimolecular self-reaction of peroxyl radicals, which finally does not result in a well-defined inhibition period. The redox potential of betanin (0.4 V) (Butera et al. 2002), would make the molecule an efficient reductant for lipid-derived peroxyl radicals (Buettner 1993). Nevertheless, kinetic solvent effects (Avila et al. 1995; Valgimigli et al. 1995), in particular polarity and hydrogen bond-accepting ability (HBA) of the solvent, may strongly affect the capacity of phenol antioxidants to transfer the hydroxylic H-atoms to radicals, because of preferential formation of a H-bonded complex between the reducing phenol-OH and a molecule of solvent (Barclay et al. 1999). Since methanol has a high HBA (Kamlet and Taft 1976), a strong interference could account for the very modest antioxidant effects of betanin in this solvent. In the absence of defined inhibition periods, Eq. 6.2 cannot be applied, then the K_{inh} for the reaction of betanin with peroxyl radicals in methanol cannot be determined. On the other hand, the hydrophilic nature of the pigment makes more apolar solvents inapplicable (Livrea and Tesoriere, unpublished data).

The interference of protic solvents on the H-atom-donating ability of *ortho*-diphenols is lower than monophenolic compounds (Foti and Ruberto 2001). Indeed, LAME autoxidation is very effectively inhibited by the betanin aglycone (betanidin) that acts as a classic chain-breaking antioxidant, with well-defined concentration-dependent inhibition periods, and total consumption at the end of the inhibition phase. According to a chain-breaking mechanism, the length of the inhibition period is determined by the number of radicals scavenged per each molecule of antioxidant (Niki 1996). Equations 6.1 and 6.2 can be then applied to calculate the stoichiometric factor n and K_{inh} of betanidin. The kinetic parameters characterizing the lipoperoxyl radical-scavenging activity of betanidin in methanol are reported in Table 6.1. Interestingly, K_{inh} of α-tocopherol (n=2) was measured $6.4 \times 10^5\ M^{-1}\,s^{-1}$ under comparable conditions (Tesoriere et al. 2009). Therefore, K_{inh} and stoichiometric factor of the reaction between betanidin and peroxyl radicals are of the same order as those of α-tocopherol.

The oxidation of phenol antioxidants by peroxyl radicals proceeds through H-atom abstraction and formation of the transient resonance-stabilized aryloxyl radical that can either undergo reactions of fast termination leading to formation of adducts, or quinones, or even self-termination reactions forming dimers or other products (Barclay 1993; Ingold 1969; Barclay et al. 1990). According to spectrophotometric and parallel high-performance liquid chromatography (HPLC) analysis, betanidin quinone, to an extent consistent with the consumed betanidin, was the only product generated during LAME peroxidation in methanol (Tesoriere et al. 2009). The stoichiometry of the reaction between betanidin and peroxyl radicals suggests that, after H-atom transfer from the *ortho*-diphenol moiety, the intermediate radical undergoes termination reactions with lipoperoxyl radicals leading to the stable betanidin quinone (Fig. 6.3, pathway A).

Other studies reported on the antioxidant activity of betanin and betanidin against peroxidation of linoleic acid in buffered detergent solution (Kanner et al. 2001). In those experiments linoleate peroxidation was induced by cyt c, metmyoglobin or lipoxygenase. Betanin acted slightly better than betanidin when cyt c or lipoxygenase were the oxidizing agents, and exhibited almost the same effect when metmyoglobin was the oxidant. Then, in aqueous micellar dispersions, the molecules were allowed to act in a nearly comparable manner. This appears to be in substantial agreement with recent observations, discussed below.

(b) *Peroxyl radical-scavenging activity of betanin and betanidin in liposomes.* Liposomes are convenient biomimetic models to study the activity of natural antioxidants. The oxidation kinetics of water-dispersed unilamellar soybean PC liposomes exposed to the hydrophilic azo-initiator AAPH can be followed by the time-course of formation of lipid hydroperoxides either in the absence or in the presence of antioxidants (Niki 1990). Both betanin and betanidin exhibit a net chain-breaking antioxidant activity in the etherogeneous aqueous-soybean phosphatidylcholine vesicular system (Fig. 6.4). The stoichiometric factors reported in Table 6.1 are calculated from the length of the relevant inhibition periods in accordance to Eq. 6.1.

Table 6.1 Kinetic parameters of the antioxidant activity of betacyanins in in vitro oxidation models

Model	Betacyanin	$10^8 \times Rp^a$ (Ms^{-1})	$10^9 \times Ri^b$ (Ms^{-1})	$10^9 \times R_{inh}$ (Ms^{-1})	kcl	t_{inh} (s)	n^c	$10^{-5} \times K_{inh}{}^d$ ($M^{-1}s^{-1}$)	K_{inh}(betacyanin)/K_{inh}(α-toc)[e]
LAME[f]	None	66	22		30[g]				
	10 μM betanidin			204	9.2[h]	900	1.98	3.75	
Liposomes[i]	None	8.6	2.77		31[g]				
	5.0 μM betanin			12.25	4.4	1841	1.02		0.53
	5.0 μM betanidin			3.90	1.4	3574	1.98		0.84

[a]Rates are expressed for total solution
[b]Measured by the duration of inhibition of 10 μM a-tocopherol
[c]Calculated by Eq. 6.1
[d]Calculated by Eq. 6.2
[e]The relative antioxidant activity of betacyanins is evaluated with respect to α-tocopherol (α-toc) by the ratio R_{inh}(betacyanin)/R_{inh}(α-toc) = nK_{inh}(α-toc)/nK_{inh}(betacyanin)
[f]AMVN (2 mM)-induced oxidation of 300 mM methyl linoleate (LAME) in methanol (Tesoriere et al. 2009)
[g]*kcl* kinetic chain length in the absence of antioxidant (Rp/Ri)
[h]kcl_{inh} kinetic chain length during the inhibition period (R_{inh}/Ri)
[i]Unilamellar soybean PC liposomes (10 mM lipid concentration) were oxidized by AAPH (2 mM) (Tesoriere et al. 2009)

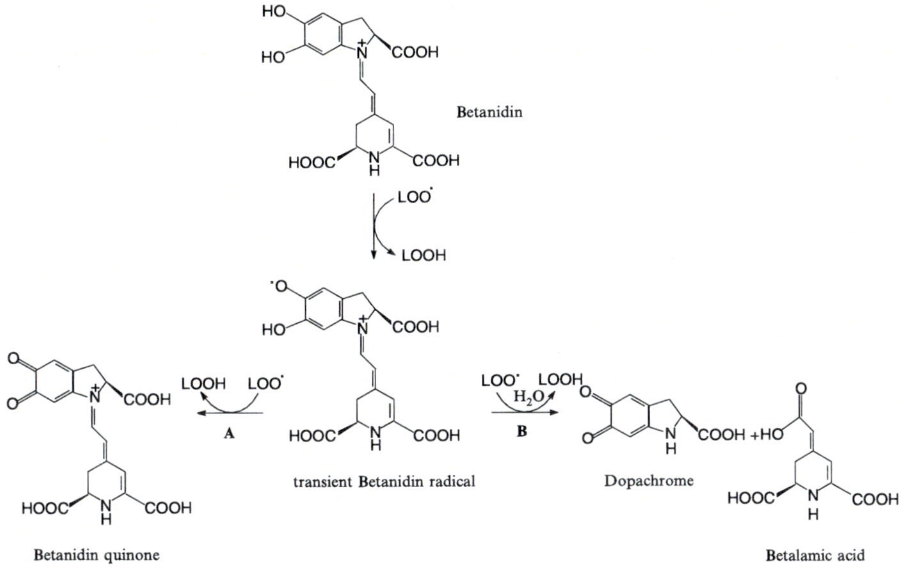

Fig. 6.3 The oxidation pathway of betanidin by lipoperoxyl radicals (LOO·) from LAME in methanol (*pathway A*), or from soybean PC in an etherogenous aqueous/vesicular system (*pathway B*)

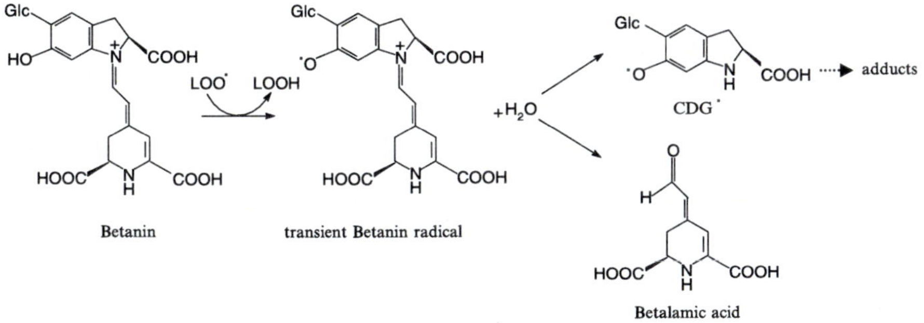

Fig. 6.4 The oxidation pathway of betanin by lipoperoxyl radicals (LOO·) in an aqueous/vesicular system. CDG· *cyclo*-DOPA 5-*O*-b-D-glucoside radical

With respect to the organic solution, an increase of the antioxidant effectiveness of betanin in the aqueous/lipid system may be expected for a number of reasons. Since the reaction medium is buffered at pH 7.4, the molecule is in a deprotonated state favoring hydrogen atom and/or electron donation (Gliszczynska-Swiglo et al. 2006; Gandia-Herrero et al. 2010). In addition, the HBA of water is lower than methanol (Kamlet and Taft 1976), thus the influence of the solvent on the H-atom-donating activity is less pronounced. Furthermore, partition between the water and lipid phase is to be considered a major factor determining the activity of antioxidant phytochemicals in membranes and lipid bilayers, with compounds partitioned more in the water phase showing less effectiveness (Rice-Evans et al. 1996; Shirai et al. 2001; Zou et al. 2005). According to other findings, betanin can partition in the lipid core of dipalmitoyl-phosphatidylcholine vesicles (Turco-Liveri et al. 2007). All these observations suggest that, despite the hydrophilic sugar substituent, location of the aromatic *cyclo*-DOPA in the membrane would allow its reducing phenol hydroxyl to easily interact with lipoperoxyl radicals floating from the membrane interior.

Partition and location of betanidin in liposomal phospholipids are not known. In comparison with betanin, the absence of the hydrophilic sugar substituent might finally enhance partition in lipid bilayers. Then, in addition to the antioxidant chemistry of its *ortho*-diphenol moiety, accessibility of lipoperoxyl radicals to the reducing hydroxyl groups could account for the effectiveness of betanidin in the liposomal model.

K_p, the rate constant for the propagation of the radical chain of phosphatidylcholine, is not known, which prevents application of Eq. 6.2 to evaluate the absolute inhibition constant of betanin and betanidin in the lipid bilayer. However, an estimate of the antioxidant activity of the pigments in liposomes can be obtained by relating the value of R_{inh} measured in the presence of either betanin or betanidin and of α-tocopherol. Taking into account Eqs. 6.1 and 6.2, R_{inh} can be expressed by

$$R_{inh} = K_p[LH]\,R_i / n\,K_{inh}[IH] \qquad (6.3)$$

Therefore, when comparable amounts of antioxidant and α-tocopherol are used, the ratio $R_{inh[betacyanin]}/R_{inh[\alpha\text{-tocopherol}]}$ will represent $nK_{inh[\alpha\text{-tocopherol}]}/nK_{inh[betacyanin]}$. Then, the effectiveness of betanin and betanidin can be calculated, which were 53% and 84%, respectively, of the effectiveness of α-tocopherol. The kinetic parameters of the inhibition of AAPH-induced peroxidation of unilamellar liposomes are summarized in Table 6.1.

In the liposomal system, the oxidation of betanidin resulted in stoichiometric amounts of dopachrome, as the oxidation product of the *cyclo*-DOPA moiety, and the chromophore betalamic acid, the yield of which was lower than the parent compound, which was explained by molecular degradation (Tesoriere et al. 2009). Then, in the heterogeneous water/lipid vesicular system, the betanidin radical generated after H-atom abstraction by lipoperoxyl radicals undergoes nucleophilic attack of water to the C adjacent to the indolic nitrogen, before being oxidized by a second lipoperoxyl radical, with final release of dopachrome and betalamic acid (Fig. 6.3, pathway B).

Betalamic acid, again to an extent not consistent with the amount of the parent compound, was found as a product from betanin during liposomal oxidation (Tesoriere et al. 2009), indicating that, similarly to betanidin, the intermediate betanin radical generated after reaction of its phenol moiety undergoes solvolytic splitting of the aldimine bond (Fig. 6.5). On the basis of spectrophotometric evidence, unidentified product(s) from the reaction has/have been considered as derivatives of the *cyclo*-DOPA 5-*O*-β-D-glucoside radical (CGD·), possibly highly conjugated structures of adducts from self-termination reactions (Tesoriere et al. 2009).

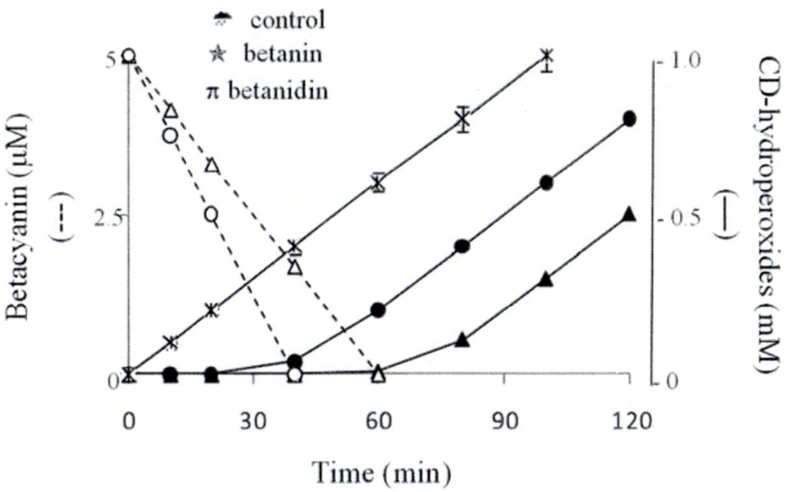

Fig. 6.5 Time course of CD-hydroperoxides formation (*filled symbols*) during the AAPH-induced soybean PC liposome oxidation in the absence (*control*) or in the presence of betacyanins and consumption of the pigments (*open symbols*) (Tesoriere et al. 2009)

The investigations reported above, showing that betanidin is a lipoperoxyl radical-scavenger better than betanin both in solution and lipid bilayers, confirm the importance of peculiar structural features conferring antiradical capacity to betalains. A recent systematic study assessed the reducing activity, as Trolox equivalence antioxidant capacity (TEAC) of 15 betalains with increasingly complex chemistry, from 1-ethylamine betaxanthin to betanin (Gandia-Herrero et al. 2010). The data support the existence of a strong "intrinsic" antiradical activity, possibly linked to the electron resonance system supported by both nitrogen atoms, which is common to all betalains. The presence of a mono/diphenol moiety in resonance with the betalamic acid moiety, plus a second cycle fused in an indoline manner, as in betacyanins, implies a significant enhancement of the radical-scavenging capacity (Gandia-Herrero et al. 2010). The formation of betanidin quinone or dopachrome from the oxidation of betanidin in methanol or liposomes, respectively (Tesoriere et al. 2009), while confirming the importance of the phenol hydroxyls, may rule out that the cyclic nitrogen is involved in the antioxidant mechanism of the molecule in the model systems considered.

6.4 Inhibition of Low-Density Lipoprotein Oxidation by Betanin

Free radical-induced oxidation of low-density lipoproteins (LDL) proceeds by a chain mechanism generating phosphatidylcholine hydroperoxides and cholesteryl ester hydroperoxides as the major primary products (Esterbauer et al. 1992).

These reactions, and the consequent internalization of oxidized LDL (ox-LDL) in macrophages, are considered key events in the progression and eventual development of atherosclerosis (Steinbrecher et al. 1987; Steinberg et al. 1989; Heinecke 1998). LDL are endowed with several lipophilic antioxidants, the most abundant being α-tocopherol (Esterbauer et al. 1992); however oxidants from endogenous and/or exogenous sources can reduce the defense, which makes the particle prone to oxidize, thus becoming an agent of damage. Under these circumstances, dietary bioavailable antioxidants that may interact with and/or partition in LDL and be involved in LDL protection have continuously been explored.

The oxidation of human LDL by transition metal ions such as iron or copper has been a model for generating knowledge of the kinetics of LDL oxidation (Esterbauer et al. 1992), and has widely been considered for assessing intrinsic activity of natural antioxidants. The biological relevance of such a model has been questioned, however. In more recent studies, oxidation of LDL in vivo has been suggested to depend on the activity of myeloperoxidase (MPO) (Daugherty et al. 1994; Heller et al. 2000), a heme-enzyme that utilizes hydrogen peroxide and a variety of co-substrates to generate reactive enzyme intermediates, namely compound I and compound II (Heinecke 1998; Daugherty et al. 1994; Klebanoff 1980). MPO activity also depends on the metabolism of nitric oxide (NO) forming nitrite, the final oxidation product of NO metabolism, a substrate for the enzyme (Burner et al. 2000; Eiserich et al. 1998; van der Vliet et al. 1997; Sampson et al. 1998). Nitrogen dioxide radical $(NO_2^{\bullet})$, the one-electron oxidation product of nitrite by MPO compound I, has been proposed as the reactive species to start massive oxidation of the LDL lipids (Byun et al. 1999; Kostyuk et al. 2003). Both these models have been used to assess whether the sensitivity of human LDL to oxidation could be altered by betanin (Tesoriere et al. 2003; Allegra et al. 2007).

The production of lipid hydroperoxides in LDL exposed to oxidative challenge does not start before all LDL antioxidants are consumed in the sequence from the most active (α-tocopherol) to the least active (β-carotene) (lag phase). After the lag period, peroxidation begins to accelerate and formation of CD hydroperoxides can be measured (propagation phase), until all lipid is oxidized. Betanin can incorporate in human LDL in vivo and in vitro (Tesoriere et al. 2004a, 2003). In ex vivo experiments, betanin-enriched LDL were isolated after spiking human plasma with pure betanin, then the resistance of these particles to copper-induced oxidation was measured in comparison with LDL obtained from the same plasma that did not undergo the spiking procedure (Tesoriere et al. 2003). Betanin-enriched LDL showed a significant elongation of the time preceding lipid oxidation, during which betanin was totally consumed (Fig. 6.6). Behaving as a lipoperoxyl radical scavenger, betanin affects the chain process of the copper-induced LDL lipid oxidation. In this system, vitamin E consumption is unaltered in the presence of betanin, whereas consumption of β-carotene is delayed. Betanin starts declining only after vitamin E depletion, and is totally consumed before β-carotene. While indicating the higher effectiveness of vitamin E in protecting LDL lipids and all LDL antioxidants, these findings show that betanin acts as a lipoperoxyl radical-scavenger better than β-carotene in the copper-oxidized LDL model.

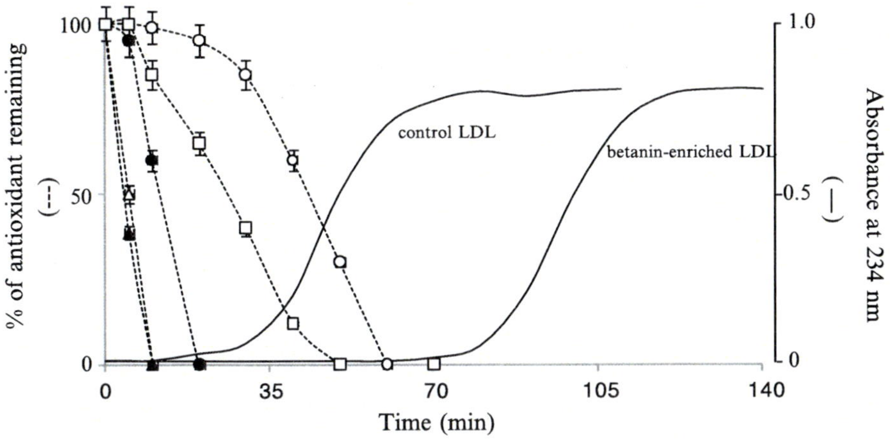

Fig. 6.6 Time course of the consumption of vitamin E (*triangle*), β-carotene (*circle*) and betanin (*square*) during the copper-induced oxidation of control (*filled symbols*) or betanin-enriched (*open symbols*) LDL. LDL oxidation is followed by the formation of CD hydroperoxides at 234 nm (Tesoriere et al. 2003)

As for the other model, betanin effectively inhibited the production of lipid hydroperoxides in human LDL submitted to a MPO/nitrite-induced oxidation (Allegra et al. 2007). In this system, the time-course of lipid oxidation follows the same phases as the copper-oxidised LDL, followed by the formation of lipid hydroperoxides. It was imperative from a number of kinetic measurements that the betalain can block the process at various levels, that betanin not only acts as a scavenger of the initiator radical nitrogen dioxide, but can also act as a lipoperoxyl radical scavenger. In addition, unidentified products from the oxidation of betanin by MPO/nitrite further inhibit LDL oxidation as effectively as the parent compound (Allegra et al. 2007), thus extending the antioxidative protection of LDL beyond the time in which betanin is consumed. It should be mentioned that other studies showed that betanin is a reducing substrate for the intermediates—compound I/II of the peroxidative MPO cycle (Allegra et al. 2005), an action potentially pro-oxidant in this LDL model. This however appears to be counteracted by the activity of betanin and possibly by its oxidized products through scavenging of NO_2. Figure 6.7 depicts the catalytic cycle of MPO/nitrite and suggests sites of action of betanin.

6.5 Interactions of Betanin and Betanidin with Vitamin E

In living organisms, antioxidants do not function individually, rather, they function cooperatively or even in synergism with each other. Since α-tocopherol is the main lipid antioxidant in membranes, exploring interactions between dietary antioxidants and α-tocopherol is considered important to envisage eventual effects and possibly mechanism of action of these molecules in vivo. For instance, either synergistic or

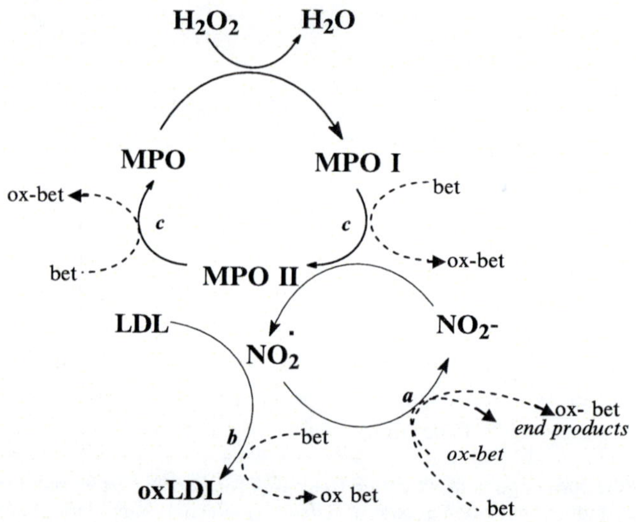

Fig. 6.7 Proposed mechanisms of antioxidant activity of betanin on myeloperoxidase-induced LDL oxidation. *a* scavenger of NO_2·, *b* lipoperoxyl radical scavenger, *c* reductant for compound I and II of MPO

additive effects or co-antioxidant action have been reported between polyphenol phytochemicals and α-tocopherol (Zou et al. 2005; Jia et al. 1998; Pedrielli and Skibsted 2002). In soybean PC liposomes, at a 1:1 betacyanin:α-tocopherol ratio, either betanin or betanidin cannot extend the inhibition period beyond the sum of the individual inhibition periods, providing evidence of merely additive effects (Tesoriere et al. 2009). On the other hand, even in a model of copper-oxidized LDL, the time-course of vitamin E consumption, either in the absence or in the presence of betanin, suggests an independent antioxidant activity of the two molecules (Tesoriere et al. 2003). The redox potential of betanin is lower than α-tocopherol (0.5 mV) (Buettner 1993), which would allow reduction of the α-tocopheroxyl radical at the membrane surface (Fukuzawa 2008), provided favorable site-specific interactions (Barclay 1993). The absence of cooperative effects may be the expression of the partition of betanin in either the lipid bilayer or LDL and of its activity in scavenging lipoperoxyl radicals.

6.6 Peroxyl Radical-Scavenging Activity of Vulgaxanthin I

Antioxidative effects of vulgaxanthin I were evaluated in an oxidation model of LAME in the presence of AMVN (Tesoriere et al. 2008). The amount of lipid hydroperoxides formed after a 30-min incubation was taken as a reference end-point, and the inhibition by vulgaxanthin I was expressed in terms of IC_{50}, that is, the amount of pigment required for a 50% inhibition. Under these conditions, vulgaxanthin I showed an IC_{50} of 0.75 μM, of the same order as betanin and α-tocopherol taken as a comparison, 1 μM and 0.56 μM, respectively.

6.7 Conclusions

The unanimously recognized dual and complex role of radical species and oxidants in the cell functioning and in pathology points to the necessity to get better knowledge of what the so-called antioxidant compounds may really do, since scavenging of reactive species and interactions with cell constituents involved in maintaining the redox homeostasis may significantly interfere with cell signal transduction. These new concepts have recently led to consider the role of antioxidant vitamins even as modulators of redox-regulated cell signaling, and must be used to investigate and interpret effects, including eventual adverse effects, of phytochemicals with redox properties at a molecular level (Leonarduzzi et al. 2010).

Phenolic hydroxyls have been repeatedly proven as efficient reducers of pro-oxidant/oxygen radicals under a wide range of conditions (Valgimigli et al. 1995; Barclay et al. 1999). In accordance, the higher the number of hydroxyl groups, the higher the antioxidant activity of polyphenol phytochemicals such as flavonoids, has been shown (Rice-Evans et al. 1996). The betacyanin pigments, betanin and betanidin, exhibit an antioxidant effectiveness linked to the presence of the glucose-substituted phenol moiety of betanin and to the *ortho*-diphenol moiety of its aglycone, the latter being a much more efficient reductant in both organic solvent and liposomal lipid bilayers. These findings may be of an even greater interest since the calculated constants characterizing the activity in solution and in liposomes have appeared of the same order as those of α-tocopherol (Tesoriere et al. 2009), the major lipid antioxidant in our body (Niki 1996). More importantly, betanin also shows antioxidant activity in a biologically relevant LDL oxidation model (Allegra et al. 2007).

Information on chemistry, reactivity in as many as possible different systems, particularly biological environments, and interactions with physiological antioxidants, are first steps to characterize dietary antioxidants. Activity in cell cultures and investigation of cell redox changes and specific signaling may further enhance our knowledge and allow hypotheses on potential health effects. However none of these studies make sense until it is proven that the compound of interest can really reach body sites and the observed in vitro actions may be accomplished in vivo. Studies in this direction have shown that betanidin, being a highly unstable molecule (Gandia-Herrero et al. 2007; Stintzing and Carle 2004), was not found after a simulated digestion of betanin-containing foods, including beet root, though the aglycone could have been generated by pancreatic amylase (Tesoriere et al. 2008). These observations make its eventual systemic activity in vivo hard to determine. Beneficial effects could be considered at the gastrointestinal level, however (Halliwell et al. 2005). Betanin, instead, has been shown to be bioavailable in humans, after ingestion of either cactus pear fruits or red beet (Kanner et al. 2001; Tesoriere et al. 2004a,

2005; Frank et al. 2005), reaching plasma concentrations sufficient to promote its incorporation in LDL and red blood cells (Tesoriere et al. 2004a, 2005). It is in light of these findings that the chemistry of the peroxyl radical-scavenging activity of betanin, and relevant parameters, deserve to be considered. It is suggested that betanin, and foods rich in betanin, such as beetroot and the fruits of the *Opuntia* cactus, may be of nutraceutical interest and contribute to maintain the natural redox homeostasis and possibly prevent disease states. With focus on the latter point, a small clinical trial carried out with eight healthy volunteers who consumed cactus pear fruit pulp for 15 days demonstrated a remarkable positive effect on the body's redox status that was reasonably attributed to betalains, and not to the fruit vitamin C (Tesoriere et al. 2004b). As a final note, current studies in the authors' laboratory show that betanin is transported through human CaCo-2 cell monolayers with an apparent permeability coefficient suggesting that dietary betanin can be absorbed quite effectively during its intestinal transit (data to be published). While these data appear to confirm the observations in humans (Kanner et al. 2001; Tesoriere et al. 2004a, 2005; Frank et al. 2005), the actual amounts recovered in vivo, quite lower than suggested by in vitro experiments, would indicate metabolism and/or bacterial degradation of the molecule in gut, which should be investigated.

References

Allegra, M., P.G. Furtmuller, W. Jantschko, M. Zederbauer, L. Tesoriere, M.A. Livrea, and C. Obinger. 2005. Mechanism of interaction of betanin and indicaxanthin with human myeloperoxidase and hypochlorous acid. *Biochemical and Biophysical Research Communications* 332: 837–844.

Allegra, M., L. Tesoriere, and M.A. Livrea. 2007. Betanin inhibits the myeloperoxidase/nitrite-induced oxidation of human low density lipoproteins. *Free Radical Research* 41: 335–341.

Avila, D.V., K.U. Ingold, J. Lusztyk, W.H. Green, and D.R. Procopio. 1995. Dramatic solvent effects on the absolute rate constants for abstraction of the hydroxylic hydrogen atom from tert-butyl hydroperoxide and phenol by the cumyloxyl radical. The role of hydrogen bonding. *Journal of the American Chemical Society* 117: 2929–2930.

Barclay, L.R.C. 1993. Model biomembranes: Quantitative studies of peroxidation, antioxidant action, partitioning, and oxidative stress. *Canadian Journal of Chemistry* 71: 1–16.

Barclay, L.R.C., K.A. Baskin, K.A. Dakin, S.J. Locke, and M.R. Vinqvist. 1990. The antioxidant activities of phenolic antioxidants in free radical peroxidation of phospholipid membranes. *Canadian Journal of Chemistry* 68: 2258–2269.

Barclay, L.R.C., C.E. Edwards, and M.R. Vinqvist. 1999. Media effects on antioxidant activities of phenols and cathecols. *Journal of the American Chemical Society* 121: 6226–6231.

Buettner, G. 1993. The pecking order of free radicals and antioxidants: Lipid peroxidation, α-tocopherol, and ascorbate. *Archives of Biochemistry and Biophysics* 300: 535–543.

Burner, U., P.G. Furtmuller, A.J. Kettle, W.H. Koppenol, and C. Obinger. 2000. Mechanism of reaction of myeloperoxidase with nitrite. *Journal of Biological Chemistry* 275: 20597–20601.

Burton, G.W., and K.U. Ingold. 1981. Autoxidation of biological molecules. 1. The antioxidant activity of vitamin E and related chain-breaking phenolic antioxidants in vitro. *Journal of the American Chemical Society* 103: 6472–6477.

Butera, D., L. Tesoriere, F. Di Gaudio, A. Bongiorno, M. Allegra, A.M. Pintaudi, R. Kohen, and M.A. Livrea. 2002. Antioxidant activities of sicilian prickly pear (*Opuntia ficus indica*) fruit extracts and reducing properties of its betalains: Betanin and indicaxantin. *Journal of Agricultural and Food Chemistry* 50: 6895–6901.

Byun, J., D.M. Mueller, J.S. Fabjan, and J.W. Heinecke. 1999. Nitrogen dioxide radical generated by the myeloperoxidase-hydrogen peroxide-nitrite system promotes lipid peroxidation of low density lipoprotein. *FEBS Letters* 455: 243–246.

Cao, G., E. Sofic, and R.L. Prior. 1997. Antioxidant and prooxidant behavior of flavonoids: Structure-activity relationships. *Free Radical Biology & Medicine* 22: 749–760.

Czapski, J., K. Mikolajczyk, and M. Kaczmarek. 2009. Relationship between antioxidant capacity of red beet juice and contents of its betalain pigments. *Polish Journal of Food and Nutrition Sciences* 59: 119–122.

Daugherty, A., J.L. Dunn, D.L. Rateri, and J.W. Heinecke. 1994. Myeloperoxidase, a catalyst for lipoprotein oxidation, is expressed in human atherosclerotic lesions. *The Journal of Clinical Investigation* 94: 437–444.

Dröge, W. 2002. Free radicals in the physiological control of cell function. *Physiological Reviews* 82: 47–95.

Echtay, K.S., T.C. Esteves, J.L. Pakai, M.B. Jekabson, A.J. Lambert, M. Portero-Otin, R. Pamplona, A.J. Vidal-Puig, S. Wang, S.J. Roebuck, and M.D. Brand. 2003. A signalling role for 4-hydroxy-2 nonenal in regulation of mitochondrial uncoupling. *EMBO Journal* 22: 4103–4110.

Eiserich, J.P., M. Hristova, C.E. Cross, A.D. Jones, B.A. Freeman, B. Halliwell, and A. van der Vliet. 1998. Formation of nitric oxide-derived inflammatory oxidants by myeloperoxidase in neutrophils. *Nature* 391: 393–397.

Escribano, J., M.A. Pedreno, F. Garcia-Carmona, and R. Munoz. 1998. Characterization of the antiradical activity of betalains from *Beta vulgaris* L. roots. *Phytochemical Analysis* 9: 124–127.

Esterbauer, H., J. Gebicki, H. Puhl, and G. Jürgens. 1992. The role of lipid peroxidation and antioxidants in oxidative modification of LDL. *Free Radical Biology & Medicine* 13: 341–390.

Foti, M., and G. Ruberto. 2001. Kinetic solvent effects on phenolic antioxidants determined by spectrophotometric measurements. *Journal of Agricultural and Food Chemistry* 49: 342–348.

Frank, T., F.C. Stintzing, R. Carle, I. Bitsch, D. Quaas, G. Strass, R. Bitsch, and M. Netzel. 2005. Urinary pharmacokinetics of betalains following consumption of red beet juice in healthy humans. *Pharmacological Research* 52: 290–297.

Fukuzawa, K. 2008. Dynamics of lipid peroxidation and antioxidation of α-tocopherol in membranes. *Journal of Nutritional Science and Vitaminology* 54: 273–285.

Gandia-Herrero, F., J.A. Escribano, and F. Garcia-Carmona. 2007. Characterization of the activity of tyrosinase on betanidin. *Journal of Agricultural and Food Chemistry* 55: 1546–1551.

Gandia-Herrero, F., J.A. Escribano, and F. Garcia-Carmona. 2010. Structural implications on color, fluorescence, and antiradical activity in betalains. *Planta* 232: 449–460.

Giles, G.I. 2006. The redox regulation of thiol dependent signaling pathways in cancer. *Current Pharmaceutical Design* 12: 4427–4443.

Giorgio, M., M. Trinei, E. Migliaccio, and P.G. Pelicci. 2007. Hydrogen peroxide: A metabolic by-product or a common mediator of ageing signals? *Nature Reviews. Molecular Cell Biology* 8: 722–728.

Gliszczynska-Swiglo, A., H. Szymusiak, and P. Malinowska. 2006. Betanin, the main pigment of red beet: Molecular origin of its exceptionally high free radical-scavenging activity. *Food Additives and Contaminants* 23: 1079–1087.

Halliwell, B., J. Rafter, and A. Jenner. 2005. Health promotion by flavonoids, tocopherols, tocotrienols, and other phenols: Direct or indirect effects? Antioxidant or not? *American Journal of Clinical Nutrition* 81: 268S–276S.

Hancock, J.T. 2009. The role of redox mechanisms in cell signaling. *Molecular Biotechnology* 43: 162–166.

Heinecke, J.W. 1998. Oxidants and antioxidants in the pathogenesis of atherosclerosis: Implications for the oxidized low density lipoprotein hypothesis. *Atherosclerosis* 141: 1–15.

Heller, J.I., J.R. Crowley, S.L. Hazen, D.M. Salvay, P. Wagner, S. Pennathur, and J.W. Heinecke. 2000. p-Hydroxyphenylacetaldehyde, an aldehyde generated by myeloperoxidase, modifies phospholipid amino groups of low density lipoprotein in human atherosclerotic intima. *Journal of Biological Chemistry* 275: 9957–9962.

Herbach, K.M., F.C. Stintzing, and R. Carle. 2006. Betalain stability and degradation. Structural and chromatic aspects. *Journal of Food Science* 71: R41–R50.

Ingold, K. 1969. Peroxy radicals. *Accounts of Chemical Research* 2: 1–9.

Jia, Z.-S., B. Zhou, L. Yang, L.-M. Wu, and Z.-L. Liu. 1998. Antioxidant synergism of tea polyphenols and α-tocopherol against free radical induced peroxidation of linoleic acid in solution. *Journal of the Chemical Society, Perkin Transactions* 2: 911–915.

Kamlet, M.J., and R.W. Taft. 1976. The solvatochromic comparison method. I. The beta-scale of solvent hydrogen-bond acceptor (HBA) basicities. *Journal of the American Chemical Society* 98: 377–383.

Kanner, J., S. Harel, and R. Granit. 2001. Betalains—A new class of dietary cationized antioxidants. *Journal of Agricultural and Food Chemistry* 49: 5178–5185.

Klebanoff, S.J. 1980. Oxygen metabolism and the toxic properties of phagocytes. *Annals of Internal Medicine* 93: 480–489.

Kostyuk, V.A., T. Kraemer, H. Sies, and T. Schewe. 2003. Myeloperoxidase/nitrite-mediated lipid peroxidation of low-density lipoprotein as modulated by flavonoids. *FEBS Letters* 537: 146–150.

Kushi, L.H., E.B. Lenart, and W.C. Willett. 1995. Health implication of Mediterranean diets in light of contemporary knowledge. 1. Plant foods and dairy products. *American Journal of Clinical Nutrition* 61(suppl 6): 1407S–1415S.

Lee, M.Y., and K.K. Griendling. 2008. Redox signaling, vascular function, and hypertension. *Antioxidants & Redox Signaling* 10: 1045–1059.

Leitinger, N. 2005. Oxidised phospholipids as triggers of inflammation in atherosclerosis. *Molecular Nutrition & Food Research* 49: 1063–1071.

Leonarduzzi, G., A. Scavazza, F. Biasi, E. Chiarpotto, S. Camandola, S. Vogel, L. Dargel, and G. Poli. 1997. The lipid peroxidation end product 4-hydroxy-2,3 nonenal upregulates transforming growth factor beta1 expression in the macrophage lineage: A link between oxidative injury and fibrosclerosis. *The FASEB Journal* 11: 851–857.

Leonarduzzi, G., B. Sottero, and G. Poli. 2010. Targeting tissue oxidative damage by means of cell signaling modulators: The antioxidant concept revisited. *Pharmacology and Therapeutics* 128: 336–374.

Lin, R.I.-S. 1995. Phytochemical and antioxidants. In *Functional foods*, ed. I.E. Goldberg, 393–441. New York: Chapman and Hall.

Livrea, M.A., and L. Tesoriere. 2004. Antioxidant activities of prickly pear (*Opuntia ficus indica*) fruit and its betalains. In *Herbal and traditional medicines*, ed. L. Packer, C.N. Ong, and B. Halliwell, 537–556. New York: Marcel Dekker.

Ma, Q. 2010. Transcriptional responses to oxidative stress: Pathological and toxicological implications. *Pharmacology and Therapeutics* 125: 376–393.

Martin, K.R., and J.C. Barrett. 2002. Reactive oxygen species as double-edged swords in cellular processes: Low-dose cell signaling versus high-dose toxicity. *Human and Experimental Toxicology* 21: 71–75.

Matsuzawa, A., and H. Ichijo. 2008. Redox control of cell fate by MAP kinase: Physiological roles of ASK1-MAP kinase pathway in stress signaling. *Biochimica et Biophysica Acta* 1780: 1325–1336.

Niki, E. 1990. Free radical initiators as source of water- or lipid-soluble peroxyl radicals. In *Methods in enzymology*, ed. L. Packer and A.N. Glazer, vol. 186, 100–108

Niki, E. 1996. α-tocopherol. In *Handbook of antioxidants*, ed. E. Cadenas and L. Packer, 3–26. New York: Marcel Dekker.

Pan, J.S., M.Z. Hong, and J.L. Ren. 2009. Reactive oxygen species: A double-edged sword in oncogenesis. *World Journal of Gastroenterology* 15: 1702–1707.

Pedrielli, P., and L.H. Skibsted. 2002. Antioxidant synergy and regeneration effect of quercetin, (−)-epicatechin, and (+)-catechin on α-tocopherol in homogeneous solutions of peroxidating methyl linoleate. *Journal of Agricultural and Food Chemistry* 50: 7138–7144.

Pryor, W.A., and L. Castle. 1984. Chemical methods for detection of lipid hydroperoxides. In *Methods in enzymology*, vol. 105, ed. L. Packer, 203–208. New york: Academic Press.

Rice-Evans, C.A., N.J. Miller, and G. Paganga. 1996. Structure-antioxidant activity relationships of flavonoids and phenolic acids. *Free Radical Biology & Medicine* 20: 933–956.

Sampson, J.B., Y. Ye, H. Rosen, and J.S. Beckman. 1998. Myeloperoxidase and horseradish peroxidase catalyze tyrosine nitration in proteins from nitrite and hydrogen peroxide. *Archives of Biochemistry and Biophysics* 356: 207–213.

Shirai, M., J.H. Moon, T. Tsushida, and J. Terao. 2001. Inhibitory effect of a quercetin metabolite, quercetin 3-O-beta-D-glucuronide, on lipid peroxidation in liposomal membranes. *Journal of Agricultural and Food Chemistry* 49: 5602–5608.

Steinberg, D., S. Parthasarathy, T.E. Carew, J.C. Khoo, and J.L. Witztum. 1989. Beyond cholesterol. Modifications of low-density lipoprotein that increase its atherogenicity. *The New England Journal of Medicine* 320: 915–924.

Steinbrecher, U.P., J.L. Wiztum, S. Parthasarathy, and D. Steinberg. 1987. Decrease in reactive amino groups during oxidation or endothelial cell modification of LDL. Correlation with changes in receptor-mediated catabolism. *Arteriosclerosis* 7: 135–143.

Stintzing, F.C., and R. Carle. 2004. Functional properties of anthocyanins and betalains in plants, food, and in human nutrition. *Trends in Food Science and Technology* 15: 19–38.

Tesoriere, L., D. Butera, D. D'Arpa, F. Di Gaudio, M. Allegra, C. Gentile, and M.A. Livrea. 2003. Increased resistance to oxidation of betalain-enriched human low density lipoproteins. *Free Radical Research* 37: 689–696.

Tesoriere, L., M. Allegra, D. Butera, and M.A. Livrea. 2004a. Absorption, excretion, and distribution in low density lipoproteins of dietary antioxidant betalains. Potential health effects of betalains in humans. *American Journal of Clinical Nutrition* 80: 941–945.

Tesoriere, L., D. Butera, A.M. Pintaudi, M. Allegra, and M.A. Livrea. 2004b. Supplementation with cactus pear (*Opuntia ficus indica*) fruit decreases oxidative stress in healthy humans: A comparative study with vitamin C. *American Journal of Clinical Nutrition* 80: 391–395.

Tesoriere, L., D. Butera, M. Allegra, M. Fazzari, and M.A. Livrea. 2005. Distribution of betalain pigments in red blood cells after consumption of cactus pear fruits and increased resistance of the cells to ex vivo-induced oxidative hemolysis in humans. *Journal of Agricultural and Food Chemistry* 53: 1266–1270.

Tesoriere, L., M. Fazzari, F. Angileri, C. Gentile, and M.A. Livrea. 2008. In vitro digestion of betalainic foods. Stability and bioaccessibility of betaxanthins and betacyanins and antioxidative potential of food digesta. *Journal of Agricultural and Food Chemistry* 56: 10487–10492.

Tesoriere, L., M. Allegra, C. Gentile, and M.A. Livrea. 2009. Betacyanins as phenol antioxidants. Chemistry and mechanistic aspects of the lipoperoxyl radical scavenging activity in solution and liposomes. *Free Radical Research* 43: 706–717.

Turco-Liveri, M.L., L. Sciascia, R. Lombardo, L. Tesoriere, E. Passante, and M.A. Livrea. 2007. Spectrophotometric evidence for the solubilization site of betalain pigments in membrane biomimetic systems. *Journal of Agricultural and Food Chemistry* 55: 2836–2840.

Uchida, K. 2007. Lipid peroxidation and redox-sensitive signaling pathways. *Current Atherosclerosis Reports* 9: 216–221.

Valgimigli, L., J.T. Banks, K.U. Ingold, and J. Lusztyk. 1995. Kinetic solvent effects on hydroxylic hydrogen atom abstractions are independent of the nature of the abstracting radical. Two extreme tests using vitamin e and phenol. *Journal of the American Chemical Society* 117: 9966–9971.

Valko, M., D. Leibfritz, J. Moncol, M.T. Cronin, M. Mazur, and J. Telser. 2007. Free radicals and antioxidants in normal physiological functions and human disease. *The International Journal of Biochemistry & Cell Biology* 39: 44–84.

van der Vliet, A., J.P. Eiserich, B. Halliwell, and C.E. Cross. 1997. Formation of reactive nitrogen species during peroxidase-catalyzed oxidation of nitrite. A potential additional mechanism of nitric oxide-dependent toxicity. *Journal of Biological Chemistry* 272: 7617–7625.

Willett, W.C., F. Sacks, A. Trichopoulou, G. Drescher, A. Ferro-Luzzi, E. Helsing, and D. Trichopoulous. 1995. Mediterranean diet pyramid: A cultural model for healthy eating. *American Journal of Clinical Nutrition* 61(suppl 6): 1402S–1406S.

Wu, W.S., R.K. Tsai, C.H. Chang, S. Wang, J.R. Wu, and Y.X. Chang. 2006. Reactive oxygen species mediated sustained activation of protein kinase C alpha and extracellular signal-regulated kinase for migration of human hepatoma cell Hepg2. *Molecular Cancer Research* 4: 747–758.

Yamamoto, Y., E. Niki, and Y. Kamiya. 1982. Oxidation of lipids: III. Oxidation of methyl linoleate in solution. *Lipids* 17: 870–877.

Zou, B., Q. Miao, L. Yang, and Z.-L. Liu. 2005. Antioxidative effects of flavonols and their glycosides against the free-radical induced peroxidation of linoleic acid in solution and in micelles. *Chemistry – A European Journal* 11: 680–691.

Chapter 7
Anticancer Effects of Red Beet Pigments

Govind J. Kapadia and G. Subba Rao

Abstract Currently, there is considerable interest in the anticancer effects of red beetroot (*Beta vulgaris* L.) pigment extract, which is used worldwide as red food color E162 and as a natural colorant in cosmetics and drugs. Of particular significance is its broad spectrum of multi-organ antitumor activity demonstrable in laboratory animal models. Further, this nontoxic plant extract, when used in combination with potent anticancer drugs such as doxorubicin (Adriamycin), has the potential to act synergistically and mitigate treatment-related drug toxicity. Betanin, the betacyanin constituent primarily responsible for red beet color, is an antioxidant with an exceptionally high free radical-scavenging activity and is a modulator of oxidative stress. Research focused on anticancer activities of beetroot extract, in animal models, has unraveled their potential benefits as chemopreventive and chemotherapeutic agents, although further progress is needed on the identification and elucidation of anticancer mechanism(s) of individual active constituent(s) in additional well-designed experimental models and clinical trials, as discussed in this chapter.

7.1 Introduction

Since ancient times, pigment extract of various cultivated forms of red beet (*Beta vulgaris* L., of family Chenopodiaceae) has been widely used as a natural colorant in food, cosmetics, decorative art, and paintings, and as a medicinal product in

G.J. Kapadia (✉)
Department of Pharmaceutical Sciences, College of Pharmacy, Howard University, Washington, DC 20059, USA
e-mail: gkapadia@howard.edu

G.S. Rao
Global Technology Resource Center, 145 Rosewood Drive, Streamwood, IL 60107, USA
e-mail: rao1938@hotmail.com

B. Neelwarne (ed.), *Red Beet Biotechnology: Food and Pharmaceutical Applications*, DOI 10.1007/978-1-4614-3458-0_7,

the management of blood, heart, liver, pancreas, digestive, neurological, and other common diseases (Nottingham 2004). At the present time, red beetroot extract is a US Food and Drug Administration (FDA)-approved and European Union (EU)-approved food color that is designated by the number E162. It is used also by the pharmaceutical industry as a coloring agent in drug formulations, both in solid and liquid forms (Pai and D'Mello 2004). Red beet color, E162, showed significant anticancer effects in animal studies by the authors (Kapadia et al. 1996, 2001, 2003, 2005) and by other investigators (Bobek et al. 2000; Klewicka et al. 2010; Lechner et al. 2010). Of particular importance is the wide spectrum of antitumor activity exhibited by E162. When administered in drinking water at very low doses (between 25 and 78 μg/ml), it consistently reduced the incidence of tumors in skin, lung, liver, colon, and esophagus in various laboratory animal models. Further, betalains, the group of chemicals responsible for color in red beet, were found to exhibit powerful antioxidative, free radical-scavenging properties with potential health benefits in humans, such as boosting immune system and prevention of cardiovascular diseases, neurodegenerative disorders, and cancer (Azeredo 2009; Kanner et al. 2001; Moreno et al. 2008; Stintzing and Carle 2007; Strack et al. 2003). These results have encouraged further research in exploring the anticancer effects of red beetroot extract with the ultimate goal of establishing its efficacy in humans.

Also, currently there is considerable interest in the use of beetroot extract and its constituents as dietary supplements in cancer prevention (Geshner et al. 1998; Lee et al. 2004; Pan and Ho 2008; Riboli and Norat 2003; Song et al. 2010; Stintzing and Carle 2004, 2008b). This is based on its perceived ability to manage oxidative stress involved in the origin and aggravation of cancer, and the prevailing consensus that long-term daily exposure to small quantities of antioxidant dietary components has cancer chemopreventive potential (Azeredo 2009; Boivin et al. 2009; Gutteridge and Halliwell 2000; Kelloff et al. 2000; Ramarathnam et al. 1997; Sporn and Suh 2000; Stanner et al. 2004). The present chapter reviews the world literature on the investigations of red beetroot extract and its constituents as novel natural products with chemopreventive and/or chemotherapeutic effects. Further, studies directed toward the identification of active components of beetroot extract, such as betalains and its red color contributing constituent, betanin (**1**), and various mechanisms that may be involved in anticancer activity are also discussed. These include investigations in laboratory animals and bioavailability in humans, and, at a cellular level, antioxidant, free radical-scavenging, antiproliferative, anti-inflammatory, pro-apoptotic, and crucial enzyme inhibitory studies with intact cancer cells and various cellular components such as DNA, low-density lipoproteins (LDL), and with redox-regulated cell processes. Finally, because beetroot extract has no known toxicity in humans, studies exploring its potential in mitigating toxic side effects of currently used potent anticancer drugs, such as doxorubicin (Adriamycin) (**2**), when used in combination for synergistic chemotherapeutic effect are also reviewed.

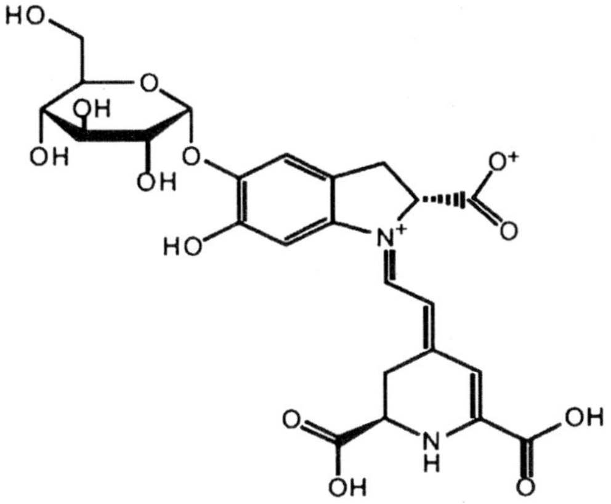

(1) Betanin

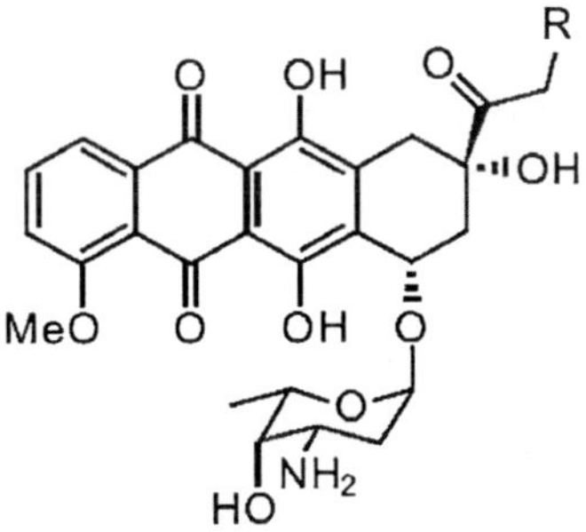

(2) Doxorubicin, R=OH
(7) Daunorubicin, R=H

7.2 Red Beet Plant

The modern beet plant is believed to have originated from the wild beet growing in prehistoric times in North Africa which then spread wild along the Mediterranean and European seashores extending quickly to British Isle, western Asian countries, India and as far east as China. In the beginning people ate beet leaves only and not roots. The ancient Romans appear to be one of the first civilizations to use beet roots as food. By the 16th century beet roots became popular throughout northern Europe. The commercial value of beets escalated upon the discovery of its high sugar content. The feasibility of commercial extraction of sugar from beets was demonstrated by the German chemist, Andreas Marggrat in 1774. This was quickly followed by setting up of the world's first commercial beet sugar factory in Poland in 1801. Before the war Poland had 110 factories with an annual output of 1,000,000 tons of

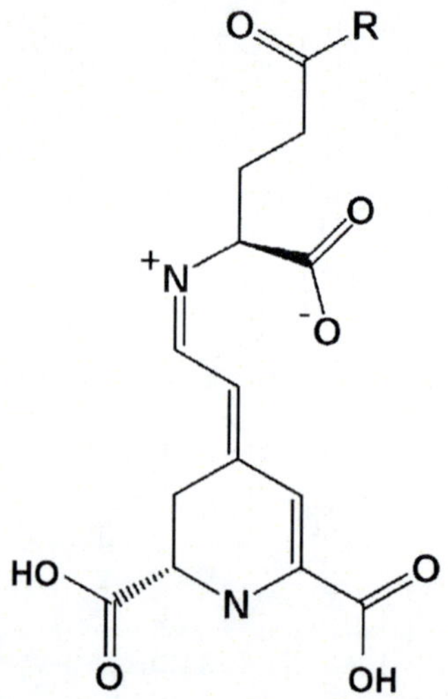

(3) Vulgaxanthin I, R = NH_2
(4) Vulgaxanthin II, R = OH

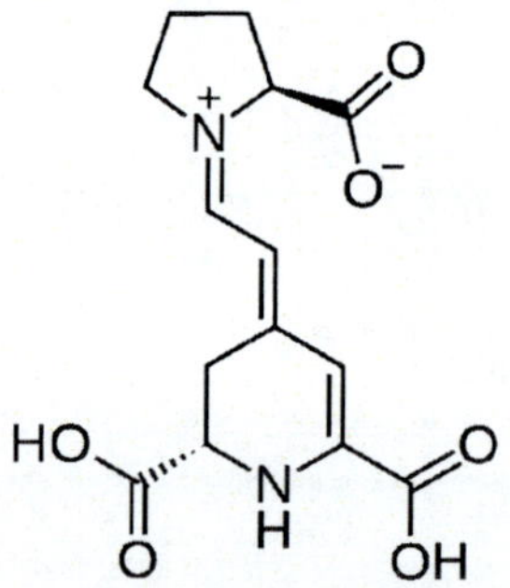

(5) Indicaxanthin

sugar. The US, the Russian Federation, France, Poland and Germany are the current leading commercial producers of beet for use in sugar manufacture and as a vegetable for general consumption (Ford-Lloyd 1995; Nottingham 2004).

The scientific name for red beet is *Beta vulgaris* L., family, Chenopodiaceae, subspecies, vulgaris, commonly known as garden beet and its five popular varieties are: 'Bonel', 'Detroit', 'Favorit', 'Nero' and 'Rubin'. Among these, the variety 'Rubin' is considered the most suitable one for food color production based on its red pigment content and composition (Gasztonyi et al. 2001).

7.3 Extraction of Red Beetroot Pigments

Almost all studies with red beet have utilized its roots as the source of red pigments and very limited investigations with beet leaves, stems and seeds are currently available (Gennari et al. 2011; Lee et al. 2009; Ninfali et al. 2007; Pyo et al. 2004). In general, beet roots are chopped into small pieces or ground prior to extraction. Only organically grown beet should be utilized to avoid synthetic fertilizer and pesticide contamination. The extraction procedure of choice is the one that achieves maximum yield of the betalain pigments responsible for red color while keeping to the minimum its loss during the process. The goal is to obtain a stable pigment extract with long shelf life. In general, pigments are extracted with water although methanol or ethanol solutions (20–50%) may be required for complete extraction (Delgado-Vargas et al. 2000). Aqueous extraction increases pigment stability and it may be further stabilized by slight acidification of the extraction medium which renders color constituents more stable and resistant to oxidation, both chemical and by endogenous polyphenoloxidases (Escribano et al. 2002; Strack et al. 2003).

Factors such as low water-content (achieved by spray drying), lower pH (4.0–6.0), presence of antioxidats (ascorbic and isoascorbic acids), and absence of metal cations (Fe, Cu, Sn and Al), heat, oxygen and light increase the stability (shelf life) of the extracted beetroot pigments (Attoe and von Elbe 1985; Azeredo et al. 2007; Haber et al. 1979; Havlikova et al. 1983; Pedreno ans Escribano 2001; Vitti et al. 2005; von Elbe et al. 1974). Reports on the stability evaluation of red beetroot color in various pharmaceutical matrices (Pai and D'Mello 2004) and soft drinks (Havlikova et al. 1985) are available. While temperature during processing and storage is the most important factor governing the stability of the beetroot pigments (Kujala et al. 2000; Patkai et al. 1997), extraction methods employing low electrical fields appear to be superior to those utilizing cryogenic freezing (Zvitov et al. 2003). Several studies comparing various extraction methods for beetroot colorants are available (Gasztonyi et al. 2001; Kujala et al. 2001b; Vitti et al. 2005). Further discussions on the stability of betalains are available in Chap. 3.

β-Glucosidases, polyphenoloxidases and peroxidases are some of the endogenous enzymes present in red beet which if not properly inactivated by blanching may breakdown betalain pigments leading to color loss during extraction process (Escribano et al. 2002; Lee and Smith 1979)). The degradation products appear to be similar to those produced by acid, alkali or heat (Escribano et al. 2002; Stintzing and Carle 2004). The red betacyanin component of beet betalain pigments is more susceptible than its yellow betaxanthin component to degradation by peroxidases since the presence of catalase efficiently suppresses the oxidation of the latter.

The following progressive losses in red betacyanin (betanin) and yellow betaxanthin (vulgaxanthin) contents during processing of beetroots, such as blanching and peeling, were observed (Patkai et al. 1997): betanin content: (1) raw material, 100%; (2) blanching, 99.8%; (3) peeling, 99.4%; (4) crushing, homogenization, 91.6%; (5) pasteurization, 50.1%; (6) storage, 60 days at 50C, 46.9% and (7) storage,

60 days at 20°C , 31.3% and vulgaxanthin content: (1) 100%, (2) 84.1%, (3) 82.9%, (4) 67.5%, (5) 42.7%, (6) 36.4% and (7) 45.2%.

Several brands of red beetroot extracts in powder form blended with dextrin added as a diluent to enhance water solubility, stabilized with ascorbic acid and rendered acidic, pH 5.4, with citric acid are currently available from various commercial sources (marketed under the trade name "Betanin" and Red Food Color E162). These commercial extracts have been utilized in recent research studies by several investigators including the authors (Kapadia et al. 1996, 2003, 2011a,b; Lechner et al. 2010).

In recent years, the development of biotechnology utilizing beet cell cultures for the commercial production of red beet pigments has been an active field of research (Georgiev et al. 2010a; O'Callaghan 1996; Pavokovic et al. 2009; also see Chaps. 9–13). In addition to higher yields (e.g., up to 250% increase in betanin content, Kriznik and Pavokovic 2010), such technology provides better quality control and pigment consistency independent of environmental changes that could alter pigment qualities in the field-grown beet, such as the vagaries of weather, pests, diseases and nutrient availability (Dornenberg and Knorr 1977; Hunter and Kilby 1990). However, there are reports of significant differences in the chemical composition of the pigments from the two sources. For example, rutin was present only in the betalain extracts from hairy root cultures, whereas chlorogenic acid was found only in the extracts from intact plants (Georgiev et al. 2010b). Also, the observed higher antioxidant activity of the hairy root extract compared to that of the field-grown beetroot extract was attributable to its increased concentrations (≈20-fold) of total phenolics compounds which may have synergistic effect with the beetroot betalain pigments. Thus, there is a need to adequately document the source of red beetroot pigment extracts, especially when they are used in biological studies, such as anticancer effects.

The economic sustainability of newer biotechnologies in competition with the traditional natural production methods of extracting red beetroot pigments remains to be ascertained. This is because beet is an abundant and inexpensive crop readily cultivable around the world. However, for technologically feasible preparation of extractives, application of the newer technologies, such as pulsed electric field treatment (Chalermchat et al. 2004; Fincan et al. 2004; Kannan 2011; Lopez et al. 2009), gamma-irradiation (Nayak et al. 2006) and ultrasound (Sivakumar et al. 2009) in conjunction with traditional extraction methods, such as solid-liquid extraction, diffusion extraction, reverse osmosis and ultrafiltration, appear promising. Since most pigments are not excreted by the cells but stored within their vacuoles (Cormier 1997), newer technologies help increase cell permeability resulting in higher pigment yields. Additional details in this regard are provided in Chaps. 14 and 15.

It is of interest to note that addition of brewer's yeast to beetroot juice helps to achieve optimum yield of pigments and thus improve nutritive quality through the action of lactic acid fermenting *Lactobacillus (L)* (Rakin et al. 2007) and sugar fermenting *Saccharomyces* bacteria (Drdak et al. 1992).

7.4 Chemical Constituents of Red Beetroot Extract

The main interesting constituents of red beet extract are its pigments. The colorants in red beetroot pigments are collectively designated as betalains, which are further divided into two chemical classes, betacyanins and betaxanthins. Betacyanin pigments are red–violet in color while betaxanthins are yellowish in color (Stintzing and Carle 2008b). Betacyanins are the dominant pigments in red beetroot and, among these, betanin (**1**) is the major constituent responsible for the red–violet color (Jackman and Smith 1996; Pucher et al. 1938; Robinson and Robinson 1932), with a greater quantity of betanin found in the peel than in the flesh (Kujala et al. 2002). Other betalain colorants found in smaller quantities in red beetroot extract are the minor red betacyanins: probetanin, isobetanin (C-15 epimer of betanin), and neobetanin, and the principal yellow xanthines: vulgaxanthin I (**3**) and vulgaxanthin II (**4**) and the minor yellow colorant, indicaxanthin (**5**) (Kujala et al. 2001b).

Betanin (**1**) was first isolated in 1918 from red beetroot by G. Schudel as part of his dissertation research in Prof. R. Willstatter's laboratory in Zurich, Switzerland. It is present in quantities of up to 200 mg in one single beetroot (0.41–0.5 g/kg) (Nollet 2004; Pucher et al. 1938) and represents up to 95% of the total betacyanins present in a typical extract, the latter accounting for 0.35–0.5% of the extract based on dry weight. The content may vary from one variety to another, forming distinctly different ratios between yellow and purple pigments, imparting different color intensity to the tissue (see Chap. 1). By breeding and recurrent selection, high-pigment beet (>310 mg/100 g fresh weight) lines have been developed (Gabelman et al. 2002).

In addition to betalain pigments, a typical beetroot extract also contains sugars (~45%, of which ~34% is sucrose), protein (~13%), and betaine (trimethylglycine ~1.0%), all based on percentage dry weight (Lechner et al. 2010).

The red betacyanins and yellow betaxanthins are iminium derivatives of betalamic acid (BA). Biosynthesis of BA involves two molecules of 3,4-dihydroxyphenylalanine (DOPA), one undergoing 4,5-extradiol oxidative cleavage of the aromatic ring and recyclization (see Chap. 2 for more details). BA is an integral part of the structural feature of all plant betalains. It is found as a natural constituent only in those plants that produce betalains, such as red beet. Condensation of BA with *cyclo*-DOPA (or its derivatives) leads to the betacyanin aglycone, betanidin, while its condensation with amino acids or amines results in betaxanthins (Delgado-Vargas et al. 2000; Roberts et al. 2010). This decisive step is believed to be spontaneous rather than an enzyme-catalyzed reaction, based upon biosynthetic studies in hairy root cultures of yellow beet (*Beta vulgaris* L. subspecies *vulgaris*, "Golden Beet") (Schliemann et al. 1999). Both betacyanins and betaxanthins have a protonated 1,7-diazaheptamethin system as a common chromophore and, in the red betacyanins, its conjugation is extended to include a substituted aromatic ring, e.g., *cyclo*-DOPA as in betanin (**1**), while in the yellow betaxanthins, the conjugation is not extended, e.g., vulgaxanthins I (**3**) and II (**4**), and indicaxanthin (**5**), which are attached to a non-aromatic amino acid or amine. All betacyanins are composed of glycosylated betanidin and some are modified with acyl groups, whereas betaxanthins are not. Detailed structural and chromatic aspects of betalains in general and betanin in particular, regarding their stability and degradation, have been investigated (Herbach et al. 2006a, b).

Also, the presence of the following constituents in red beetroot extract is reported employing high-performance/pressure liquid chromatography (HPLC) in conjunction with mass spectrometry (MS): 5,5′,6,6′-tetrahydroxy-3,3′biindolyl, *cyclo*-DOPA, BA, L-trytophan, p-coumaric acid and ferulic acid; ferulylglucose, β-D-fructofuranosyl-α-D-[6-*O*-(*E*)-feruloylglucopyranoside] and glucosides of *N*-formyl*cyclo*-DOPA and dihydroxyindolcarboxylic acid; flavonoids: betagarin, betavulgarin, cochliophilin A, and dihydroisorhamnetin; and phenolic amides: *N*-*trans*-feruloyltyramine and *N*-*trans*-feruloylhomovanillylamine (Kujala et al. 2001a, b, 2002). In addition, the occurrence of *cyclo*-DOPA-5-*O*-glucoside in red beet (*B. vulgaris* var. *rubra* L.) has been reported (Wyler et al. 1984). These various structurally diverse compounds, pigments, and their intermediary and degradative products probably contribute for cumulative anticancer effects observed when whole red beet extracts are used.

7.5 Anticancer Effects of Red Beetroot Extract

Data on properly designed, definitive anticancer clinical studies with red beetroot extract or any other beet-derived products, including isolated compounds, such as betanin (**1**), are currently not available. To date, all antitumor studies in laboratory animals have been conducted exclusively with beetroot extract utilizing only two murine species: mouse (HOS:HR-1, ICR and SENCAR species) and rat (F344 Harlan Spargue-Dawley and Wistar species).

7.5.1 Anticancer Studies in Laboratory Animals

The authors, in collaboration with their Japanese colleagues, were the first to report that drinking water containing red beetroot extract (red food color E162) exhibited significant anticancer activity by inhibiting the induction of lung and skin tumors in chemical carcinogen-treated mice (Kapadia et al. 1996). Encouraged by such observation, a systematic search was made for cancer chemopreventive agents among the natural colorants and related products of biologic and synthetic origin (Kapadia et al. 1997, 1998; Konoshima and Takasaki 2003). Subsequent studies have established the multi-organ tumor chemopreventive effects of red beetroot extract in multiple species (Kapadia et al. 2001, 2003; Bobek et al. 2000; Klewicka et al. 2010; Lechner et al. 2010).

In all animal models employed by the authors and other investigators for chemopreventive studies, the administration of red beetroot extract in drinking water exhibited no toxic effect in any of the test animals. There was no noticeable difference in weight gain between the control and experimental groups throughout the experimental period (20 weeks in mice and 35 weeks in rats). In addition, the mean water consumption per animal per day was also comparable in both groups. The low doses of beetroot extract employed in these studies (25 μg/ml in mice and 78 μg/ml in rats) were equivalent to a daily intake of about 6–20 μg of beetroot betacyanins/kg body weight of the test animals, and the chemopreventive effects of red beetroot extract were clearly demonstrable as a statistically significant reduction in tumor incidence and multiplicity and an accompanying delay in tumor latency period.

7.5.1.1 Anticancer Studies in Mice

In the anticancer studies in mice summarized below, some of the references cited utilize the commercial term "betanin" to describe the red beetroot extract tested and it should not be mistaken for betanin (**1**), the chemically pure compound isolated from the extract.

Inhibition of Skin Tumors

Administration of red beetroot extract (25 μg/ml) in drinking water *ad libitum* produced 37% reduction in 7,12-dimethylbenz[*a*]anthracene (DMBA)-induced, 12-*O*-tetradecanoylphorbol-13-acetate (TPA)-promoted skin tumors in female ICR strain mice (Kapadia et al. 1996). In a subsequent study (Kapadia et al. 2003), similar doses of red beetroot extract in drinking water of female HOS:HR-1 strain mice significantly delayed the appearance of skin papillomas (by 2 weeks) and inhibited the DMBA-induced, UV-B (280–320 nm, 3,430 J/m^2) light-promoted skin tumors by 70%. When the tumor data were evaluated for tumor multiplicity (cumulative number of tumors per group or number of tumors per mouse), beginning with the first tumor appearance up to the termination of the experiment at 20 weeks, beetroot extract showed a highly significant protection against UV-B-promoted skin cancer. Further, examination of the spleen indicated that, surprisingly, beetroot extract also protected (100%) all mice against splenomegaly (enlarged spleen) caused by DMBA–UV-B treatment.

In the female HOS:HR-1 strain mouse skin carcinogenesis model by topical applications of (±)-(E)-4-methyl-2-[(E)-hydroxyamino]-5-nitro-6-methoxy-3-hexanamide (NOR-1) as tumor initiator and TPA as promoter, the tumor modulation by *ad libitum* oral administration of 25 μg/ml red beetroot extract was also significant (Kapadia et al. 2003). There was a 1-week delay in the first tumor appearance and a 40% inhibition of the number of papillomas formed per mouse in the beetroot extract-treated group compared with the non-treated control mice.

Advanced glycation end products (AGE), generally recognized as products of multiple complications from various diseases including diabetes, have been considered as possible human carcinogens (Yamagishi et al. 2005). During the authors' studies on chemical carcinogens and chemopreventive agents of natural origin (Kapadia et al. 1997, 1998; Tokuda and Iida 2006), human albumin and glucose natural mixture (HAGE) was identified as a possible human carcinogen with significant experimental tumor-initiating potency in the two-stage mouse skin tumor model described above. In this new anticancer test model utilizing HAGE as initiator and TPA as promoter of skin tumors in SENCAR mice, red beetroot extract exhibited significant chemopreventive activity (Kapadia et al. 2005). Thus, oral administration of beetroot extract (25 μg/ml) for 2 weeks decreased the tumor incidence and tumor burden by 20% and 45%, respectively. The treatment also extended the tumor latency by 10% compared with the control animals. Apart from establishing beetroot extract as an effective broad-spectrum cancer chemopreventive agent, these results may also explain the increase in risk of neoplasm formation in individuals with diabetes.

Inhibition of Lung Tumors

A 60% reduction in 4-nitroquinoline-1-oxide (4NQO)-induced, glycerol-promoted lung tumors in male ICR strain mice drinking *ad libitum* water containing 25 μg of red beetroot extract/ml was reported in the authors' 1996 study (Kapadia et al. 1996) and in their subsequent US patent (Kapadia et al. 2001). In these investigations, lung tumors (adenomas) had been taken into account after separation of each pulmonary lobe in all experimental animals.

Inhibition of Liver Tumors

In the female ICR strain mouse model, *N*-nitrosodiethylamine (DEN) initiation followed by phenobarbital promotion resulted in 67% tumor formation in the liver (Kapadia et al. 2003). However, oral administration of red beetroot extract (25 μg/ml, *ad libitum* access) reduced the gross tumor incidence to 40%, with a tumor inhibition rate of 40%. Concurrently, there was a 51% drop in the total number of tumor nodules formed in the livers of the animals treated with beetroot extract.

7.5.1.2 Anticancer Studies in Rats

Experiments using rat as animal model have proven to be valuable for the study of chemopreventive potentials of red beetroot extract in chemical-induced carcinogenesis (Bobek et al. 2000; Klewicka et al. 2010; Lechner et al. 2010). Like in the mouse model, beetroot extract was found to exhibit antitumor effects also in multiple organs in the rat model. Prior to these anticancer studies, a metabolic investigation in 1980 reported that orally administered betanin (**1**) was poorly absorbed and established the gut as the major site for its degradation in rats (Krantz et al. 1980).

Inhibition of Esophageal Tumors

Oral consumption of red beetroot extract (red food color E162) was found to inhibit *N*-nitrosomethylbenzylamine (NMBA)-induced tumors in the esophagus of the male rat (F344 strain) (Lechner et al. 2010). Rats treated with NMBA and given free access to water containing 78 μg/ml of beetroot extract had a reduced esophageal papilloma formation of 45%. The treatment also resulted in reduced rates of cell proliferation in both precancerous esophageal lesions and in papillomas of NMBA-treated rats. Additionally, angiogenesis and inflammation were reduced, and the apoptotic rate was increased.

Inhibition of Colon Tumors

The effect of diet supplemented with 15% in red beet (*B. vulgaris* var. *rubra*) fiber on dimethylhydrazine (DMHZ)-induced colon carcinogenesis was studied in male Wistar rats (Bobek et al. 2000). The animals were sacrificed 14 weeks after the application of DMHZ, at which point they were on a beet fiber diet for a total of 26 weeks. The presence of both higher cellulose (5% and 15%) and red beet fiber (15%) in the diet significantly reduced the incidence of precancerous lesions (aberrant crypt foci [ACF]) in rat colon. The diet containing red beet fiber did not significantly affect the incidence of colon tumors, although the number of animals bearing tumors was reduced by 30%.

Another study (Klewicka et al. 2010) investigated the effects of beetroot juice fermented by *L. brevis* 0944 and *L. paracasel* 0920 (FBJ) on the induction of ACF by the carcinogen *N*-nitroso-*N*-methylurea (NMU) in rat colon. FBJ significantly reduced the number of ACF in NMU-treated rats (from 55 ± 18 to 21 ± 6). Moreover, the number of extensive aberrations (more than four crypts in a focus) decreased from 45 ± 21 to 7 ± 4. Fecal water obtained from rats fed with the NMU-containing diet induced significant cytotoxic and genotoxic effects in control Caco 2 cells. Both of these toxic effects were abolished in rats on FBJ supplementation.

7.6 Studies in Humans

Several studies on health benefits of diets rich in fresh fruits and vegetables have implied that red beet extracts and juices may reduce the risk of cancer in humans (Boivin et al. 2009; Kelloff et al. 2000; Nottingham 2004; Pan and Ho 2008; Riboli and Norat 2003; Stanner et al. 2004). However, no properly designed clinical studies to establish such anticancer effects of red beet products or their constituents are currently available. Bioavailability and pharmacokinetics of betanin (**1**), the main red beetroot pigment, in humans have gained attention in recent years. Also, the incidence of beeturia, the red or pink coloration of urine in beetroot consumers, has been the subject of clinical research in humans over the past many years.

Since chemical studies with the betalains isolated from red beet (*B. vulgaris* various subspecies) indicated their similarity to the betalain pigments isolated from other plant species, e.g., from the edible fruits of cactus *Opuntia ficus-indica* L. Mill., and its various cultivars (Castellanos-Santiago and Yahia 2008; Chavez-Santoscoy et al. 2009; Livrea and Tesoriere 2006), and those of the Amaranthaceae family (Cai et al. 2003, 2005), relevant studies with these non-beet betalains are also included in this review for comparative purpose.

7.6.1 Bioavailability and Pharmacokinetic Studies

The urinary pharmacokinetics of the betalain constituents after the consumption of red beet juice in healthy human volunteers have been studied. In general, the urinary excretion of 60–363 mg of betalains in 300–500 ml of beet juice was studied over a 24-h period. The red beet betacyanins (betanin and isobetanin) were found to be quickly absorbed from gut and began to appear in urine after 2 h of ingestion in a group of six healthy volunteers (Netzel et al. 2005). However, only 0.28% of the ingested dose of betacyanins was accounted for in the 24-h urine with a half-life of 7.43 ± 1.47 h and a terminal elimination rate constant of 0.097 ± 0.021 h^{-1} (Frank et al. 2005). Compared with the water controls, red beet juice consumption resulted in a significantly increased urinary excretion of total phenolics (51% of the administered phenolics). Similar low recovery (0.01–0.6%) of intact betalains excreted in the urine were observed in an earlier study with a larger population of 100 healthy subjects after an 8-h period of beetroot ingestion (Watts et al. 1993).

The plasma kinetics and urinary excretion of the betalains from *Opuntia ficus-indica* were studied in eight healthy volunteers after a single ingestion of 500 g of its fruit pulp, which provided 16 mg of betanin (**1**) and 28 mg of indicaxanthin (**5**) (Tesoriere et al. 2004). The two betalain constituents reached their maximum plasma concentrations 3 h after the fruit meal and declined according to first-order kinetics. The half-life of betanin (0.94 ± 0.07 h) was shorter than that of indicaxanthin (2.36 ± 0.17 h). Both compounds had disappeared from plasma by 12 h after intake and their urinary excretion over a 12-h period represented $76.0 \pm 3.0\%$ and $3.7 \pm 0.2\%$ of ingested indicaxanthin and betanin, respectively. These results indicate higher bioavailability of indicaxanthin than betanin and that their renal clearance is a minor route of systemic elimination for these phytochemicals.

The above studies indicate that the urinary excretion rates of unmetabolized betalains were fast and appeared to be monoexponential, suggesting a one-compartment model. In order to get a more complete picture of the pharmacokinetics of the red beet betalains, quantitative data on their bioavailability that includes measurements of both unchanged betalains and their metabolites in plasma, urine, and bile are needed. Further, in spite of the similarity in their chemical structures, the betalains containing common betanin constituent from different plant species may have different pharmacokinetic profiles for their absorption, distribution, and excretion in humans. This may be attributable to the differences in the non-betanin constituents and their composition in different plant betalains.

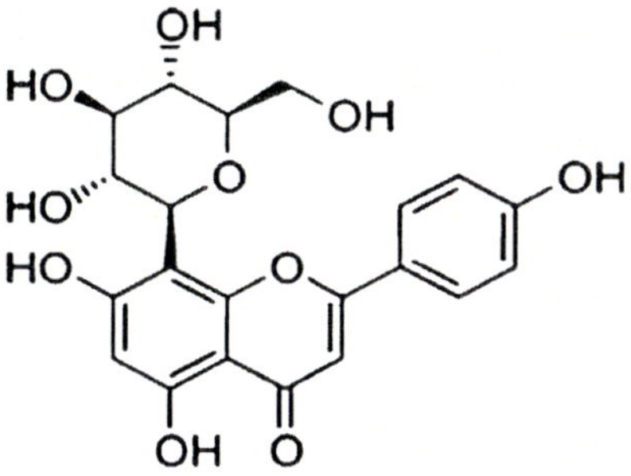

(6) Vitexin

7.6.2 Studies of Beeturia

Consumption of red beetroot extract produces red or pink coloration of urine and feces, known as beeturia (also called betaninuria, named after betanin, the major constituent of the extract) in 10–14% of the population (Mitchell 2001; Watson et al. 1963; Watts et al. 1993). It appears to be more prevalent among those with pre-existing iron deficiency and malabsorption conditions. The susceptible individuals are unable to completely break down the red pigment in their stomach and colon, and thus more of it becomes available for re-absorption and bodily re-circulation. Two major contributing factors for this condition are higher gastric pH and a longer emptying rate, which increase the absorption rate of the red colorant from the ingested beetroot extract (Pavlov et al. 2005; Watts et al. 1993).

To study the potential role of genetic traits in beeturia occurrence, 99 adult like-sexed twin pairs have been studied for the presence of beeturia in urine after the ingestion of beetroot juice (Forrai et al. 1982). However, the hereditary factor of the intensity of beeturia could not be definitely established, although some circumstances pointed to an effect of a genetic factor on the phenomenon. A similar study of the frequencies of beeturia among different racial groups failed to establish a hereditary trait (Saldanha et al. 1960), while another study suggested polymorphism in inborn patterns of the beetroot pigment metabolism and excretion (Saldanha 1962).

A clinical investigation reported the effect of colonic oxalic acid on beeturia in beeturic and non-beeturic subjects (Eastwood and Nyhlin 1995). Since betalains act as redox indicators, their color is protected by reducing agents. Betalains are decolorized by hydrochloric acid, ferric ions, and colonic bacteria preparations. In animals, oral administration of betalains does not produce beeturia, but its injection into the peritoneum does. Oral ingestion of betalain and 1 g of oxalic acid produced beeturia in non-beeturic healthy subjects, but passed into lieostomies without beeturia. This suggests that beeturia results from colonic absorption of betalains. Oxalic acid preserves the red color in the colon; otherwise it is decolorized in non-beeturic individuals by a non-enzymatic process in their stomach and colon.

Studies on the stability and bioaccessibility of betalains under simulated oral, gastric, and small intestine digestive conditions suggest that digestive stability controls bioaccessibility of betaxanthins, whereas additional factors relevant to the food matrix and its processing affect betacyanin bioaccessibility (Tesoriere et al. 2008).

7.7 Mechanisms of Anticancer Effects

In an effort to unravel various biochemical mechanisms that may be involved in the anticancer effects of red beetroot extract observed in animal models, studies are being conducted in various tumor cell lines and subcellular models, such as microsomal membranes, enzymes, and DNA. These investigations have provided important information on the nature of bioactive constituents, their metabolic fate, and the cell components involved, which should help elucidate the potential mechanism(s) of anticancer activity exhibited by red beetroot pigment extract. Additionally, these studies help identify the areas that need further investigations to establish the clinical usefulness of beetroot extract and its constituent(s) as chemopreventive and/or chemotherapeutic agents.

It should be noted that betacyanins, among the beetroot betalain constituents, are generally recognized as the most important anticancer component (Kanner et al. 2001; Kapadia et al. 2011a; Lechner et al. 2010; Stintzing and Carle 2007, 2008a, b; Pan and Ho 2008). Currently, there is growing recognition that betanin (**1**), the major constituent of the red beet-derived betalains, is most likely the primary and potent compound responsible for the observed anticancer effects (Gliszczynska-Swiglo et al. 2006; Kapadia et al. 2011a). Accordingly, anticancer and other biological studies involving red beet extract and its constituents, betalains, betacyanins, and betanin, in various biological models, are reviewed below. Also, such studies with non-beet betalains from other plant species are included for comparison.

7.7.1 Studies in Tumor Cell Lines of Human Origin

Protocols using cultured human tumor cell lines as model systems have been found useful for elucidating potential mechanisms involved in cancer initiation and progression, and suppression by known anticancer compounds of natural origin and their synthetic analogs (Langdon 2003). Thus, betanin (**1**), isolated from red beetroot, exhibited a dose-dependent growth inhibition of human breast (MCF-7), lung (NCI-H460), colon (HCT-116), stomach (AGS), and central nervous system (CNS) tumor cell lines (Reddy et al. 2005). A subsequent study (Boivin et al. 2009) reported the antiproliferative activity of red beetroot extract against the following tumor cell lines of human origin: breast (MCF-7), lung (A549), kidney (Caki-2), prostate (PC-3), pancreatic (Panc-1), stomach (AGS), and medulloblastoma (DAOY). Results with the colon HCT-116 cell line were recently confirmed with beetroot juice, which reduced cancer cell proliferation and increased apoptosis but did not exhibit cytotoxicity (Kannan 2011).

Another study reported that betanin (**1**), isolated from the fruits of *Opuntia ficus-indica*, induced apoptosis in the human chronic myeloid leukemia cell K562 line (Sreekanth et al. 2007). This in-depth investigation demonstrated that betanin entered K562 cells and altered mitochondrial membrane integrity, which led to cytochrome *c* leakage, activation of caspases, and nuclear disintegration. These biochemical alterations were reflected in ultrastructural changes typical of cells undergoing apoptosis. Also, juices containing betacyanins and betaxanthins along with other phenolic compounds from nine species of *Opuntia* prickly pears were found to be active against human breast, hepatic, colon, and prostate cancer cells (Chavez-Santoscoy et al. 2009).

More recently, the authors studied the cytotoxic effect of red beetroot extract in human prostate (PC-3), pancreas (PaCa), and breast (MCF-7) cancer cells, comparing it with equivalent doses of an anticancer drug, doxorubicin (Adriamycin), and in combinations of the two in different proportions (Kapadia et al. 2011a, b). This red-colored anticancer antibiotic was selected for the comparative cytotoxic study because its chemical structure (**2**) with a planar configuration of an aromatic chromophore attached to a six-membered sugar molecule, is remarkably similar to that of betanin (**1**). Moreover, like beetroot extract, doxorubicin also causes red coloration of urine in patients taking this drug (Carvalho et al. 2009). Beetroot extract and doxorubicin showed synergistic cytotoxic effects in the three human tumor cell lines tested. Further, the results suggested that betanin, the major betacyanin constituent, may play an important role in the cytotoxicity exhibited by beet extract.

It is of interest to note that the leaves and seeds of Swiss chard (*B. vulgaris* subsp. *cycla*, also known as *cicla*) contain flavonol glycosides, identified as apigenin-8-glucoside (vitexin) (**6**) derivatives of rhamnose and xylose, which were found to inhibit human breast cancer MCF-7 (Ninfali et al. 2007) and colon cancer RKO cell proliferation (Gennari et al. 2011). Further, vitexin-2″-O-rhamnoside strongly inhibited DNA synthesis in MCF-7 cells, while xylosylvitexin exhibited the strongest antiproliferative activity against RKO cells, with an enhancement of the apoptosis as well as an increase of cells in the G(1) phase and a reduction of cells in the S phase. Based on these results, it was concluded that xylosylvitexin in the seeds of Swiss chard appears to be the main and more efficient chemopreventive compound.

7.7.2 Studies of Antioxidative, Antiradical, and Anti-Inflammatory Activities

Recent epidemiological and laboratory studies provide ample evidence for oxidative stress leading to biochemical changes that contribute to degenerative diseases, such as aging and cancer (Ames et al. 1993, 1995; Lee et al. 2004; Sporn and Suh 2000; Stanner et al. 2004). This involves pro-oxidant factors that increase formation of free radicals and other reactive oxygen species, which then attack the body's defense systems. The increased production of oxidants and free radicals during inflammatory disorders is also recognized as an integral part of cell and tissue injury (Gutteridge and Halliwell 2000; Stanner et al. 2004). Oxidative balance is achievable by reducing pro-oxidant factors and maintaining endogenous protective enzymes, such as glutathione peroxidase, by dietary intake of antioxidant radical scavenging compounds.

The antioxidant properties of betalains from red beet are currently under active investigations (Kanner et al. 2001; Moreno et al. 2008; Stintzing and Carle 2007; Strack et al. 2003). Beets are now ranked among the ten most potent vegetables, and their total content of phenolic compounds is the highest among the common vegetables tested (Vinson et al. 1998), and they show the highest cellular antioxidant activity (Song et al. 2010). Betanin (**1**), the main red beet pigment constituent, has exceptionally high free radical-scavenging activity (Gliszczynska-Swiglo et al. 2006). The ability of betalains to induce quinone reductase (QR), a potent phase II detoxification enzyme and an indicator of cancer chemoprevention (Talalay 1989), has been reported (Lee et al. 2005). In murine hepatoma (Hep 1c1c7) cells *in vitro*, beetroot extract induced QR. A comparison of four strains of beetroot (*Beta vulgaris* L.) with different pigmentation (white, orange, regular red, and high-pigment red) indicated that only the red strains were the most capable free radical scavengers and reducers of radical cations as well as inducers of QR (Wettasinghe et al. 2002). Also, betanin isolated in pure form was confirmed to be an efficient QR inducer in the bioassay employed. However, no such activity was observed in rats administered a betalain-enriched diet (Lee et al. 2005). This lack of response to betanin in the betalain-containing solid diet may be related to its stability, absorptive capacity and/or insufficient level of enrichment, or to difficulties in assessing such activity *in vivo*.

In contrast, another study reported a protective effect of red beetroot juice against carbon tetrachloride- and *N*-nitrosodiethylamine-induced oxidative stress in rats (Kujawska et al. 2009). Similarly, liver- and duodenum-protecting effects of table beets during ischemia–reperfusion were observed in rats (Vali et al. 2006, 2007). The beetroot diet resulted in increased levels of glutathione peroxidase and superoxide dismutase in the liver, indicating a positive effect on its redox state. Also, the free radical-scavenging property in plasma increased significantly. This was accompanied by similar increases in other antioxidant parameters and reducing power, such as H-donating ability and free SH-group concentration. Also, mice on a high-fat, high-cholesterol diet supplemented with 8% freeze-dried red beet leaf had a reduced lipid peroxidation and an improved antioxidant status (Lee et al. 2009). Thus, together with results from the multi-organ tumor chemopreventive studies in rodents described earlier (Sect. 7.5.1), it could be concluded that pretreatment with beetroot extract or juice has the potential to counteract xenobiotic-induced stress in animals.

A study of beetroot juice on stimulated and unstimulated human peripheral blood mononuclear cells found it to exhibit immunosuppressive and anti-inflammatory activities (Winkler et al. 2005). The data showed that the extract counteracted pro-inflammatory cascades in the cells. Additionally, an *in vitro* study of various beetroot products, both in liquid and solid forms, on oxidative metabolism and apoptosis in neutrophils of obese individuals, demonstrated their antioxidant and anti-inflammatory capacities (Zielinska-Przyjemska et al. 2009). Also observed were pro-apoptotic effects of beetroot products on stimulated neutrophils. Because inflammation is intimately involved in the development and progression of several clinical conditions including obesity and cancer, the beneficial effect of beetroot may relate to this anti-inflammatory capacity.

Hydrochlorous acid (HOCL) is the most powerful oxidant produced by human neutrophils and is largely responsible for the damage caused by the inflammatory cells. It is produced by the heme enzyme myeloperoxidase (MPO) from hydrogen peroxide and chloride. Because the beetroot betalains are antioxidant and anti-inflammatory phytochemicals, the mechanism of interaction of betanin (**1**), the major betalain constituent, with MPO was studied (Allegra et al. 2005). This study showed that betanin is indeed a good peroxidase substrate for MPO, functioning as one-electron reductant of its redox intermediates. In addition, betanin could effectively scavenge HOCL. Thus, the chemopreventive properties of the betalain pigments of beet may depend on its betanin content as well as its antioxidant capacity to act as a free radical scavenger in quenching radical oxygen species and preventing them from causing tissue damage (Pedreno and Escribano 2000).

Spiking human LDL with betanin (**1**) yielded betanin-enriched LDL with a maximum binding of 0.51 nmoles/mg LDL protein (Tesoriere et al. 2003, 2009). The enriched LDL was more resistant to a copper-induced oxidation. But the kinetic studies did not provide evidence of pro-oxidant effects over a large concentration range of betanin. Similar studies with indicaxanthin (**5**), another constituent of red beetroot betalain, indicated that it had twice the effectiveness of betanin in increasing the length of the lag phase. One pathway for LDL oxidation may involve MPO (which is secreted by activated phagocytes as a part of the defense process against invading pathogens) and nitrite (the final oxidation product of nitric oxide metabolism) (Burner et al. 2000) by the lipoprotein acting as a substrate for the enzyme. Other details of the lipoperoxyl radical-scavenging activity of red beet pigments are presented in Chap. 6.

Although low concentrations were attained in plasma by betanin in human subjects after ingestion of *Opuntia* cactus fruit, it effectively inhibited the production of lipid hydroperoxides in human LDL submitted to MPO/nitrite-induced oxidation (Allegra et al. 2005, 2007). Here, kinetic measurements suggested scavenging of the initiator radical nitrogen dioxide and lipoperoxyl radicals. Also, possible involvement of unidentified oxidation product(s) of betanin by MPO/nitrite in the inhibition of LDL oxidation, potentially as effective as the parent betanin, was noted. In the earlier investigation (Tesoriere et al. 2004), LDL isolated from plasma at 3 and 5 h after a cactus fruit meal was found to incorporate betalains at concentrations of 100.5 ± 11 and 50 ± 7.2 pmol/mg LDL protein, respectively. In addition, the LDL appeared more resistant to *ex vivo*-induced oxidative injury than did the samples isolated from the same volunteers before fruit ingestion. It was noted that the higher the amount of betalains incorporated, the higher was the resistance. These results suggest that *Opuntia* cactus pear fruit could be a source of bioavailable betalains and its constituents, betanin (**1**) and indicaxanthin (**5**), may be involved in the observed protection of LDL against *ex vivo*-induced oxidative modifications.

Further, under *in vitro* conditions, low concentrations of betanin (170 μmole) were found to inhibit lipid peroxidation and cyclooxygenase enzymes, COX-1 and COX-2, which catalyze the conversion of arachidonic acid to generate chemical mediators of inflammation (Reddy et al. 2005). This study also found betanin to inhibit breast and other tumor cell growth as described earlier (Sect. 7.7.1). It was noted that when betanin was tested in the presence of anthocyanin, the mixture did not enhance the COX enzyme inhibitory activity and failed to inhibit tumor cell growth. However, this observation may not be relevant in the case of beetroot extract since anthocyanins are not present in red beet (Jackman and Smith 1996).

Among the red beet betalain colorants, betacyanins exhibit greater antiradical activity than betaxanthins under *in vitro* test conditions (Escribano et al. 1998). Studies focused on the relationship between structures of compounds present in betalains and their radical scavenging activity showed that, in betacyanins, e.g., betanin (**1**), glycosylation reduced their antioxidant activity while acylation generally increased it (Cai et al. 2005; Gandia-Herrero 2009, 2010). Further, 6-*O*-glycosylated betacyanins had higher antioxidant capacity than 5-*O*-glycosylated ones and additional glycosylation reduced activity. In betaxanthins, e.g., vulgaxanthins I (**3**) and II (**4**), the free radical-scavenging activity usually increased with the number of hydroxyl and imino moieties as well with the location of hydroxyl groups. Also, the presence of phenolic hydroxyl groups enhanced the antiradical activity for compounds with two hydroxyl groups, based upon studies with synthetic analogs (Gandia-Herrero et al. 2009).

Also, radical scavenging activity and stability of betalains in simulated conditions of human gastrointestinal tract have been studied (Pavlov et al. 2005). Under the conditions of pH below 3 and bile salts concentration of up to 4%, betalains were relatively stable with radical scavenging activity ranging from 75% to 38%. Similarly, antiradical activity and stability of betanin under different pH, temperature, and light conditions have been investigated (Pedreno and Escribano 2001). Further details in this regard are available in Chap. 3.

It is of interest to note that the phenolic extracts of leaves, stems (Pyo et al. 2004), and seeds (Gennari et al. 2011) of Swiss chard (*B. vulgaris* subsp. *cycla*) also exhibit antioxidant activities. This gives credence to the practice of binding wounds with beet leaves advocated by the ancient Greek physician, Hippocrates, in 400 BC. It was thought at the time that juice from the leaf bindings aided healing process (Nottingham 2004).

7.7.3 Interactions with Tumor Cell DNA

Currently, limited studies are available on the interaction of betanin (**1**) and other constituents of red beetroot extract with DNA. Betanin induced DNA fragmentation in human chronic myeloid leukemia K4562 cells when they were treated at a concentration of 40 μmoles for 24 h, which resulted in 50% decrease in cell growth (Sreekanth et al. 2007). Vitaxin-2″-O-rhamnoside, the primary constituent of the phenolic fraction of Swiss chard leaf extract, is found to be a strong inhibitor of DNA synthesis in human breast cancer MCF-7 cells (Ninfali et al. 2007). Also, betanin is a weak inhibitor of DNA methyltransferase (DNAMT) in MCF-7 tumor cells (Paluszczak et al. 2010). The significance of these results in elucidating the mechanism of anticancer effects exhibited by red beetroot extract is yet to be delineated. Dietary polyphenols, including betanin and other betalains, are known to inhibit DNAMT in human esophageal, colon, prostate, and breast cancer cell lines (Fong et al. 2007). Thus, prevention or reversal of hypermethylation-induced inactivation of key tumor suppression genes or receptor genes by DNAMT inhibitors could be an effective approach for cancer prevention.

The authors' recent *in vitro* studies comparing cytotoxic effects of red beetroot extract and the anticancer drug, doxorubicin, in human cancer cells suggested a possible common mechanism of action between this drug and betanin, the constituent primarily responsible for red color of the extract (Kapadia et al. 2011a, b). This is based on the fact that doxorubicin and betanin are both red-colored compounds with remarkably similar planar aromatic chromophore attached to a six-membered sugar molecule in their chemical structures, (**1**) and (**2**), respectively. A proposed mechanism of cytotoxic activity of doxorubicin involves its ability to intercalate with cancer cell DNA to form adducts that interfere with its unwinding (Pigram et al. 1972; Formari et al. 1994). Thus, the cytotoxic effect attributable to betanin may also involve such adduct formation with cancer cell DNA. Additionally, the inhibition of DNA topoisomerase I and II enzyme activities exhibited by doxorubicin (Bodley et al. 1989; Foglesong et al. 1992) may also play a role in its anticancer effect. However, evidence for such enzyme inhibitions by either betanin or red beetroot extract is currently not available.

It is noteworthy that recent studies in humans have noted decreased oxidative damage to lymphocyte DNA when consuming a diet containing red pigment extract of *Opuntia ficus-indica* cactus fruits (Siriwarchana et al. 2006) and antioxidants from other fruits and vegetables (Duthie et al. 1996). Similar results were obtained in mice on a high-fat, high-cholesterol diet when supplemented with 8% freeze-dried red beet leaf (Lee et al. 2009).

7.8 An Overview

Although, over the centuries, red beetroot extract has been utilized in the prevention and treatment of a variety of human diseases, only in the past few years the potential health benefits of the extract have gained renewed attention by the scientific community as well as by the general public. This is largely due to the recent finding that the betalain pigments in red beetroot extract act as natural antioxidants, exhibiting powerful free radical-scavenging properties with potentials for prevention and cure of a variety of diseases associated with oxidative stress (Azeredo 2009; Gliszczynska-Swiglo et al. 2006; Kanner et al. 2001). Currently, there is general consensus (at least in theory) that diseases, such as cancer, that originate in or are aggravated by oxidative stress, may be mitigated by antioxidant compounds found in diet or taken as supplements (Ames et al. 1995; Carlsen et al. 2010; Dorai and Aggarwal 2004; Geshner et al. 1998; Lee et al. 2004; Stanner et al. 2004; Staruchova et al. 2006). Beetroot extract and its betalain constituents, e.g., betanin (**1**), may fit into such a category, as discussed in various sections of this and other chapters of this book.

In recent years, the food, beverage, and nutritional supplements industries have shown great interest in the antioxidant betalains found in red beetroot extract to capitalize on their marketable potential health benefits (Moreno et al. 2008). Much attention is now focused on beetroot extract as an antioxidant nutrient in lowering oxidative stress (Kanner et al. 2001; Strack et al. 2003). Additionally, the betalain-rich beet extract appears to provide protection against radiation as reported in a recent study in mice exposed to gamma ray irradiation (Lu et al. 2009) and to exhibit a prophylactic effect against experimental influenza infection in mice (Prahoveanu et al. 1986). These effects are attributable to the extract's antioxidant activity and potential modulation of the immune system. Interestingly, beetroot extract also acts as corrosion inhibitor for metals in corrosive media, such as carbon steel in well water, by providing a protective pigment film that is environmentally friendly (Raja and Sethuraman 2008; Selvi et al. 2009).

The betalain-rich beetroot extract has a very intense red color and it is more potent than most synthetic red colorants currently used by the food and beverage industries (Henry 1996). Betanin (**1**), the principal red pigment, is more stable in beetroot extract than when isolated in its pure chemical form (Degenhardt and Winterhalter 2001; Pedreno and Escribano 2001). Therefore, the chemically pure betanin is not recommended for use as food colorant. For commercial use, betanin in beetroot extract is stabilized with ascorbic acid and rendered acidic with citric acid. Although it is susceptible to degradation when exposed to heat, light, and oxygen (Jackman and Smith 1996; Pedreno and Escribano 2001; von Elbe et al. 1974), beetroot extract is well suited for coloring frozen, dried, and short shelf life products, such as ice cream, yogurt, powdered beverage mixes, and confectionary fruit and cream fillings. Beverages enriched with antioxidant beetroot extract are currently gaining much popularity among the health conscious younger as well as older population. It should be noted that, while epidemiological studies suggest potential health benefits of dietary intake of betalains, such as those found in red beetroot extract, no direct cause and effect relationship has been established to date.

However, a broad spectrum of antitumor activity is demonstrated in laboratory animals receiving red beetroot extract in drinking water *ad libitum* (Kapadia et al. 1996, 2001, 2003; Lechner et al. 2010). The extract also exhibits protective activity against chemically induced genotoxicity (Edenharder et al. 1994, 2002; Platt et al. 2010). It is used around the world as a natural red food color E162 and considered safe by both the US FDA and the EU for human consumption, with no known toxicity (Downham and Collins 2000). It is non-carcinogenic (Schwartz et al. 1983) as well as non-mutagenic (von Elbe and Schwartz 1981).

The fact that beetroot extract did not initiate or promote hepatocarcinogenesis in rats (Schwartz et al. 1983) was further confirmed by the authors' studies in a mouse model, which showed a delay in the DEN-induced tumor onset by both short interval and prolonged treatment with the extract (Kapadia et al. 2003). This delay in tumor onset could involve inhibition of intracellular DEN–DNA adduct formation by the beetroot constituent(s). To achieve this, there is a need to transport the antitumor constituent(s) through the plasma membrane, which may involve metabolic activation by cytochrome P450 monooxygenase and reductase enzymes. Based on these observations, the following three mechanisms of cancer chemoprevention by the beetroot extract constituent(s) may be postulated in the case of DEN-induced liver tumors: (1) by blocking the passage of the chemical carcinogen, DEN, through the plasma membrane, (2) by competing with enzyme(s) involved in DEN activation, and/or (3) by competitively inhibiting DNA adduct formation with DEN.

In diverse *in vitro* studies in a variety of cancer cells of human origin, both red beetroot extract and its betalain constituents exhibited significant cytotoxic effects. The cancer cell lines tested include those from human breast, lung, liver, colon, prostate, pancreas, kidney, leukemia, and central nervous system tumors (Boivin et al. 2009; Kapadia et al. 2011a, b; Reddy et al. 2005). By studying cell biology and biochemical events at extra- and intra-cellular levels, a number of potential mechanisms of anticancer activity attributable to red beetroot extract constituents may be proposed. Thus, involvement of processes, such as antioxidative, antiradical, anti-proliferative, anti-inflammatory, and anti-angiogenesis in tumor cells as well as pro-apoptosis, and enzymes, such as LDL, MPO, DMNT, phase-II, COX-1, and COX-2, and DNA adduct formation are some of the possible mechanisms of anticancer effects exhibited by red beetroot constituent(s) (Azeredo 2009; Kanner et al. 2001; Strack et al. 2003).

In addition to polyphenolic betalains in red beetroot extract, betaine (trimethylglycine, present at a concentration level of up to 1.5% on dry weight basis) may act as a cytotoxic agent by methylating DNA in cancer cells (Lechner et al. 2010). Animals on a methyl donor-deficient diet are known to develop hypomethylation and cancers (Pogribny et al. 2006). Thus, it has been suggested that betaine in beetroot extract may antagonize cancer development by augmenting *S*-adenosylmethionine levels, which, in turn, antagonize development of hypomethylation. Additionally, betaine is known to reduce inflammation (Detopoulou et al. 2008), which may be one of the mechanisms of chemoprevention by beetroot extract. However, currently there is no consensus on any correlation between dietary consumption of betaine (together with choline) in humans and cancer mortality rate based on large-scale epidemiological studies (Xu et al. 2009; Cho et al. 2010; Lee et al. 2010).

It is perhaps fortuitous that the naturally occurring anticancer red beet pigments of diverse chemical structures, such as betanin (**1**) and vitexin (**6**), have the potential to act synergistically with potent antitumor drugs like doxorubicin (**2**), which itself is an aliphatic hydroxy derivative of the natural anthracycline antibiotic, daunorubicin (**7**) isolated from red pigment produced by *Streptomyces peucetius*, a soil bacteria found in Andria, Southern Italy (Minotti et al. 2004). A common feature in the chemical structures of these anticancer red pigments is a planar configuration of an aromatic chromophore attached to a six-membered sugar molecule, which suggests a possible common mechanism of cytotoxic action among them (Kapadia et al. 2011a, b). The authors' studies demonstrated synergistic cytotoxic effects of beetroot extract with doxorubicin in human prostate, pancreas, and breast cancer cells, and postulated that dietary supplementation of the extract may have potential to lower side effects associated with this widely used anticancer drug. The synergistic anticancer activity of red beetroot pigments with anthracycline cancer drugs and other plant betalain pigments, such as those from *Opuntia* cactus fruits (Castellanos-Santiago and Yahia 2008), deserves further attention. Recent *in vivo* studies with red beet juice in mice with leukemia L1250 tumors (Lukowicz et al. 2010) and in *in vitro* studies with grape seed extract in human breast cancer cells (Sharma et al. 2004) have demonstrated drug toxicity mitigation and synergistic anticancer activity, respectively, when used in combination with doxorubicin. Thus, beetroot extract may act as a cytoprotective agent capable of reducing the treatment-related toxicity of anticancer drugs (Griggs 1998). Such abatement of the therapy-related toxicity may also be achievable through reduction of the effective dosage of the potent tumor drugs when co-administered with beetroot extract.

7.9 Conclusions and Future Directions

Anticancer effects of red beetroot extract and its constituents continue to command both research and commercial interests around the globe. Approval of the extract as red food color E162 for human use has boosted its standing among the phytochemicals with potential health benefits. Its broad spectrum antitumor activity demonstrated in various animal models when administered in drinking water *ad libitum* (Kapadia et al. 1996, 2001, 2003; Lechner et al. 2010) meet the following criteria of an efficient chemopreventive agent and should qualify it to be considered as a natural chemopreventive food product: (1) sufficiently potent to provide protection at low doses, (2) effective when taken orally with regular diet, (3) easy to administer, (4) inexpensive, (5) shows prevention of multiple-organ tumor incidences or reduction of damaged cell replication from a variety of carcinogen insults, (6) shows inhibition of both initiation and promotion steps of chemical carcinogenesis, and (7) nontoxic, both on short-term and long-term usage.

Prolonged exposure to the sun is known to cause skin cancer (IARC/WHO 1997). The UV-B (280–320 nm) component of sunlight is harmful to skin and causes DNA damage (Ley and Reeve 1997). Most topically used sunscreen agents contain compounds that absorb the harmful UV-B radiation and prevent the skin DNA damage. Since beetroot extract is orally active in preventing mouse skin carcinogenesis (Kapadia et al. 2003), its use as an oral and/or topical sunscreen with potential for preventing skin cancer in humans needs to be investigated. The finding that beetroot extract prevents splenomegaly in carcinogen-treated mice needs further studies to understand its implication in cancer development and prevention. Also, the potential relationship between skin carcinogenesis and immunoglobulin-regulating activity of the extract deserves to be explored (Kuramoto et al. 1996).

The question of which constituent is responsible for the anticancer effects of red beetroot extract remains to be resolved. The current data does not rule out the presence of more than one cytotoxic constituent in the extract. Since the betalain constituents exhibit strong antioxidant activity, betanin (**1**), the principal betalain component, is being considered as a leading candidate among the potential anticancer constituents of beetroot extract (Gliszczynska-Swiglo et al. 2006; Kapadia et al.

2011a; Pedreno and Escribano 2000). Further studies are needed to confirm this assertion.

In conclusion, this review found research on anticancer effects of red beetroot pigment extract to be at its early stage with no definitive clinical data. Rapid advances are achievable through systematic studies designed to: (1) isolate, identify, and delineate mechanism(s) of action of anticancer constituent(s) of the extract, (2) conduct clinical studies to establish the efficacy of the extract or its active constituent(s) as chemopreventive and/or chemotherapeutic agent(s), and (3) study combination therapy with currently used potent anticancer drugs to achieve synergistic effect and mitigation of treatment-related drug toxicity with potential for lowering effective dosage.

Acknowledgments The authors are profoundly grateful to Prof. Harukuni Tokuda and his colleagues in Japan as well as all of Prof. Kapadia's collaborators in the US who contributed to their research on red beetroot extract cited in this review.

References

Allegra, M., P.G. Furtmuller, W. Jantschko, M. Zederbauer, L. Tesoriere, M. Livrea, and C. Obinger. 2005. Mechanism of interaction of betanin and indicaxanthin with human myeloperoxidase and hypochlorous acid. *Biochemical and Biophysical Research Communications* 332: 837–844.

Allegra, M., L. Tesoriere, and M.A. Livrea. 2007. Betanin inhibits the myeloperoxidase/nitrite-induced oxidation of human low-density lipoproteins. *Free Radical Research* 41: 367–375.

Ames, B., M.K. Shigenaga, and T.M. Hagen. 1993. Oxidants, antioxidants and the degenerative disease of aging. *Proceedings of the National Academy of Sciences of the United States of America* 90: 7915–7922.

Ames, B.N., L.S. Gold, and W.C. Willet. 1995. The causes and prevention of cancer. *Proceedings of the National Academy of Sciences of the United States of America* 92: 5258–5265.

Attoe, I., and J.H. von Elbe. 1985. Oxygen involvement in betanin degradation: Effect of antioxidants. *Journal of Food Science* 50: 106–110.

Azeredo, H.M.C. 2009. Betalains: Properties, sources, applications, and stability – A review. *International Journal of Food Science and Technology* 44: 2365–2376.

Azeredo, H.M.C., C.R.A. Santos, C.B.K. Mendes, and R.A.M. Iranilde. 2007. Betacyanin stability during processing and storage of a microencapsulated red beetroot extract. *American Journal of Food Technology* 2: 307–312.

Bobek, P., S. Galbavy, and M. Mariassyova. 2000. The effect of alimentary hypercholesterolemia and chemically induced colon carcinogenesis in rats by dimethylhydrazine. *Nahrung* 44: 184–187.

Bodley, A., L.F. Liu, M. Israel, R. Seshadri, Y. Koseki, F.C. Giuliani, S. Kirschenbaum, R. Silber, and M. Potmesil. 1989. DNA transisomerase II-mediated interaction of doxorubicin and daunorubicin congeners with DNA. *Cancer Research* 49: 5969–5978.

Boivin, D., S. Lomy, S. Lord-Dufour, J. Jackson, E. Beauliou, M. Cote, A. Mograbi, S. Barrette, D. Gingas, and R. Beliveau. 2009. Antiproliferative and antioxidant activities of common vegetables: A comparative study. *Food Chemistry* 112: 374–380.

Burner, U., P.G. Furtmuller, A.J. Kettle, W.H. Koppenol, and C. Obinger. 2000. Mechanism of reaction of myeloperoxidase with nitrite. *Journal of Biological Chemistry* 275: 20597–20601.

Cai, Y.Z., M. Sun, and H. Corke. 2003. Antioxidant activity of betalains from plants of Amaranthaceae. *Journal of Agricultural and Food Chemistry* 51: 2288–2294.

Cai, Y.Z., M. Sun, and H. Corke. 2005. Characterization and application of betalain pigments from plants of the Amaranthaceae. *Trends in Food Science and Technology* 16: 370–376.

Carlsen, M.H., B.L. Halvorsen, K. Holte, S.K. Bohn, S. Dragland, L. Sampson, C. Willey, H. Senoo, Y. Umezono, C. Sanada, I. Barikmo, N. Berthe, W.C. Willett, K.M. Phillips, D.R.

Jacobs Jr., and R. Blomhoff. 2010. The antioxidant content of more than 3100 foods, beverages, spices, herbs and supplements used worldwide. *Nutrition Journal* 9: 3.

Carvalho, C., R.X. Santos, S. Cardoso, S. Correia, and P.J. Oliveira. 2009. Doxorubicin: The good, the bad and the ugly effect. *Current Medicinal Chemistry* 16: 3267–3285.

Castellanos-Santiago, E., and E.M. Yahia. 2008. Identification and quantification of betalains from the fruits of 10 Mexican prickly pear cultivars by high-performance liquid chromatography and electronspray ionization mass spectrometry. *Journal of Agricultural and Food Chemistry* 56: 5758–5764.

Chalermchat, Y., P. Dijmek, and M. Fincan. 2004. Pulsed electric field treatment for solid-liquid extraction of red beetroot pigment: mathematical modeling of mass transfer. *Journal of Food Engineering* 64: 146–152.

Chavez-Santoscoy, R.A., J.A. Gutierrez-Uribe, and S.O. Sema-Saldivar. 2009. Phenolic composition, antioxidant capacity and in vitro cancer cell cytotoxicity of nine prickly pear (*Opuntia* spp) juices. *Plant Foods for Human Nutrition* 64: 146–152.

Cho, E., M.D. Holmes, S.E. Hankinson, and W.C. Willett. 2010. Choline and betaine intake and risk of breast cancer in post-menopausal women. *British Journal of Cancer* 10: 489–494.

Cormier, F. 1997. Food colorants from plant cell cultures. In *Recent advances in phytochemistry*, Functionality of phytochemicals, vol. 31, ed. J.T. Romeo and T. Johns, 201–222. New York: Plenum.

Degenhardt, A., and P. Winterhalter. 2001. Isolation of natural pigments by high speed CCC. *Journal of Liquid Chromatography and Related Technologies* 24: 1745–1764.

Delgado-Vargas, F., A.R. Jimenez, and O. Paredes-Lopez. 2000. Natural pigments: Carotenoids, anthocyanins, and betalains – Characteristics, biosynthesis, processing, and stability. *Critical Reviews in Food Science and Nutrition* 40: 173–289.

Detopoulou, P., D.B. Panagiotakos, S. Antonopoulou, C. Pitsavos, and C. Stefanadis. 2008. Dietary choline and betaine intakes in relation to concentrations of inflammatory markers in healthy adults: The ATTICA study. *American Journal of Clinical Nutrition* 87: 424–430.

Dorai, T., and B.B. Aggarwal. 2004. Role of chemopreventive agents in cancer therapy. *Cancer Letters* 215: 129–140.

Dornenberg, H., and D. Knorr. 1977. Challenges and opportunities of metabolite production from plant cell and tissue culture. *Food Technology* 51: 47–54.

Downham, A., and P. Collins. 2000. Coloring our foods in the last and next millennium. *International Journal of Food Science and Technology* 35: 5–22.

Drdak, M., R.C. Altamirano, A. Rajniakova, P. Simko, J. Karovicova, and D. Benkovska. 1992. Red beet pigment composition. Effects of fermentation by different strains of *Saccharomyces cerevisiae*. *Journal of Food Science* 57: 935–936.

Duthie, S.J., A. Ma, M.A. Ross, and A.R. Collins. 1996. Antioxidant supplementation decreases oxidative DNA damage in human lymphocytes. *Cancer Research* 56: 1291–1295.

Eastwood, M.A., and H. Nyhlin. 1995. Beeturia and colonic oxalic acid. *Quarterly Journal of Medicine* 88: 711–717.

Edenharder, R., P. Kurz, K. Joh, S. Burgard, and K. Seeger. 1994. In vitro effect of vegetable fruit juices on the mutagenicity of 2-amino-3-methylimidazole[4,5-f]quinoline, 2-amino-3,8-dimethylimide-3-ol[4,5-f]quinoxaline. *Food and Chemical Toxicology* 32: 443–459.

Edenharder, R., J.W. Sagr, H. Glatt, E. Muckel, and K.L. Platt. 2002. Protection by beverages, fruits, vegetables, herbs, and flavonoids against genotoxicity of 2-acetylaminofluorene and 2-amino-1-methyl-6-phenylimodazole[4,5-b]pyridine (PhIP) in metabolically competent V79 cells. *Mutation Research* 521: 57–72.

Escribano, J., M.A. Pedreno, F. Garcia-Carmona, and R. Munoz. 1998. Characterization of the anti-radical activity of betalains from *Beta vulgaris* L. roots. *Phytochemical Analysis* 9: 124–127.

Escribano, J., F. Gandia-Herrero, N. Caballero, and M.A. Pedreno. 2002. Subcellular localization of isoenzyme pattern of peroxidase and polyphenoloxidase in beetroot (*Beta vulgaris* L.). *Journal of Agricultural and Food Chemistry* 50: 6123–6129.

Foglesong, P.D., C. Rockford, and S. Swink. 1992. Doxorubicin inhibits human DNA topoisomerase I. *Cancer Chemotherapy and Pharmacology* 30: 123–125.

Fong, M., D. Chen, and C.S. Yang. 2007. Dietary polyphenols may affect DNA methylation. *Journal of Nutrition* 137: 223S–228S.

Ford-Lloyd, B.V. 1995. Sugar beet and other cultivated beets (*Beta vulgaris*). In *Evaluation of crop plants*, 2nd ed, ed. J. Smartt and N.W. Simmonds, 35–40. London: Longman Scientific.

Formari, F.A., J.J. Randolph, J.C. Yalowich, M.K. Ritke, and D.A. Gewirtz. 1994. Interference of doxorubicin with DNA unwinding in MCF-7 breast tumor cells. *Molecular Pharmacology* 45: 649–656.

Forrai, G., G. Bankovi, and D. Vagujfalvi. 1982. Betaninuria: A genetic trait? *Acta Physiologica Academiae Scientiarum Hungaricae* 59: 265–282.

Frank, T., F.C. Stintzing, R. Carle, R. Bitsch, and M. Netzel. 2005. Urinary pharmacokinetics of betalains following consumption of red beet juice in healthy humans. *Pharmacological Research* 52: 290–297.

Gabelman, W.H., I.L. Goldman, and N.D. Breitbach. 2002. High pigment beet. US Patent No 6353156.

Gandia-Herrero, F., J. Escribano, and F. Garcia-Carmona. 2009. The role of phenolic hydroxyl groups in the free radical scavenging activity of betalains. *Journal of Natural Products* 72: 1142–1146.

Gandia-Herrero, F., J. Escribano, and F. Garcia-Carmona. 2010. Structural implications on color, fluorescence, and antiradical activity in betalains. *Planta* 232: 449–460.

Gasztonyi, M.N., H. Daood, M.T. Hajos, and P. Biacs. 2001. Composition of red beet (*Beta vulgaris* var. *Conditiva*) varieties on the basis of their pigment components. *Journal of the Science of Food and Agriculture* 81: 932–933.

Gennari, L., M. Felletti, M. Blasa, C. Celeghini, A. Corallini, and P. Ninfali. 2011. Total extract of *Beta vulgaris* var. *Cicla* seeds versus its purified phenolic components: Antioxidant activities and antiproliferative effects against colon cancer cells. *Phytochemical Analysis* 22: 272–279.

Georgiev, V., M. Ilieva, T. Bley, and A. Pavlov. 2010a. Review: Betalain production in plant in vitro systems. *Acta Physiologiae Plantarum* 30: 581–593.

Georgiev, V.G., J. Weber, E.-M. Kneschke, P.N. Denev, T. Bley, and A.I. Pavlov. 2010b. Antioxidant activity and phenolic content of betalain extracts from intact plants and hairy root cultures of the red beetroot *Beta vulgaris* cv. Detroit Dark Red. *Plant Foods for Human Nutrition* 65: 105–111.

Geshner, A., U. Pastorino, S.M. Plummer, and M.M. Manson. 1998. Suppression of tumor development by substances from the diet: Mechanisms and clinical implications. *British Journal of Clinical Pharmacology* 45: 1–12.

Gliszczynska-Swiglo, A., H. Szymusiak, and P. Malinowska. 2006. Betanin, the main pigment of red beet: Molecular origin of its exceptionally high free radical-scavenging activity. *Food Additives and Contaminants* 11: 1079–1087.

Griggs, J.J. 1998. Reducing the toxicity of anticancer therapy: New strategies. *Leukemia Research* 22(Suppl 1): S27–S33.

Gutteridge, J.M., and B. Halliwell. 2000. Free radicals and antioxidants in the year 2000. A historical look to the future. *Annals of the New York Academy of Sciences* 899: 136–147.

Haber, G.J., C.T. Tan, and J. Wu. 1979. Stable beet color composition. US Patent No 4132793.

Havlikova, L., K. Mikova, and V. Kyzlink. 1983. Heat stability of betacyanins. *Zeitschrift fur Lebensmittel Untersuchung und Forshung* 177: 247–250.

Havlikova, L., K. Mikova, and V. Kyzlink. 1985. Red beet pigments as soft drink colorants. *Nahrung* 29: 723–770.

Henry, B.D. 1996. Natural food colors: Beetroot. In *Natural food colorants*, 2nd ed, ed. G.A.F. Hendry and J.D. Houghton, 59–64. Glasgow: Blackie Academic Professional.

Herbach, K.M., F.C. Stintzing, and R. Carle. 2006a. Betalain stability and degradation. Structural and chromatic aspects. *Journal of Food Science* 71: R41–R50.

Herbach, K.M., F.C. Stintz, and R. Carle. 2006b. Stability and color change of thermally treated betanin, phyllocactin and hylocerenin solutions. *Journal of Agricultural and Food Chemistry* 54: 390–398.

Hunter, C.S., and N.J. Kilby. 1990. Betanin production and release in vitro from suspension cultures of *Beta vulgaris*. *Methods in Molecular Biology* 6: 545–554.

IARC. 1997. *IARC monographs on the evaluation of carcinogenic risks to humans*, Solar and ultraviolet radiation, vol. 55. Geneva: WHO.

Jackman, R.L., and J.L. Smith. 1996. Anthocyanins and betalains. In *Natural food colorants*, 2nd ed, ed. G.A.F. Hendry and J.D. Houghton, 244–310. Glasgow: Blackie Academic Professional.

Kannan V. 2011. Extraction of bioactive compounds from whole red cabbage and beetroot using pulsed electric fields and evaluation of their functionality. Master of Science thesis in food science and technology, University of Nebraska, Lincoln. http://digitalcommons.unl.edu/foodscidiss/11. Accessed 4 Feb 2011.

Kanner, J., S. Harel, and R. Granit. 2001. Betalains – A new class of dietary cationized antioxidants. *Journal of Agricultural and Food Chemistry* 49: 5178–5185.

Kapadia, G.J., H. Tokuda, T. Konoshima, M. Takasaki, and H. Nishino. 2001. Inhibitory effect of synthetic and natural colorants on carcinogenesis. US Patent No 6284224.

Kapadia, G.J., H. Tokuda, T. Konoshima, and H. Nishino. 1996. Chemoprevention of lung and skin cancer by *Beta vulgaris* (beet) root extract. *Cancer Letters* 100: 211–214.

Kapadia, G.J., V. Balasubramanian, H. Tokuda, A. Iwashima, and H. Nishino. 1997. Inhibition of 12-*O*-tetradecanoylphorbol-13-acetate induced Epstein-Barr virus antigen activation by natural colorants. *Cancer Letters* 115: 173–178.

Kapadia, G.J., H. Tokuda, R. Sridhar, V. Balasubramanian, J. Takayasu, P. Bu, F. Enjo, M. Takasaki, T. Kinoshima, and H. Nishino. 1998. Cancer chemopreventive activity of synthetic colors used in foods, pharmaceuticals and cosmetic preparations. *Cancer Letters* 129: 87–95.

Kapadia, G.J., M.A. Azuine, R. Sridhar, Y. Okuda, A. Tsuruta, E. Ichiishi, T. Mukainake, M. Takasaki, N.H. Konoshima, and H. Tokuda. 2003. Chemoprevention of DMBA-induced UV-B promoted, NORI-induced TPA promoted skin carcinogenesis, and DEN-induced phenobarbital promoted liver tumors in mice by extract of beetroot. *Pharmacological Research* 47: 141–148.

Kapadia, G.J., M.A. Azuine, T. Konoshima, M. Takasaki, F. Enjo, H. Nishino, and H. Tokuda. 2005. Chemopreventive activity of the natural colorant, betanin against advanced glycation endproducts induced carcinogenesis. In 46th annual meeting of the American Society of Pharmacognosy, Corvallis, Abstr No P241.

Kapadia, G.J., M.A. Azuine, G.S. Rao, T. Arai, A. Iida, and H. Tokuda. 2011a. Cytotoxic effect of the red beetroot (*Beta vulgaris* L.) extract compared to doxorubicin (Adriamycin) in the human prostate (PC-3) and breast (MCF-7) cancer cell lines. *Anticancer Agents in Medicinal Chemistry* 11: 280–284.

Kapadia, G.J., G.S. Rao, A. Iida, N. Suzuki, and H. Tokuda. 2011b. Synergistic cytotoxic effect of red beetroot (*Beta vulgaris* L.) food colorant E162 with anticancer drug doxorubicin (Adriamycin) against human prostate, breast and pancreatic tumor cells. In 52nd annual meeting of the American Society of Pharmacognosy, San Diego, Abstr No P241.

Kelloff, G.J., J.A. Crowell, V.E. Steele, R.A. Lubert, W.A. Malone, C.W. Boone, L. Kopelovich, E.T. Hawk, R. Lieberman, A. Lawrence, I. Ali, J.L. Viner, and C.C. Sigman. 2000. Progress in chemoprevention: Development of diet-derived chemopreventive agents. *Journal of Nutrition* 130((2S Suppl)): 467S–471S.

Klewicka, E., A. Nowak, Z. Zdunczyk, B. Cukrowska, and J. Blasiak. 2010. Protective effect of lactofermented beetroot juice against aberrant crypt foci and genotoxicity of fecal water in rats. *Experimental and Toxicologic Pathology* 63. http://dx.doi.org/10.1016/j.etp.2010.12001

Konoshima, T., and M. Takasaki. 2003. Anticarcinogenic activities of natural pigments from beetroot and saffron. *Foods and Food Ingredients Journal of Japan* 208: 615–622.

Krantz, C., M. Monier, and B. Wahlstrom. 1980. Absorption, excretion, metabolism and cardiovascular effects of beetroot extract in the rat. *Food and Cosmetics Toxicology* 18: 363–366.

Kriznik, B., and D. Pavokovic. 2010. Enhancement of betanin yield in transformed sugar beet (*Beta vulgaris* L.). *Acta Botanica Croatia* 69: 173–182.

Kujala, T.S., J.M. Loponen, K.D. Klika, and K. Pihlaja. 2000. Phenolics and betacyanins in red beetroot (*Beta vulgaris*) root: Distribution and effect of cold storage on the content of total phenolics and three individual compounds. *Journal of Agricultural and Food Chemistry* 48: 5338–5342.

Kujala, T., K. Klika, V. Ovcharenko, J. Laponen, M. Vienola, and K. Pihlaja. 2001a. 5,5′-6,6′-tetrahydroxy-3,3′-biindolyl from beetroot (*Beta vulgaris*) peel extract. *Zeitschrift fur Naturforschung C* 56: 714–718.

Kujala, T., J. Laponen, and K. Pihlaja. 2001b. Betalains and phenolics in red beetroot (*Beta vulgaris*) peel extract. *Zeitschrift fur Naturforschung C* 56: 343–348.

Kujala, T.S., M.S. Vienola, K.D. Klika, J.M. Loponen, and K. Pihlaja. 2002. Betalain and phenolic composition of four beetroot (*Beta vulgaris*) cultivars. *European Food Research and Technology* 214: 505–510.

Kujawska, M., E. Ignatowicz, M. Murias, M. Ewertowska, K. Mikolajczyk, and J. Jodynis-Liebert. 2009. Protective effect of red beet against carbon tetrachloride- and *N*-nirosodiethylamine-induced oxidative stress in rats. *Journal of Agricultural and Food Chemistry* 57: 2570–2575.

Kuramoto, Y., K. Yamada, O. Tsuruta, and M. Sugano. 1996. Effect of natural food colorings on immunoglobulin production in vitro by rat spleen lymphocytes. *Bioscience, Biotechnology, and Biochemistry* 60: 1712–1713.

Langdon, S.P. (ed.). 2003. *Cancer cell culture: Methods and protocols*. New York: Humana Press.

Lechner, J.F., L.-S. Wang, C.M. Rocha, B. Larue, C. Henry, C.M. McIntyre, K.M. Riedl, S.J. Schwartz, and G.D. Stoner. 2010. Drinking water with red beetroot food color antagonizes esophageal carcinogenesis in *N*-nitrosomethylbenzylamine-treated rats. *Journal of Medicinal Food* 13: 1–7.

Lee, C.Y., and N.L. Smith. 1979. Blanching effect on polyphenoloxidase activity in table beets. *Journal of Food Science* 44: 82–86.

Lee, J., N. Koo, and D.B. Min. 2004. Reactive oxygen species, aging and antioxidant neutraceuticals. *Comprehensive Reviews in Food Science and Food Safety* 3: 21–33.

Lee, C.-H., M. Wattasinghe, B.W. Bolling, L.-L. Ji, and K.L. Parkin. 2005. Betalains, phase II enzyme-inducing components from red beetroot (*Beta vulgaris* L.) extracts. *Nutrition and Cancer* 53: 91–103.

Lee, J.H., C.W. Son, M.Y. Kim, M.H. Kim, H.R. Kim, E.S. Kwak, S. Kim, and M.R. Kim. 2009. Red beet (*Beta vulgaris* L.) leaf supplementation improves antioxidant status in C57BL/6 J mice fed high fat high cholesterol diet. *Nutrition Research and Practice* 3: 114–121.

Lee, J.E., E. Giovannucci, C.S. Fuchs, W.C. Willett, S.H. Zeisel, and E. Cho. 2010. Choline and betaine intake and the risk of colorectal cancer in men. *Cancer Epidemiology, Biomarkers and Prevention* 19: 884–887.

Ley, R.D., and V.E. Reeve. 1997. Chemoprevention of ultraviolet radiation-induced skin cancer. *Environmental Health Perspectives* 105(Suppl 4): 981–984.

Livrea, M.A., and L. Tesoriere. 2006. Health benefits and bioactive components of the fruits from *Opunita ficus-indica* [L.] Mill. *Journal of the Professional Association for Cactus Development* 2006: 73–90.

Lopez, N., E. Puertolas, S. Condon, J. Raso, and I. Alvarez. 2009. Enhancement of the extraction of betaine from red beetroot by pulsed electric fields. *Journal of Food Engineering* 90: 60–66.

Lu, X., Y. Wang, and Z. Zhang. 2009. Radioprotective activity of betalains from red beets in mice exposed to gamma irradiation. *European Journal of Pharmacology* 615: 223–227.

Lukowicz, J., G. Peszynska-Sularz, A. Cieslak, A. Piasek, and W. Popadiuk. 2010. Dietary intervention with red beet juice during cancer chemotherapy with doxorubicin as means of combating toxic side effects resulting from oxidative stress. www.Science24.com/paper/7039. Accessed 9 Dec 2010.

Minotti, G., P. Menna, E. Salvatorelli, and L. Gianni. 2004. Anthracylines: Molecular advances and pharmacologic developments in antitumor activity and cardiotoxicity. *Pharmacological Reviews* 56: 185–229.

Mitchell, S.C. 2001. Food idiosyncrasies. Beetroot and asparagus. *Drug Metabolism and Disposition* 29: 539–543.

Moreno, D.A., C. Garcia-Veguera, J.I. Gil, and A. Gil-Izuierdo. 2008. Betalains in the era of global agri-food science, technology and nutritional health. *Phytochemistry Reviews* 7: 261–280.

Nayak, C.A., S. Chethana, N.K. Rastogi, and K. Raghavarao. 2006. Enhanced mass transfer during solid-liquid extraction of gamma-radiated red beetroot. *Radiation Physics and Chemistry* 75: 173–178.

Netzel, M., F.C. Stintzing, D. Quaas, G. Straub, R. Carle, R. Bitsch, I. Bitsch, and T. Frank. 2005. Renal excretion of antioxidative constituents from red beet in humans. *Food Research International* 38: 1051–1058.

Ninfali, P., M. Bacchiocca, A. Antonelli, E. Biagiotti, A.M. Gioacchino, G. Piccoli, V. Stocchi, and G. Brandi. 2007. Characterization and biological activity of main flavonoids from Swiss chard (*Beta vulgaris* subspecies cycla). *Phytomedicine* 14: 216–221.

Nollet, L.M.L. 2004. *Handbook of food analysis: Physical characterization and nutrient analysis*, 852. Boca Raton: CRC Press.

Nottingham, S. 2004. Beetroot e-book- 2004. The Times, London. http://stephennottingham.co.uk. Accessed 28 Jan 2011.

O'Callaghan, M.C. 1996. Biotechnology in natural food colours: The role of bioprocessing. In *Natural food colorants*, 2nd ed, ed. G.A.F. Hendry and J.D. Houghton, 80–111. Glasgow: Blackie Academic Professional.

Pai, S.R., and P. D'Mello. 2004. Stability evaluation of beetroot colour in various pharmaceutical matrices. *Indian Journal of Pharmaceutical Sciences* 66: 696–699.

Paluszczak, J., V. Krajka-Kuzniak, and W. Baer-Dobowska. 2010. The effect of dietary polyphenols on the epigenetic regulation of gene expression on MCF7 breast cancer cells. *Toxicology Letters* 192: 119–125.

Pan, M.H., and C.T. Ho. 2008. Chemopreventive effects of natural dietary compounds on cancer development. *Chemical Society Reviews* 37: 2558–2574.

Patkai, G.Y., J. Bara, and I. Varsanyi. 1997. Decomposition of anticarcinogen factors of the beetroot during juice and nectar production. *Cancer Letters* 114: 105–106.

Pavlov, A., P. Kovatcheva, D. Tuneva, M. Ilieva, and T. Bley. 2005. Radical scavenging activity and stability of betalains from *Beta vulgaris* hairy root culture in simulated conditions of human gastrointestinal tract. *Plant Foods for Human Nutrition* 60: 43–47.

Pavokovic, D., G. Rusak, V. Besendorfer, and M. Krsnik-Rasol. 2009. Light-dependent betanin production by transformed cells of sugar beet. *Food Technology and Biotechnology* 47: 153–158.

Pedreno, M.A., and J. Escribano. 2000. Studying the oxidation and the antiradical activity of betalain from beet root. *Journal of Biological Education* 35: 49–51.

Pedreno, M.A., and J. Escribano. 2001. Correlation between antiradical activity and stability of betanine from *Beta vulgaris* L. roots under different pH, temperature and light conditions. *Journal of the Science of Food and Agriculture* 81: 627–631.

Pigram, W.J., W. Fuller, and L.D. Hamilton. 1972. Stereochemistry of interactions: Interaction of daunomycin with DNA. *Nature: New Biology* 235: 17–19.

Platt, K.L., R. Edenharder, S. Aberhold, S. Muckel, and H. Glatt. 2010. Fruits and vegetables protect against genotoxicity of heterocyclic aromatic amines activated by human xenobiotic-metabolizing enzymes expressed in immortal mammalian cells. *Mutation Research* 703: 90–98.

Pogribny, I.P., S.A. Ross, C. Wise, M. Pogribna, E.A. Jones, V.P. Tryndyak, S.J. James, Y.P. Dragan, and L.A. Poirier. 2006. Irreversible global DNA hypomethylation as a key step in hepatocarcinogenesis induced by dietary methyl deficiency. *Mutation Research* 593: 80–87.

Prahoveanu, E., V. Easanu, G. Anton, and S. Frunzulica. 1986. Prophylactic effect of a *Beta vulgaris* extract on experimental influenza infection in mice. *Virologie* 37: 121–123.

Pucher, G.W., L.C. Curtis, and H.B. Vickery. 1938. The red pigment of the root of the beet (*Beta vulgaris*) 1. The preparation of betanin. *Journal of Biological Chemistry* 123: 61–70.

Pyo, Y.-H., T.-C. Lee, L. Logendra, and R.T. Rosen. 2004. Antioxidant activity of phenolic compounds of Swiss chard (*Beta vulgaris* subspecies cycla) extracts. *Food Chemistry* 85: 19–26.

Raja, P.B., and M.G. Sethuraman. 2008. Natural products as corrosion inhibitor for metals in corrosive media – A review. *Materials Science* 62: 113–116.

Rakin, M., M. Vukasinovic, S. Silver-Marinkovic, and M. Maksimovic. 2007. Contribution of lactic acid fermentation to improve nutritive quality of vegetable juices enriched with brewer's yeast autolysate. *Food Chemistry* 100: 599–602

Ramarathnam, N., H. Ochi, and M. Takeuchi. 1997. Antioxidant defense system in vegetable extracts. In *Natural antioxidants: Chemistry, health effects, and applications*, ed. F. Shahidi, 76–87. Champaign: AOCS Press.

Reddy, M.K., R.L. Alexander-Lindo, and M.G. Nair. 2005. Relative inhibition of lipid peroxidation, cyclogenase enzymes and human tumor proliferation by natural food colors. *Journal of Agricultural and Food Chemistry* 53: 9268–9273.

Riboli, E., and T. Norat. 2003. Epidemiologic evidence of the protective effect of fruit and vegetables on cancer risk. *American Journal of Clinical Nutrition* 78: 559S–569S.

Roberts, M.F., D. Strack, and M. Wink. 2010. Biosynthesis of alkaloids and betalains. In *Annual plant reviews*, Biochemistry of plant secondary metabolites, vol. 40, 2nd ed, ed. M. Wink, 20–91. Chichester: Wiley. http://www.onlinelibrary.wiley.com. Accessed 20 Apr 2011.

Robinson, A.M., and R. Robinson. 1932. Synthetic experiments on the nature of betanin and related nitrogenous anthocyanins. Part I. *Journal of the Chemical Society* 1932: 1439–1445.

Saldanha, P.H. 1962. On the genetics of betanin excretion. *Journal of Heredity* 53: 296–298.

Saldanha, P.H., L.E. Magalhaes, and W. Horta. 1960. Race differences in the ability to excrete beetroot pigment (betanin). *Nature* 187: 806–807.

Schliemann, W., N. Kobayashi, and D. Strack. 1999. The decisive step in betaxanthin biosynthesis is a spontaneous reaction. *Plant Physiology* 119: 1217–1232.

Schwartz, S.J., J.H. von Elbe, M.W. Pariza, T. Goldsworthy, and H.C. Pitot. 1983. Inability of red beet betalain pigments to initiate or promote hepatocarcinogenesis. *Food and Chemical Toxicology* 21: 531–535.

Selvi, J.A., S. Rajendran, V.G. Sri, A.J. Amaraj, and B. Narayanaswamy. 2009. Corrosion inhibition by beetroot extract. *Portugaliae Electrochimica Acta* 27: 1–11.

Sharma, G., A.K. Tyagi, R.P. Singh, D.C. Chan, and R. Agrawal. 2004. Synergistic anti-cancer effects of grape seed extract and conventional cytotoxic agent doxorubicin against human breast carcinoma cells. *Breast Cancer Research and Treatment* 85: 1–12.

Siriwarchana, N., F. Shahidi, and Y. Jeon. 2006. Potential antioxidative effects of cactus pear fruit (*Opuntia ficus-indica*) extract of radical scavenging and DNA damage reduction in human peripheral lymphocytes. *Journal of Food Lipids* 13: 445–458.

Sivakumar, V., J.L. Anna, J. Vijayeeswarri, and G. Swaminathan. 2009. Ultrasound assisted enhancement in natural dye extraction from beetroot for industrial applications and natural dyeing of leather. *Ultrasonics Sonochemistry* 16: 782–789.

Song, W., C.M. Derito, X. Liu, X. He, M. Dong, and R.H. Liu. 2010. Cellular antioxidant activity of common vegetables. *Journal of Agricultural and Food Chemistry* 58: 6621–6629.

Sporn, M.B., and N. Suh. 2000. Chemoprevention of cancer. *Carcinogenesis* 21: 525–530.

Sreekanth, D., M.K. Arunasree, and K.R. Roy. 2007. Betanin, a betacyanin pigment purified from fruits of *Opuntia ficus-indica* induces apoptosis in human chronic myeloid leukemia cell line-K562. *Phytomedicine* 14: 739–746.

Stanner, S.A., J. Hughes, C.N. Kelly, and J. Buttriss. 2004. A review of the epidemiological evidence for the 'antioxidant hypothesis'. *Public Health Nutrition* 7: 407–422.

Staruchova, M., K. Volkova, A. Lajdova, C. Mislanova, A. Collins, L. Wsolova, L. Staruch, and M. Dusinska. 2006. Importance of diet in protection against oxidative damage. *Neuro Endocrinology Letters* 27(Suppl 2): 112–115.

Stintzing, F.C., and R. Carle. 2004. Functional properties of anthocyanins and betalains in plants and human nutrition. *Trends in Food Science and Technology* 15: 19–38.

Stintzing, F.C., and R. Carle. 2007. Betalains: Emerging prospects for food scientists. *Trends in Food Science and Technology* 18: 514–525.

Stintzing, F.C., and R. Carle. 2008a. Analysis of betalains. In *Food colorants: Chemical and functional properties*, ed. C. Socaciu, 507–520. Boca Raton: CRC Press.

Stintzing, F.C., and R. Carle. 2008b. *N*-Heterocyclic pigments: Betalains. In *Food colorants: Chemical and functional properties*, ed. C. Socaciu, 87–93. Boca Raton: CRC Press.

Strack, D., T. Vogt, and W. Schliemann. 2003. Recent advances in betalain research. *Phytochemistry* 62: 247–269.

Talalay, P. 1989. Mechanisms of induction of enzymes that protect against chemical carcinogenesis. *Advances in Enzyme Regulation* 28: 237–250.

Tesoriere, L., D. Butera, D. D'Arpa, F. DiGuadio, M. Allegra, C. Gentile, and M.A. Livrea. 2003. Increased resistance to oxidation of betalain-induced human low density lipoproteins. *Free Radical Research* 37: 689–696.

Tesoriere, L., M. Allegra, D. Butera, and M. Livrea. 2004. Absorption, excretion, and distribution of dietary antioxidant betalains in LDLs: Potential health effects in humans. *American Journal of Clinical Nutrition* 80: 941–945.

Tesoriere, L., M. Fazzari, F. Angileri, C. Gentile, and M.A. Livrea. 2008. In vitro digestion of betalainic foods. Stability and bioaccessibility of betaxanthins and betacyanins, and oxidation potential of food digesta. *Journal of Agricultural and Food Chemistry* 56: 10487–10492.

Tesoriere, L., M. Allegra, C. Gentile, and M.A. Livrea. 2009. Betacyanins as phenol antioxidants. Chemistry and mechanistic aspects of the lipoperoxyl radical-scavenging activity in solution and liposomes. *Free Radical Research* 43: 706–717.

Tokuda, H., and A. Iida. 2006. Chemopreventive activity of natural compounds against advanced glycation endproduct induced carcinogenesis. In 5th AACR international conference in cancer prevention research, Boston, Abstr No A59.

Vali, L., H. Feber, E. Stefanovits-Banyai, E. Sardi, A. Lugasi, K. Szentmihalyi, and A. Blazovics. 2006. Duodenum protecting effects of table beet (*Beta vulgaris* L. ssp. *esculata* var. *rubra*) during hepatic eschemia-reperfusion. *Acta Alimentaria* 35: 445–453.

Vali, L., E. Stefanovits-Banyai, K. Szentmihalyi, H. Febel, E. Sardi, A. Lugasi, I. Kocsis, and A. Blazovics. 2007. Liver-protecting effects of table beet (*Beta vulgaris* var. *rubra*) during ischemia-reperfusion. *Nutrition* 23: 172–178.

Vinson, J.A., Y. Hao, C. Su, and L. Zubic. 1998. Phenol antioxidant quantity and quality in foods: Vegetables. *Journal of Agricultural and Food Chemistry* 46: 3630–3634.

Vitti, M.C.D., L.K. Yamamoto, F.F. Sasaki, J.S. Aguila, R.A. Kluge, and A.P. Jacomimo. 2005. Quality of minimally processed beet roots stored in different temperatures. *Brazillian Archives of Biology and Technology* 48: 503–510.

von Elbe, JH. 2001. Betalains: Spectrophotometric determination of betacyanins and betaxanthins: In *Current Protocols in Food Analytical Chemistry,* F3.1.1-F3.1.7, New York: Wiley.

von Elbe, J.H., and S.J. Schwartz. 1981. Absence of mutagenic activity and a short-term study of beet pigments as food colorants. *Archives of Toxicology* 49: 93–98.

von Elbe, J.H., I.-Y. Maing, and C.H. Amudson. 1974. Color stability of betanin. *Journal of Food Science* 39: 334–337.

Watson, W.C., R.G. Luke, and J.A. Inall. 1963. Beeturia: Its incidence and a clue to its mechanism. *British Medical Journal* 2: 971–973.

Watts, A.R., M.S. Lennard, S.L. Mason, G.T. Tucker, and H.F. Woods. 1993. Beeturia and the biological fate of beetroot pigments. *Pharmacogenetics* 3: 302–311.

Wettasinghe, M.B., B. Bolling, L. Plhak, H. Xiao, and K. Parkin. 2002. Phase II enzyme-inducing antioxidant activities of beetroot (*Beta vulgaris* L.) extracts from phenotypes of different pigmentation. *Journal of Agricultural and Food Chemistry* 50: 6704–6709.

Winkler, C., B. Wirleitner, K. Schroecksnadel, H. Schennach, and D. Fuchs. 2005. In vitro effects of beetroot juice on stimulated and unstimulated peripheral blood mononuclear cells. *American Journal of Biochemistry and Biotechnology* 1: 180–185.

Wyler, H., U. Meuer, J. Bauer, and L. Stravs-Mombelli. 1984. Cyclodopa glucoside (= (2 S)-5-(β-D-glucopyranosyloxy)-6-hydroxyindoline-2-carboxylic acid) and its occurrence in red beet (*Beta vulgaris* var. *rubra* L.). *Helvetica Chimica Acta* 67: 1348–1355.

Xu, X., M.D. Gammon, S.H. Zeisel, P.T. Bradshaw, J.G. Wetmur, S.L. Teitelbaum, A.I. Neugut, R.M. Santella, and J. Chen. 2009. High intakes of choline and betaine reduce breast cancer mortality in a population-based study. *The FASEB Journal* 23: 4022–4028.

Yamagishi, S., K. Nakamura, H. Inoue, S. Kikuchi, and M. Takeuchi. 2005. Possible participation of advanced glycation endproducts in the pathogenesis of colorectal cancer in diabetic patients. *Medical Hypotheses* 64: 1208–1210.

Zielinska-Przyjemska, M., A. Olejnik, A. Dobrowolska-Zachwieja, and W. Grajek. 2009. In vitro effects of beetroot juice and chips on oxidative metabolism and apoptosis in neutrophils from obese individuals. *Phytotherapy Research* 23: 49–55.

Chapter 8
Anti-diabetic Potentials of Red Beet Pigments and Other Constituents

Kotamballi N. Chidambara Murthy and Shivapriya Manchali

Abstract Diabetes mellitus is a group of metabolic disorders characterized by elevated sugar levels in blood, caused by defects in insulin secretion and further signaling cascades. Among the approaches to overcome diabetes, diet plays a key role in the maintenance of blood sugar level, in the hyper-glycosylation of biomolecules associated with diverse metabolisms, and in the prevention of pathologies thus associated. In this connection, a number of bioactive molecules found in fruits, vegetables, dietary constituents, and other natural sources are being continuously explored for their direct or indirect benefits in preventing and/or management diabetes. Consumption of red beet (beetroot) is associated with numerous health benefits, attributed to its wide-ranging array of bioactive molecules. The major bioactive molecules in beet are polyphenols, flavonoids, betalains, therapeutic enzymes, ascorbic acid, and dehydroascorbic acid (DHAA). Among the phenolic compounds, betalains are the major bioactive molecules, probably due to their water-soluble nature, nitrogen content, intactness during assimilation, and stability in the circulatory system. The health benefits of red beet demonstrated by in vitro and pre-clinical studies include hypoglycemic, anti-inflammatory, antiproliferative, antitumor, antimicrobial, anti-acetylcholinesterase, antimutagenic, and lipid-lowering benefits, protection from cardiovascular disease, prevention of peripheral capillary fragility, induction of phase II enzymes activity, etc. All of these pharmacological properties

K.N.C. Murthy (✉)
Triesta Sciences, HealthCare Global Oncology Hospital, Sampangiram Nagar,
Bangalore 560 027, India

Fruit and Vegetable Improvement Centre, Texas A&M University,
College Station, TX 77843-2119, USA
e-mail: kncmurthy@gmail.com

S. Manchali
Department of Biotechnology and Crop Improvement, University of Horticultural sciences,
Bagalkot, GKVK campus, Bengaluru 560 065, India
e-mail: priyachidambara@gmail.com

B. Neelwarne (ed.), *Red Beet Biotechnology: Food and Pharmaceutical Applications*,
DOI 10.1007/978-1-4614-3458-0_8,

are said to be associated directly or indirectly with free radical-scavenging abilities of bioactive molecules, which are abundant in red beet. Extract of red beet is known to reduce glycemic value and serum lipid in diabetic animals, and similar results were found in human intervention studies. The current chapter provides an insight into the role of phytochemicals in the prevention of diabetes, with an emphasis on red beet constituents.

8.1 Incidence of Diabetics

According to the World Health Organization (WHO), diabetes mellitus, or simply diabetes, is a condition of metabolic disorder in which the fasting blood glucose level in humans is ≥7.0 mmol/l (126 mg/dl) and 2-h post-prandial plasma glucose is ≥11.1 nmol/l (200 mg/dl). It is projected that by 2020 there will be approximately 366 million people with diabetes (Wild et al. 2004). The hyperglycemic condition of people with diabetes is known to induce microvascular damage, leading to retinopathy, nephropathy, and neuropathy. Diabetes is a physiological disorder causing impaired production of insulin or an inefficiency of the body to utilize insulin. Based on the involvement of insulin, there are two types of people with diabetes, namely, insulin dependent (type I) and non-insulin dependent (type II), commonly known as IDDM and NIDDM, respectively. According to the WHO database, there are roughly more than 285 million people with diabetes, corresponding to 6.4% of the world population. This number is expected to grow to 438 million, accounting for 7.4% of the population, by 2030. The prevalence is as high as 10.2% of the population in the Western Pacific region and as low as 3.8% in the African region (http://www.worlddiabetesfoundation.org/composite-35.htm). According to the American Diabetes Association's statistics of 2011, 25.8 million children and adults have diabetes in the USA, accounting for 8.3% of the total population. Among the leading causes of diabetes, obesity, stress, physical inactivity, and diet are most important factors. Several clinical studies have demonstrated that physical exercise and diet regulation is helpful in the prevention of type II diabetes (Sartor et al. 1980; Astrup 2001). These studies suggest that *ad libitum* consumption of foods that are low in fat and high in protein, fiber, and complex carbohydrates with a low glycemic index can help in management and prevention of diabetes. In addition, a number of bioactive molecules found in diet and other botanicals are known to help in preventing both type I and type II diabetes (Dembinska-Kiec et al. 2008). An inverse relation between the consumption of fruits and vegetables and the incidence of diabetes was observed in a clinical investigation conducted in the USA, where significant prevention of diabetes in women compared with men was observed (Ford and Mokdad 2001). Another prospective study of women in the USA suggested that high intakes of green leafy and dark yellow vegetables is inversely associated with incidences of type II diabetes in overweight women (Liu et al. 2004). Based on such information, a number of bioactive molecules of fruit and vegetable origin are being studied to understand their role in the prevention of diabetes in both in vitro and in vivo models.

8.2 Red Beet- and Other Botanicals-Derived Molecules in the Control of Diabetes by Glucose Regulation

Numerous botanicals, including red beet, and their products are claimed to normalize blood glucose level in humans and prevent the complications of hyperglycemia (Fig. 8.1). Many of these and their specific extracts and concentrates are consumed as supplements in regular diet by individuals with diabetes to maintain a normal range of blood sugars. Table 8.1 lists some of the major botanicals that have been claimed to prevent diabetes and associated complications. The utilization of botanicals is either by virtue of ancient knowledge or by scientific evidence provided through modern science. Among the botanicals, gymnema (*Gymnema sylvestra*), fenugreek, bitter melon, ginseng, nopal, aloe, bilberry, and milk thistle (*Silybum marianum*) are the products being clinically evaluated for their antidiabetic efficacies (Shane-McWhorter 2001).

Consumption of 200-mg gymnema capsules by people with type I diabetes for 6–30 months resulted in a decrease in mean HbA1c value by 12%. Furthermore, in this study, there were mean decreases in the fasting blood glucose from 232 to 177 mg/dl after 6–8 months, to 150 mg/dl after 16–18 months, and to 152 mg/dl after 20–24 months, and a relative decrease in the mean insulin dose was observed(Shanmugasundaram et al. 1990). In another study of type II diabetic individuals, consumption of 400 mg of gymnema powder along with sulfonylurea resulted in a 28% decrease in the HbA1c value compared with baseline, and mean fasting blood sugar was decreased from 174 to 124 mg/dl. A decrease in blood lipid was also observed with treatment of gymnema (Baskaran et al. 1990). On similar lines, fenugreek has also been demonstrated to impart hypoglycemic effect in people with type I and type II diabetes when used at a level of 50 g/day. Biochemical analysis has demonstrated that fenugreek can act through delaying the gastric emptying,

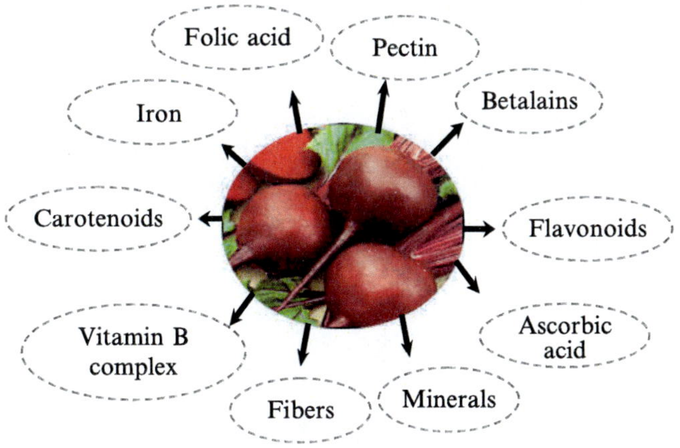

Fig. 8.1 Major class of bioactive molecules found in red beet associated with diabetes prevention and other health benefits in people with diabetes

Table 8.1 Botanicals useful for the prevention/management of diabetes

Name of botanical	Part/preparation used for study	Study design	Major anti-diabetic efficacy	Reference
Gymnema sylvestre	Water soluble acidic fraction of ethanol extract of leaves	Type II human subjects	Reduction in blood glucose, glycosylated hemoglobin, and plasma proteins	Baskaran et al. (1990)
Trigonella foenum graecum (Fenugreek)	Water-soaked whole seeds	Type II human subjects	Fall in blood and urine glucose, serum total cholesterol, LDL and VLDL cholesterol and triglyceride levels	Sharma and Raghuram (1990)
Momordica charantia (bitter gourd)	Fruits juice	Hyperglycemic individuals	Improved glucose tolerance	Welihinda et al. (1986)
Panax ginseng (Ginseng)	Powder of root	Newly diagnosed type II human subjects (double-blind placebo-controlled study)	Reduction in blood glucose and body weight, improved glycated hemoglobin, serum amino-terminal propeptide	Sotaniemi et al. (1995)
Aloe vera	Leaf phosphate-buffered saline (PBS) extract Leaf gel PBS extract	Type I and II rat models	Leaf extract was effective in both types and leaf gel extract was effective only in type II	Okyar et al. (2001)
Silybum marianum (milk thistle)	Seed extract	Type II randomized control trail	Significant decreases in HbA1c, fasting blood sugar, total cholesterol, LDL, triglyceride SGOT and SGPT levels in silymarin-treated patients compared with placebo control	Huseini et al. (2006)
28 plants selected based on the ancient knowledge from Mexico and other parts of world (Please see reference for details of plants)	Water decoctions of dried plants	Temporarily hyperglycemic rabbits	Eight plants of 28 were effective in reducing blood glucose: *Guazuma ulmifolia*, *Tournefortia hirsutissima*, *Lepechinia caulescens*, *Rhizophora mangle*, *Trigonella foenum graceum*, *Musa sapientum*, *Turnera diffusa*, and *Euphorbia prostrate*	Alarcon-Aguilara et al. (1998)
12 edible plants used in Mexico (Please see reference for details of plants)	Water decoctions of dried plants	Temporarily hyperglycemic rabbits	*Cucurbita ficifolia, Phaseolus vulgaris, Opuntia streptacantha, Spinacea oleracea, Cucumis sativus* and *Cuminum cyminum* significantly decreased the area under the glucose tolerance curve and the hyperglycemic peak. *Brassica oleracea* var. *botrytis, Allium cepa* and *Allium sativum* only decreased the hyperglycemic peak	Roman-Ramos et al. (1995)
Murraya koenigii	Leaves	STZ-induced rats	Reduction in blood glucose level	Yadav et al. (2002)
Ocimum sanctum	Aqueous extract of leaves	C57BL/KsJ db/db diabetic mice	Increase in insulin release and reduction in blood glucose	Nyarko et al. (2002)

carbohydrate absorption, and inhibition of glucose transport (Raghuram et al. 1994). Other mechanisms by which the botanicals are known to act as hypoglycemic are discrimination of sweet taste, modulation of enzyme activity responsible for glucose uptake, stimulation and increase in the number of pancreatic β-cells, increasing erythrocyte insulin receptors, improving peripheral glucose utilization, increasing tissue glucose uptake, elevation in glycogen synthesis, enhancing glucose oxidation, modulation of insulin secretion, antioxidant activity, and increasing serum insulin levels (Shane-McWhorter 2001). Table 8.2 presents a list of the hypoglycemic effects of some of the molecules commonly found in fruits and vegetables.

Stevia (*Stevia rebaudiana*) is another botanical, also known as natural sweetener, that helps in prevention of diabetes. The activity of stevia is attributed to its glycosides. One of the streptozotocin (STZ)-induced animal studies suggested that steviosides can help in diabetes through regulation of blood glucose levels by enhancing insulin secretion and insulin utilization in insulin-deficient rats. Furthermore, enhanced insulin utilization was mainly due to the ability of steviosides to decreased Phosphoenol Pyruvate Carboxy Kinase (PEPCK) gene expression in rat liver, leading to a slowing down of gluconeogenesis (Chen et al. 2005). Stevioside is known to act directly on β-cells of the pancreas to release insulin in a cyclic adenosine monophosphate (cAMP)- and ATP-independent mechanism (Jeppesen et al. 2000).

8.3 Intervention with Energy Metabolism

Both extracts and bioactive molecules of dietary constituents are known to alter energy balance though their interference with the metabolism of lipids and carbohydrates by virtue of their antioxidant and physico-chemical effects and their ability to act on metabolic enzymes. Bioactive compounds from dietary and other natural sources are endowed with an ability to help obesity through an increase in energy expenditure by inhibiting the phosphodiesterase (PDE)-induced degradation of intracellular cyclic adenosine monophosphate (cAMP). Caffeine is said to help in the prevention of obesity by increasing energy expenditure (Racotta et al. 1994). Another mode of action is by thermogenic effect through sympathetic neuronal release of norepinephrine (NE) and epinephrine, as was observed in the case of alkaloid ephedrine (Astrup et al. 1992). Other bioactive molecules known to increase the energy expenditure in pre-clinical and clinical modes are epigallocatechin gallate (EGCG, chiefly present in green tea) and capsaicin, the alkaloid of *Capsicum*. These compounds influence catecholamine secretion from the adrenal medulla and inhibit the activity level of catechol-*O*-methyl transferase (Borchardt and Huber 1975; Kawada et al. 1986). Another mechanism of diabetes prevention is through suppression of appetite by botanicals, which leads to decrease in the availability of glucose as energy, diverting toward lipid utilization seeking for energy. Steroidal glucoside, hydroxycitric acid, extracts of Ephedra, *Citrus aurantium*, *Caralluma fimbriata*, and *Phaseolus vulgaris* are some of the bioactive compounds and botanicals known to suppress the appetite (Rayalam et al. 2008). Compounds or extracts capable of

Table 8.2 Selected molecules from fruits and vegetables known for hypoglycemic effects

Molecule	Commonly found in	Study design	Major outcome of the study	Reference
Quercetin	Onion, pepper	STZ-induced animal model	Dose-dependent decreases in blood glucose, plasma lipids, and triglycerides	Vessal et al. (2003)
Rutin		STZ-induced animal model	Decreased lipid peroxidation, fasting plasma glucose, and glycosylated hemoglobin, and increased insulin	Kamalakkannan and Prince (2006)
Lycopene	Tomato, pepper	Type II human volunteers	Consumption of lycopene, tomato and tomato related products will help decrease blood glucose	Wang et al. (2006)
Anthocyanin	Grapes	STZ-induced animal model	Decreased blood and urine glucose levels, decreased lipid peroxidation	Jankowski et al. (2000)
Green tea polyphenols epicatechin (EC), epigallocatechin (EGC), epicatechin gallate (ECG), and epigallocatechin gallate (EGCG)	Green tea	Alloxan-induced rats	Increased antioxidant enzymes, reduced serum glucose and lipid peroxidation	Sabu et al. (2002)
Terpenoid-type quinones	*Pycnanthus angolensis*	Type II diabetic mouse	Increased plasma insulin content, lower plasma glucose concentration	Luo et al. (1999)
Corosolic acid	*Lagerstroemia speciosa* L. leaf	Animal and human clinical trails	Acts similar to insulin, lower blood glucose	Sivakumar et al. (2009)
Phyllanthin	*Phyllanthus niruri* L	Alloxan-induced rats	Dose-dependent reduction in blood glucose, cholesterol, and triglycerides	Okoli et al. (2010)
Curcumin	Rhizomes of *Curcuma* Sp	STZ-induced animal model	Reduced blood glucose and cholesterol, improved renal function	Suresh Babu and Srinivasan (1995)
[6]-gingerol	Ginger	db/db mice	Reduction in blood glucose, cholesterol, triglycerides, free fatty acid, c-LDL	Singh et al. (2009)
Resveratrol	Grapes and groundnuts	STZ-induced animal model	Reduced blood glucose, lipid peroxidation product MDA, and copper concentrations; increased catalase activity, nitric oxide level, and zinc concentrations	Aribal-Kocatürk et al. (2007)
Berberine	Phylantus and other plants	Alloxan-induced rats	Significantly decreased fasting blood glucose levels, serum content of TC, TG, LDL-c, and effectively increased HDL-c, and NO levels in diabetic rats. Increased SOD and GSH-px levels in diabetic rats	Tang et al. (2006)
		Type II human subjects	Lowered fasting blood glucose, postprandial blood glucose, hemoglobin A_{1c}; decreased total cholesterol and LDL cholesterol	Yin et al. (2008)

blocking the breakdown of starch through inhibiting the activity of salivary amylase and pancreatic amylase, known as starch blockers, are also capable of reducing carbohydrate utilization and help in the prevention of obesity and the symptoms of diabetes (Bo-Linn et al. 1982). Botanicals are also known to decrease glucose sequestration and increase lipid metabolism by functioning through adipocyte-specific and non-specific mechanisms through their interventions in the process of adipogenesis and lypolysis. Bioactive compounds like polyphenols, plant sterols, organosulfur compounds, and extracts of garlic are known to increase lipid metabolism in animals and human subjects (Rayalam et al. 2008).

8.4 Suppression of α-/β-Glycosidase in the Intestine

Glycosidases are class of enzymes responsible for the hydrolytic breakdown of glycosides, mainly complex molecules, like starch and glycogen, resulting in the formation of glucose, a direct source of metabolic energy. Compounds/extracts capable of inhibiting the activity of glycosidases are known to prevent breakdown of complex molecules, leading to reduced glucose concentrations in blood. The activity of both α- and β-glycosidase are closely associated in diabetes. Nine commonly edible plants of Thailand, namely Beijing Grass, Sweet leaf, Pennywort, Safflower, Gingko, Cat's Whiskers, Senna, Grape seed, and Jiaogulan were screened for their for efficiency of inhibiting α-glycosidase, and all of the extracts demonstrated significant inhibition of enzyme activity. The activity of these extracts is attributed to their polyphenol, including flavonoids and tannins, content (Adisakwattana et al. 2010). Baicalein (5,6,7-trihydroxyflavone) is a flavone responsible for the α-glycosidase inhibition activity of *Scutellaria baicalensis*, *Rheum officinale*, and *Paeonia suffruticosa* plant extracts in rat intestinal sucrase (Nishioka et al. 1998).

8.5 Red Beet and Diabetes

8.5.1 History of Use of Red Beet for Hypoglycemic Effect

The ancestor of today's beets, sea beet, is said to have originated from the Mediterranean Sea. The red beet is believed to have originated from the third dynasty Saqqara pyramid at Thebes, Egypt, and four chard beetroots were found in the Neolithic site of Aartswoud in the Netherlands. According to Zohary and Hopf, the earlier written documents on red beet come from the eighth century BC (Zohary and Hopf 2000). During the sixteenth century, both red and white beet juices and leaves were used as cleansing agents for the liver and spleen. Information available in the literature explains the ability of red beet juice to prevent headache and toothache (Crellin and Philpotte 1989).

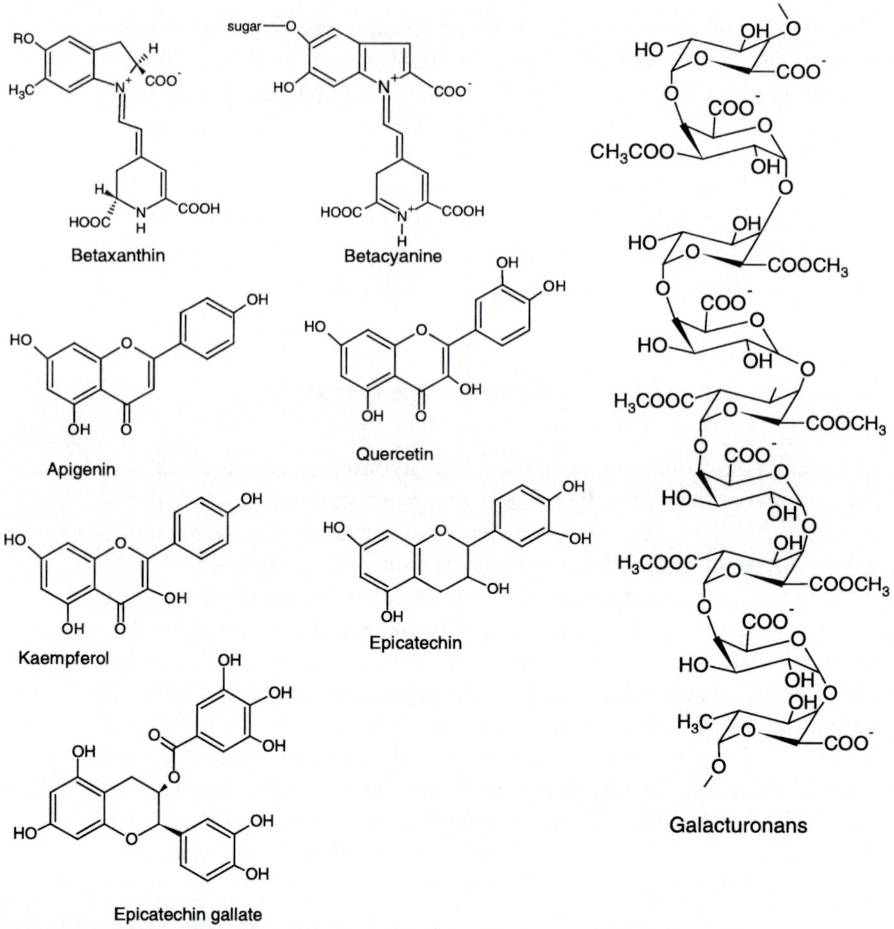

Fig. 8.2 Chemical structure of selected bioactive molecules of red beet responsible for the prevention/management of diabetes

8.5.2 Major Bioactive Molecules of Red Beet Responsible for Its Biological Activity

The major bioactive molecules responsible for the anti-diabetic and other biological activities of red beet are betalains, flavonoids, pectin, and fibers. Beet root is known to contain 20–25 mg/g (on a dry weight basis) of polyphenols (Gallic acid equivalent), 10–15 mg/g (dry weight) of betalains (Kujala et al. 2002), and $2.5 \pm 0.3\%$ fiber (Jenkins et al. 1981), and the flavonoids content of red beet is up to 50 mg/100 g of edible portion (Bhagwat et al. 2003). Chemical structures of the major health beneficial compounds found in red beet are shown in Fig. 8.2.

Betalains are the major pigments of red beet, and include both betacyanins and betaxanthins. Betacyanins (red–violet pigments) consist of red color-imparting pigments, such as betanin, betanidine, prebetanin, isobetanin, and neobetanin. Chemically, betalains are immonium conjugates of betalamic acid with *cyclo*-DOPA and amino acids. Betanidin and indicaxanthin are the first betalains identified by chemical means. The major enzymes involved in the biosynthesis of betalains in plants are polyphenoloxidase (PPO)-type tyrosinase and extradiolic DOPA dioxygenase, betalain-specific glucosyl and hydroxycinnamoyl transferases (Roberts and Strack 1999). Betacyanins have recently been extensively researched due to their high hue color for use as a natural source for food applications, and red beet is the major source of betalains, although a few other sources, such as cacti fruit, may also be useful in prevention of diabetes.

Betaxanthins are conjugates of betalamic acid with amino acids, which are yellow to orange in color and found in most of the Caryophyllales order of plants. Betaxanthin is the major pigment responsible for the fluorescence of red beet color (Gandía-Herrero et al. 2005). The biosynthesis of the pigments in red beet is presented in Chap. 2 of this book and elsewhere (Vincent and Scholz 1978; Tanaka et al. 2008). The content of betaxanthin in red beet varies from 9.6 to 21.9 mg/100 g of beet and the ratio of betaxanthin to betacyanin ranges from 0.41 to 0.66 (Sapers and Hornstein 1979).

Flavonoids are major biologically active phytochemicals in red beet, consisting mainly of flavones, flavonols, and flavan-3-ols. Apigenin and luteolin are the major flavones; quercetin, myricitin, and kaempferol are the major flavonols, and (−)-epicatechin and (−)-epicatechin gallate are the other flavonols found in red beet (Bhagwat et al. 2003). The presence of two major biologically active flavonoids, quercetin and kaempferol, with a mean content of <1 and <2 mg/kg, respectively, has been reported in red beet (Hertog et al. 1992). Flavonoids namely, vitexin-2″*O*-rhamnoside, its demethylated derivative, 2″-xylosylvitexin, isorhamnetin 3-gentiobioside, and rutin were identified in extracts of red beet (Ninfali et al. 2007).

Pectin is a structural heteropolysaccharide found in most of the plants cell walls and contains 1,4-linked α-D-galactosyluronic acid residues. The major three pectic polysaccharides characterized from the alcohol-soluble fraction of red beet are homogalacturonans, substituted galacturonans, and rhamnogalacturonans. The major pectins found in red beet are galacturonans, and breakage of these has led to formation of glucose, galactose, mannose, rhamnose, arabinose, and xylose as major monosaccharides (Dongowski 2001). Another paper from this research group reported on the pectins from alcohol-insoluble fractions of red beet, which consist of galacturonic acid, rhamnose, and final residues of sugars as apiose, 2-*O*-methyl-xylose, 2-*O*-methyl-fucose, 3,4-linked fucose, 2,3,4-linked rhamnose, 2-linked glucuronic acid, aceric acid, 3-deoxy-D-manno-2-octulosonic acid (Kdo), and 3-deoxy-D-lyxo-2-heptulosonic acid (DHA), which are characteristic of rhamnogalacturonan II (RG-II), containing two major groups with molecular weights of 4,100 and 4,300 kDa (Strasser and Amadò 2002).

8.5.3 Diabetes-Associated Biological Activities of Red Beet and Its Major Bioactive Molecules

Betalains are known for a number of health benefits, including antioxidant activity (Cao et al. 1996), anti-inflammatory activity (Reddy et al. 2005), and induction of apoptosis (Sreekanth et al. 2007). There are numerous reports on the radical scavenging ability of betalains. Extracts from garden red beets and cultured hairy roots have shown significant radical scavenging activities as measured by the 2,2-dyphenyl-1-picrylhydrazyl (DPPH) radical and oxygen radical absorbance capacity (ORAC) assay (Georgiev et al. 2010). Betanin has shown inhibition of lipid hydroperoxide production in human low-density lipoprotein (LDL) subjected to an MPO/nitrite-induced oxidation. This ability of betanin is attributed to its radical scavenging activity and betanin is known to inhibit hydroperoxides through inhibition of nitrogen dioxide and lipoperoxyl radical-scavenging activity (Allegra et al. 2007). Supplementing with 8% freeze-dried red beet leaf for 4 weeks demonstrated its extensive antioxidant property and protection against DNA damage in C57BL/6 J mice fed a high-fat and high-cholesterol diet. Elevated levels of HDL and decreased triglycerides were observed in subjects with red beet leaf feeding in comparison with control subjects. Furthermore, lipid peroxidation was also found to be decreased with red beet leaf supplementation to a high extent, of up to 50% in plasma and 20–40% in other vital organs. Because of the higher oxidative and lipoperoxidative risk status of people with diabetes than people without diabetes (Lee et al. 2009), these anti-oxidant properties of red beet are of great relevance to diabetes.

The effect of red beet on oxidative stress was also demonstrated in a chemically induced in vivo study. Treatment with 8 ml/kg/day of red beet juice for 4 weeks followed by challenge with *N*-nitrosodiethylamine (NDEA) or CCl_4 was found to protect animals from oxidative stress and related damage. The key markers measured in these studies were lipid peroxidation and activity of antioxidant enzymes (catalase, SOD, glutathione peroxidase, and glutathione reductase). Additionally, treatment with red beet juice has also protected leukocyte DNA against damage, suggesting multiple roles for red beet phytochemicals (Kujawska et al. 2009).

There is a direct association between the phase II detoxification enzymes and diabetes, as evidenced from impaired levels of glutathione peroxidase (GPx), quinine reductase (QR), and other enzymes in animals models of diabetes and -individuals with diabetes (Aydına et al. 2001). Betalains are also known to induce phase II detoxification enzymes in cultured cells. Aqueous (5%) ethanol extracts of freeze-dried *Beta vulgaris* L. has shown an up to 4-fold increase in QR activity. The aqueous extract was found to be a more effective inducer of QR compared with acetonitrile extract, suggesting that the water-soluble betalains are more effective inducers of phase II enzymes (Wettasinghe et al. 2002).

Red beet extracts have also shown anticancer properties in both cell culture and in vivo studies, as discussed in detail in Chap. 7 of this book. However, a few major anti-cancer functionalities are mentioned here. Administration of 0.025% red beet juice for several weeks followed by administration of 7,12-dimethylbenz (a) anthracene (DMBA) and exposure to 3,430 J/m^2 of ultraviolet light B (UV-B) significantly protected animals from skin cancer. Protection was also observed against other chemical inducers, namely, (±)-(E)-4-methyl-2-[(E)-hydroxyamino]-5-nitro-6-methoxy-3-hexanamide (NOR-1) in 100 µl of acetone and 12-*O*-tetradecanoylphorbol-13-acetate (TPA) (Kapadia et al. 2003). The extracts also showed inhibition of lung cancer induced by 4-nitroquinoline l-oxide (4NQO) (Kapadia et al. 1996). Flavonoids namely, vitexin-2″ *O*-rhamnoside, its demethylated derivative 2″-xylosylvitexin, isorhamnetin 3-gentiobioside, and rutin have inhibited human estrogen-positive breast (MCF-7) cancer cell proliferation through induction of apoptosis. Further study on DNA synthesis by these flavonoids has shown that vitexin-2″ *O*-rhamnoside strongly inhibited DNA synthesis and 2″-xylosylvitexin and isorhamnetin 3-gentiobioside activated DNA. The net effect of these compounds are said to be responsible for inhibition of MCF-7 cells (Ninfali et al. 2007).

8.6 Red Beet and Diabetes

8.6.1 Anti-diabetic Properties of Red Beet Pigments and Whole Root

Traditionally red beet is used as a hypoglycemic agent in Turkey and other countries. Red beet, popularly known as chard, has been traditionally used as medicine by people with diabetes. Treatment by chard (2 g/kg extract for 28 days) has been shown to increase the number of beta-cells and secretary granules in islet of Langerhans in STZ-induced mammals. Interestingly, the extract did not show any effect on the body weight and blood sugar of normal animals (Bolkent et al. 2000). Daily consumption of 100–200 g of red beet along with horseradish was found to be helpful in the drastic reduction of blood sugar in people with diabetes (Josef et al. 1973). Red beet is also known to increase the glucose tolerance and exhibit hypoglycemic effect only in people with diabetes, without affecting the glucose of healthy subjects. The extract (1 mg/kg) has reduced blood glucose through regeneration of beta-cells in STZ-induced hyperglycemic animals (Feugang et al. 2006). Treatment of four compounds isolated from red beet, namely, betavulgaroside I, II, III, and IV, when administered to male Wistar rats at 5 ml/kg after 30 min of glucose (0.5 g/kg) resulted in significant hypoglycemic activity. The relative activity strengths were betavulgaroside II > betavulgaroside IV > betavulgaroside III; betavulgaroside I did not show activity (Yoshikawa et al. 1996). Cactus pear (*Opuntia* spp.), rich in betalains, which is also the major pigment of red beet, have been traditionally used to treat people with diabetes in Mexico and Italy. Both animal and human studies have demonstrated an anti-diabetic role of Cactus pear. In an animal study, treatment with extract of cactus pear was found to be effective in reducing blood sugar compared to insulin treatment (Feugang et al. 2006; Butterweck et al. 2011).

One of the red beet cultivars, Detroit red, has shown reduction of fatty liver symptoms in a chemically induced diabetic animal model. Feeding with 2 g/kg of beet power for 10 days resulted in significant changes in serum cholesterol levels, and alkaline phosphatase and alanine-aminotransferase activities. This activity of red beet was attributed to betanin and the high polyphenol content of the beet (Sárdi et al. 2009). An in vivo feeding study on ALS strain mice demonstrated an anti-diabetic property of red beet. In this study, red beet solution was fed to mice with and without induction of diabetes using alloxan. Treatment using 36 ppm of red beet extract prevented induction of diabetes by alloxan in animals with ED_{50} of 19 and 76.4, for female and male, respectively. Furthermore, feeding of 36 ppm red beet extract for 3–18 weeks also resulted in a delayed onset of diabetes (Yamashita and Sato 2008).

Extracts of red beet were found to decrease the extent of diabetes by increasing the levels of liver protection markers. Treatment with 2-g extracts/kg body weight for 28 days followed by induction of diabetes using STZ resulted in increases in the levels of serum alanine and aspartate transaminase and alkaline phosphatase activities, total lipid levels, and sialic acid and uric acid levels. In addition, red beet extract also decreased blood glucose level, liver lipid peroxidase (LPO), and non-enzymatic glycosylation (NEG), demonstrating the anti-diabetic potential of its constituents (Ozsoy-Sacan et al. 2004). In vitro screening of red beet extract for angiotensin-converting enzyme (ACE) inhibition demonstrated the ability of extracts to inhibit this enzyme, suggesting a possible role in reducing hypertension and other ACE-related cardiovascular health problems (Inoue et al. 2011). Cyanidin 3-glucoside, found in red beet, is known to down-regulate retinol binding protein 4 (RBP4) expression, which is known to reduce the sensitivity to insulin in diabetic mice (Sasaki et al. 2007).

In animals fed high-fat diets, feeding with cellulose and fiber isolated from red beet was found to significantly reduce serum cholesterol, triglycerides, and aortic cholesterol. In addition, it has also reduced symptoms of dimethylhydrazine-induced colon cancer, suggesting the multifaceted benefits of red beet constituents (Bobek et al. 2000).

8.6.2 Anti-diabetic Properties of Red Beet Flavonoids

Apigenin and luteolin are the major flavones; quercetin, myricitin, kaempferol are the flavonols; and (–)-epicatechin and (–)-epicatechin gallate are the flavonols found in red beet. Apigenin has been implicated in the regulation of hyperglycemia, lipid peroxidation, and thyroid dysfunction in an alloxan-induced animal model. Treatment of 0.78 mg/kg of apigenin for 10 days showed increased serum insulin content and decreased content of thyroxin (T_4) and triiodothyronine (T_3). Furthermore, alloxan-induced higher serum cholesterol, decreased activity of cellular antioxidants, such as catalase (CAT) and superoxide dismutase (SOD), and

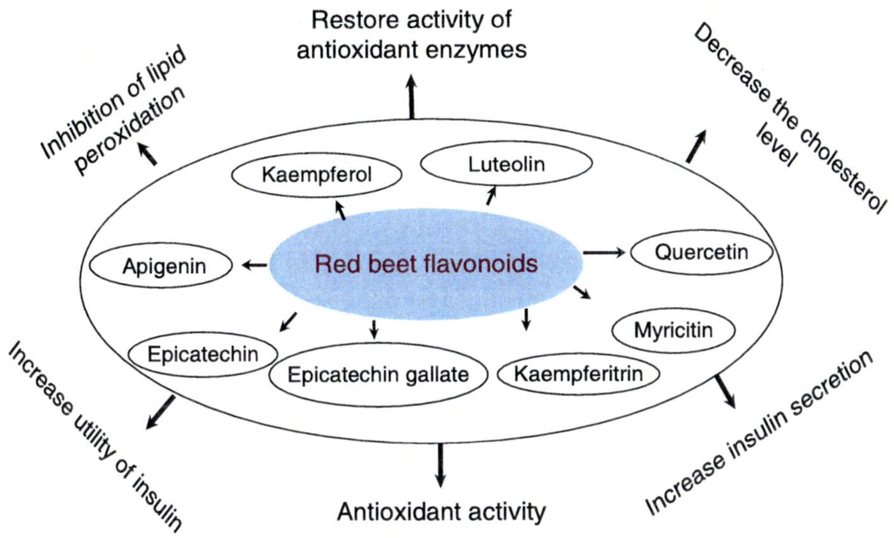

Fig. 8.3 Major flavonoids found in red beet and their mode of supporting people with diabetes

glutathione (GSH) content were reversed by apigenin, suggesting the anti-diabetic potential of the compound (Panda and Kar 2007). Treatment with 2 mg/kg of luteolin 5-rutinoside for 20 days resulted in increased pancreatic insulin and DNA content. Treatment with luteolin did not influence the plasma insulin level, but reduced glycemia by 22%. Furthermore, treatment of luteolin with glibenclamide (1.0 mg/kg) significantly reduced glycemia and serum insulin levels by 2.5 fold, suggesting synergistic activity (Zarzuelo et al. 1996). Quercetin, the other flavone present in red beet, is also known for anti-diabetic activity. Feeding of 1.0 g/kg of quercetin to STZ-induced diabetic rats up to 6 weeks resulted in significant reductions in blood sugar levels and sugar excretion in urine (Shetty et al. 2004). Measurement of lipid peroxidation, antioxidant defense is said to be one of the possible mode of action (Sanders et al. 2001). However, the exact mechanism of action is not clear. Kaempferitrin, chemically, kaempferol-3, 7-*O*-(α)-dirhamnoside, has been shown to have antioxidant and hypoglycemic effects both in vitro and in vivo. The effects were found to be significant at 50 mg/kg kaempferitrin. Surprisingly, high doses (200 mg/kg) caused hypoglycemia in both healthy and alloxan-induced diabetic rats (de Sousa et al. 2004). Figure 8.3 illustrates the major flavonoids found in red beet and the modes of action to help prevention of diabetes.

Epicatechin treatment has resulted in regeneration of pancreatic cells in diabetic animals. Treatment with 30 mg/kg of epicatechin for 4–5 days brought the sugar level down to normal and the regeneration of beta cells in islet region of pancreas was similar to the control group (Chakravarthy et al. 1982). However, a later study using both STZ-induced diabetic rats and the spontaneously diabetic BB/E rat model showed that epicatechin failed to halt the progress of disease in either model and a caution was given for consideration of the results (Bone et al. 1985). A clinical study conducted later showed that (–)-epicatechin can act similar to insulin in controlling diabetes by increasing the acetylcholinesterase (AChE) at a concentration of 1 mM (Rizvi and Zaid 2001). (–)-Epicatechin gallate, another flavonol of red beet, has been shown to prevent oxidative stress in the pancreas of diabetes-induced animals. Treatment with 5 mg/kg/day for 4 days resulted in recovery of loss of islet cells and insulin immunoreactivity in beta cells. Results showed the efficiency of the compound was <100 nM (Yun et al. 2006), indicating potential for applications in diabetes and pre-diabetes. Similar observations were made in studies with type II diabetes in experimental animals (Wolfram et al. 2006). In addition, epicatechin gallate from tea is known to protect animals from oxidation induced by secondary damages of diabetes (Rizvi et al. 2005).

8.6.3 *Anti-diabetic Properties of Red Beet Fiber*

Fibers, the complex carbohydrates, have been repeatedly found to play multiple roles in the prevention of diabetes. Fibers, due to their physico-chemical nature, act by interfering with the microenvironment of the gut. The presence of dietary fiber in the gut is known to decrease the uptake of nutrients in the intestine, and soluble fibers are known to form a gel/viscous solution, resulting in slow and steady absorption of nutrients. Fiber is also known to reduce the transit time in the intestine, resulting in greater frequency of defecation. In addition, fiber helps in the maintenance of gut microflora (Brownlee 2011). A clinical study has shown that high consumption of dietary fibers rich in soluble fractions will help in glycemic control and decrease hyperinsulinemia and lower plasma lipid concentrations in people with type II diabetes (Chandalia et al. 2000).

A 6-year follow-up cohort study of non-insulin-dependent women with diabetes suggested that a diet with a high glycemic load low in fiber increased the risk of diabetes (Salmerón et al. 1997b). Similar observations were made from another study consisting of 42,759 men, of which, 523 were reported to have diabetes at the end of 6 years (Salmerón et al. 1997a). High fiber consumption, rich in the soluble type, was demonstrated to improve glycemic control, decrease hyperinsulinemia, and lower blood lipids (decrease in cholesterol by 6% and triglyceride by 10%) (Chandalia et al. 2000).

8.7 Benefits of Red Beet and Its Constituents in Prevention of Toxicity

Extracts of red beet rich in betalains has been shown to protect against gamma radiation (^{60}Co gamma, 6.0 Gy, 1.5 Gy min^{-1}) in experimental animals. The exact mechanism is unclear, however, the protection may be attributed to antioxidant activity because the antioxidant activity was restored with treatment with extracts at different concentrations (Lu et al. 2009). The antioxidant enzymes measured in the study were catalase, lipid peroxidase, SOD, and glutathione peroxidase.

A study to evaluate betalain (fermented solution of betalain) safety in hepatectomized rats that were treated with *N*-nitrosodiethylamine (NDEA) demonstrated that pigments of red beet fed at 100 ppm for a period of 8 months neither initiated nor promoted chemically induced hepatocarcinoma (Schwartz et al. 1983). Additionally, treatment of table red beet and freeze-dried red beet (2 g/kg/day) for 10 days resulted in liver protection in an 45-min ischemia–reperfusion injury model. Treatment with red beet significantly increased the antioxidant status. The parameters measured in the treated animals were H-donating ability, reducing power, and free SH group concentration (global parameters); and glutathione peroxidase and superoxide dismutase content (Váli et al. 2007). Another animal feeding study, in which freeze-dried solvent extracts of seven fruits and vegetables (red beet being one of the vegetables) were fed to animals challenged with cyclophosphamide and benzo[*a*]pyrene showed that extracts are capable of protecting animals against chemical clastogenicity (Edenharder et al. 1998).

Other, less explored, health benefits of red beet include support of sex hormone formation due to the presence of boron, and mood modification due to the presence of betaine. However, the exact mechanisms of these effects is not clear (Nottingham 2004).

8.8 Concluding Remarks

Bioactive molecules and chemical constituents of red beet vegetables have the ability to reduce diabetes and associated complications. This fact is well supported by in vitro, pre-clinical, and clinical investigations using various vegetables and fruits. Among the bioactive compounds in the diet, polyphenols, terpenoids, and sterols play major roles in the reduction of serum glucose levels. The dietary constituents and the bioactive molecules are known to prevent diabetes mainly by increases in insulin production/sensitivity, increases in energy metabolism, alerting the activity of digestive/metabolic enzymes, activation of detoxification (phase II) enzymes, and decreasing LDL, cholesterol, and triglyceride levels. Red beet is known to help in preventing diabetes by its fiber, flavonoids, betacyanin, betaxanthin, pectin, and other cell compositions. The antioxidant abilities of bioactive molecules in red beet and other botanicals are said to be the key mode of action. Although research on direct association of beet root per se is scanty, the convincing information on anti-diabetic properties of many botanicals similar to beet, and the existing scientific knowledge on their association with diabetes is convincing for its functionalities. The fact that red beet and its juice help in the reduction of blood pressure through NO formation and mood elevation through its DOPA and NO actions (see Chap. 1) also support that this vegetable imparts an overall health improvement status to the individual who consumes it. Further research toward understanding the role of individual constituents and their combined effects on people with diabetes and related pathologies needs to be pursued both using clinical and pre-clinical models, and may be of great importance for efficient utilization of this wonderful vegetable for natural management of diabetes.

References

Adisakwattana, S., P. Jiphimai, et al. 2010. Evaluation of α-glucosidase, α-amylase and protein glycation inhibitory activities of edible plants. *International Journal of Food Sciences and Nutrition* 61(3): 295–305.

Alarcon-Aguilara, F.J., R. Roman-Ramos, et al. 1998. Study of the anti-hyperglycemic effect of plants used as antidiabetics. *Journal of Ethnopharmacology* 61(2): 101–109.

Allegra, M., L. Tesoriere, et al. 2007. Betanin inhibits the myeloperoxidase/nitrite-induced oxidation of human low-density lipoproteins. *Free Radical Research* 41(3): 335–341.

Aribal-Kocatürk, P., G. Özelçi Kavas, et al. 2007. Pretreatment effect of resveratrol on streptozotocin-induced diabetes in rats. *Biological Trace Element Research* 118(3): 244–249.

Astrup, A. 2001. Healthy lifestyles in Europe: Prevention of obesity and type II diabetes by diet and physical activity. *Public Health Nutrition* 4(2b): 499–515.

Astrup, A., L. Breum, et al. 1992. The effect and safety of an ephedrine/caffeine compound compared to ephedrine, caffeine and placebo in obese subjects on an energy restricted diet. A double blind trial. *International journal of obesity and related metabolic disorders/journal of the International Association for the Study of Obesity* 16(4): 269.

Aydına, A., H. Orhanb, et al. 2001. Oxidative stress and nitric oxide related parameters in type II diabetes mellitus: Effects of glycemic control. *Clinical Biochemistry* 34(1): 65–70.

Baskaran, K., B.K. Ahamath, et al. 1990. Antidiabetic effect of a leaf extract from *Gymnema sylvestre* in non-insulin-dependent diabetes mellitus patients. *Journal of Ethnopharmacology* 30(3): 295–305.

Bhagwat, S., and J. Holden., et al. 2003. USDA database for the flavonoid content of selected foods.

Bobek, P., Š. Galbavý, et al. 2000. The effect of red beet (*Beta vulgaris* var. rubra) fiber on alimentary hypercholesterolemia and chemically induced colon carcinogenesis in rats. *Food/Nahrung* 44(3): 184–187.

Bo-Linn, G.W., C.A.S. Ana, et al. 1982. Starch blockers-their effect on calorie absorption from a high-starch meal. *The New England Journal of Medicine* 307(23): 1413–1416.

Bolkent, S., R. Yanardag, et al. 2000. Effects of chard (*Beta vulgaris L.* var. cicla) extract on pancreatic B cells in streptozotocin-diabetic rats: A morphological and biochemical study. *Journal of Ethnopharmacology* 73(1–2): 251–259.

Bone, A.J., C.S.T. Hii, et al. 1985. Assessment of the antidiabetic activity of epicatechin in streptozotocin-diabetic and spontaneously diabetic BB/E rats. *Bioscience Reports* 5(3): 215–221.

Borchardt, R.T., and J.A. Huber. 1975. Catechol O-methyltransferase. 5. Structure-activity relations for inhibition by flavonoids. *Journal of Medicinal Chemistry* 18(1): 120–122.

Brownlee, I.A. 2011. The physiological roles of dietary fibre. *Food Hydrocolloids* 25(2): 238–250.

Butterweck, V., L. Semlin, et al. 2011. Comparative evaluation of two different *Opuntia* ficus-indica extracts for blood sugar lowering effects in rats. *Phytotherapy Research* 25(3): 370–375.

Cao, G., E. Sofic, et al. 1996. Antioxidant capacity of tea and common vegetables. *Journal of Agricultural and Food Chemistry* 44(11): 3426–3431.

Chakravarthy, B.K., S. Gupta, et al. 1982. Functional beta cell regeneration in the islets of pancreas in alloxan induced diabetic rats by (–)-epicatechin. *Life Sciences* 31(24): 2693–2697.

Chandalia, M., A. Garg, et al. 2000. Beneficial effects of high dietary fiber intake in patients with type 2 diabetes mellitus. *The New England Journal of Medicine* 342(19): 1392–1398.

Chen, T.-H., S.-C. Chen, et al. 2005. Mechanism of the hypoglycemic effect of stevioside, a glycoside of *Stevia rebaudiana*. *Planta Medica* 71(2): 108–113.

Crellin, J.K., and J. Philpotte (eds.). 1989. *A referance guide to Medicinal plants. Herbal medicine past and present*. Duke: Duke University Press.

de Sousa, E., L. Zanatta, et al. 2004. Hypoglycemic effect and antioxidant potential of kaempferol-3,7-O-(α)-dirhamnoside from *Bauhinia forficata* leaves. *Journal of Natural Products* 67(5): 829–832.

Dembinska-Kiec, A., O. Mykkänen, et al. 2008. Antioxidant phytochemicals against type 2 diabetes. *British Journal of Nutrition* 99(E-S1): ES109–ES117.

Dongowski, G. 2001. Enzymatic degradation studies of pectin and cellulose from red beets. *Food/Nahrung* 45(5): 324–331.

Edenharder, R., J. Frangart, et al. 1998. Protective effects of fruits and vegetables against in vivo clastogenicity of cyclophosphamide or benzo[a]pyrene in mice. *Food and Chemical Toxicology* 36(8): 637–645.

Feugang, J.M., P. Konarski, et al. 2006. Nutritional and medicinal use of cactus pear (*Opuntia* spp.) cladodes and fruits. *Frontiers in Bioscience* 11: 2574–2589.

Ford, E.S., and A.H. Mokdad. 2001. Fruit and vegetable consumption and diabetes mellitus incidence among U.S. adults. *Preventive Medicine* 32(1): 33–39.

Gandía-Herrero, F., J. Escribano, et al. 2005. Betaxanthins as pigments responsible for visible fluorescence in flowers. *Planta* 222(4): 586–593.

Georgiev, V., J. Weber, et al. 2010. Antioxidant activity and phenolic content of betalain extracts from intact plants and hairy root cultures of the red beetroot *Beta vulgaris* cv. detroit dark red. *Plant Foods for Human Nutrition* 65(2): 105–111.

Hertog, M.G.L., P.C.H. Hollman, et al. 1992. Content of potentially anticarcinogenic flavonoids of 28 vegetables and 9 fruits commonly consumed in the Netherlands. *Journal of Agricultural and Food Chemistry* 40(12): 2379–2383.

Huseini, H.F., B. Larijani, et al. 2006. The efficacy of *Silybum marianum* (L.) Gaertn. (silymarin) in the treatment of type II diabetes: A randomized, double-blind, placebo-controlled, clinical trial. *Phytotherapy Research* 20(12): 1036–1039.

Inoue, K., M. Kitade, et al. 2011. Screening assay of angiotensin-converting enzyme inhibitory activity from complex natural colourants and foods using high-throughput LC-MS/MS. *Food Chemistry* 126(4): 1909–1915.

Jankowski, A., B. Jankowska, et al. 2000. The effect of anthocyanin dye from grapes on experimental diabetes. *Folia Medica Cracoviensia* 41(3–4): 5.

Jenkins, D., T. Wolever, et al. 1981. Glycemic index of foods: A physiological basis for carbohydrate exchange. *American Journal of Clinical Nutrition* 34(3): 362.

Jeppesen, P.B., S. Gregersen, et al. 2000. Stevioside acts directly on pancreatic [beta] cells to secrete insulin: Actions independent of cyclic adenosine monophosphate and adenosine triphosphate – sensitivie K+–channel activity. *Metabolism* 49(2): 208–214.

Josef, V., Schusselbergstrasse, et al. 1973. Antidiabetic composition of beet roots and horseradish. USA: Google Patents.

Kamalakkannan, N., and P.S.M. Prince. 2006. Antihyperglycaemic and antioxidant effect of Rutin, a polyphenolic flavonoid, in streptozotocin-induced diabetic wistar rats. *Basic & Clinical Pharmacology & Toxicology* 98(1): 97–103.

Kapadia, G.J., H. Tokuda, et al. 1996. Chemoprevention of lung and skin cancer by *Beta vulgaris* (beet) root extract. *Cancer Letters* 100(1–2): 211–214.

Kapadia, G.J., M.A. Azuine, et al. 2003. Chemoprevention of DMBA-induced UV-B promoted, NOR-1-induced TPA promoted skin carcinogenesis, and DEN-induced phenobarbital promoted liver tumors in mice by extract of beetroot. *Pharmacological Research* 47(2): 141–148.

Kawada, T., T. Watanabe, et al. 1986. Capsaicin-induced beta-adrenergic action on energy metabolism in rats: Influence of capsaicin on oxygen consumption, the respiratory quotient, and substrate utilization. *Proceedings of the Society for Experimental Biology and Medicine. Society for Experimental Biology and Medicine (New York)* 183(2): 250–256.

Kujala, T., M. Vienola, et al. 2002. Betalain and phenolic compositions of four beetroot (*Beta vulgaris*) cultivars. *European Food Research and Technology* 214(6): 505–510.

Kujawska, M., E. Ignatowicz, et al. 2009. Protective effect of red beetroot against carbon Tetrachloride- and *N*-Nitrosodiethylamine-induced oxidative stress in rats. *Journal of Agricultural and Food Chemistry* 57(6): 2570–2575.

Lee, J.H., C.W. Son, et al. 2009. Red beet (*Beta vulgaris* L.) leaf supplementation improves antioxidant status in C57BL/6 J mice fed high fat high cholesterol diet. *Nutrition Research and Practice* 3(2): 114–121.

Liu, S., M. Serdula, et al. 2004. A prospective study of fruit and vegetable intake and the risk of type 2 diabetes in women. *Diabetes Care* 27(12): 2993–2996.

Lu, X., Y. Wang, et al. 2009. Radioprotective activity of betalains from red beets in mice exposed to gamma irradiation. *European Journal of Pharmacology* 615(1–3): 223–227.

Luo, J., J. Cheung, et al. 1999. Novel terpenoid-type quinones isolated from pycnanthus angolensis of potential utility in the treatment of type 2 diabetes. *Journal of Pharmacology and Experimental Therapeutics* 288(2): 529–534.

Ninfali, P., M. Bacchiocca, et al. 2007. Characterization and biological activity of the main flavonoids from Swiss Chard (*Beta vulgaris* subspecies cycla). *Phytomedicine* 14(2–3): 216–221.

Nishioka, T., J. Kawabata, et al. 1998. Baicalein, an α-glucosidase inhibitor from *Scutellaria baicalensis*. *Journal of Natural Products* 61(11): 1413–1415.

Nottingham, S. 2004. Beetroot health and nutrition S. Nottingham

Nyarko, A.K., H. Asare-Anane, et al. 2002. Extract of *Ocimum canum* lowers blood glucose and facilitates insulin release by isolated pancreatic [beta]-islet cells. *Phytomedicine* 9(4): 346–351.

Okoli, C., A. Ibiam, et al. 2010. Evaluation of antidiabetic potentials of *Phyllanthus niruri* in alloxan diabetic rats. *African Journal of Biotechnology* 9(2): 248–259.

Okyar, A., A. Can, et al. 2001. Effect of *Aloe vera* leaves on blood glucose level in type I and type II diabetic rat models. *Phytotherapy Research* 15(2): 157–161.

Ozsoy-Sacan, O., O. Karabulut-Bulan, et al. 2004. Effects of chard (*Beta vulgaris* L. var cicla) on the liver of the diabetic rats: A morphological and biochemical study. *Bioscience Biotechnology and Biochemistry* 68(8): 1640–1648.

Panda, S., and A. Kar. 2007. Apigenin (4′,5,7-trihydroxyflavone) regulates hyperglycaemia, thyroid dysfunction and lipid peroxidation in alloxan-induced diabetic mice. *Journal of Pharmacy and Pharmacology* 59(11): 1543–1548.

Racotta, I.S., J. LeBlanc, et al. 1994. The effect of caffeine on food intake in rats: Involvement of corticotropin-releasing factor and the sympatho-adrenal system. *Pharmacology Biochemistry and Behavior* 48(4): 887–892.

Raghuram, T., R. Sharma, et al. 1994. Effect of fenugreek seeds on intravenous glucose disposition in non insulin dependent diabetic patients. *Phytotherapy Research* 8(2): 83–86.

Rayalam, S., M.A. Della-Fera, et al. 2008. Phytochemicals and regulation of the adipocyte life cycle. *The Journal of Nutritional Biochemistry* 19(11): 717–726.

Reddy, M.K., R.L. Alexander-Lindo, et al. 2005. Relative inhibition of lipid peroxidation, cyclooxygenase enzymes, and human tumor cell proliferation by natural food colors. *Journal of Agricultural and Food Chemistry* 53(23): 9268–9273.

Rizvi, S.I., and M.A. Zaid. 2001. Insulin-like effect of (–) Epicatechin on erythrocyte membrane acetylcholinesterase activity in type 2 diabetes mellitus. *Clinical and Experimental Pharmacology and Physiology* 28(9): 776–778.

Rizvi, S.I., M.A. Zaid, et al. 2005. Protective role of tea catechins against oxidation-induced damage of type 2 diabetic erythrocytes. *Clinical and Experimental Pharmacology and Physiology* 32(1–2): 70–75.

Roberts, M.J., and D. Strack. 1999. *Biochemistry and physiology of alkaloids and betalains*. Sheffield: Academic.

Roman-Ramos, R., J.L. Flores-Saenz, et al. 1995. Anti-hyperglycemic effect of some edible plants. *Journal of Ethnopharmacology* 48(1): 25–32.

Sabu, M., K. Smitha, et al. 2002. Anti-diabetic activity of green tea polyphenols and their role in reducing oxidative stress in experimental diabetes. *Journal of Ethnopharmacology* 83(1–2): 109–116.

Salmerón, J., A. Ascherio, et al. 1997a. Dietary fiber, glycemic load, and risk of NIDDM in men. *Diabetes Care* 20(4): 545–550.

Salmerón, J., J.E. Manson, et al. 1997b. Dietary fiber, glycemic load, and risk of non–insulin-dependent diabetes mellitus in women. *JAMA: The Journal of the American Medical Association* 277(6): 472–477.

Sanders, R.A., F.M. Rauscher, et al. 2001. Effects of quercetin on antioxidant defense in streptozotocin-induced diabetic rats. *Journal of Biochemical and Molecular Toxicology* 15(3): 143–149.

Sapers, G.M., and J.S. Hornstein. 1979. Varietal differences in colorant properties and stability of red beet pigments. *Journal of Food Science* 44(4): 1245–1248.

Sárdi, É., É. Stefanovits-Bányai, et al. 2009. Effect of bioactive compounds of table beet cultivars on alimentary induced fatty livers of rats. *Acta Alimentaria* 38(3): 267–280.

Sartor, G., B. Schersten, et al. 1980. Ten-year follow-up of subjects with impaired glucose tolerance: Prevention of diabetes by tolbutamide and diet regulation. *Diabetes* 29(1): 41–49.

Sasaki, R., N. Nishimura, et al. 2007. Cyanidin 3-glucoside ameliorates hyperglycemia and insulin sensitivity due to downregulation of retinol binding protein 4 expression in diabetic mice. *Biochemical Pharmacology* 74(11): 1619–1627.

Schwartz, S.J., J.H. von Elbe, et al. 1983. Inability of red beet betalain pigments to initiate or promote hepatocarcinogenesis. *Food and Chemical Toxicology* 21(5): 531–535.

Shane-McWhorter, L. 2001. Biological complementary therapies: A focus on botanical products in diabetes. *Diabetes Spectrum* 14(4): 199–208.

Shanmugasundaram, E., G. Rajeswari, et al. 1990. Use of *Gymnema sylvestre* leaf extract in the control of blood glucose in insulin-dependent diabetes mellitus. *Journal of Ethnopharmacology* 30(3): 281–294.

Sharma, R.D., and T.C. Raghuram. 1990. Hypoglycaemic effect of fenugreek seeds in non-insulin dependent diabetic subjects. *Nutrition Research* 10(7): 731–739.

Shetty, A.K., R. Rashmi, et al. 2004. Antidiabetic influence of quercetin in streptozotocin-induced diabetic rats. *Nutrition Research* 24(5): 373–381.

Singh, A.B., N.S. Akanksha, et al. 2009. Anti-hyperglycaemic, lipid lowering and anti-oxidant properties of [6]-gingerol in db/db mice. *International Journal of Medicine and Medical Sciences* 1(12): 536–544.

Sivakumar, G., D.R. Vail, et al. 2009. Plant-based corosolic acid: Future anti-diabetic drug? *Biotechnology Journal* 4(12): 1704–1711.

Sotaniemi, E.A., E. Haapakoski, et al. 1995. Ginseng therapy in non-insulin-dependent diabetic patients. *Diabetes Care* 18(10): 1373–1375.

Sreekanth, D., M. Arunasree, et al. 2007. Betanin a betacyanin pigment purified from fruits of *Opuntia ficus*-indica induces apoptosis in human chronic myeloid leukemia Cell line-K562. *Phytomedicine: International Journal of Phytotherapy and Phytopharmacology* 14(11): 739.

Strasser, G.R., and R. Amadò. 2002. Pectic substances from red beet (*Beta vulgaris* L. var. conditiva). Part II. Structural characterisation of rhamnogalacturonan II. *Carbohydrate Polymers* 48(3): 263–269.

Suresh Babu, P., and K. Srinivasan. 1995. Influence of dietary curcumin and cholesterol on the progression of experimentally induced diabetes in albino rat. *Molecular and Cellular Biochemistry* 152(1): 13–21.

Tanaka, Y., N. Sasaki, et al. 2008. Biosynthesis of plant pigments: Anthocyanins, betalains and carotenoids. *The Plant Journal* 54(4): 733–749.

Tang, L.-Q., W. Wei, et al. 2006. Effects of berberine on diabetes induced by alloxan and a high-fat/high-cholesterol diet in rats. *Journal of Ethnopharmacology* 108(1): 109–115.

Váli, L., É. Stefanovits-Bányai, et al. 2007. Liver-protecting effects of table beet (*Beta vulgaris* var. rubra) during ischemia-reperfusion. *Nutrition* 23(2): 172–178.

Vessal, M., M. Hemmati, et al. 2003. Antidiabetic effects of quercetin in streptozocin-induced diabetic rats. *Comparative Biochemistry and Physiology Part C, Toxicology & Pharmacology* 135(3): 357–364.

Vincent, K.R., and R.G. Scholz. 1978. Separation and quantification of red beet betacyanins and betaxanthins by high-performance liquid chromatography. *Journal of Agricultural and Food Chemistry* 26(4): 812–816.

Wang, L., S. Liu, et al. 2006. The consumption of lycopene and tomato-based food products is not associated with the risk of type 2 diabetes in women. *The Journal of Nutrition* 136(3): 620–625.

Welihinda, J., E.H. Karunanayake, et al. 1986. Effect of *Momordica* charantia on the glucose tolerance in maturity onset diabetes. *Journal of Ethnopharmacology* 17(3): 277–282.

Wettasinghe, M., B. Bolling, et al. 2002. Phase II enzyme-inducing and antioxidant activities of beetroot (*Beta vulgaris* L.) extracts from phenotypes of different pigmentation. *Journal of Agricultural and Food Chemistry* 50(23): 6704–6709.

Wild, S.H., G. Roglic, et al. 2004. Global prevalence of diabetes: Estimates for the year 2000 and projections for 2030. *Diabetes Care* 27(10): 2569.

Wolfram, S., D. Raederstorff, et al. 2006. Epigallocatechin gallate supplementation alleviates diabetes in rodents. *Journal of Nutrition* 136(10): 2512.

Yadav, S., V. Vats, et al. 2002. Hypoglycemic and antihyperglycemic activity of *Murraya koenigii* leaves in diabetic rats. *Journal of Ethnopharmacology* 82(2–3): 111–116.

Yamashita, T., and K. Sato. 2008. Effects of beet red inhibiting experimental diabetes in vivo-studies using ALS strain mice. *Foods and Food Ingedients Journal of Japan* 213(2): 122.

Yin, J., H. Xing, et al. 2008. Efficacy of berberine in patients with type 2 diabetes mellitus. *Metabolism* 57(5): 712–717.

Yoshikawa, M., T. Murakami, et al. 1996. Medicinal foodstuff. III. Sugar beet.(1): Hypoglycemic oleanolic acid oligoglycosides, betavulgarosides I, II, III, and IV, from the root of *Beta vulgaris* L.(*Chenopodiaceae*). *Chemical and Pharmaceutical Bulletin* 44(6): 1212.

Yun, S.-Y., S.-P. Kim, et al. 2006. Effects of (–)-epigallocatechin-3-gallate on pancreatic beta-cell damage in streptozotocin-induced diabetic rats. *European Journal of Pharmacology* 541(1–2): 115–121.

Zarzuelo, A., I. Jiménez, et al. 1996. Effects of luteolin 5-O-[beta]-rutinoside in streptozotocin-induced diabetic rats. *Life Sciences* 58(25): 2311–2316.

Zohary, D., and M. Hopf. 2000. *Domestication of plants in the old world: The origin and spread of cultivated plants in West Asia, Europe, and the Nile Valley*. Oxford, USA: Oxford University Press.

Chapter 9
Cell and Tissue Culture Studies in *Beta vulgaris* L.

Bhagyalakshmi Neelwarne

Abstract The presence of intensely colored hydrophilic nitrogenous betalain pigments, abundance of health-promoting compounds, growth-linked pigment-laden vacuoles in their cells and the possibility of quantitative regulation of pigment molecules have all made red beet cultured cells both an experimental model for studying cellular physiologies and a valuable source of various nutraceutical compounds. Compared with many other fruit/vegetable crops and sugar beets, studies targeted at developing in vitro regeneration protocols for varietal improvements and exploiting cell cultures for other biotechnological applications have not received much importance in red beet. With a focus on classic attempts made towards using cultured cells and organs for understanding growth, development and pigment biosynthesis-related physiologies, this chapter discusses aspects of newer technologies where cultured cells of *Beta vulgaris* may find applications.

Abbreviations

2,4-D	2,4-Dichlorophenoxy acetic acid
B5	Gamborg's B5 medium
BA	Benzyl adenine (6-benzylamino-purine)
NAA	Naphthalene acetic acid
IBA	Indole butyric acid
Kn	Kinetin
LS	Linsmaier and Skoog
MS	Murashige and Skoog
TIBA	2,3,5-Triiodobenzoic acid

B. Neelwarne (✉)
Plant Cell Biotechnology Department, Central Food Technological Research Institute
(Council of Scientific and Industrial Research), Mysore 570 020, India
e-mail: bneelwarne@cftri.res.in

B. Neelwarne (ed.), *Red Beet Biotechnology: Food and Pharmaceutical Applications*,
DOI 10.1007/978-1-4614-3458-0_9,

9.1 Introduction

Ever since the scientific understanding that each cell in an organism carries essentially the same set of genes, our fore-biologists were driven to conceptualize that higher plants could be reduced to single cells, grown in culture under human-controlled environment and produce specific compounds as an alternative strategy for agriculture. This idea of treating the higher plant cells almost like microbial cells triggered interest among many researchers to take up cell culture research. Among the earliest such trials is a success story of culturing red beet cells by Constabel and Nassif-Makki (1971), who raised cell cultures of red beet in vitro and demonstrated their pigment biosynthetic ability. With an inherent interest of unraveling the pigment biosynthetic routes, these researchers tried to advance our understanding of the function and regulation of betalain biosynthesis. Moreover, it is still barely known what actual roles are played by betalains and many other such secondary metabolites in the beetroot plant (Escribano et al. 1998; Grassmann et al. 2002).

9.2 Micropropagation

Clonal propagation is one of the oldest applications among plant in vitro culture techniques, which has paved the path for modern plant biotechnology industry. Micropropagation is based on obtaining genetically uniform (clonal) plant propagules through vegetative means, commercially used to mass-produce identical somatic clones. This technology has significantly complemented the growth of agriculture and horticulture sectors. Inexpensive cultivation methods and short life cycles of beet cultivars in fields have sidelined the application of micropropagation technology for this crop. Owing to the commercial importance, the sugar beet was the first among many varieties of *B. vulgaris* to experience mass propagation technology, starting with meristems and floral buds of sub-apical inflorescences (Margara 1977) and axillary shoot proliferation from in vitro germinated seedlings (Hussey and Hepher 1978), several valuable breeding lines and dihaploids, aneuploids and genetic transformants were multiplied in vitro. However, red beet and leaf beets have also been occasionally researched for establishing their regeneration potential, and more so for the production of betalains and for other technological purposes. Notable is the fact that many explants from some cultivars of beet have shown high totipotency in vitro, while certain cultivars have shown high recalcitrancy.

9.2.1 Culture Initiation and Role of Explants

For presumed final application, selection of consistently well-performing cultures is important. If clonal propagation is considered, rapid shoot multiplication of shoot propagule without an interphase of callus is preferred. When cultures are induced for the selection of variants, highly totipotent cell lines or embryogenic cultures are preferred. In case of cell cultures for secondary metabolites, cell lines with high growth rate, consistent high yields of desirable compounds and sensitivity to treatments such as elicitation and amenability to product recovery are important technological parameters. Prior growth conditions of the mother plants, selection of explants and their pre-treatments are known to profoundly alter cellular physiologic conditions which in turn influence the culture establishment as well as the type and fate of cultures produced. Most of the successful experiments with *B. vulgaris* have used sterile seedlings as explants for culture initiation, be it for cell cultures, callus with regeneration capacities or hairy roots. Table 9.1 lists types of cell/callus or other cultures induced from various explants in different beetroot cultivars and foliage beet.

9.2.2 Propagation from Pre-Existing Meristems

Pre-existing meristems such as shoot tips, axillary buds, flower buds and young inflorescences have been found useful for continuous multiplication of shoots. Clonal multiplication in beets is applied for rare genotypes, unfertile plants and homozygous haploids/di-haploids and cybrids for technical reasons. In red beet ssp. *vulgaris* var. *conditiva* ALEF, shoot tips were used to develop an efficient propagation method by Harms et al. (1983). These shoot-tip explants were very sensitive to wounding formed during cutting of shoot tips, which leaked betacyanins into the medium, which rapidly oxidized, causing growth inhibition, which subsided upon repeated transfers to fresh medium (Harms et al. 1983). In the same study, petiole explants were found to produce a large number of shoot buds originating from the upper surface, which were later perpetually multiplied in vitro. In another study using wild beets (*B. maritima*), a large number of shoots were found to originate from receptacle (the non-essential part of flower bud) when immature inflorescences were grown in Murashige and Skoog (MS) medium (Murashige and Skoog 1962) with 1 mg/L of benzyl adenine (6-benzylamino-purine) (BA). These shoots grew into whole plants, rooting efficiently on medium with 1.00 mg/L of naphthalene acetic acid (NAA) and established easily in green house (Zhong et al. 1993a).

9.2.3 Regeneration de Novo of Adventitious Shoots and Somatic Embryos

De novo or spontaneous shoot development has also received importance in the micropropagation industry because of the possibility of obtaining a large number of shoots from an explant. Theoretically, each cell in an explant is totipotent, capable of entering into morphogenic cycle through adventive embryony. However, practically various genetic and environmental as well as physiological gradients occur in and around each cell/tissue, due to micro-local variations from one cell to another,

Table 9.1 Explants and media used for raising cell/callus and other cultures in different beetroot cultivars and forage beets

Variety/cultivar of beetroot	Explant	Medium	Type of cells/callus	Reference
Detroit dark red	Seedling root tips	MS+NAA+Kn	Red and yellow callus	Weller and Lasure (1982)
Conditiva	Beetroot	B5+2,4-D+Kn	Callus and cell	Bokern et al. (1991)
Detroit dark red extra early flat Egyptian	Hypocotyls	LS+2,4-D	Callus and cell suspension	Akita et al. (2000)
Cicla	Various parts from 30-d-old sterile seedlings	MS	Organogenic/non-organogenic red/white callus	Xu et al. (2009)
Bikores monogerm	Cotyledon	LS	Green callus	Girod and Zryd (1987)
Bikores monogerm	Cotyledonary cell line	MS+2,4-D+BA	Cell suspension	Khlebnikov et al. (1995).
Cosby Egyptian	Hypocotyl	B5+2,4-D+Kn	Cell suspension	Rodríguez-Monroy et al. (1994)
				Trejo-Tapia et al. (2001)
B. vulgaris (monosomatic hybrids)	Inflorescence	MS+BA MS+BA MS+IBA	Shoot formation and multiplication Rooting of shoots	Miedema (1982)
B. vulgaris	Shoot tips Petioles		Shoot multiplication Adventitious shoot buds	Harms et al. (1983)
B. maritima (wild beet)	Inflorescence	MS+BA	Adventitious shoots	Zhong et al. (1993a)
B. vulgaris hybrid	Ovary		Embryogenesis	Bossoutrot and Hosemans (1985)
Bikores monogerm				Girod and Zryd (1987)
Forage beet:	Shoot meristem	–	Regeneration of somaclones	Botău and Chiş (2008)
Monogal				
Ursus				
Tetra Roşu Tetra 181				
Ruby Queen	Cotyledon	MS+2,4-D	Red callus	In author's laboratory (Fig. 9.1)
Red Ace Cylindra	Cotyledon	MS+BA+NAA	Embryogenesis Adventitious shoots	Livadariu et al. (2009)
B. Vulgaris	Leaf	MS+BA	Callus	Yu (1989)
B. macrocarpa	Leaf	MS+BA	Adventitious shoots	
B. maritima				
B. intermedia 9 other Beta spp.	Leaf	MS+BA	Callus	Yu (1989)
B. vulgaris (6 lines) B. maritime (3accn.)	Leaf	MS+BA	Direct shoot regeneration	Mikami et al. (1989)
Red Ball	Seedling stem	B5		Yang et al. (2003)
B. vulgaris var. Altissima (sugar beet transformed)	Leaf	PG0	Callus	Pavokovic et al. (2009)

which in turn influence the quality and number of adventitious organ/embryo formation. This being the case, the regenerability in red beet cultures was found to vary mainly on the basis of initial explants, although growth regulators were also found to significantly influence morphogenesis. Five different types of explants (cotyledons, leaves, hypocotyls, petioles, and shoot tips) were used to optimize and choose the best explant, along with any pre-treatments, if required, for improving their performance in vitro (Xu et al. 2009). Explants from aerial vegetative portions were found highly regenerable in the case of leaf of red beet seedlings that had previously been grown on medium with BA for 30 days, where over 81% shoot tip explants, 17% of petiole explants and only 6% of leaf explants formed transparent nodular structures, which later developed into adventitious shoot buds. On the medium supplemented with different concentrations of BA, 2,4-dichlorophenoxy acetic acid (2,4-D), indole butyric acid (IBA) and 2,3,5-triiodobenzoic acid (TIBA), all types of explants from sterile seedlings, irrespective of whether pre-treated with BA or not, formed different callus types, such as red compact (RC), red loose (RL), translucent nodular (TN), white compact (WC) and/or white loose (WL) callus, after 4 weeks of incubation at either 25 or 16°C (Xu et al. 2009). These responses were similar to those observed in three sugar beet lines, where pre-treatment by germinating seedlings on medium containing different concentrations (1.0, 3.0 or 5.0 mg/l) of BA (Gurel et al. 2003a) or TDZ (Gurel et al. 2003b) for 5 weeks and then culturing excised petiole explants on two different regeneration media containing 0.5 or 1.0 mg/l BA combined with 0.1 or 0.3 mg/l NAA, respectively, resulted in good regenerative responses. Since the responses in red leaf beet were similar to those observed in sugar beet (Zhang et al. 1993; Wei and Zhang 1998), one may take clues from the extensive studies conducted on morphogenic studies with sugar beet (Gurel et al. 2008) for further work on red beet.

In cases where regenerable globular calluses were formed, they arose from and around the leaf midrib regions, from the cut ends of hypocotyls and petioles, and from shoot tips, particularly in the presence of BA (Table 9.1). This indicates the probable involvement of highly competent cells such as xylem and/or shoot tip parenchyma cells in the formation of pro-embryogenic cultures. It has been suggested that a high level of cytokinin reservoir was needed to boost the undetermined cells to regain meristematic activity and shoot morphogenesis (Veit 2009). Thus, the proembryogenic cultures from var. *cicla*, upon continuous culturing on high cytokinin-supplemented medium regenerated adventitious buds. In these cultures, temperature also influenced the pattern of organ formation, where, despite providing the right medium components, incubation at 16°C played a crucial role (Xu et al. 2009). In this study it was observed that all the globular tissues originating from an explant went through adventitious bud formation where each explant unit produced adventitious buds or shoots, with the highest average number of nearly nine developing from each shoot tip and petiole.

In red beet var. *conditiva* ALEF, the primary shoot tip explants often formed tuber-like hard compact neoplasms at their bases. Such globular structures were made up of highly organized cells, unlike calluses. After cutting off such structures upon each transfer, the tuber-forming tendency diminished, yielding multifoliar rosette plantlets suitable for repeated sub-culturing (Harms et al. 1983). Whenever a non-globular type of callus was formed from explants, it remained in an undifferentiated state, although cultured under similar conditions as for regenerable cultures, indicating that pre-destined regenerable genetic status is altered in these cells.

9.2.4 *Regeneration from Callus and Cell Suspension Cultures*

Naturally endowed with fast growth and short life cycles, growing red beet cells under aseptic condition has been successful in a few laboratories. It has been well established through experimentation that, in most of the higher plants, cellular organizations and regeneration are the result of interactions among genetic factors and environmental conditions. Thus by creating alterations in the latter, mainly by making various alterations in the supply of various phytohormones, it is possible to regulate the pattern of regeneration. For example auxins and cytokinins are known to play major roles in controlling plant development. During in vitro culture, an excess of cytokinin over auxin promotes shoot formation in callus, while auxin in excess over cytokinin induces root formation (Veit 2009).

The first report of Constabel and Nassif-Makki (1971) using red beet callus cultures, reported by Rodríguez-Monroy et al. (1994), who established red beet cells in vitro aiming to understand the nutritional requirements for their continuous growth and pigment synthesis. Later, callus cultures were initiated from the hypocotyls of aseptically raised seedlings on LS agar-gelled medium with 10^{-7}or 3×10^{-7} M of 2,4-D (Akita et al. 2000). In most of the studies that involved prolonged culture of pigmented calluses and cell suspensions of red beet, further regeneration into either buds or somatic embryos was not observed, except for one study where both red and white compact or lose-type calluses regained regeneration potential when transferred to high BA (15 mg/L)-containing medium and incubated at low temperature (Xu et al. 2009). Such responses observed in various plant culture systems have been the subject of intense study, where physiological basis of shift from one biochemical status to another have been unraveled. Based on the agreement with the competence-determination theory of Christianson and Warnick (1983; 1985), who postulated that organogenesis proceeds through three sequential stages: (1) the acquisition of competence (which is believed to be predetermined at an early stage of development) to respond to a particular inductive signal, (2) induction, and (3) morphogenic differentiation and development. In accordance with this theory, it is imperative that the *B. vulgaris* explants obtained from BA-pre-treated seedlings might have gained the competence in most of the cells at the seedling development period, so that their cells respond readily toward either direct shoot bud induction (organogenesis) or somatic embryogenesis. Such responses may differ according to the degree of cell differentiation, which in turn is governed by genetic factors, thus producing different regenerative responses (Meng et al. 2010). For instance, preculture of explants or long-time culture of callus in medium supplemented with a particular growth regulator is known to alter the ratio of endogenous hormones.

Accordingly, increasing the level of the endogenous cytokinin pool by pre-treating explants on BA-supplemented medium has also been proved to improve regeneration efficiency in sugar beet protoplasts (Dovzhenko and Koop 2003).

The growth regulator used for subsequent culture, after the explant pre-treatment, is also very important. In case of leaf beet, *B. vulgaris* var. *cicla*, culturing BA-pre-treated seedling explants in an auxin-containing medium, particularly 0.1–3 mg/l 2,4-D, formed calluses from all seedling parts. However, such calluses resulted in root formation with no further ability to re-differentiate into shoots upon exposure to any single or combination of growth regulators (Xu et al. 2009), which the investigators linked to probable high levels of endogenous auxin, indicative of a strong tendency to re-synthesize this hormone. To support this view, when the root-forming calluses were cultured in a medium with 1–2 mg/L of TIBA, an auxin transport inhibitor, the expected regenerative type of calluses were formed, supporting that a higher cytokinin over auxin ratio is needed for regeneration in beets. The fact that the endogenous level of auxin remains high in *B. vulgaris* is also supported by the observation in sugar varieties where only BA was adequate for the induction of an organogenic callus (Saunders 1982).

Apart from pre-treatment, the concentration of BA in the callus induction medium was also important for obtaining cultures composed of small nodular structures or globular cell masses that eventually organized into shoot buds. When different explants obtained from 30-day-old sterile seedlings were pre-treated with BA and cultured at 16°C on MS medium with a range of BA concentrations (0.1, 1, 2, 5, 10 and 20.0 mg/l), several types of calluses, appearing either compact or non-compact with or without red pigmentation, were formed. However, regenerable compact calluses were observed only on the MS medium supplemented with 15 mg/l BA at 16°C, with an adverse effect when BA level increased further (Xu et al. 2009).

Cell suspensions have always been raised from callus cultures. Although there was no change in mineral composition from that of callus cultures, higher sucrose requirements (50 g/L) and changing the level of auxin were found beneficial to maintain cultures in cell suspension form (Akita et al. 2000). For these cultures, genetic composition of the source material played a crucial role. For instance, the variety Detroit Dark Red produced a larger amount of callus than the variety Extra Early Flat Egyptian in a span of time. Different explants of red beet have been successfully grown in MS medium (Table 9.1) and sometimes in LS (Akita et al. 2000, 2001, 2002) or B5 media.

9.2.5 *Anther, Ovary and Ovule Culture*

Culturing these organs in vitro are of great relevance to commercial varieties, mainly for applications targeted at cultivar improvements, therefore, one can expect a great body of research data on ovary culture in sugar beet cultivars. The long period required for developing breeding lines through conventional crossing has prompted the regeneration of haploid plants from unfertilized ovary/ovule cultures of sugar beet and the derived doubled haploids were integrated into breeding programs (Bossoutrot and Hosemans 1985; Marylise et al. 1989). Gynogenic haploid regeneration or haploid parthenogenesis using the un-pollinated female gametophytics (ovules, ovaries or floral buds) is a method for haploid induction used in several agronomically important sugar beets (Bohanec 2009; Germana 2011). Such protocols appear applicable to red beets also, as both are the same species and their somatic cells have shown similar responses in vitro.

9.2.6 Protoplast Cultures

Protoplasts are plant cells without their cellulosic cell walls, which are obtained in vitro by enzymatic hydrolysis of cellulose. Each protoplast is capable of re-entering into its normal cellulosic cell form and then dividing to form cell colonies and develop into plants in vitro. Protoplasts have been enormously useful for many physiological and genetic engineering studies. Red beet protoplasts have been used as a model system to isolate vacuoles and for studies related to membrane-transport mechanisms studies (Pradedova et al. 2011) (see Chap. 4). Through electroporation or treatment with chemicals that change electric potential of cell membranes (e.g., polyethylene glycol), protoplasts are made to take up macromolecules, plasmids or naked DNA fragments, or fuse either with similar or dissimilar other protoplasts resulting in cytoplasmic hybrids or polyploids. These properties make protoplasts important for direct gene transfer or for somatic hybridization of cells from distantly related plants, thereby circumventing several incompatibility barriers. While most efforts with protoplast-derived callus resulted in only callus formation, there are reports where protoplasts of *B. vulgaris* were found to generate somaclonal variations (Steen et al. 1986; Krens et al. 1989, 1990; Lenzner et al. 1995; Jazdzewska et al. 2000). Protoplast technology has high relevance in the characterization of the cell wall. In regenerating leaf-derived protoplasts of *B. vulgaris*, several pectin oligosaccharides such as arabinogalactan proteins (AGPs) were visualized in situ by a series of immunocytochemical reactions and compared with those in tobacco protoplast. Conspicuous differences between the two species were found; *B. vulgaris* cell walls had a lower content of pectin side-chains bearing (1→4)-β-D-galactose residues as compared with *N. tabacum* (Wiśniewska and Majewska-Sawka 2008).

9.2.7 Temperature Effects

Temperature appears to affect both callus induction and shoot formation in red beet. In a specific study with leaf beet var. *cicla* explant, the responses as well as their further regenerative performances were found sensitive to temperature treatments. Transparent nodular regenerative cells were formed at 16°C, which ultimately formed adventitious shoots, whereas cultivation of same explants at 25°C resulted in non-regenerative calluses (Xu et al. 2009). The red beet var. *conditiva* ALEF, however, was micropropagated successfully at 25°C (Harms et al. 1983), and almost all cell cultures performed well at 25°C (Akita et al. 2000; Trejo-Tapia et al. 1999, 2001). For the induction of embryos from unfertilized ovules in vitro, 27°C was preferable over 24°C in sugar beets (Doctrinal et al. 1989). Thus, large number of studies indicated embryogenic potential at temperatures at or nearer to 25°C for tuberous varieties, which are stored at lower temperatures to suppress bolting, which indirectly means growth suppression.

9.2.8 Cellular Differentiation and Organogenesis

Studies focused on regeneration events in *B. vulgaris* adopted histological techniques to keep track of the mode of differentiation. In almost all the studies, whether red beet or sugar beet, the adventive embryos, or meristematic calluses that later regenerated into shoots, developed from tissues at the vicinity of vascular tissue, probably from the vascular parenchyma, which is known to possess a higher capacity for regeneration (Skoog and Miller 1957; Sreedhar et al. 2008). In many plant species, juvenile explants were found more regenerative in vitro than mature ones (Ibrahim and Pierre 2001; Wang et al. 2006). However embryogenic cell suspensions and the similar cultures from protoplasts need more nurturing environments for progressing toward embryogenesis apparently because of the absence of maternal tissues or supportive cells, which may offer nutritional support to presumptive embryogenic cells/ protoplasts. In case of protoplasts from shoot meristems that are rich in meristematic cells pre-destined to form shoots, the shoot tip formed globular shoot meristemoids, subsequently forming shoots when exposed to the right combination of growth regulators. BA was the preferable cytokinin for shoot regeneration in almost all the cases, with or without TIBA, indole butyric acid, or NAA (Xu et al. 2009; Gurel et al. 2011). The in vitro regeneration of red leaf beet could also be similar.

9.2.9 Acclimatization

Irrespective of whether red beet or sugar beet, rooting and acclimatization have been easy in *B. vulgaris*. In a recent study, Xu et al. (2009) observed that the use of half-strength MS was suitable for inducing roots in all the adventitiously regenerated shoots in 30 days, however, with a reduction in the level of sucrose (10 g/L), where the presence of NAA did not make a big difference. However, increasing sugar level to 30 g/L, as is used for other types of in vitro cultures, was found to inhibit root formation in regenerated red beet shoots. A good 80% survival rate of regenerated shoots has been recorded for leaf beets (Xu et al. 2009) and between 65% and 95% for tuberous red beet (Trejo-Tapia et al. 1999). In var. *Red Ace*, 82% of in vitro plants were acclimatized successfully under ex vitro conditions, whereas in var. *cylindra* only 73% showed normal growth cycle (Livadariu et al. 2009). In case of var. *cicla*, all nine shoots derived from TN callus from petiole explants could easily root on half-strength MS medium containing 10 g/L sucrose with or without NAA (Xu et al. 2009).

9.2.10 Regeneration Studies in Sugar Beet

Sugar beet, which is a variety derived from red *B. vulgaris*, has been extensively studied for its regeneration potential with a main intent of widening the array of variants for breeding purposes and to facilitate the application of transgene technology. Therefore, several sugar beet breeding lines and commercial varieties were cultured in vitro. Regeneration through adventitious shoot organogenesis or via an intervening callus phase was obtained from cotyledons, hypocotyls, root/hypocotyl/shoot transition zone tissues, leaf petiole and lamina. As in the case of red beet, in sugar beet also, preconditioning seedlings on medium containing BA (Jacq et al. 1993; Zhong et al. 1993a, b; Zhang et al. 2001; Gurel et al. 2003a), TDZ (Zhang et al. 2001; Gurel et al. 2003b) or 2,3,5-triiodobenzoic acid (TIBA) alone or with a combination of either BA (Jacq et al. 1993) or L-proline (Moghaddam et al. 2000) was found to improve regeneration efficiency from different types of explants. As a sole growth regulator, cytokinin–BA was used for routine regeneration of adventitious shoots from petiole explants and the method was easily adoptable for the propagation of elite genotypes (Atanassov 1986; Detrez et al. 1988; Ritchie et al. 1989; Zhong et al. 1993a; Grieve et al. 1997; Zhang et al. 2001). Callus induction and adventitious shoot regeneration in sugar beet was dependent on genotype and combinations of plant growth regulators. Somatic embryogenesis was observed to occur indirectly (with the intervention of a callus phase) from shoot base tissue or seedling explants treated with BA or directly from young zygotic embryos treated with 2,4-D (Pedersen et al. 1993; Kubalakova 1990; Tenning et al. 1992; Zhang et al. 2001, and unpublished results). However, these approaches find applications in a limited number of genotypes and the frequency of regeneration is generally low. The procedures involved are complicated due to the fact that several regimes of induction and subculture are needed. Cotyledon and hypocotyl explants of sugar beet lines SDM 3 and 10 showed higher response toward adventitious shoot regeneration in medium containing BA and TIBA or NAA than SDM 11, 5, and 9. Shoot regeneration was obtained both from hypocotyl and mesocotyl explants in SDM 9, 10, and HB 526 when grown on medium supplemented with BA, where the regeneration frequency was 25%. Leaf explants were also capable of forming adventitious shoots in breeding lines SDM 3 and 9 when cultured on a first medium containing NAA for callus induction and then in medium with BA and NAA to induce shoot regeneration, whereas, in lines SDM 10 and CMS 22003, culturing explants on medium containing BA produced a callus which eventually, on the same medium, regenerated into shoots (Zhang and Lemaux 2004).

In many cell cultures, white friable calluses were found non-regenerative since such white friable calluses consisted of large highly vacuolated cells with the least cell-to-cell contacts. As an exception, such cells were able to produce both roots and shoots in case of sugar beets (Saunders and Daub 1984; Konwar and Coutts 1990; Shimamoto et al. 1993), while the chlorophyllous compact calluses were non-organogenic (Ritchie et al. 1989; Tetu et al. 1987). Because of the prevalence of high heterozygosity as a consequence of outcrossing, sugar beets display a high degree of genotypic variation (Gurel 1997; Gurel et al. 2001; Zhang et al. 2008), which is a serious problem when the optimization of their regeneration potential is considered. In view of improving regeneration frequencies in several breeding lines of sugar beet, Gurel et al. (2011) screened the effects of explants and pre-treatments and found that petiole explants produced significantly more shoots than lamina explants. A strong influence of genotype was evident, since some breeding lines, such as the line ELK345, were endowed with high shoot regenerative capacity from both petiole and lamina explants. Like in many other studies, these breeding lines also performed well when medium was supplemented with low levels of BA and IBA, although a great difference was noticeable between any two breeding lines (Gurel et al. 2011). A review of regeneration potential of various sugar beet lines by Gurel et al. (2008) indicates that although several attempts were made by several different research groups; the protocols available are not always suitable for direct adoption since sugar beets also display high recalcitrance, as in the case of red beet.

9.3 Production of Secondary Metabolites

In vitro technology often offers an array of benefits: (a) the production of novel products not found in respective filed plants, (b) independence from geographic/climatic factors, (c) simpler extraction/purification free from interfering microbes, enzymes and inhibitors, (d) better control over biosynthetic routes for obtaining the most desired compounds in shorter production cycles and (e) easier fulfillment of quality demands. The most important of all is the possibility for genetic engineering avoiding legal restrictions against GMO introduction into the natural environment. Cultured cells are also exploited for several enzymatic chain reactions of biotransformation of simple compound to a specific complex biomolecule, which cannot be accomplished either outside the cell or through a synthetic route. All these biotechnological interceptions have been possible if the plant system has most of the features of an experimental model, for which red beet has been researched for several years. Accordingly different varieties and cultivars of *Beta vulgaris* are continuously researched for exploring newer compounds and to enhance desirable known compounds.

9.3.1 Initiation of Betalains Producing Cultures

Although agriculturally produced red beets are the cheaper sources for commercial extraction of betalains, there are various technical reasons for using cultured red beet cells. Different varieties and cultivars of *Beta vulgaris* are continuously researched because of the ease with which cultures can be maintained in vitro where colorful betalains serve as visual markers of biosynthetic potential of cell types. Therefore, red beet cell cultures serve as a model system for studying the synthesis

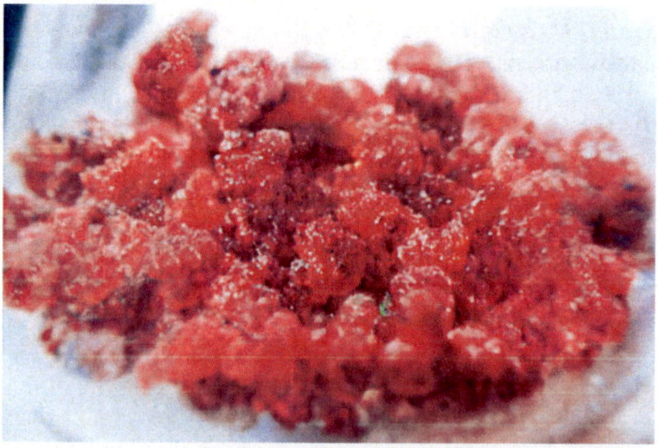

Fig. 9.1 Callus culture of red beet var. Ruby Queen grown on Murashige and Skoog's medium with 2,4-D and kinetin

of secondary metabolite, and hence serve as tools for technology development. The nutritional and health benefits of red beet extract, particularly betalain pigments, are discussed in other chapters. Callus cultures induced from cotyledonary leaves of the cv. Ruby Queen resulted in a deep red callus (Fig. 9.1) interspersed rarely with clusters of green cells, whereas other varieties of red beet produced yellow (betaxanthin-rich) and/or red (betacyanin-rich) callus masses in vitro. Akita et al. (2000) initiated calluses from cv. Detroit Dark Red that displayed both yellow and red clusters, and from cv. Extra Early Flat Egyptian (EEFE) that displayed cell clusters of violet color, indicating their capability to solely synthesize pure betacyanin. These cells were subsequently used for establishing suspension cultures for betacyanin production. *B. vulgaris* cv. Bikores Monogerm was found to initially produce a white/green non-betalainous callus, which upon repeated culture on medium with different ratios of auxin (2,4-D) and cytokinin (BAP) grew into secondary calluses of green, yellow, orange, red and violet phenotypes (Girod and Zryd 1991). In this study, the suspension cultures of pure yellow and pure violet cell lines, which predominantly synthesized betaxanthins and betacyanins, respectively, were successfully established. The violet cell culture showed significantly higher betacyanin content (28.01 mol/g dry weight [DW]) than that found in the whole plant, which was 21.18 mol/g DW (Leathers et al. 1992).

9.3.2 Culture Line Selection

Choosing appropriate cultivars is important for obtaining highly productive cell lines, since intra-specific variations may occur in terms of growth rates and product formation. As in the case of other higher plants, initial studies on red beet callus cultures were also found to perform poorly, with much lower productivities of betalains than in conventionally grown beet. Owing to such characteristics, initial research emphasized screening and selection of highly productive cell lines. Careful selection of *B. vulgaris* cells during subsequent subcultures nearly doubled the betalain yield (Akita et al. 2000). Both callus and cell suspension cultures were capable of producing an array of betalains in the presence of growth regulator, particularly 2,4-D, a synthetic auxin. A vast number of reports indicate that secondary metabolites are suppressed by synthetic auxins, particularly, 2,4-D, betalain pigments are also secondary metabolites that have been exceptionally and highly expressed in the presence of 2,4-D (Stafford et al. 1985). The callus raised from Detroit Dark Red displayed distinct yellow and red clusters indicating differences in their pigment biosynthetic capacity. Repeated selection and subculturing of yellow and red cell clusters resulted in the establishment of two separate cell culture lines. However, the callus induced from another cultivar, EEFE, showed uniform violet cells, indicative of the predominant presence of betacyanin, which, upon spectral analysis, were also found to contain small amounts of betaxanthins (Akita et al. 2000). Table 9.2 summarizes different types of beet cell cultures and contents of pigment(s) in each culture type.

9.3.3 Optimization of Biomass and Yield of Betalains

Since the yield of betalains from in vitro systems is lower than from crops, several research groups have attempted to optimize the pigment production both by the selection of highly productive cell lines or by modification of growth conditions. While careful selection of pigmented clusters of *B. vulgaris* cells during their subcultures resulted in nearly two-fold increases in betalain yield (Akita et al. 2000), other studies have sought to increase betalain yield by modifying growth conditions, where nutrients and growth regulators have played significant roles, as discussed in the following section.

9.3.4 Influence of Nutrients

Betalain synthesis was found to alter when the level of sucrose in the medium and the ratio of ammonium:nitrate were modified, resulting in 200 mg/L of betalains (Akita et al. 2000). Although zinc-deficient medium further enhanced betalain production in cell suspensions to a highest level of 590 mg/L (in 21 days) (Akita et al. 2001), it may not be possible to continuously maintain red beet cells on zinc-deficient medium, since zinc is needed for the formation of hundreds of proteins (Mustafa et al. 2011). In fact it is the only metal ion in certain important enzymes (oxidoreductases, transferases, hydrolases, lyases, isomerases and ligases). Root cortical cell vacuoles, cell walls and cytoplasm contain $\leq 3.4 \times 10^8$, 1.4×10^8 and $\leq 1.5 \times 10^8$

Table 9.2 Levels of different betalain pigments in cell and callus cultures of Beta vulgaris

Variety/cultivar of beetroot	Type of culture	Type of pigment	Pigment content in 1 g DW	Reference
Extra early flat Egyptian	Cell suspension	Betacyanins	7 mg	Akita et al. (2000)
Detroit dark red			22 mg	Akita et al. (2002)
B. vulgaris	Cell suspension	Betanin Iampranthin-II	2.6 μmol	Bokern et al. (1991)
			5.1 μmol	
Crosby Egyptian	Cell suspension	Betalains	4.2 mg	Rodríguez-Monroy et al. (1994)
Bikores monogerm	Cell suspension	Betalains	12 to 35 mg	Khlebnikov et al. (1995)
Bikores monogerm	Callus	Vulgaxanthin-II	1.813 μmol	Girod and Zryd (1991)
	Yellow phenotype	Miraxanthin-V	1.177 μmol	
		Total betaxanthins	4.278 μmol	
	Orange phenotype	Vulgaxanthin-II	5.519 μmol	
		Miraxanthin-V	3.061 μmol	
		Total betaxanthins	12.210 μmol	
	Red phenotype	Betanin	8.688 μmol	
		Isobetanin	0.721 μmol	
		Total betacyanins	11.22 μmol	
	Violet phenotype	Betanin	22.63 μmol	
		Isobetanin	2.903 μmol	
		Total betacyanins	28.02 μmol	
Bikores monogerm	Cell suspension Orange phenotype	Vulgaxanthin-I	10 mg	Leathers et al. (1992)
	Violet phenotype	Betanin	28 mg	
Altissima	Genetically transformed callus	Betanin	13.8 mg	Pavokovic et al (2009)

atoms of Zn, respectively (Broadley et al. 2007) and hence they are invariably involved in both tissue growth and betalain synthesis in red beet. It is also well established that the deficiency of one micronutrient leads to the over-accumulation of another. For instance zinc-deprived sugar beets accumulated enormous amounts of iron (Henriques 2001), Hajiboland and Amirazad 2010). The fact that zinc enhances betalain accumulation only transiently has been confirmed in a subsequent study by the same research group (Akita et al. 2001).

Among the macronutrients, the concentration of nitrate ions has been found to profoundly influence the biosynthesis of betalains in cell cultures of *B. vulgaris* cv. Bikores Monogerm, which exhibits different color phenotypes (Leathers et al. 1992). Revising the LS medium (Linsmaier and Skoog 1965) for vars. Extra Early Flat Egyptian and Detroit Dark Red, by reducing the nitrogen concentration from 60 to 30 mM and changing the ratio of ammonium to nitrate from 1:2 to 1:4, resulted in an increase in betacyanin accumulation in cells from 163 to 248 mg/L when additional sucrose (50 g/L) was added (Akita et al. 2000). These authors later found that the productivities were not cumulative when only nitrates were altered, and hence the medium was further revised. Subsequently they reported that keeping the lower level of nitrates (30 mM/L), some alterations in the combination of microelements were made that efficiently enhanced specifically the reddish-purple betacyanin production up to 550 mg/L in a 14-day batch and had a productivity of 40 mg/L/day in red beet cell suspensions (Akita et al. 2002).

A few more studies have also focused on studying the influence of different microelements on betalain production. Increasing the iron concentration from 0.1 to 0.2 mM in LS medium was found to markedly increase the betacyanin content without causing drastic effects on biomass formation of *B. vulgaris* var. Extra Early Flat Egyptian cell suspension (Akita et al. 2000). The complete removal of zinc from cultivation medium resulted in an increase in betacyanin content (591 mg/L) in a 21-day batch culture (Akita et al. 2001). Removal of boron, iodine, manganese and molybdenum was found to adversely affect both the cell growth and betacyanin content, whereas complete absence of copper and cobalt did not show any negative effect on red beet cell suspension. While these effects of microelements were confirmed in three different types of *B. vulgaris* L. cell suspensions (Akita et al. 2002), such effects were not long lasting since certain enzymes require these elements to complete their structural integrities. For example tyrosinase involved in betalain synthesis (see Chap. 2) is a copper-containing enzyme (Moreno et al. 2008), and similarly polyphenoloxidase has many versions with different metal ions. Copper ions are also considered to act as cofactors imparting various effects in different effects indifferent culture systems (Leathers et al. 1992). Therefore, when experiments were conducted on the effects of cobalt, copper, iron, manganese, molybdenum and zinc on the betalain production by *B. vulgaris* cv. Crosby Egyptian cell suspension grown in Gamborg's B5 medium (Gamborg et al. 1968), it was observed that higher concentrations of all microelements except manganese showed a positive effect on betalain production (Akita et al. 2002). A 5-fold increase of cobalt concentration provoked the highest betalain accumulation (<17.75 mg/g DW), which was 60% higher than that obtained in standard B5 medium

(Trejo-Tapia et al. 1999, 2001). All in all, a careful optimization of microelements is essential, which should be based on a thorough knowledge about the target metabolite(s) (betalains in the present case) expected from the cell cultures.

9.3.5 Carbon Source

The type and concentration of carbon source have been found to significantly influence the type of cell cultures induced, their further growth and betalain synthesis. Initial studies with cell suspensions of *B. vulgaris* cv. Crosby Egyptian grown in B5 medium supplemented with 20 g/L of one among sucrose, lactose, galactose and glucose, was found to utilize sucrose for growth and betalain synthesis. Although this low sugar supplementation (20 g/L) was utilized only to a 38% extent, the betalain content was as high as 26 mg/g DW. These cells could also utilize lactose, but glucose was partially utilized with almost no growth in galactose. Further optimization of sucrose level resulted in the finding that the use of 10 g/L of sucrose was utilized to a 98% extent with a high betalain content of 48 mg/g DW (Rodríguez-Monroy et al. 1994). In other studies, a high sucrose requirement for cell growth up to 60 g/L was recorded, which was efficiently utilized; although the highest betalain content in cells (18 mg/g DW) was observed at 20 g/L of sucrose (Akita et al. 2000).

9.3.6 Effect of Growth Regulators

Growth regulators are essential for keeping continuous cell multiplication as well as for betalain synthesis. Growth regulators 2,4-D, indole-3-acetic acid (IAA), NAA, IBA (auxins), BA and kinetin (cytokinins) have been used for *B. vulgaris* callus and cell suspensions (Table 9.1). The biosynthesis of betacyanins in suspension culture of *B. vulgaris* cv. Extra Early Flat Egyptian increased when auxins were added. Maintaining cultures as cell suspension required continuous use of a stronger synthetic auxin, such as 2,4-D and IBA, and rarely low levels of cytokinins, particularly BA, (Akita et al. 2000). Use of growth regulators in different ratios provoked shifts in pigment types, indicating their influence on different regulatory enzymes of betalain biosynthesis. Visible color changes occurred from white/green calluses to a variety of colors, ranging from yellow, orange and red to deep violet (Girod and Zryd 1991; Akita et al. 2000). Use of auxins was found to increase polysomaty through endopolyploidization, causing drastic changes in ploidy levels in different organs of *Beta vulgaris*, calluses in vitro and rarely hairy roots (Weber et al. 2010). Such auxins and auxin-type of compounds may also affect the type and yield of betalains due to the presence of multiple copies of genes formed after endoreduplication (Weber et al. 2010). However, to what extent the duplicate copies of genes are involved in the biosynthetic pathways of these pigments has not so far been elucidated.

9.3.7 *Elicitation of Betalains*

Elicitors are traditionally used to stimulate cellular response, which induces the synthesis, and sometimes the release of valuable secondary metabolites. While cell cultures of other plant species producing betalains have been subjected for elicitor treatments (Bhuiyan and Adachi 2003), cell cultures of red beet have not been researched much in this direction. However, a large number of trials were done to elicit betalains in hairy root cultures of red beet (see Chap. 10), the applicability of such treatments to cell cultures needs well-designed experimental evidence.

9.3.8 *Recycling of Cell Cultures*

The stringent steps involved in culture initiation and cell line selection for high yields, as well as the long growth cycles of plant cell cultures indicate the need for smarter methods for the recovery of metabolites and recycling of the biomass. Thus suitable techniques of significant commercial value would be the continuous production and simultaneous recovery of secondary metabolites from plant cells, without causing much loss to their viabilities and biosynthetic capacities (see Chap. 13). Although betalains are stored in the vacuoles within cells, they can be forced out of the vacuole and further into the cell exterior by subjecting cells/tissues to a temporary stress phase (transient stress). Stress brings about changes in the normal physiology of cells, making cells release mainly the defense-related compounds from their vacuoles. Certain types of stress alter the electrical charges of lipid membranes, including that of vacuoles and plasmalemma, allowing permeabilization of cell membranes. Although most of the cell permeabilization experiments in red beet have been conducted with hairy roots, knowledge from such trials is often applicable to cell cultures of the same plant species. Permeabilization of hairy roots and pigment recovery techniques are discussed in Chap. 13.

9.4 Bioreactors

Shake flask cultures have always been good models for betalain production in cultured red beet cells. However, different types of bioreactors have been tried, particularly because of the clear visibility of the colorful betalains synthesized by red beet cells, these cell cultures are preferred as model system for exploring different types of bioreactors for reaping higher productivities. Most of the cultures obtained from mesophytes perform rarely well under submerged conditions, as far as secondary metabolites are concerned. The problem is enhanced further when cells are exposed to large medium volumes, where cells undergo hydrodynamic stress. In stirred tank bioreactors, *B. vulgaris* cells were found to undergo higher hydrodynamic stress than in shake flasks, causing a high accumulation of extracellular proteins and polysaccharides, which changed the properties of the medium from Newtonian to pseudoplastic (Sánchez et al. 2002). To overcome this problem, an air lift bioreactor with lower hydrodynamic stress than stirred tanks was used where the yield of betalains ranged widely from 4.2 to 35.0 mg/g DW, probably due to poor mass transfer caused by high variations in the pattern of mixing and aeration of culture medium within the growth chamber (Sánchez et al. 2002).

Fluidized bed bioreactors of 5 and 50 L were used to grow suspension cultures of *B. vulgaris* cv. Bikores Monogerm; high betalain yields of 14–17 mg/g DW/day were achieved, which were almost similar to the yields obtained in shake flasks (13–24 mg/g DW/day) (Khlebnikov et al. 1995). Contrarily, cell cultures of *B. vulgaris* cv. Crosby Egyptian grown in a stirred tank bioreactor resulted in a 62.5% lower betalain content than in shake flasks (Rodríguez-Monroy and Galindo 1999). Like many other stressed cells, the cell cultures of *B. vulgaris* cv. Bikores Monogerm were found to secrete high amounts of polysaccharides and proteins into the culture medium, which resulted in continuous changes in broth rheology, causing hydrodynamic stress in the bioreactor. Therefore, in a subsequent trial, this problem was overcome by adopting an air-lift bioreactor (10 L) which imparted low shear stress on the biomass. Although the problem of protein secretion was lowered, this model resulted in low betalain productivities (Sánchez et al. 2002). A special type of bioreactor with an inorganic membrane, namely, an electrophoretic tubular membrane reactor, was used by Yang et al. (2003) for accomplishing electro-permeabilization of red beet cells of var. Red Ball. This system allowed simultaneous release, transport and collection of ionic metabolic products – the betalains. Batch and continuous operations of reactor allowed red beet cell suspensions to grow well in this reactor, releasing betalains into the buffer, which also resulted in retaining cell viability (Yang et al. 2003).

9.5 Production of Betalains from Hairy Root-Derived Callus

Red beet hairy roots, obtained as a result of transfection of beet leaves of var. Detroit Dark Red with *Agrobacterium rhizogenes* ATCC 15834 bacteria, were cultured to induce callus by incubation in the dark on MS medium supplemented with various concentrations (0, 0.1, 0.5 mg/L) of IAA, IBA and BAP. The hairy root explants formed calluses after 3 weeks on MS medium containing 0.5 mg/L IBA and 0.5 mg/L BAP, where bright red and pure yellow cultures could be obtained by repeated manual selection and were maintained as respective colored cultures for over a year (Weber et al. 2010). Wide variations in the ratio of betacyanin:betaxanthin were observed in these cultures. DNA content analyses of in vivo leaves, hairy roots and callus indicated that leaf cells undergo more endopolyploidization, whereas hairy roots consistently retained normal diploid level. However, the callus cells of hairy roots showed the highest DNA content, meaning the highest endo-reduplication, which the authors linked to the presence of auxin in the culture medium (Weber et al. 2010), since auxin has been reported to enhance polysomaty (Mishiba et al. 2001). In red beet calluses, the degree of endo-reduplication correlated with the endopolyploidy of the explant (D'Amato 1990). The fact that primary callus from hairy roots had a greater polysomaty (up to 16×) than the secondary calluses is indicative of some unknown interactions between the red and the yellow cells that trigger high endo-reduplication, which also reflects their altered metabolic profiles.

9.6 Production of Betalains in Sugar Beet Cultures

As indicated in Chap. 1, the non-pigmented high sugar-producing beet cultivars originated from red beet, which were improved by selection and breeding for high sugar biosynthesis. Apparently, betalain biosynthetic pathways in these plants remain redundant, recent studies have shown that they also display pigments under stress and artificial culture conditions. Large variations in cellular metabolic profiles were observed when sugar beet (*Beta vulgaris* L. var. *altissima*) was transformed using a wild octopine *Agrobacterium tumefaciens* strain, B6S3. Infection of leaf explants and cultivation in vitro of the responsive tumerous type of cultures resulted in the selection of two culture lines: one appearing pale green in color and the other red–violet. The latter type of culture was found to synthesize betanin (betacyanin – the purple pigment), which was strongly regulated by light (Pavokovic et al. 2009). To enhance the betanin yield, nutrient media with different carbohydrates like sucrose, glucose and fructose, either singly or in combination were tested. These sugars profoundly influenced pigment biosynthesis, the highest betanin pigment formation being 13.5 mg/g DW in fructose-containing medium, which was 20–40% higher than in medium with sucrose or the combination of glucose and fructose. Since sucrose supported high biomass, the productivity of betanin was higher in this medium when cultures were grown for 14 days. Because the pigment was purely betanin, which has more preference for food applications owing to its high stability, these transformed cultures hold great promise for commercial production of betanin, provided the T-DNA of Agrobacterium has legal approval.

The red violet callus cultures of sugar beet were found to lose pigmentation when grown in complete darkness. Such non-pigmented dark-grown culture lines regained the capacity to synthesize red pigments when repeatedly cultured in light. This is an interesting phenomenon worth studying at biochemical level, in view of contrasting behavior of other cell cultures of beet such as dark-grown cell cultures of red beet (Akita et al. 2000) and light-insensitive beet hairy root (also transformed) cultures (Thimmaraju et al. 2003a, b), which are profusely pigmented. However, the transformed sugar beet cultures were stable, retaining green, white and red color, and containing, respectively, chloroplasts, amyloplasts and chromoplasts under appropriate conditions for over 10 years (Pavokovic et al. 2009).

Aiming at betanin yield enhancement, cell suspensions were initiated from the friable callus cells and grown as 21-day batch cultures (Křižnik and Pavoković 2010). The biomass accumulation and the content of betanin were monitored under functions of sucrose levels versus mineral modifications and elicitor treatment. Results showed that sequential increase of sucrose from 3% to 6% (w/v) significantly enhanced biomass, increasing the betanin yield up to 250%, whereas mineral modifications enhanced 20% betanin. Elicitation at early exponential phase by ten-fold increase in calcium or yeast extract (0.25%) was ineffective. Contrarily, the latter elicitor caused leaching of pigments. Thus certain beet cell cultures display much divergent behaviors in vitro.

9.7 Concluding Remarks

The high cellular plasticity of the genus *Beta* has opened a wide panorama for its application as a research model for the development of cell culture-related technologies. So far such opportunities are exploited mostly for improving sugar beet varieties, although red beet cells are more interesting for basic research, starting from cell cycle studies to genetics of totipotency and secondary metabolite (betalains) synthesis followed by their sequestration into the vacuoles. A few studies used cultured red beet cells as a model system for protoplast and vacuole isolation techniques; however, the knowledge thus generated has not been applied for characterizing various biochemical events associated with cell membrane, wall synthesis and cell-to-cell communication signals, which are hot topics of current interest. Beet cells in culture have shown great versatality such as high genetic stabilities with consistent production of betalains for over 10 years (Akita et al. 2002) as well as the formation of variants, the waking-up of the sleeping betalain biosynthetic pathway, as in sugar beet (Pavokovic et al. 2009) and yellow–red–purple callus cultures from hairy roots (Weber et al. 2010). The pigment biosynthetic capacity of cultured red beet cells, even in the absence of photosynthesis, is an interesting phenomenon, not fully characterized so far. The advent of modern genomics, proteomics and metabolomics techniques offers tremendous opportunities to dissect and understand the uncharacterized biosynthetic steps and relevant gene regulation mechanisms. The acquired knowledge may unravel the mystery behind the non-reversible and stable pigment biosynthesis by certain cell lines as those of transformed sugar beet callus cultures, and pure betanin synthesising cell lines of Extra Early Flat Egyptian cultivar used by Akita et al. (2000, 2002).

References

Akita, T., Y. Hina, and T. Nishi. 2000. Production of betacyanins by a cell suspension culture of table beet (*Beta vulgaris* L.). *Bioscience, Biotechnology, and Biochemistry* 64: 1807–1812.

Akita, T., Y. Hina, and T. Nishi. 2001. Effect of zinc deficiency on betacyanin production in a cell suspension culture of table beet (*Beta vulgaris* L.). *Bioscience, Biotechnology, and Biochemistry* 65: 962–965.

Akita, T., Y. Hina, and T. Nishi. 2002. New medium composition for high betacyanin production by a cell suspension culture of table beet (*Beta vulgaris* L.). *Bioscience, Biotechnology, and Biochemistry* 66: 902–905.

Atanassov, A.I. 1986. Sugar beet. In *Handbook of plant cell culture*, ed. D.A. Evans, W.R. Sharpe, and P.V. Ammirato, 652–680. New York: Macmillan.

Bhuiyan, N.H., and T. Adachi. 2003. Stimulation of betacyanin synthesis through exogenous methyl jasmonate and other elicitors in suspension-cultured cells of *Portulaca*. *Journal of Plant Physiology* 160: 1117–1124.

Bohanec, B., A. Touraev, B. Forster, and M. Jain. 2009. Doubled haploids via gynogenesis. In *Advances in haploid production in higher plants*, 35–46. Heidelberg: Springer.

Bokern, M., S. Heuer, V. Wray, L. Witte, T. Macek, T. Vanek, and D. Strack. 1991. Ferulic acid conjugates and betacyanins from cell cultures of *Beta vulgaris*. *Phytochemistry* 30: 3261–3265.

Bossoutrot, D., and D. Hosemans. 1985. Gynogenesis in *Beta vulgaris* L.: From in vitro culture of unpollinated ovules to the production of doubled haploid plants in soil. *Plant Cell Reports* 4: 300–303.

Botău, D., and S. Chiş. 2008. The production of some forage beet somaclones. *Botău* 40: 263–266.

Broadley, M.R., P.J. White, J.P. Hammond, I. Zelko, and L.A. Alexander. 2007. Zinc in plants – Tansley review. *New Phytologist* 173: 677–702.

Christianson, M.L., and D.A. Warnick. 1985. Temporal requirement for phytohormone balance in the control of organogenesis in vitro. *Developmental Biology* 12: 494–497.

Christianson, M.L., and D.A. Warnik. 1983. Competence and determination in the process of in vitro shoot organogenesis. *Developmental Biology* 95: 288–293.

Constabel, F., and H. Nassif-Makki. 1971. Betalain bildung in beta-callus kulturen. *Berichte der Deutschen Botanischen Gesellschaft* 84: 629–636.

D'Amato, F. 1990. Somatic nuclear mutations in vivo and in vitro in higher plants. *Caryologia* 43: 191–204.

Detrez, C., T. Tetu, R.S. Sangwan, and B.S. Sangwan-Norrel. 1988. Direct organogenesis from petiole and thin cell layer explants in sugar beet cultured in vitro. *Journal of Experimental Botany* 39: 917–926.

Doctrinal, M., R.S. Sangwan, and B.S. Sangwan-Norreel. 1989. In vitro gynogenesis in *Beta vulgaris* L.: Effect of plant growth regulators, temperature, lines and season. *Plant Cell Tissue and Organ Culture* 17: 1–12.

Dovzhenko, A., and H.U. Koop. 2003. Sugar beet (*Beta vulgaris* L.): Shoot Regeneration from callus and callus protoplasts. *Planta* 217: 374–381.

Escribano, J., M.A. Pedrenõ, F. García-Carmona, and R. Munõz. 1998. Characterization of the antiradical activity of betalains from *Beta vulgaris* L. roots. *Phytochemical Analysis* 9: 124–127.

Gamborg, O.L., R.A. Miller, and K. Ojima. 1968. Nutrient requirements of suspension cultures of soybean root cells. *Experimental Cell Research* 50: 151–158.

Germana, M.A. 2011. Gametic embryogenesis and haploid technology as valuable support to plant breeding. *Plant Cell Reports* 30: 839–857.

Girod, P.A., and J.P. Zryd. 1987. Clonal variability and light induction of betalain synthesis in red beet cell cultures. *Plant Cell Reports* 6: 27–30.

Girod, P.A., and J.P. Zryd. 1991. Secondary metabolism in cultured red beet (*Beta vulgaris* L.) cells: Differential regulation of betaxanthin and betacyanin biosynthesis. *Plant Cell, Tissue and Organ Culture* 25: 1–12.

Grassmann, J., S. Hippeli, and E.F. Elstner. 2002. Plant's defence and its benefits for animal and medicine: role of phenolics and terpenoids in avoiding oxygen stress. *Plant Physiology and Biochemistry* 40: 471–478.

Grieve, T.M., K.M.A. Gartland, and M.C. Elliot. 1997. Micropropagation of commercially important sugar beet cultivars. *Plant Growth Regulation* 21: 15–18.

Gurel, E. 1997. Callus and root development from leaf explants of sugar beet (*Beta vulgaris* L.): Variability at variety, plant and organ level. *Turkish Journal of Botany* 21: 131–136.

Gurel, S., E. Gurel, and Z. Kaya. 2001. Callus development and indirect shoot regeneration from seedling explants of sugar beet (*Beta vulgaris* L.) cultured in vitro. *Turkish Journal of Botany* 25: 25–33.

Gurel, E., E. Topal, and S. Gurel. 2003a. The effect of pretreating seedlings with BA on direct shoot regeneration from petiole explants of sugar beet (*Beta vulgaris* L.). *Biotechnol Biotechniques* 17: 89–96.

Gurel, S., E. Topal, and E. Gurel. 2003b. The effect of pretreating seedlings with TDZ on direct shoot regeneration from petiole explants of sugar beet (*Beta vulgaris* L.). *Asia Pacific Journal of Molecular Biology* 11: 57–62.

Gurel, E., S. Gurel, and P.G. Lemaux. 2008. Biotechnology applications for sugar beet. *Critical Reviews in Plant Sciences* 27: 108–140.

Gurel, S., M. Cengiz Baloglu, E. Gurel, H. Avni Oktem, and M. Yucel. 2011. A two-stage pretreatment of seedlings improves adventitious shoot regeneration in sugar beet (*Beta vulgaris* L.). *Plant Cell, Tissue and Organ Culture*. DOI 10.1007/s11240-011-9916-8.

Hajiboland, R., and E. Amirazad. 2010. Growth, photosynthesis and antioxidant defense system in Zn-deficient red cabbage plants. *Plant, Soil and Environment* 56: 209–217.

Harms, C.T., I. Baktir, and J.J. Oertli. 1983. Clonal propagation in vitro of red beet (*Beta vulgaris* ssp.) by multiple adventitious shoot formation. *Plant cell, tissue and organ culture* 2: 93–102.

Henriques, F.S. 2001. Loss of blade photosynthetic area and of chloroplasts' photochemical capacity account for reduced CO2 assimilation rates in zinc-deficient sugar beet leaves. *Journal of Plant Physiology* 158: 915–919.

Hussey, G., and A. Hepher. 1978. Clonal propagation of sugar beet plants and the formation of polyploids by tissue cultures. *Annals of Botany* 42: 447–478.

Ibrahim, R., and C.D. Pierre. 2001. Factors controlling high efficiency adventitious bud formation and plant regeneration from in vitro leaf explants of roses (*Rosa hybrida* L.). *Scientia Horticulturae* 88: 41–57.

Jacq, B., T. Tetu, R.S. Sangwan, A. De Laat, and B.S. Sangwan-Norreel. 1993. Efficient production of uniform plants from cotyledon explants of sugar beet (*Beta vulgaris* L). *Plant Breeding Reviews* 110: 185–191.

Jazdzewska, E., Z. Sadoch, A. Niklas, and A. Majewska-Sawka. 2000. Plant regeneration from sugar beet leaf protoplasts: Analysis of shoots by DNA fingerprinting and restriction fragment length polymorphism. *Canadian Journal of Botany* 78: 10–18.

Khlebnikov, A., B. Dubuis, O.M. Kut, and J.E. Prenosil. 1995. Growth and productivity of *Beta vulgaris* cell culture in fluidized bed reactors. *Bioprocess Engineering* 14: 51–56.

Konwar, B.K., and R.H.R. Coutts. 1990. Rapid regeneration of sugar beet (*Beta vulgaris* L.) plants from in vitro cultures. In *Progress in plant cellular and molecular biology*, ed. H.J.I. Nijkamp, L.H.W. Van Der Plas, and J. Van Aartrijk, 114–118. The Netherlands: Kluwer Academic.

Krens, F.A., and D. Jamar. 1989. The role of explants source and culture conditions on callus induction and shoot regeneration in sugar beet (*Beta vulgaris* L.). *Journal of Plant Physiology* 134: 651–655.

Krens, F.A., D. Jamar, G.J.A. Rouwendal, and R.D. Hall. 1990. Transfer of cytoplasm from new *Beta* CMS sources to sugar beet by asymmetric fusion. I. Shoot regeneration from mesophyll protoplasts and characterization of regenerated plants. *Theoretical and Applied Genetics* 79: 390–396.

Križnik, B., and D. Pavoković. 2010. Enhancement of betanin yield in transformed cells of sugar beet *(Beta vulgaris* L.). *Acta Botanica Croatica* 69: 173–182.

Kubalakova, M. 1990. Somatic embryogenesis and cytoplasmic sterility in *Beta vulgaris* L. var. saccharifera. *Biologia Plantarum* 32: 414–419.

Leathers, R.R., C. Davin, and J.P. Zryd. 1992. Betalain producing cell cultures of *Beta vulgaris* L var. Bikores monogerm (Red beet). *In Vitro Cellular & Development Biology Plant* 28: 39–45.

Lenzner, S., K. Zoglauer, and O. Schieder. 1995. Plant regeneration from protoplasts of sugar beet *(Beta vulgaris)*. *Physiologia Plantarum* 94: 342–350.

Linsmaier, E.M., and F. Skoog. 1965. Organic growth factor requirements of tobacco tissue cultures. *Physiologia Plantarum* 18: 100–127.

Livadariu, O., N. Băbeanu, O. Popa, and I. Ionescu-Mălăncu. 2009. Study regarding in vitro propagation through somatic embryogenesis in *Beta vulgaris* L. *Bulletin UASVM Animal science biotechnology* 66: 368–373.

Margara, J. 1977. La multiplication vegetative de la betterave (*Beta vulgaris* L.) en culturein vitro. *Comptes-Rendu Acadimie des Sciences* 285: 1041–1044.

Marylise, D., R.S. Sangwan, and B.S. Sangwan-Norreel. 1989. In vitro gynogenesis in *Beta vulgaris* L.: Effects of plant growth regulators, temperature, genotypes and season. *Plant cell, tissue and organ culture* 17: 1–12.
Meng, L., S. Zhang, and P.G. Lemaux. 2010. Toward molecular understanding of in vitro and in planta shoot organogenesis. *Critical Reviews in Plant Sciences* 29: 108–122.
Miedema, P. 1982. A tissue culture technique for vegetative propagation and low temperature preservation of *Beta vulgaris* L. *Euphytica* 31: 635–643.
Mikami, T., R. Sudoh, E. Nagao, and T. Kinoshita. 1989. Genotypic variation in the in vitro morphogenesis from leaf explants of *Beta vulgaris* L. and *B. maritima* L. *Euphytica* 40: 271–273.
Mishiba, K., T. Okamoto, and M. Mii. 2001. Increasing ploidy level in cell suspension cultures of *Doritaenopsis* by exogenous application of 2,4-dichlorophenoxyacetic acid. *Physiologia Plantarum* 112: 142–148.
Ehsani, M.B., M. Mahmoud, and Y. Nasrin. 2000. The effect of *in planta* TIBA and proline treatment on somatic embryogenesis of sugar beet (*Beta vulgaris* L.). *Euphytica* 112: 151–156.
Moreno, D.A., C. García-Viguera, J.I. Gil, and A. Gil-Izquierdo. 2008. Betalains in the era of global agri-food science, technology and nutritional health. *Phytochemistry Reviews* 7: 261–280.
Murashige, T., and F. Skoog. 1962. A revised medium for rapid growth and bioassay with tobacco tissue cultures. *Physiologia Plantarum* 15: 473–497.
Mustafa, N.R., W. de Winter, F. van Iren, and R. Verpoorte. 2011. Initiation, growth and cryopreservation of plant cell suspension cultures. *Nature Protocols* 6: 715–742.
Pavokovic, D., G. Rusak, V. Besendorfer, and M. Krsnik-Rasol. 2009. Light-dependent betanin production by transformed cells of sugar beet. *Food Technology and Biotechnology* 47: 153–158.
Pedersen, H.C., R.D. Hall, and F.A. Krens. 1993. Petioles as the tissue source for isolation and culture of *Beta vulgaris* and *B. maritima* protoplasts. *Plant Science* 95: 89–97.
Pradedova, E.V., O.D. Isheeva, and R.K. Salyaev. 2011. Antioxidant defense enzymes in cell vacuoles of red beet roots. *Russian Journal of Plant Physiology* 58: 36–44.
Ritchie, G.A., K.C. Shoort, and M.R. Davey. 1989. In vitro shoot regeneration from callus, leaf axils and petioles of sugar beet (*Beta vulgaris* L.). *Journal of Experimental Botany* 40: 277–283.
Rodríguez-Monroy, M., A. Jiménez-Aparicio, G. Dávila-Ortíz, and G. Sepúlveda-Jiménez. 1994. Effect of carbon source on cell growth and betalain production in cell suspension culture of *Beta vulgaris*. *Biotechnology Letters* 16: 853–858.
Rodriguez-Monroy M. and E. Galindo. 1999. Broth rheology, growth and metabolite production of *Beta vulgaris* suspension culture: a comparative study between cultures grown in shake flasks and in a stirred tank. *Enzyme and Microbial Technology* 24: 687–693.
Sánchez, M.J., A. Jiménez-Aparicio, G.G. López, G. Trejo-Tapia, and M. Rodríguez-Monroy. 2002. Broth rheology of *Beta vulgaris* cultures growing in an air lift bioreactor. *Biochemical Engineering Journal* 12: 37–41.
Saunders, J.W. 1982. A flexible *in vitro* shoot culture propagation system for sugar beet that induces rapid floral induction of ramets. *Crop Science* 22: 1102–1105.
Saunders, J.W., and M.E. Daub. 1984. Shoot regeneration from hormone-autonomous callus from shoot cultures of several sugar beet (*Beta vulgaris* L.) genotypes. *Plant science Letter* 34: 219–223.
Shimamoto, Y., H. Hayakawa, J. Abe, H. Nakashima, and T. Mikami. 1993. Callus induction and plant regeneration of *Beta* germplasm. *Journal of Sugarbeet Research* 30: 317–319.
Skoog, F., and C.O. Miller. 1957. Chemical regulation of growth and organ formation in plant tissues culture in vitro. *Symposia of the Society for Experimental Biology* 11: 118–131.
Sreedhar, R.V., L. Venkatachalam, R. Thimmaraju, N. Bhagyalakshmi, M.S. Narayan, and G.A. Ravishankar. 2008. Direct organogenesis from leaf explants of *Stevia rebaudiana* and cultivation in bioreactor. *Biologia Plantarum* 52: 355–360.
Stafford, A., L. Smith, and M.W. Fowler. 1985. Regulation of product synthesis in cell cultures of *Catharanthus roseus* (L) G. Don. *Plant cell, tissue and organ culture* 4: 83–94.
Steen, P., B. Keimer, K. D'Halluin, and H.C. Pedersen. 1986. Variability in plants of sugar beet (*Beta vulgaris* L.) regenerated from callus, cell-suspension and protoplasts. In *Genetic manipulation in plant breeding*, ed. W. Horn, C.J. Jensen, W. Odenbach, and O. Schieder, 633–635. Berlin: Walter de Gruyter.

Tenning, P., E.W. Weich, U.B. Kjarsgaard, M.A. Lelu, and M. Nihlgard. 1992. Somatic embryogenesis from zygotic embryos of sugar beet (*Beta vulgaris* L.). *Plant Science* 81: 103–109.

Tetu, T., R.S. Sangwan, and B.S. Sangwan-Norreel. 1987. Hormonal control of organogenesis and somatic embryogenesis in *Beta vulgaris* callus. *Journal of Experimental Botany* 38: 506–517.

Thimmaraju, R., N. Bhagyalakshmi, M.S. Narayan, and G.A. Ravishankar. 2003a. Kinetics of betalaine release under the influence of physical factors. *Proceedings of Biochemical* 38: 1067–1074.

Thimmaraju, R., N. Bhagyalakshmi, M.S. Narayan, and G.A. Ravishankar. 2003b. Food grade chemical and biological agents assist the release of betalains from hairy root cultures of *Beta vulgaris*. *Biotechnology Progress* 19: 1274–1282.

Trejo-Tapia, G., A. Jimenez-Aparicio, M. Rodriguez, G. Sepulveda, G. Salcedo, B. Martinez, et al. 1999. Influence of medium constituents on growth and betalain production in cell suspension cultures of *Beta vulgaris*. *Asia-Pacific Journal Molecular Biology Biotechnol* 7: 167–172.

Trejo-Tapia, G., A. Jimenez-Aparicio, M. Rodriguez-Monroy, A. De Jesus-Sanchez, and G. Gutierrez-Lopez. 2001. Influence of cobalt and other microelements on the production of betalains and the growth of suspension cultures of *Beta vulgaris*. *Plant cell, tissue and organ culture* 67: 19–23.

Veit, B. 2009. Hormone mediated regulation of the shoot apical meristem. *Plant Molecular Biological* 69: 397–408.

Wang, H.M., Y.G. Zu, W.J. Wang, and F.L. Dong. 2006. Establishment of *Camptotheca* acuminata regeneration from leaf explants. *Biologia Plantarum* 50: 725–728.

Weber, J., V. Georgiev, C. Haas, T. Bley, and A. Pavlov. 2010. Ploidy levels in *Beta vulgaris* (red beet) plant organs and *in vitro* systems. *Engineering Life Science* 10: 139–147.

Wei, Q., and J.F. Zhang. 1998. Tissue culture and plant regeneration with petiole of six different genotypes of sugar beet. *Sugar Crops China* 4: 21–24.

Weller, T.A., and L. Lasure. 1982. Betalains in beet root tissue culture. *Journal of Food Science* 47: 162–163.

Wiśniewska, E., and A. Majewska-Sawka. 2008. The differences in cell wall composition in leaves and regenerating protoplasts of *Beta vulgaris* and *Nicotiana tabacum*. *Biologia Plantarum* 52: 634–641.

Xu, Q.L., Y.H. Xie, H. Ru, X. Hu, Ch-Y Wang, and W.X. Yu. 2009. Efficient plant regeneration in vitro from red leaf beet via organogenesis. *Russian Journal Plant Physiological* 56: 546–550.

Yang, R.Y.K., O. Bayraktar, and H.T. Pu. 2003. Plant-cell bioreactors with simultaneous electropermeabilization and electrophoresis. *Journal of Biotechnology* 100: 13–22.

Yu, M.H. 1989. Callus induction and differentiation from leaf explants of different species of the genus *Beta*. *Crop Science* 29: 205–209.

Zhang, S., and P.G. Lemaux. 2004. Molecular analysis of in vitro shoot organogenesis. *Critical Reviews in Plant Sciences* 23: 325–335.

Zhang, J.E., X.R. Wang, Q. Wei, and M.H. Li. 1993. Tissue culture and plant regeneration in *Beta maritime* L. *China Beet Sugar* 2: 9–12.

Zhang, C.L., D.F. Chen, M.C. Elliott, and A. Slater. 2001. Thidiazuron-induced organogenesis and somatic embryogenesis in sugar beet (*Beta vulgaris* L). *In Vitro Cellular & Development Biology Plant* 37: 305–310.

Zhang, C.L., D.F. Chen, M. Kubalakova, J. Zhang, N.W. Scott, M.C. Elliott, and A. Slater. 2008. Efficient somatic embryogenesis in sugar beet (*Beta vulgaris* L.) breeding lines. *Plant Cell, Tissue and Organ Culture* 93: 209–221.

Zhong, Z., H.G. Smith, and T.H. Thomas. 1993a. Micropropagation of wild beet (*Beta maritima*) from inflorescence pieces. *Plant growth regulation* 12: 53–57.

Zhong, Z., H.G. Smith, and T.H. Thomas. 1993b. In vitro culture of petioles and intact leaves of sugar beet (*Beta vulgaris*). *Plant growth regulation* 12: 59–66.

Chapter 10
Red Beet Hairy Root Cultures

Bhagyalakshmi Neelwarne

Abstract The invention of hairy root induction from various higher plants has provided a new set of hopes for using such organs in vitro as an alternative to cell cultures, chiefly because of their genetic stabilities that impart them high biochemical consistencies. Several research groups characterized different types of red beet hairy root clones, their scaled-up performances as well as their applications for various basic studies. Rather than cell cultures, hairy roots of red beet have attracted the attention of more number of research groups, probably because of their genetic aspects, spectacular colors, various morphologies, clonal stability and for other technological challenges associated with their scale-up and product recovery. The astonishingly variable responses of a cultivar of red beet (Ruby Queen) to different strains of *Agrobacterium rhizogenes* and the variable spectrum of morpho-physiological responses to nutrient components make red beet an ever interesting material for genetic as well as physiological studies, as discussed in this chapter. The fact that higher plants can transcribe animal genes and perfectly translate them into functional proteins has attracted newer opportunities for obtaining boundless number of important therapeutic proteins from cultured plant cells and organs. Hairy roots are often chosen for such purposes because of their consistent performances under fully manageable in vitro conditions. Many opportunities could also be realized with red beet hairy roots.

B. Neelwarne (✉)
Plant Cell Biotechnology Department, Central Food Technological Research Institute
(Council of Scientific and Industrial Research), Mysore 570 020, India
e-mail: bneelwarne@cftri.res.in

B. Neelwarne (ed.), *Red Beet Biotechnology: Food and Pharmaceutical Applications*, 199
DOI 10.1007/978-1-4614-3458-0_10, © Springer Science+Business Media New York 2013

10.1 Introduction

The new market trend for natural compounds has concomitantly increased the demand for natural compounds from plants. Over 20,000 plant species are used for food and pharmaceutical purposes, of which, the majority of the pharmaceutically important compounds are sourced from roots. Hence there is a great scope for the application of cultured roots for the production of important bioactive compounds. Hairy roots of higher plant species, including trees, often produce the same metabolites that are found in the normal roots. Hairy roots in culture are genetically very stable, capable of rapid and indefinite growth with prolific lateral root formation on a hormone free medium, amenable for desirable genetic manipulation and hence form an attractive alternative to suspension cultures and normal roots, which frequently exhibit biochemical variability. In tuberous crops like red beet, the hairy roots do not form tubers, instead they grow continuously like non-storage roots with tremendous branching, producing similar levels of almost all compounds present in normal tubers. Brief descriptions of the induction and genetic characteristics of hairy roots are provided below, which are indicative of the enormous potential inherent in red beet hairy roots for various technological applications.

10.2 General Features of Hairy Roots

The infection in dicotyledonous plants with the soil-inhabiting bacterium *A. rhizogenes* and the consequent root formation is commonly known as "hairy root". In nature, the formation of hairy root disease or hairy root syndrome was first reported by Riker in 1930 and subsequently by Hildebrand in 1934. Since then the process of infection (Fig. 10.1), the resulting hairy roots and the associated cellular genetics have been intensely researched. Consequently, several clones of *A. rhizogenes* have been continuously induced from different food and pharmaceutically important plants. Further, application of genetic engineering techniques have resulted in isolating root clones for their various beneficial applications, which involve using Agrobacterium as a means of transferring desired genes into plants as well as to induce hairy roots for the over-production of root-based secondary metabolites (Flores 1986; Tepfer 1984; Gelvin 2000; Hilton and Rhodes 1990; Granicher et al. 1995; Christen et al. 1989; Fukui et al. 1998; Pavlov and Bley 2005; Pavlov et al. 2005; Savitha et al. 2006; Thimmaraju et al. 2008; Neelwarne and Thimmaraju 2009). The ploidy and karyotypes of hairy roots in a number of plant species were found to be identical to donor plants (Aird et al. 1988; Baiza et al. 1999; Weber et al. 2010), which was not the case in respective cell cultures, indicating a high genetic stability of hairy roots. Many key genetic factors are associated with the infecting bacterium as well as with the host plant cell, which together predominantly influence the quality of hairy roots, their growth performance and the quality/quantity of secondary metabolites produced. Therefore, the following section describes the genetic aspects of *Agrobacterium* infection process as well as the host genome effects on

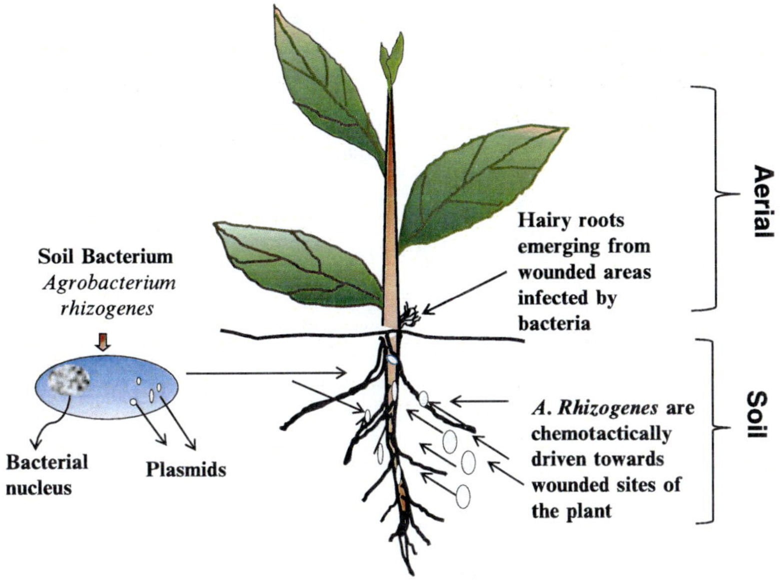

Fig. 10.1 The process of infection by *A. rhizogenes.* The hairy roots synthesize opines and shed them to the soil, where they are used up by Agrobacterium

the quality of hairy roots, which probably are the main reasons for wide differences in red beet hairy root clones reported by various research groups. The genetic interactions described below will also provide insights into the vast differences in the characteristics of red beet hairy roots discerned from each infection process.

10.2.1 Genetic Aspects of Hairy Roots

Major developments in plant genetic engineering have been possible due to an upsurge of information on the functioning of plasmids present in the two bacterial species of *Agrobacterium.* Plasmids in *A. tumefaciens* (Tm) cause crown gall disease (Seishiro and Kunihiko 2000) and plasmids in *A. rhizogenes* (Ri) cause rooty teratomas in plants, the latter is commonly known as "hairy roots", deriving the name from their morphological appearance. The bacterium performs the function of transferring a portion of the plasmid genome, namely, the T-DNA region and integrates into the genome of most of the dicotyledonous plants, including red beet and a few monocotyledonous plants. For the integration of T-DNA, left-border and right-border sequences, which are *cis*-acting 25-bp direct repeats, are crucial (Wang et al. 1984). The T-DNA contains a number of genes (Fig. 10.2) that are transferred, integrated and expressed in the transformed higher plant cells. The genes mainly expressed are those coding for synthesis as well as alterations of phytohormones: the synthesis of cytokinin (*ipt*), auxin indoleacetic acid (*tms* or *tmr* and *iaaM*) and

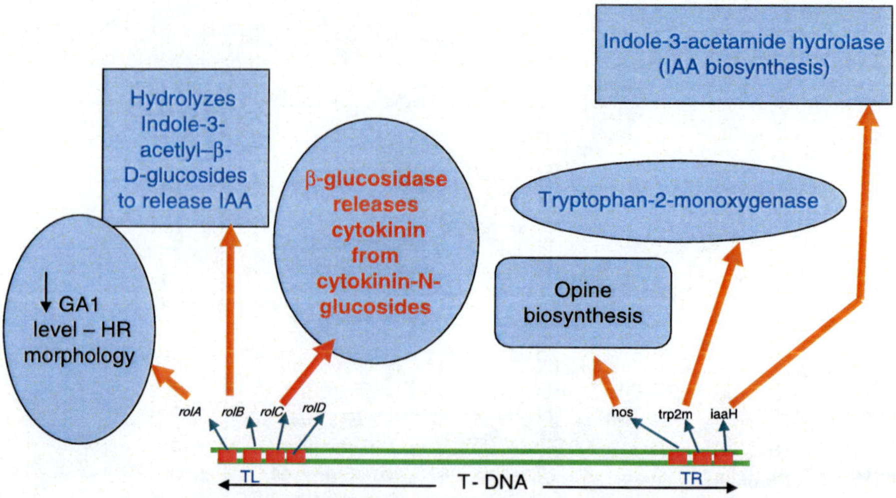

Fig. 10.2 The T-DNA of plasmid lodged within *A. rhizogenes*, showing some of the important genes that instigate hairy root formation and their further growth. *TR* right border of T-DNA, *TL* left border of T-DNA, *rol* rooty locus, *nos* opine synthesis gene, *trp* gene for tryptophan synthesis, *iaaH* gene coding for IAA biosynthesis

indoleacetamide (*iaaH*), which explains why the transformed cells produce either tumors or hairy roots in vivo and in vitro even in the absence of externally supplied phytohormones. There are a number of different types of tumor-inducing (Ti) and hairy root-inducing (Ri) plasmids. They are classified based on the type of opines (non-protein amino acids) produced in the tumors or hairy roots. Opines are formed via enzymes that are encoded by the T-DNA only after its integration into the chromosome of higher plant cell. Since opines are not generally detected in most of the normal higher plants, their presence in a tissue or organ can be considered as a reliable indication of transformation via *Agrobacterium* of the respective tissue or organ of the plant (Hooykaas and Schilperoort 1992).

10.2.2 *Agrobacterium rhizogenes* and Its Plasmids

T-DNA of the Ri plasmid is known to harbor two important categories of genes, one set is involved in phytohormone biosynthesis for stimulating cell division at the site of infection resulting in root growth (Morris 1986; Walden et al. 1993; Nilsson and Olsson 1997; Tanaka et al. 1985, 1994, 2001). The other set of genes are responsible for hairy root morphology. The T-DNA is divided into left T-DNA (T_L-DNA) and right T-DNA (T_R-DNA) regions (De Paolis et al. 1985; Jounin 1984), the T_L-DNA of *A. rhizogenes* is known to contain genes that control hormonal synthesis, altering the balance of hormones in the host cell in such a way that such cells are driven to form hairy roots (Cardarelli et al. 1987); wherein, each transformed cell results in the formation of one root clone, which may be entirely different from that formed from the adjacent cell. T_R-DNA contains two genes, *iaaM* and *iaaH*, responsible for the biosynthesis of auxins, as well as genes coding for the synthesis of opines mannopine (*mas1'* and *mas2'*) and agropine (*ags*). T_L-DNA carries 18 open-reading frames (ORF), of which four play vital roles in hairy root inductio: ORF10, ORF11, ORF12 and ORF15, which correspond to *rolA*, *rolB*, *rolC* and *rolD* genes, respectively (White et al. 1985) (Fig. 10.2). Of these, the *rolB* gene is absolutely essential for the induction of hairy roots, since it induced hairy roots even when expressed alone (Nilsson and Olsson 1997).

Although Ri and Ti plasmids are compatible and indicate divergent replication systems, they were found to have retained homology in the replication region. A clone containing the replication region of pTiA6 was found to hybridize to the *Hin*dIII fragment 3 of pRiA4b-derived clone pGH35 (Huffman et al. 1984). The homologous region within the plasmid T-DNA region of *A. rhizogenes* clone, pRiA4b and *A. tumefaciens* clones pTiA6 or pTiT37, was found to extend to a short region of *Hin*dIII-11 to *Hin*dIII-16, i.e., about 15–20 kb (Yoshikawa and Furuya 1987). Therefore homology between these regions of either Ti plasmid or Ri plasmid is limited to the *Hin*dIII homologous segment of T-DNA from pTiT37. Homology of Ri plasmid to the *tms* loci of the Ti plasmid pTiA6 mapped to more than 15 kb to the right region of earlier identified T-DNA. This information indirectly suggests that there could be additional T-DNA regions existing in hairy root teratomas. Therefore, one can argue that *tms* homology must occur in the T-DNA of Ri plasmids or *tmr* homology should be lacking. The observation that the rooty teratomas induced by *A. rhizogenes* is comparable to that of tumors induced by *tmr* mutants of *A. tumefaciens* suggests that both the phenotypes have the same underlying mechanism (White et al. 1985). Nevertheless, sequencing of the entire T_L-DNA of an agropine type Ri plasmid of *A. rhizogenes* revealed no extensive homology in a study where Ri T_L-DNA coding and non-coding regions were compared with Ti plasmid pTi15955 (Slightom et al. 1986).

10.2.3 Functions of rol genes

Although the *aux* genes are necessary for auxin-autotrophic cell division of the infected plant cell, and the *rolABCD* genes are not directly involved in auxin autotrophy (Nemoto et al. 2009a, b), hairy root morphology is mostly conferred by *rol* genes. Therefore, more information about these genes is contextual, especially when different morphotypes have been reported in hairy roots of red beet (Thimmaraju et al. 2008). The morphological changes that were recorded when the *rolA* gene of T_L-DNA was transferred and expressed in tobacco were promoter dependent. Tobacco with *rolA* driven by its own promoter resulted in normal-sized plants (Schmulling et al. 1988), whereas plants with *rolA* driven by the *CaMV35S* promoter (p35S) showed stunted growth, leading to dwarf plants with dark wrinkled leaves (Dehio et al. 1993). Other variations recorded were altered length-to-width ratio, condensed inflorescence, retarded onset of flowering, reduced number of flowers and flowers with shortened styles (Dehio et al. 1993). These altered pleiotropic growth patterns were found to have a distinct link with altered hormonal status observed in the immunoreactive levels of indole-3-acetic acid (IAA), cytokinins, abscisic acid (ABA), gibberellic acid (GA) and the ethylene precursor 1-aminocyclopropane (ACC), where multiple tissue-specific alterations of hormone contents were found to be the consequences of *rolA* gene activity. All the roots formed in tobacco leaf discs were found to harbor *rolA* genes, and infection with *Agrobacterium* having plasmid with *rolA* alone could also induce hairy roots (Dehio et al. 1993). The *rolA* transcripts accumulated in an organ-specific fashion in tobacco plants transformed with *rolA* as well as with *rolB* and *rolC* (Schmulling et al. 1988, 1989; Spena et al. 1987).

Specific combinations of *rol* genes chiefly influenced root formation efficacy in various plants. In carrot, abundant roots could be induced, only in the presence of auxin, by a single copy of *rolABC* or *rolB* alone. A reduced rooting capability was caused by the inclusion of *rolC* with *rolB* in tobacco. Rooting was also elicited if rol constructions were co-inoculated with a strain carrying T_L-DNA genes, *ORF13+ORF14*, instead of the T_R-DNA strain. These roots were shown to contain both the *rol* genes and the *ORF13+ORF14* genes (discussed later). Roots containing different complements of T_L-DNA genes exhibited striking differences in growth patterns with a high growth rate, branching, and, most noticeably, the absence of geotropism, as shown by roots containing *rolB* alone, while roots with *rolA+B+C* were found to be geotropic like normal roots. Hairy root morphology was imparted to roots in the presence of *ORF13*+14 with *rolABC* and by the addition of auxin to the culture medium (Capone et al. 1989b), where the *rolB* promoter was found to exhibit differential tissue-specific expression (Capone et al. 1989a, 1991). These results clearly indicate a strong interaction among these genes in altering the endogenous hormone levels and influencing the morphology of the roots.

Foreign gene insertion into a cell depends on the specific stage of cell cycle, as it increases the chance of foreign DNA integration into the host genome. To establish this phenomenon, *rolABC* genes with their own promoters were moved into carrot cells under suspension culture where the cell cycle had been synchronized by adding 2-fluoro-de-oxy-uridine. Plants were then allowed to develop from transformed somatic embryos, PCR analysis of six morphologically different groups derived from the above experiment revealed that the original gene combination, i.e., *rolABC*, was not integrated in every transgenic plant in total, instead, a single *rol* gene or combinations of two *rol* genes could be detected. Since the expected molecular weight bands were not detectable by Southern blot analysis, one may expect the possibility of DNA methylation or rearrangement leading to altered *rol* gene sequences. These observations showed a very limited influence of cell cycle synchronization on the frequency of foreign DNA insertion (Claudia 1991).

The coding sequences of *ORFs* show various levels of homology and code for proteins involved in auxin-response modulation or IAA metabolism. For example, *ORF8* showed limited homology of about 31.5 and 37.5% with the *iaaM* genes of *A. tumefaciens* and *Pseudomonas savastoni*, respectively (Levesque et al. 1988;

Slightom et al. 1986), which code for tryptophan-2-monooxygenase (t2m), which further catalyzes the formation of indole-3-acetamide (IAM) from tryptophan. The N-terminal domain of *ORF8*, homologous to *rolB* protein, may modulate auxin responsiveness of host cells, whereas the C terminal had 38% homology to t2m (Lemcke and Schmulling 1998a, b, c). *ORF8* over-expressing plants were found to contain a 5-fold higher concentration of IAM than untransformed plants. In addition to this, the protein extracts of *rolB*-transformed seedlings of tobacco and those from *E. coli* over-expressing *ORF8* showed significantly higher turnover rates of tryptophan to IAM than the respective negative controls, thereby indicating that the *ORF8* gene product has significant t2m activity (Lemcke et al. 2000).

The coding regions of pRi1724 with other plasmids such as *pRiA4b* (agropine type) (Slightom et al. 1986) and a mannopine-type plasmid, pRi8196 (Hansen et al. 1991), were observed having strong homology with each other, especially at the 3′ end. Using primers for *rol* genes, the expression of *rolA*, *rolB* and *rolC* from pRiA4b was detected, whereas *rolA* expression from pRi1724 was not detectable. Therefore it was predicted that morphological differences resulting after transformation were probably because of the expression of the *rolA* gene from pRiA4b. The function of the *ORF13* protein is related to cytokinin regulation, because the *rolC* and *ORF13* proteins have a consensus sequence extending over 49 amino acids (Prinsen et al. 1994).

10.2.4 Production of Secondary Metabolites

The expression of individual *rol* genes in transgenic plants not only results in rhizogenesis (Spena et al. 1987), but also affects the development and physiology of the entire plant (Schell et al. 1993). It has been demonstrated that hairy root cultures of a number of plants synthesized higher levels of secondary products (e.g., nicotine, tropane alkaloids, anabasine, cytisine, anagyrine, hyocyamine, scopolamine, ginsenosides, thiarubrin, anthraquinones (AQ), polyacetylenes and betalains) than the normal-type plants (Bulgakov et al. 1998; Bulgakov et al. 2004; Ahn et al. 1996; Mateus et al. 2000; Thimmaraju and Bhagyalakshmi 2002; Pavlov et al. 2002), and released substantial levels into the medium, accounting for higher overall productivity than normal roots and other organ cultures (Thimmaraju et al. 2003a, b, 2004; Mukundan et al. 1998; Georgiev et al. 2008).

Hairy roots are induced in most cases from aerial parts of the plant, however, in many instances, the resultant roots fail to synthesize the compounds produced in the aerial portions. For example vindoline and its derivative, vinblastine, the anti-cancer drug synthesized in aerial parts, were produced at very low levels in hairy roots of *Catharanthus roseus* (Parr 1989). Contrarily, there are reports where certain compounds stored in the aerial portions were found at high levels, such as the quinine found at high levels in the bark of *Cinchonia ledgeriana*, which was synthesized at appreciable levels in the hairy roots of the species (Hamill et al. 1986, 1987, 1989). Since *rol* genes are involved in the overall process of hairy root formation via alteration of endogenous hormone levels (Bulgakov 2008), their involvement in specifically modifying the metabolic pathway in different systems remain obscure, red beet could be a good research model for this. Nevertheless, certain studies establish the direct involvement of *rol* genes in altering the secondary metabolism in roots that are discussed below.

The most extensively studied root secondary metabolites are ginsenosides, the saponin glycosides from *Panax ginseng*. Ginsenosides are of high economic value as invigorators and longevity drugs. The *rolC*-transformed ginseng roots consistently produced high levels of ginsenosides for over 6 months, where saponins exceeded 1.8- to 3-fold higher levels than in untransformed cultures. The production of ginsenosides varied from one subculture to another in the 1c *rolC*–II cultures maintained over 2 years. The total ginsenoside production in hairy roots ranged from 6.76 to 65.83 mg/g dry weight (DW), whereas in untransformed culture, it varied between 5.12 and 8.92 mg/g DW. On an average, the transformed root cultures 1c *rolC*-II had three times more ginsenosides than control 1c cells. When *rol* genes were individually transferred, much variation between *rolA*, *rolB* and *rolC* cultures in terms of their ability to produce ginsenosides was observed (Bulgakov et al. 1998), where cultures obtained from the insertion of *rolC* alone-accumulated 2–3-fold higher glycosides, whereas the cultures with *rolA* and *rolB* produced traces of ginsenosides. To assess whether stimulation of ginsenosides in *rolC* roots is due to the direct action of *rolC* gene or is a phenomenon associated with rhizogenesis (caused by *rolC* gene integration), the metabolite contents in tumors and hairy roots obtained after *rolC* transformation were tested, where both the organs showed similar levels, indicating that *rolC* is, of course, involved, but not the root morphology induced by the same gene (Bulgakov et al. 1998). In general, most of the studies ascribed the increased level of secondary metabolite synthesis to rhizogenesis caused by T-DNA integration. In a study involving callus cultures of *Rubia cordifolia*, the content of anthraquinone (AQ) phytoalexins was significantly higher in cultures transgenic for 35 S-*rolB* and 35 S-*rolC* as compared with that of non-transformed cultures (Bulgakov et al. 2002). The ratio of the two major anthraquinones (AQ) of *R. cordifolia*, munjistin and purpurin (Mischenko et al. 1999), was significantly altered in *rolC*- and *rolB*-transformed cultures, where the different clones of *rolC* alone showed wide variations. In general, *rolB* caused higher levels of purpurin and munjistin, whereas total AQ was highest (293 mg/L) in one of the *rolC*-expressing clones, which was 1.8-times higher than the untransformed callus (Bulgakov et al. 2002). Production of increased tropane alkaloid synthesis was observed in tobacco transformed with *rol* genes (Palazón et al. 1998a, b). Further, in the case of *Atropa belladona*, transgenic roots containing *rolC* and *rolABC* showed a much higher content, of about 0.37% DW for *rolC* and 0.33% DW for *rolABC*, as compared with control untransformed roots, which produced about 0.03% DW of tropane alkaloids (hyoscyamine and scopolamine) (Bonhomme et al. 1999). In a similar study in cell cultures of *R. cardifolia*, higher AQ content was observed in *rolB* transformants (Shkryl et al. 2008). Individually, *rolA* or *rolB* or *rolC* genes were found to increase the biosynthesis of AQ in transformed callus cells of *Rubia cordifolia* (Shkryl et al. 2008) and, in a different study, *rol* genes were found involved in altering the sensitivity of transformed cells to auxin (Maurel et al. 1991).

The altered array of polyamines due to the transformation by *A. rhizogenes* in *Hyoscyamus muticus* was linked to phenotypic changes. The plants regenerated via protoplasts from *A. rhizogenes*-transformed root cultures showed the integration of *rol A*, *B* and *C* genes, with variations in hyoscyamine scopolamine ratio and a range of calystegins (A_3, B_1 and B_2) (Sevon et al. 1997). Calystegins are tropane alkaloid-related hydrophilic compounds causing specific inhibition of glycosidases and galactosidases (Molyneux et al. 1993; Mitra et al. 1993). While total alkaloid production with *rol* genes was clearly reduced, the pattern of calystegins varied with a total disappearance of calystegin B_1 and highly enhanced levels of calystegin B_2, followed by calystegin-A_3. The level of calystegins was similar in all the plants that had lost *rol* genes. Contrarily, the scopolamine and hyoscyamine contents during the same growth period were much higher in such plants that had lost *rol* genes or non-transformed ones, indicating that *rol* genes suppress alkaloid accumulation in *H. muticus* (Sevon et al. 1997).

The multipurpose neem tree, *Azadirachta indica* is well known in both traditional medicine and for its insecticidal properties (Allan et al. 1999). Hairy roots of neem were found to synthesize ten-fold higher levels of azadiractin (0.007% on a DW basis) having higher antifeedant bioactivity against locust (*Schistocerca gregaria*), than that obtained from the callus cultures, nearly 30-fold higher (Allan et al. 2002) than in vivo roots (Sundaram 1996). In addition, other compounds such as nimbin, salannin, 3-acetyl-1-tigloylazadirachtinin and 3-tigloylazadirachtol were detected in hairy roots by supercritical fluid chromatography. Some of the nimboids, such as 3-acetyl-1-tigloylazadirachtinin and 3-tigloylazadirachtol and deacetylsalannin, were found to leach into the medium, particularly at the later stages between the 30th and 36th days of culture period, by which time the azadirachtin in roots disappeared. Since 3-acetyl-1-tigloylazadirachtinin is a re-arrangement product of azadirachtin (Ley et al. 1989), it is suspected to have been formed from the azadirachtin either before or after leaching into the medium. Thus the leaching of secondary metabolites has been a common feature in hairy roots, for which the part played by *rol* genes is yet to be established. All these studies clearly establish that *rol* genes strongly influence the secondary metabolism, which correlates with rhizogenesis caused by T-DNA integration (Kamada et al. 1986; Mano et al. 1989).

10.2.5 Elucidating Phytohormone Signaling Mechanisms

It is very clear that *rol* genes are directly involved in the alteration of plant hormones. As widely established, plant hormone synthesis and their interactions depend on the environmental and physiological status of the plant cell/tissue and organ. Therefore, one can expect different effects by each *rol* gene at various stages of the cell cycle in each organ of different plant species. Altered phenotypes are generally linked to the alteration in the flux of hormones at the molecular level, which generally occurs due to signaling, which has been confirmed in specific point mutation studies (Frank et al. 2000). In another study, to differentiate the hormonal role from the signaling effect, the extent of involvement of *rolC* in the alteration of cytokinin flux was evaluated. For this, the *rolC* gene was expressed in tobacco under the transcriptional control of *CaMV35S*, and the effect on endogenous cytokinin conjugates (ECC) was monitored, as ECC is the proven preferential substrate for *rolC* protein (Frank et al. 2000). The use of the *CaMV35S* promoter here was expected to rule out possible interference with secondary effects or plant homeostatic mechanisms that may mask primary in vivo events when transgenes are expressed constitutively. Since no changes in endogenous pools of different cytokinin glucosides were found following chemical induction of the *rolC* gene, and since the level of free cytokinins remained unchanged even when the cytokinin-synthesizing *ipt* gene as well as the *rolC* gene were expressed, it was evident that the altered *rolC* phenotype is mediated through a signaling pathway different from those of cytokinins (Faiss et al. 1996; Frank et al. 2000).

The genetic transformation of higher plant cell via *Agrobacterium* would be of limited practical application if the process had only resulted in either tumors or hairy roots with a loss of regenerative potential of the transformed plant cell. Fortunately it has been found that the elimination of the genes responsible for phytohormone production (commonly known as disarming) leading to tumor/hairy root formation does not affect the process of transfer of T-DNA to the plant cells. This finding in fact has been the rationale behind the development of a number of *Agrobacterium*-based vector systems used for the genetic transformation of higher plants.

10.3 Hairy Roots of Red Beet

10.3.1 Induction

Of several factors that influence successful cell transformation and hairy root induction, the important ones are the amenability of the plant species/cultivar, age and type of explants, the *Agrobacterium* strain and the density of the bacterial suspension. In red beet, the first requirement for establishing highly productive hairy root cultures is to start with a variety with desirable qualities, such as high content of betalains. Since different cultivars of colored beets (including chard and table beets) produce an array of pigments resulting from various combinations of betacyanins and betaxanthins, desirable shades of natural pigments may be produced from hairy roots by choosing an appropriate cultivar. On the contrary, this may not guarantee that the same pigments may be predominantly synthesized in hairy root lines, as has been observed in cell cultures, since both transformation-associated and intra-specific variations may occur, which are discussed in a forthcoming section of this chapter. Even after the selection of a desirable cultivar of red beet for hairy root induction, the explant selection has been shown to chiefly influence the frequency and the quality of root clones. In almost all the studies, aseptically raised seedlings have been used for raising hairy roots (Fig. 10.3), except in one study where leaf explants from 4-week-old field grown plants of cv. Detroit Dark Red were used for the induction of hairy roots (Pavlov et al. 2002). Table 10.1 lists the different cultivars of red beet and

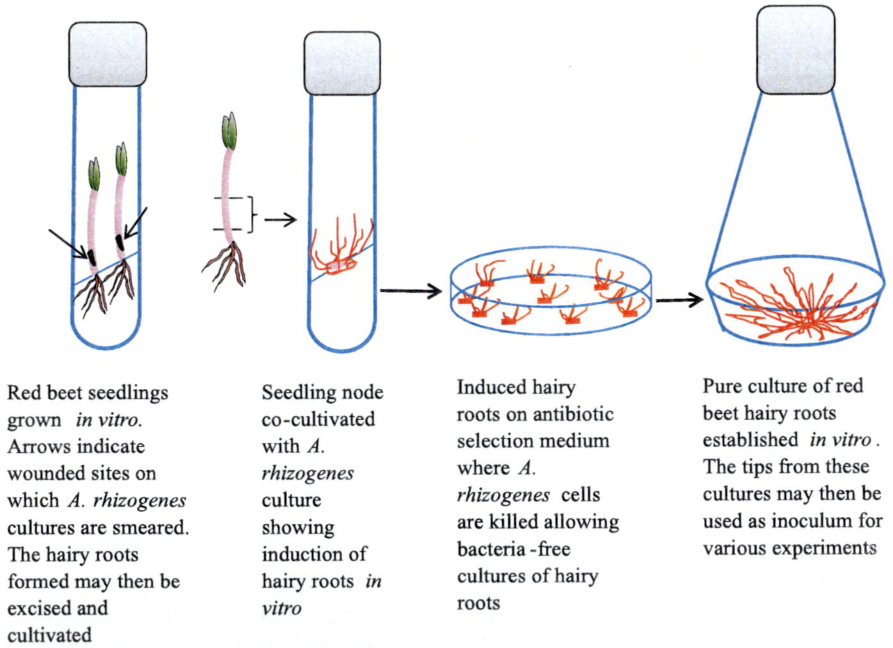

Fig. 10.3 Establishment of hairy root cultures of red beet

Table 10.1 Red beet cultivars/varieties from which hairy roots have been induced for the production of betalains.

B. vulgaris variety/cultivar	*A. rhizogenes* strain	Type of pigments	Quantity (mg/g DW)	Reference
Boltardy	LBA9402	Betacyanin	7	Hamill et al. (1986)
		Betaxanthin	13	
Detroit Dark Red	A4	Betacyanin	6.1	Taya et al. (1992)
		Vulgaxanthin-I	9.3	
Detroit Dark Red Brado Detroit 2 Egyptian	ATCC 15834	Betacyanin Betaxanthin	10 14.7	Pavlov et al. (2002, 2003)
Detroit Dark Red Leaf	A4	Betacyanin	14.6	Shin et al. (2002)
		Betaxanthin	9.2	
Detroit Dark Red	Not specified	Betalains	2.9	Weathers and Zobel (1992)
Mahyco Red	Not specified	Betacyanin	1.64	Mukundan et al. (1998)
		Betaxanthin	1.08	
Lutea (yellow beet)	Not specified	Portulaxanthin-II Vulgaxanthin-I	Not specified	Hempel and Böhm (1997)
Ruby Queen	LMG-150	Betacyanin	18.1	Thimmaraju et al. (2008)
		Betaxanthin	13.6	
Ruby Queen	A4 A2/83 A20/83	Betaxanthin Betaxanthin	Variable	Thimmaraju et al. (2008)

A. rhizogenes strains used for hairy root induction in red beet and the pigments produced by transformed roots.

Thimmaraju et al. (2008) characterized red beet hairy root clones of cv. Ruby Queen after the genetic transformation with several strains of *A. rhizogenes*. Among the different seedling explants, it was the cotyledonary leaf explant or the first node that showed the highest response, resulting in the initiation of several hairy root clones with transformation frequency ranging from 5% to 20% for each strain of *A. rhizogenes* tested. Other explants, such as mature leaf and hypocotyl showed 10% and 8% transformation, respectively, with the strain *A. rhizogenes* A2/83. When several cotyledonary leaf explants were re-examined, the best result obtained was 20% transformation frequency with the strain A2/83.

Certain chemicals, particularly the acetosyringone (4′-hydroxy-3′,5′-dimethoxy acetophenone) and similar other acetophenones, are also known to promote transformation processes (Sheikholeslam and Weeks 1987). However, in red beet cv. Ruby Queen, use of acetosyringone did not improve transfection efficiency of any of the bacterial clones used for the induction of hairy roots. Contrarily, there was a complete suppression of hairy root induction by the strain LMG-150. Of the several hairy root clones induced from different strains of *A. rhizogenes*, only ten clones obtained from cotyledonary leaf explants survived during the selection process in medium with antibiotic (Thimmaraju et al. 2008). The ten hairy root clones were induced from red beet using *A. rhizogenes*: one from strain LMG-150; three from strain A4, A4(1), A4(2), A4(3); three were from strain A2/83 viz., A2/83(1), A2/83(2), A2/83(3); and three from the strain A20/83 viz., A20/83(1), A20/83(2), A20/83(3) (Fig. 10.4). Although strain A15834 consistently induced hairy roots in beet cv. Detroit Dark Red (Pavlov et al. 2005), in cv. Ruby Queen, the roots obtained from strain A15834 were very weak, being unable to survive through the antibiotic selection process (Thimmaraju et al. 2008). Such high variations in response to *Agrobacterum* infection have also been observed in other non-beet hairy root systems (Santarem et al. 1998).

10.3.2 Morphology

Within the same cultivar, after transfection with *A. rhizogenes*, each hairy root clone was found to perform differently from other clones in terms of growth, morphology and secondary metabolite production, mainly due to the difference in physiological status of the host cell (Lemcke and Schmulling 1998a, b; Doran 2002) and other factors explained earlier. Astonishingly, different characteristics were observed in hairy roots clones generated from a single red beet var. Ruby Queen (Thimmaraju et al. 2008). There were significant differences in morphological features (Fig. 10.4) such as color, thickness, hairiness, length of the primary root, the length from tip up to the first branch point, number of laterals per centimeter of length and the length of the laterals (Table 10.2). The clone LMG-150 produced the highest number of laterals, about 9.20 lateral branches/cm length of the

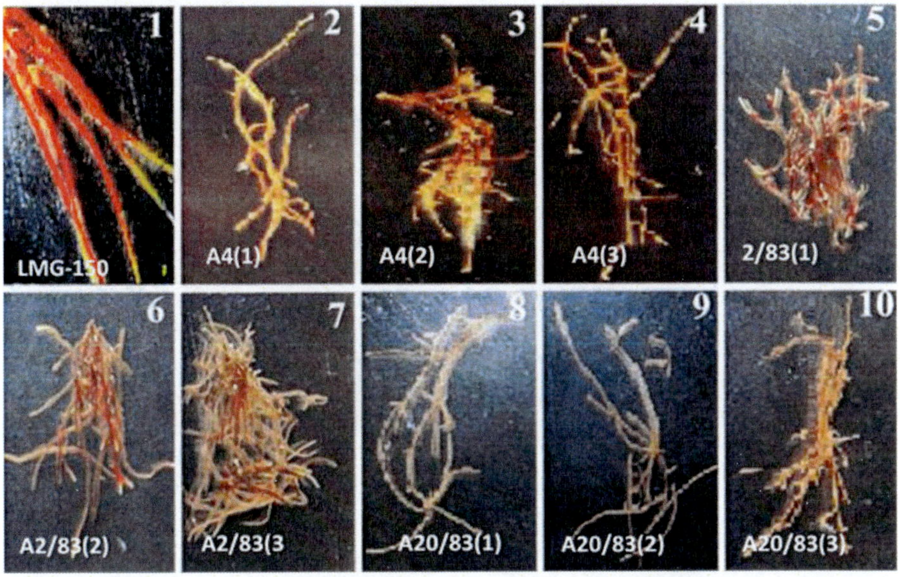

Fig. 10.4 Ten different clones of hairy roots obtained from red beet var. Ruby Queen using different *A. rhizogenes* strains: LMG-150, A4, A2/83 and A20/83. Clones grown in MS liquid medium show wide variations among the clones with respect to pigment content and other morphological properties

Table 10.2 Morphologic differences in ten different clones of hairy roots* obtained from red beet cultivar Ruby Queen

HR-Clone	Length (cm)	Length from tip to branch point (cm)	No. of laterals/cm	Length of the laterals
LMG150	7.2	2.90	9.20	2.36
A2/83(1)	1.20	0.25	2.25	0.35
A2/83(2)	1.11	0.35	8.25	0.95
A2/83(3)	1.0	0.3	6.0	0.80
A20/83(1)	1.6	0.5	3.5	0.63
A20/83(2)	0.93	0.3	5.25	0.35
A20/83(3)	2.7	1.03	4.0	0.23
A4(1)	3.0	1.15	2.75	0.8
A4(2)	2.13	0.67	4.50	0.93
A4(3)	2.22	1.08	3.50	0.43

*Hairy root clones were grown for 20 days in hormone-free MS liquid medium

primary root, with bright red pigmentation (Fig. 10.5), whereas the clone A2/83(2) produced a highest number of about 8.0 lateral branches/cm length of the root followed, by A2/83(3) and A20/83(2). The number of laterals in the rest of the clones ranged from 2.25 to 4.50. Except for the clone LMG-150 that was deep red in color, the appearance of the other clones ranged from white to pale yellow to dull red due to extreme differences in their pigment contents. Even the thickness and hairiness

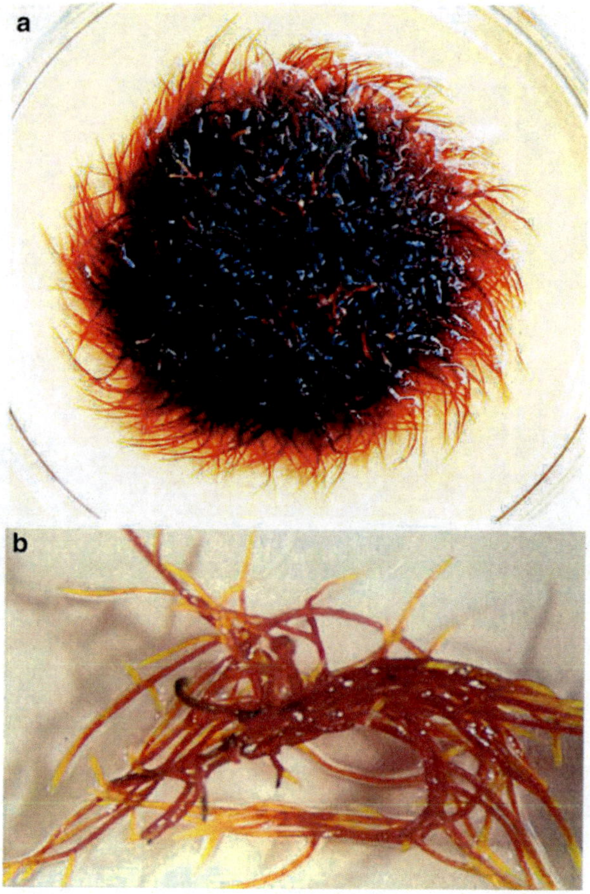

Fig. 10.5 Fully grown hairy roots of clone LMG-150. (**a**) Total biomass formed in 40 mL medium in a 150-mL flask (transferred to a 10-cm petri dish before photography). (**b**) Enlarged section from culture as in (**a**). Note that the tips first synthesized yellow betaxanthin, whereas the older portions were more loaded with betacyanins

were much different from each other. Variations in morphology, growth rates and branching patterns were also observed in hairy roots clones of *B. vulgaris* cv. Detroit Dark Red (Taya et al. 1992). This indicates that a single cultivar can display extreme variations in hairy root characteristics, making these cultures interesting for basic physiologic and genetic studies.

10.3.3 Growth Pattern and Accumulation of Betalains

Growth characteristics of hairy roots obtained from different varieties of red beet have been characterized by many research groups (Hamill et al. 1986; Mukundan et al. 1998; Taya et al. 1992; Pavlov and Bley 2005). In Ruby Queen, there was a clear cut difference in growth patterns between any two hairy root clones, where rapid growth was observed in clone LMG-150, which produced maximum biomass (Fig. 10.6a) on the 20th day of the culture period, with an exponential phase ranging

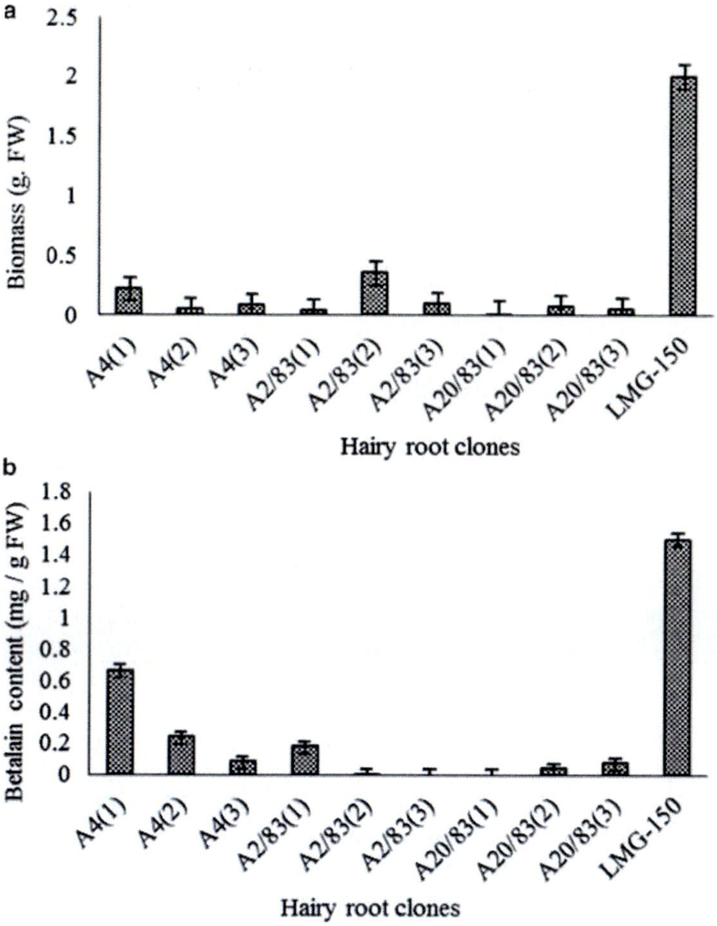

Fig. 10.6 Biomass (**a**) and contents of total betalains (**b**) in ten different clones of hairy roots obtained after 3 weeks in 40 ml MS medium (With permission from Thimmaraju et al. (2008))

from the 5th to the 20th day, reaching a stationary phase thereafter. The clones A2/83(1), A2/83(3), A20/83(2) and A20/83(3) produced maximum biomass on the 15th day, reaching a stationary phase thereafter (Thimmaraju et al. 2008). Such large differences in morphology, growth pattern and betalain biosynthesis were also observed in hairy roots induced in Detroit Dark Red (Taya et al. 1992; Takahashi et al. 2001). In another study hairy roots induced from different varieties of red beet, Detroit Dark Red, Egypt, Bordo and Detroit-2, accumulated betalains respectively 13.2, 5.0, 3.1 and 3.7 mg/g DW (Pavlov et al. 2002), indicating diverse responses among the cultivars of red beet. Even within the same cultivar, large variations in total betalains were observed in each root clone of red beet var. Ruby Queen (Fig. 10.6b), which further varied upon supplementation with different auxins.

Although the highest betalain content was recorded in the clone A4 (2), followed by A4 (3) (Thimmaraju et al. 2008), overall productivity in these clones was much less than LMG-150 due to their very low biomass output. Hairy roots of yellow beet var. Lutea contained predominantly portulaxanthin II and vulgaxanthin I, and the presence of similar several minor pigments: muscaaurin VII, indicaxanthin, dopaxanthin and vulgaxanthin II, III and IV, were suggested (Hempel and Böhm 1997; Böhm and Mack 2004). Improved response to auxins has been observed to improve membrane permeability in the *rolB*-transformed plant cells (Barbier-Brygoo et al. 1990), indicating *rol*B's involvement in bringing about the changes in the auxin perception pathway (Delbarre et al. 1994; Sedira et al. 2005; Faiss et al. 1996). Since the set of red beet hairy root clones of Ruby Queen were harboring *rol* genes, the increased sensitivity of all the clones to auxins and tryptophan may presumably be linked to the action of *rol* genes, particularly *rolB*.

10.3.4 Nutrients for Hairy Root Growth

Owing largely to differences in the functional genetic status of the cell from which a hairy root clone originates, great differences occur in their growth and metabolic profiles, as described above, and, thus, in their requirements of nutrients and hormones. A large number of studies used MS medium without any growth regulators (Hamill et al. 1986; Taya et al. 1994; Pavlov et al. 2005). Medium B5 was used successfully for the cultivation of yellow and red beets var. Burpee (golden yellow) and Mahyco-red (Mukundan et al. 1998). For another yellow beet (var. Lutea), Gamborg's B50 nutrient medium was useful for maintaining hairy root growth, where the betaxanthin biosynthesis varied depending on the amino acids present within/outside their roots (Hempel and Böhm 1997; Böhm and Mack 2004).

10.3.4.1 Carbon Source

The habitat of growing below the soil surface makes roots normally non-autotrophic, requiring the external supply of carbon-rich organic compounds, particularly sugars. Accordingly hairy root cultures need a continuous supply of sugar source, although autotrophic hairy root cultures induced under high-light conditions have also been reported (Liu et al. 2002). In the case of red beet, both normal tuberous roots and hairy root cultures are completely heterotrophic. Under normal in vivo conditions, the photosynthetic foliage synthesizes sugars that are partially transported into tubers, where they are stored as various forms of carbohydrates and as components of other primary and secondary metabolites. Contrarily, although hairy roots grown in vitro do not form tubers, their tremendous growth rate demands a continuous supply of sugars. In ascertaining the suitability of carbon source required for the growth and betalain synthesis, when red beet hairy root cultures were supplied with different sugars (3% w/v), sucrose was rapidly utilized, next was maltose, and there was a very limited use of glucose. The other sugars: fructose, lactose, xylose and galactose or glycerol suppressed growth and betalain synthesis (Bhagyalakshmi et al. 2004). Repeated culture in medium with the respective sugars did not habituate the cultures to the specific sugar, nor did the cultures synthesize higher levels of betalain pigment under such stressful conditions. Glycerol was not taken up alone by red beet hairy roots but most of these non-sucrose carbon sources were utilized when traces of sucrose was present in the medium. Yellow beet hairy roots predominantly synthesize betaxanthin from the condensation of betalamic acid with an amino acid. Such roots were found to contain gamma-aminobutyric acid (GABA), and its concentration reached a maximum level when (S)-glutamate (a GABA precursor) was provided or when the carbon source was changed from sucrose to glucose. Additional inclusion of (S)-glutamate along with glucose in the growth medium enhanced the GABA concentration to a level that exceeded that of all other amino acids. Interestingly, the level of the main betaxanthin, miraxanthin V, the pigment formed from the condensation of betalamic acid and dopamine, was markedly reduced by the replacement of sucrose with glucose (Böhm and Mack 2004). These observations indicate that pigmented beet hairy roots appear useful as a model system to study sugar metabolism/signaling that may also be directly linked to their morphologies, and probably auxin sensing, discussed in a subsequent section of this chapter. In almost all other studies, sucrose has also been found most suitable for red beet hairy roots, unlike the cell cultures of red beet, where different sugars were found to profoundly affect biomass and pigment profiles (Akita et al. 2000, 2002).

10.3.4.2 Macronutrients

Red beet displays a high sensitivity to phosphate ions, as do many other vegetable crops. Such a property was beneficially applied to regulate betalain biosynthesis in cultured hairy roots. Hairy roots of cv. Detroit Dark Red were found to consume 90% of this ion in a short growth period (Pavlov and Bley 2005); however, phosphate depletion resulted in nearly 5-fold raise in betalain content (Taya et al. 1994). Further, it was observed that phosphate limitation specifically enhanced betacyanin (nearly 8-fold) rather than betaxanthin (nearly 3.5-fold) formation (Mukundan et al. 1998).

Nitrate is the major mineral nutrient in the culture medium, changing its supply to cell cultures of *B. vulgaris* cv. Bikores Monogerm caused differences in color phenotypes (Leathers et al. 1992) and a 50% reduction in nitrogen concentration; while also changing other microelements increased betacyanin productivity up to 40 mg/L/day (Akita et al. 2002). However, while working with hairy root cultures of cv. Detroit Dark Red, Taya et al. (1994) did not find any significant effect of nitrate ions on betalain accumulation. Among the different forms of nitrates, ammonium ions were found preferentially and more rapidly utilized than other forms of nitrate ions by red beet hairy roots (Shin et al. 2003), with only 350 mg/L remaining from an initial level of 2,443 mg/L in cv. Detroit Dark Red (Pavlov and Bley 2005).

10.3.4.3 Micronutrients

A few studies investigated the influence of microelements on growth and betalain content in hairy roots of red beet. In hairy root cultures of cv. Detroit Dark Red, which were more sensitive to changes in the macroelements of MS medium (Murashige and Skoog 1962), changes in microelement composition had almost no effect on betalain production (Taya et al. 1994). Shin et al. (2003) found that the uptake of manganese and zinc started from the 6th day of sub-cultivation in fresh medium and they were completely utilized by the end of culture period, whereas iron and copper were very slowly utilized by hairy root culture. However, the cell cultures of the same cultivar of red beet were highly sensitive to fluctuations in micronutrients (see Chap. 9). For Ruby Queen red beet hairy roots, any deviation from normal levels of micronutrients in MS medium showed negative effects on growth and betalain content. Therefore, the optimal concentrations vary for each root clone, demanding specific optimizations for every root clone.

10.3.5 Effects of Growth Regulators

10.3.5.1 Levels of Endogenous Auxins

As documented, the endogenous IAA content largely determines the growth and development of transformed root cultures (Tanaka et al. 2001). The genes present on the T-DNA of *A. rhizogenes* also alter endogenous IAA levels (Schmulling et al. 1993) (also see Sect. 10.2). In red beet cv. Ruby Queen, the highest correlation of endogenous IAA of 1.41 ng/g fresh weight (FW) (Fig. 10.7a) was observed in the clone with highest biomass, which also synthesized highest level of betalain (Fig. 10.6b). Regarding kinetics, the endogenous IAA content in the roots of clone LMG-150 was highest on the 10th day of cultivation when the roots were progressing toward exponential growth phase (Fig. 10.7b). Although there was a direct correlation between root IAA content and root length (from tip to branching point), the least correlation was found between IAA and the number of laterals, root morphology and their branching patterns (Thimmaraju et al. 2008).

10.3.5.2 Exogenous Growth Regulators

External supplementation of auxins, each at levels of 0.5 and 1.0 mg/L, to the best clone, LMG-150, drastically suppressed biomass to less than 15 g/L; LMG-150 normally produced biomass of over 130 g/L FW in control MS (without additional IAA) medium (Fig. 10.8). When another auxin, naphthalene acetic acid (NAA), was used, the biomass was 100 g/L, indicating a lesser growth-suppressive effect than IAA. The highest biomass in A4(1) was 80 g/L on MS with 1 mg/L of NAA, which was followed by the clone A2/83(2), with 60 g/L on medium with IAA.

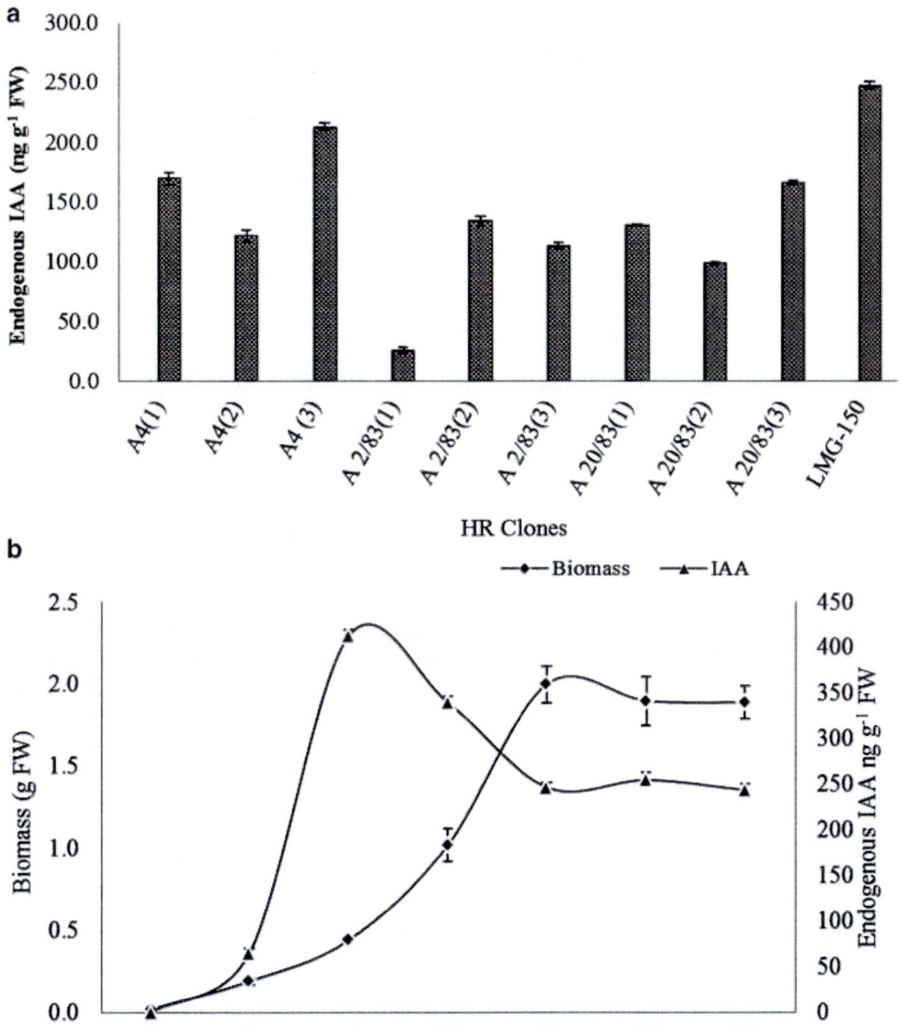

Fig. 10.7 (**a**) Endogenous IAA content in 10 different hairy root clones of red beet. (**b**) The pattern of biomass accumulation vs. endogenous IAA content during growth of root clone LMG-150 (Thimmaraju et al. (2008))

The highest biomass, of about 135 g/L, accounting for the productivity of nearly 200 mg/L of betalains, was observed for clone LMG-150 (Fig. 10.8). Less than 50% of this level was observed in the next highest betalain-producer clone, A2/83(1), with about 90 mg/L of betalains, particularly in the presence of NAA (1.0 mg/L), while most of the other root clones produced traces of betalains. The clone A2/83(3) produced the highest betalain content (20 mg l/L) in the presence of tryptophan.

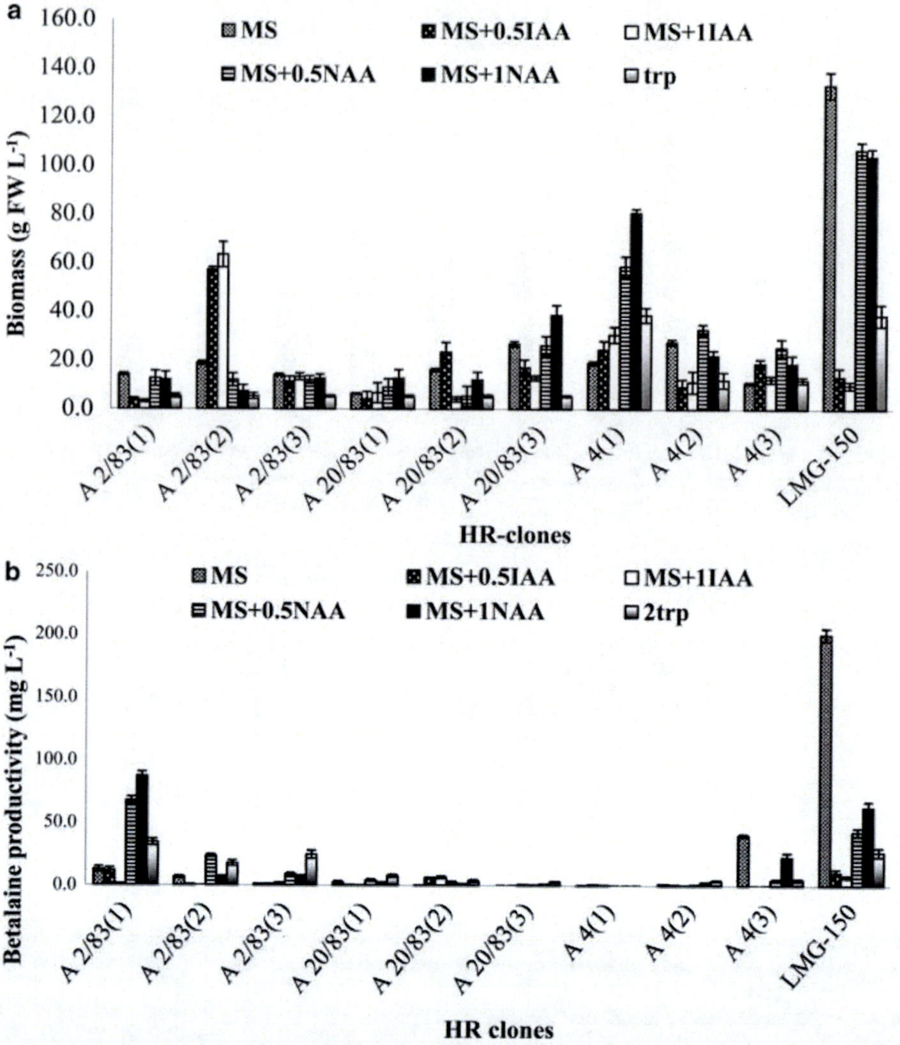

Fig. 10.8 Biomass accumulation (**a**) and the productivities of betalains (**b**) in different clones of hairy roots grown on hormone-free MS liquid medium (MS) and MS with 0.5 and 1 ppm IAA, MS with 0.5 and 1 ppm NAA and MS with 2 ppm tryptophan (Adapted from Thimmaraju et al. (2008) with permission)

Interestingly, the least IAA-containing clones showed somewhat higher betalain synthesis under the influence of external growth regulators. Since *rol* genes of T-DNA play a role in synthesizing and mobilizing endogenous IAA (Estruch et al. 1991a, b; Schmulling et al. 1993), supplementing tryptophan (the immediate precursor of IAA) to hairy roots presumably would alter endogenous IAA, cascading its effects on the root growth and morphology. However, when 2 mg/L of tryptophan was supplied to red beet hairy root clones, no significant positive effects were noticed (Thimmaraju et al. 2008), which was also the case in chicory hairy root cultures (Bais et al. 1999). Nevertheless, the clone A2/83(3), as an exception, produced much higher betalain levels, of about 20 mg/L, when grown in MS liquid medium containing tryptophan (2 ppm), when the control medium produced negligible pigments (Fig. 10.8b).

10.3.6 Effects of Physical Parameters

10.3.6.1 Light

Presence or absence of light did not make any difference in case of root clones of cv. Ruby Queen red beet (Thimmaraju et al. 2008). Red beet hairy roots are chiefly grown in the dark in various laboratories (Pavlov et al. 2005; Thimmaraju et al. 2008; Bhagyalakshmi et al. 2004; Neelwarne and Thimmaraju 2009). Effects of red (peak emission of 660 nm) and blue (peak emission of 470 nm) light, emitted by light emission diode to obtain 50 μmol/m^2/s, on growth of hairy roots and betalain accumulation were studied by Shin et al. (2003). The highest biomass and betalain pigments accumulated under a combination of blue and far red lights, although the values were not significantly higher than in white florescent light. Hairy roots of certain plant species were observed to develop chloroplast when exposed to light (Liu et al. 2002; Maldonado-Mendoza and Loyola-Vargas 1995; Kino-Oka et al. 2001), whereas red beets have never displayed any photosynthetic ability.

10.3.6.2 Hydrodynamic Stress

When hairy roots are sub-cultured, they are shifted from an environment of dense biomass–low nutrient–low volume to sparse biomass–high nutrient–high volume. Because of such a sudden shift into a different environment with high ionic and osmotic potential, presumably cells experience pressure on their cell walls, causing various biochemical shifts in membrane properties, chiefly causing membrane overloading (Clark and Blanch 1996). In red beet hairy roots cultured in shaking flasks, when an inoculum is transferred to fresh medium, a period for acclimation was needed during which lateral roots were not formed. It was always the root tip meristem cells that actively divided to add biomass. The root tip meristems suffered damage under shear stress. Experiments conducted by subjecting hairy roots to shear stress revealed that the sensitive root tips acquired tolerance to shear stress, when they were allowed to acclimate to the new environment by two-stage cultivation, which resulted in their retaining a relatively high viability of growing tips of up to 0.6 N/m^2 of loaded shear stress (Hitaka et al. 2000).

In red beet hairy root clone LMG-150 obtained from cv. Ruby Queen, the medium volume and other kinetic parameters, such as shear stress and volume, directly and significantly affected the biomass yield rather than the ratio of betacyanin: betaxanthin and peroxidase enzyme activity. A slight increase in hydrodynamic pressure was sufficient to cause a drastic reduction in biomass. The hydrodynamic stress created on the roots by large culture volume and shear stress was minimized by pulse feeding of medium instead of providing a single large volume. Based on such trials, a bioreactor was designed by creating a separate reservoir from which aerated medium was circulated through anchored biomass in the growth chamber, resulting in improved productivities of all the compounds (Neelwarne and Thimmaraju 2009).

10.4 Genetic Analyses of Red Beet Hairy Roots

While almost all studies concerned with hairy roots have confirmed the transformed nature of hairy roots by testing for the presence of plasmid DNA (T-DNA), one study focused on the characterization of different clones of hairy roots obtained from a single cultivar of red beet (Thimmaraju et al. 2008). In this study infection with *A. rhizogenes* clones A4, A2/83, A20/83 each resulted in three clones, and, from LMG-150, one hairy root clone was formed. The root clone A2/83(1) showed two inserts of T-DNA, whereas the rest had a single insert; the root clones largely differed in their growth characteristics, pigment content and levels of endogenous auxin, probably due to the physiological status of the host cell rather than the T-DNA copy number. Although a single copy transformation has been the best choice to produce a stable transgene effect (Lacroix et al. 2006), a few researchers conceptualized that the higher the number of T-DNA copies, the higher the number of genes that would be involved in the alterations of cellular growth/metabolic signaling, presumably resulting in either higher biomass formation and/or biosynthesis of desirable metabolites in such hairy root clones (Inze et al. 1984; Camilleri and Jouanin 1991). In the study of Thimmaraju et al. (2008), where one clone, 2/83(1), showed two T-DNA inserts, there was no linearity in performance, i.e., neither higher biomass nor higher metabolite synthesis occurred. This observation clearly indicates that the mere presence of a higher number of T-DNA inserts would not cause any increase in growth and production of secondary metabolites.

10.5 Re-transformation of Hairy Root Clones

The double transformation experiments with hairy root cultures reported describe the transfer of more than one gene to the seedling explants, or inducing hairy root cultures from the explants obtained from transformed plants (Zhang et al. 2004). As explained in Sect. 10.2, upon integration into the plant genome, *rol* genes of *A. rhizogenes* cause profound and variable effects on plant morphology and metabolism, which, in red beet, includes induction of hairy roots with various types of morphological features, their growth, regulation of betalain synthesis and regulation of various enzymes. To check the complementary effects of additional *rol* genes, the

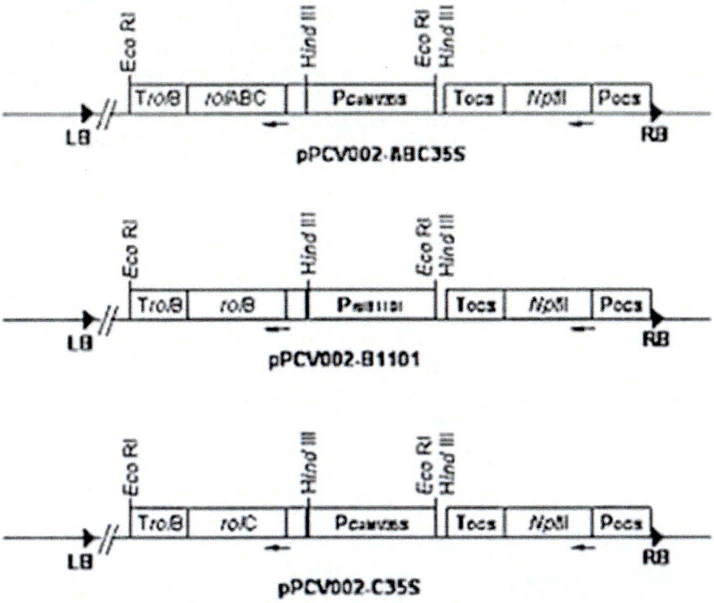

Fig. 10.9 The ready to use constructs present in *A. tumefaciens* (GV3101) containing the *rol* gene used for the re-transformation of hairy root clone LMG-150 (Adapted from Thimmaraju et al. (2008), with permission)

best-performing red beet hairy root clone obtained after infection with LMG-150 (Thimmaraju et al. 2008) was re-transformed using *A. tumefaciens* strain GV3101. To assist the selection of doubly transformed clones, the clone LMG-150 was first screened for its sensitivity to various levels of kanamycin, and 100 ppm kanamycin was found to be ideal. The re-infecting strain GV3101 harbored plasmid constructs pPCV002-A (*rolA* under the control of its own native promoter) and pPCV002-ABC (*rolABC* under the control of their native promoters); pPCV002-CaMVBT (*rolB* under the 35 S CaMV promoter) and pPCV002-CaMVC (*rolC* under the 35 S CaMV promoter) (Spena et al. 1987) (Fig. 10.9). The clones showed double transformation with significant morphological differences (Fig. 10.10) harboring additional *rolABC* and *rolC* genes, confirmed by PCR analysis for NptII (Fig. 10.11). The growth and pigment productivity were higher in clones re-transformed with additional *rolABC* (LMG-ABC) genes (Fig. 10.12). This is similar to what was found in non-beet plants, where single *rolC* and *rolB* genes were able to stimulate secondary metabolite production in transgenic plant tissues; e.g., nicotine production in *Nicotiana tabacum* root cultures, indole alkaloid production in *Catharanthus roseus* cultures (Palazón et al. 1998a, b; Bulgakov et al. 1998, 2004), ginsenoside production in cultures of *Panax ginseng* (Bulgakov et al. 1998), tropane alkaloid production in *Atropa belladonna* hairy root cultures (Bonhomme et al. 2000) and AQ production in callus cultures of *Rubia cordifolia* (Shkryl et al. 2008). The improvement in growth and pigment productivity in the LMG-ABC clone was

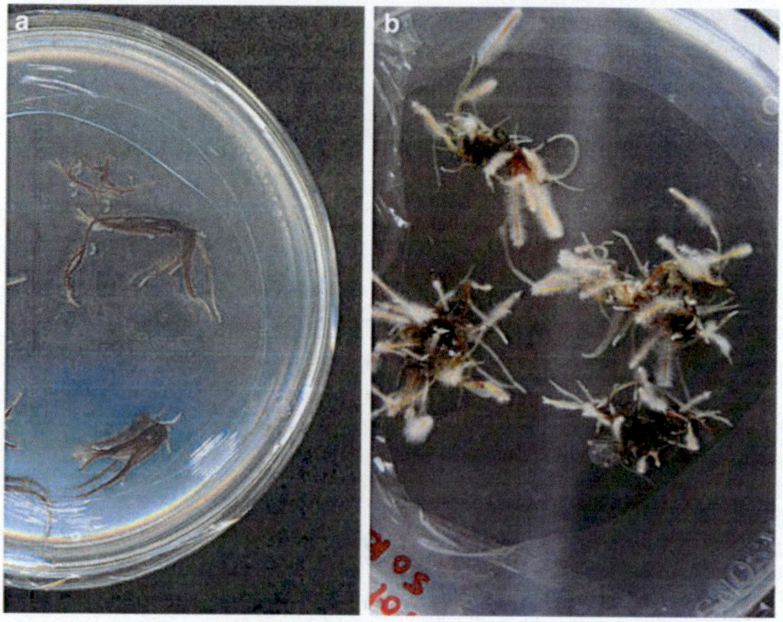

Fig. 10.10 (**a**) Original clone LMG-150 (control), and (**b**) putative double transformants of LMG-150 with additional *rolABC* genes, both cultures grown on MS medium with 100 mg/L of kanamycin

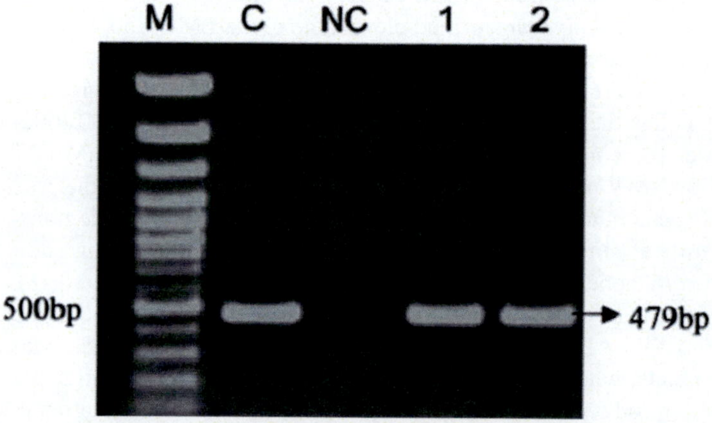

Fig. 10.11 Confirmation of double transformation and integration of additional rol genes into the clone LMG-150. PCR analysis was carried out using NPT-II specific primer. Lanes: *M* marker, *C* vector control, *NC* LMG-150 (single transformant), *1* LMG-ABC (LMG-150 transformed with additional *rolABC* genes), *2* LMG-C (LMG-150 transformed with additional *rolC* gene)

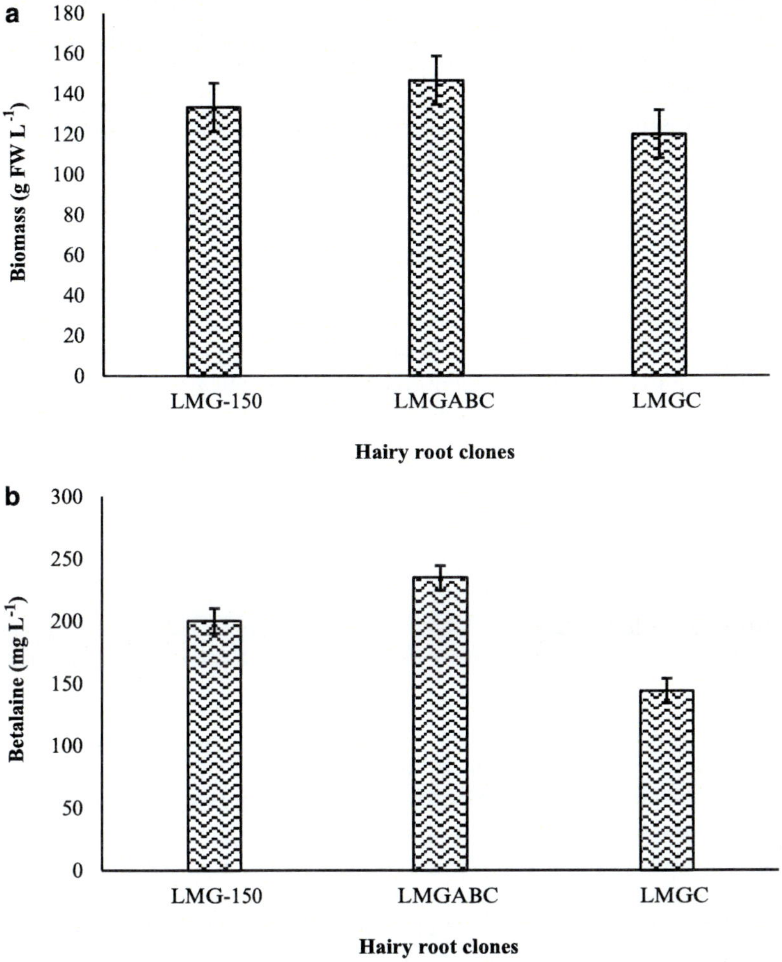

Fig. 10.12 Biomass (**a**) and betalain production (**b**) in hairy root clone (LMG-150) re-transformed with *rol*ABC (LMG-ABC) and *rol*C genes (LMG-C) (Adapted from Thimmaraju et al. (2008), with permission)

possibly due to the increase in the length of the primary roots with a concomitant increase in branching (Table 10.3) leading to overall biomass improvement, and hence productivity. However, similar effects were not observed in LMG-C. The *rol A*, *-B*, *-C*, genes, when present together, might act co-coordinately to bring about improved biomass and pigment synthesis, whereas the *rolC* gene might antagonize the function of already existing *rolABC* effects of LMG-150, resulting in the suppression both of growth and betalains. Contrarily, in a similar study, increased tropane

Table 10.3 Growth and morphology of single- and double-transformed hairy root clones in hormone free MS liquid medium (20 days after inoculation)

HR-Clone	Root length (cm)	Length from tip to branch point (cm)	No. of laterals/cm	Length of laterals (cm)	Color	Thickness	Hairiness
MS							
LMG-150	7.2	2.90	9.20	2.36	Red	Thick	Dense
LMG-ABC	8.6	3.10	13.1	2.51	Red	Thick	Dense
LMG-C	6.6	1.90	8.3	1.91	Red	Thick	Dense

alkaloid production was observed when hairy root cultures of *Brugmansia candida* were regenerated and re-transformed with additional *rol* genes (Zhang et al. 2004). All in all, it has long been demonstrated, through several studies, that the number of copies of T-DNA (mainly from Ti plasmids) is not an important factor for the manifestation of transformed phenotypes. A single copy transformation is a best choice to produce a stable transgene effect (Lacroix et al. 2006).

10.6 Metabolic Engineering

Genetically engineered root cultures have been used as a model system to study the metabolic modulation of several natural product pathways, such as phenylpropanoid biosynthesis. The lignification process, which otherwise enhances the quantity of waste material, may be suppressed in beetroots by diverting the carbon-sequestering pathway toward the biosynthesis of commercially useful end-products such as free/soluble sugars and other secondary metabolites. Accordingly, re-routing of the carbon destined for lignin production into the production industrial polymer feedstock, p-hydroxybenzoic acid (pHBA), was demonstrated by Dong et al. (2001). The synthesis of pHBA in transgenic plants has been made possible by expressing the gene coding for p-hydroxycinnamoyl-CoA hydratase/lyase (HCHL), which performs hydration of p-hydroxycinnamoyl-CoA and lysis of its substrates that participate in the phenylpropanoid pathway leading to lignin, instead forming C6–C1 metabolites, particularly p-coumaroyl-CoA-derived pHBA (Mayer et al. 2001; Merali et al. 2007; Mitra et al. 2002; McQualter et al. 2005). Since *B. vulgaris* is one of the most abundant (ca. 0.5–1% DW basis) sources of the p-hydroxycinnamic acid and ferulic acid present in cell wall material (Clifford, 1999; Walton et al. 2003), the HCHL gene under the control of a CaMV 35 S promoter was introduced into *B. vulgaris* using *A. rhizogenes* LBA 9402. Among the resultant hairy roots, one hairy root clone accumulated the glucose ester of pHBA at a very high level of 14% on a DW basis (Rahman et al. 2009).

Phenyl-alanine ammonia lyase (PAL) and cinnamate-4-hydroxylase (C4H) are the key enzymes in flavonoid biosynthetic pathway, resulting in several medicinally important flavonoids and other small phenolic molecules in plants (Weisshaar and Jenkins 1998). These enzymes are also known for their involvement in controlling the carbon flux, thereby regulating carbon allocation for the shikimic acid pathway (Howles et al. 1996; Singh et al. 2009). Even in case of betalain biogenesis, the basic phenyl group is derived from the same shikimic acid pathway (Delgado-Vargas et al. 2000). Genes for these key enzymes (PAL and C4H) were engineered into T-DNA of plasmids of *A. rhizogenes*, which were further used for hairy root induction for enhancing specific secondary metabolites where only C4H transgenic roots produced more decurcinol angelate, the neuroprotective anti-cancer drug in *Angelica gigas* (Park et al. 2011). Similarly the over-expression of the chalcone-isomerase gene caused enhanced production of epigenin (a compound that efficiently suppresses skin tumorogenesis and thyroid cancer) in *Saussurea medusa* hairy roots (Li et al. 2006). Such examples indicate the large possibilities for obtaining specifically engineered beet hairy roots for enhancement of desirable products, that have not yet been tried.

10.7 Production of Therapeutic Proteins

Foreign genes cloned between the two T-borders of the T-DNA can be transferred into the plant genome through the functions of molecular machinery inherent in *A. rhizogenes*. One of the most appealing areas of such genetically engineered plants is their ability to synthesize therapeutic proteins of human relevance at very low costs. The presence of post-translational modification machinery in plant cells can accomplish protein synthesis with more human-like modifications than can be accomplished with cultured mammalian cells and microbes. As a consequence of the enormous restrictions imposed on transgenic plant cultivation due to the potential gene transfer to wild species, the possibility of expressing functional animal proteins in plant cells was proven, which in turn opened up enormous potential for recruiting cultured plant cells for the purpose. As has been indicated earlier, the recalcitrance of cell cultures has often been a discouraging trend for stable production as well as for maintaining the quality of the product. Therefore, as an alternative, organ cultures are preferred for the purpose. Hairy roots have been found to carry out all the post-translational protein modifications found in eukaryotic systems, and there have recently been a few experiments that demonstrate the expression of foreign proteins in cultured hairy roots. Hairy roots have been proven appropriate as well as advantageous because of their ease of management, faster growth than cell cultures, simple culture conditions and high biochemical/genetic stability (Sivakumar et al. 2006; Srivastava and Srivastava 2007). They serve a strong alternative to field-cultured transgenic plants by growing in a confined space. Production of engineered proteins from hairy roots has been reported (Guillon et al. 2006). For example, functional antibodies (Putalun et al. 2003) and human secreted alkaline phosphatase (Gaume et al. 2003) have been produced in tobacco hairy roots. While the extraction and purification of complex proteins from plant tissues requires a laborious and costly process, animal proteins expressed in hairy roots are often secreted in the culture medium (see peroxidase excretion in Chap. 12). For example, the non-toxic lectin subunit ricin B fused to green fluorescence marker protein (GFP) was expressed in tobacco hairy roots and secreted in the culture medium (Medina-Bolivar et al. 2003). This murine-functional antigenic immunomodulatory fusion protein, which binds mostly to ricin B (a mucosal adjuvant in mammalian immune responses), was efficiently produced by hairy roots. Since Ricin B was found to be sensitive to the proteases that are often present in the medium used for hairy root culture, a two-phase extraction system was applied, allowing Ricin B to be concentrated in the organic phase to improve its stability (Medina-Bolivar et al. 2003). Another protein, the murine interleukin-12 (Liu et al. 2009) and monoclonal antibodies (Sharp and Doran 2001; Doran 2006) have been reported and can also be cloned and expressed through red beet hairy roots. Several strategies for production of cholinesterases (ChEs) were evaluated. Of the two ChEs in humans, only the serum enzyme butyrylcholinesterase (BChE) can be obtained from natural sources, and large-scale purification efforts from outdated blood-banked human plasma were demonstrated. Human ChEs can be used as bioscavengers of organophosphate toxins used as pesticides and chemical warfare nerve agents (Evron et al. 2007). The practical feasibility of this approach depends on the availability of the human enzymes, but because of inherent supply and regulatory constraints, a suitable production system is yet to be identified.

Human tissue plasminogen activator (TPA) is a thrombolytic protein, useful for lysing fibrin clots in blood vessels by converting the plasminogen into easily degradable plasmin. Human TPA consists of 527 amino acid residues forming a single-chain polypeptide with 17 disulfide bonds, imparting five distinct structural domains to the protein (Pennica et al. 1983), of which some domains specifically bind to fibrin clots, while others function in the conversion of plasminogen into plasmin, leading to fibrin homeostasis. Therefore, correct folding via correct pairing of multiple disulfide bridges is crucial for the functioning of TPA (Pennica et al. 1983). The gene coding for human TPA was cloned and expressed under the transcriptional control of rooting loci promoter, *rolD*, in hairy roots of Oriental melon. Abundant transcripts as well as the expression of high levels of TPA protein (0.17–0.15 μg/mg) was eventually measurable through enzyme-linked immunosorbent assay in the presence of single and dual *rolD* promoters as compared with triple and quadruple *rolD* (0.07–0.03 μg/mg) promoters. It was also confirmed that the protein was functional through an assay for fibrinolytic activity. This demonstration confirmed that hairy roots could serve as alternatives to transgenic plants for the expression of mammalian genes and the production of human therapeutic proteins (Kang et al. 2011).

Since red beet hairy roots have been found to spontaneously secrete certain proteins such as peroxidase isozyme, and can also be permeabilized, the non-destructive rhizosecretion process suggested by Guillon et al. (2006) may also be feasible for trapping high yields of recombinant proteins. Thus, red beet hairy root cultures serve as a serious alternative to whole plants for producing pharmacologically functional proteins, and probably other commercial proteins.

10.8 Production of Antioxidants

Ethanolic extracts from hairy roots of four different cultivars of red beet, Detroit Dark Red, Brado, Detroit-2 and Egyptian, at different stages of their growth were analyzed for antioxidant potential (Pavlov et al. 2002; Georgiev et al. 2010). In all the cultivars, the radical scavenging activity steadily increased along with their growth, reaching maxima on the 15th day and declining later, except in the case of cv. Egyptian, which showed the highest antioxidant potential even on the 20th day of culture, which also coincided with the highest pigment content. The highest pigment (13.2 mg/g DW) as well as radical scavenging (80% inhibition) was observed in cv. Detroit Dark Red, with cv. Egyptian showing slightly lesser pigment content and efficacy. In cv. Brado, which displayed low growth and pigment content, the antioxidant potential of the extract was also low (Pavlov and Bley 2006). These observations indicate the involvement of pigments in imparting such efficacies, as has been demonstrated with purified pigments of red beet. Although these cultivars were found to synthesize more betaxanthin than betacyanin, the fact that there was a drastic reduction in betacyanin in cv. Egyptian with a parallel reduction in antioxidant efficacy is indicative that the antioxidant efficacy displayed by these roots are mainly imparted by betacyanin. This argument has support from other studies where the betacyanin-containing fraction was found to possess higher antioxidant activity than the betaxanthin fraction (Escribano et al. 1998), and the activity was linked to the chemical structure of betacyanin (Fig. 10.13), having a cyclic amine, with similarity to the structure of ethoxyquin (Fig. 10.13), a strong antioxidant (E324) used in food preservation (Tesoriere et al. 2004). The amine in the molecule is considered to be the reactive group that gives betalains reducing properties. Using the modified 1,1-diphenyl-2-pi-crylhydrazyl method of measuring antioxidant capacity from cactus pear with betacyanins (3.7 mM) and betaxanthins (4.2 mM), Cai et al. (2003) demonstrated 3- to 4-fold greater antioxidant activity with these pigments than with ascorbic acid (13.9 mM), which was also greater than with polyphenols, rutin (6.1 mM) and catechin (7.2 mM). In another study, radical scavenging activity of betacyanin in an assay of Trolox equivalent antioxidant capacity (determined at pH 4.0 and above) showed that betacyanin was about 1.5- to 2.0-fold more potent than some anthocyanins, the latter are considered very good free-radical scavengers (Gliszczyňska-Świglo et al. 2006).

In red beet three different forms of superoxide dismutase (SOD), a strong antioxidant involved in scavenging superoxides and peroxidases, which function as oxido-reductases, were isolated (Pradedova et al. 2011). More details about the kinetics of free radical-scavenging properties of betalains are available in Chap. 6.

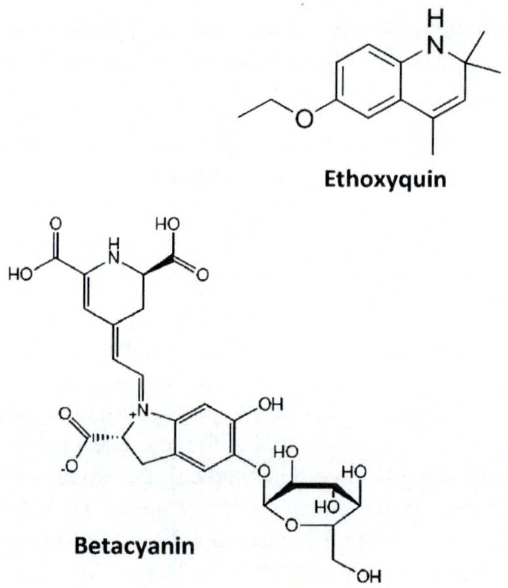

Fig. 10.13 Molecular structures of ethoxyquin and Figure is invisible

10.9 Production of Enzymes from Red Beet Hairy Roots: See Chap. 12

10.10 Scale-up and Cultivation in Bioreactor: See Chap. 11

10.11 Pigment Enhancement by Elicitation

Elicitors are either physical or chemical in nature and affect plant secondary metabolite production by a series of functional modulations from the rates of biosynthesis, accumulation and/or sequestration into vacuoles, to further turnover and release during defense followed by degradation; the overall process is termed elicitation. Since elicitors generally affect total secondary metabolism (Barz and Koster 1990), they are effective right from induction to incremental progression of constitutive secondary metabolite(s). Thus defense responsive compounds, phytoalexins, various other secondary metabolites and their precursors and enzymes may also accumulate upon elicitor treatment. The elicitation processes are executed involving cellular signal transduction networks, available precursors and physical conditions (Angelova et al. 2006; Vasconsuelo and Boland 2007; Zhang et al. 2009; Shilpa et al. 2010).

Since betalains are among the secondary metabolites that play a defense role in the species, such compounds are elicited under stress and pathogen interference. Neosynthesis of betacyanin in sugar beets was observed when exposed to stress conditions (Pavoković et al. 2009). This is indicative of the fact that betalains can be enhanced by exposing cultured red beet hairy roots to stress inducers, for which biotic and abiotic elicitors have been applied. Biotic elicitors are certain microbials that the plant cell recognizes as a danger signal and to which the plant cell enhances its defense network in response. These could be natural whole live organisms, as in vivo, whereas in culture conditions, their cell wall polysaccharides are used, which are known to efficiently elicit a defense response in a manner that particularly enhances betalain synthesis. Therefore, a large number of microbial components need to be screened for their efficient use. A few workers have attempted to elicit betalain pigment content in hairy root cultures of red beet. Among abiotic elicitors, copper sulfate (5 mM) added to 2-week-old hairy root cultures caused a 2-fold increase in pigment accumulation within 24 h (Mukundan et al. 1999) and use of the polyamines (each 0.75 mM) putrescine and spermidine, respectively, enhanced 1.63- and 2-fold higher productivities of betalains in red beet hairy roots, mainly through enhancement of biomass (Bais et al. 2000). These polyamines, when supplied to red beet hairy roots in a bubble column bioreactor, showed 1.3-fold higher betalain content after 24 days of cultivation (Suresh et al. 2004). Use of very high levels (100 mg/mL medium) of freeze-dried algae powders of *Hematococcus pluvialis* and *Spirulina platensis* resulted in pigment increases of 2.28-fold on the 15th day and 1.16-fold on the 25th day, respectively (Rao et al. 2001). In a more precisely conducted experiment using red beet hairy root clone LMG-150, almost all elicitors significantly enhanced pigment biosynthesis, culminating in higher productivities of betalains that were associated with growth suppression (Savitha et al. 2006). In cultures treated with dry cell powder (158 mg/L) of *Penicillium notatum*, after 7 days of elicitor addition, there were 2-fold higher levels of betalain than in control cultures. However, treatment with 200 mg/L of pullulan, a glycan from *Pullularia pullulans*, produced 202 mg/L of betalain on the 10th day after its addition, although it largely arrested further growth of hairy roots whereas control cultures that continued growth produced 88.4 mg/L on the 10th day. Since most of the elicitors caused early elicitation (on the 7th day) and suppressed biomass, resulting in reduced overall productivity, a strategy of using elicitor at the late exponential growth phase was worked out and found to be adoptable to treating cultures in a bubble-column bioreactor, effecting a 47% higher productivity of betalains (Savitha et al. 2006). Certain elicitors enhanced specific proteins, particularly the peroxidase enzyme (Thimmaraju et al. 2005, 2006, 2007) (see Chap. 12). These observations made in red beet hairy root cultures clearly establish that betalains, and probably other metabolites of red beet, can be substantially enhanced through the activation of genes involved in the respective metabolic pathways. Research inputs from metabolomics have facilitated the discovery of genes involved in several metabolic routes, offering new opportunities to design metabolic engineering strategies to circumvent problems linked to precursor availability or negative regulatory loops (Memelink 2005). These genetic resources are of tremendous use to modify the enzymatic routes leading to the biosynthesis of valuable metabolites produced by hairy roots. Because elicitation-mediated product enhancement can be easily accomplished in hairy roots, the elicitation technique has enormous potential for exploring differentially responsive genes involved in the metabolic processes.

10.12 Embryognesis from Beet Hairy Roots

Since genes of interest can be integrated to plasmids of *A. rhizogenes*, not only may desirable metabolites be expressed in the hairy roots, but different agronomical and other traits can be imparted to plants if hairy roots are made to regenerate shoots capable of becoming established under normal greenhouse/field conditions. However, despite intense efforts, somatic embryo induction from hairy root cultures is not very common, and only a few successful regeneration studies have been reported (Cabrera-Ponce et al. 1996; Dicola et al. 1997; Gutierrez-Pesce et al. 1998; Yang and Choi 2000; Ishizaki et al. 2002; Balen et al. 2004). The high recalcitrance of callus cultures of *B. vulgris* has diverted research interest toward the use of hairy roots, owing to the ease with which hairy roots are induced in beets. A large number of transformed root clones were obtained in 4-week-old beet seedlings using *A. Rhizogenes* strain A4M70GUS. Fast-growing hairy root clones were induced to form calluses using different combinations of growth regulators, the desirable results were achieved in MS nutrient medium containing both 1 mg/L 2,4-dichlorophenoxyacetic acid and 1 mg/L thidiazuron (TDZ). After a long cultivation period, globular embryo-like structures were formed from friable callus in MS medium supplemented with TDZ and gibberellic acid (1 mg/L each), followed by incubation medium with 1% glucose and 0.5 mg/L 2,3,5-triiodobenzoic acid. Histological analysis established that densely cytoplasmic cells interspersed in friable callus were able to gain embryogenic potential spontaneously and underwent the same sequence of steps toward embryo formation as that of a zygotic embryo, although root primordial were more than one per shoot primordium in some (Ninković et al. 2010).

Betalain production in callus cultures obtained from red beet hairy roots cv. Detroit Dark Red is discussed in Chap. 9.

10.13 Improvement of Agronomic Traits

Since the T-region of *A. rhizogenes* can be easily engineered with insertions and deletions of genes coding for different traits, allowing for the insertion of desirable genes, and the resultant hairy roots express those gene products under completely controlled heterotrophic conditions, hairy root cultures are recognized as excellent model systems to study the effects of transformation on physiological and morphological parameters. Accordingly, as summarized in Sect. 10.1, many studies have explored the relevance of individual *rol* genes in the development of transformed plants. While several important agronomical traits have been investigated using hairy root cultures, the *Hslpro*-1 gene from sugar beets was studied for plant nematode resistance (Cai et al. 1997). The nematode resistance gene (*Hslpro*-1) was introduced into *B. vulgaris* (Kifle et al. 1999), and stable expression and mitotic stability of the foreign genes were demonstrated, which was complemented by an in vitro nematode resistance test. When exposed to an engineered nematode-resistant hairy root line, the beet cyst nematode *Heteroderaschachtii* was unable to complete its life cycle. More investigations through gene expression studies revealed that *Hs1pro-1* was found to be up-regulated transcriptionally under both biotic/abiotic stress and after nematode infection in resistant *B. vulgaris* roots (Thurau et al. 2003).

In addition to the known nematode resistant gene *Hslpro-1*, another new gene potentially involved in beet cyst nematode resistance was investigated using beet hairy root cultures. When a large number of transcripts obtained from both nematode-infected resistant and susceptible hairy roots were screened, a few fragments were obtained that could be specifically linked to resistance (Samuelian et al. 2004) where a single transcript-derived fragment resulted from nematode infection, conferring resistance to such roots. The new transcript was further analyzed, and when its cDNA was transferred into the nematode-susceptible hairy roots clones, nematodes were significantly reduced.

In plants, complex defense mechanisms against herbivory have evolved, and studies have used aerial tissues, for all practical reasons. Since the aerial and underground environments are likely to differ enormously, cultured normal roots and hairy roots have been found handy for investigating root defense responses, although altered defense profiles are likely (Steeghs et al. 2004; Vaughan et al. 2011).

10.14 Beet Hairy Root as a Model System

10.14.1 Model for Bioreactor Design

Amazing genetic stability, pigment visibility and their heterotrophic growth and high sensitivity to changes in their environment have attracted researchers to use red beet hairy root as a model system for designing organ bioreactors. Although the phenomenon may not be applicable to all other hairy root systems, red beet hairy roots are ideal model systems for studying kinetic interactions between the root cells and the surrounding culture medium, chiefly because of their high sensitivity to shear stress, which is reflected both in their morphology and pigment synthesis. Changes in root tip morphology as a response to shear stress (Hitaka et al. 2000), explained in an earlier section, is one such example demonstrating the suitability of red beet as a model system for designing bioreactor systems for organ cultures. A combination of shear and hydrodynamic stress on red beet hairy roots was studied, where the slightest change in hydrodynamic stress caused extensive growth suppression, causing a 65% reduction in biomass (Neelwarne and Thimmaraju 2009); this also indicates the high sensitivity and suitability of hairy roots for various cellular membrane-loading and other physical stress studies. Driven by various types of sensitivities of red beet hairy roots, these authors designed a bioreactor model that reduced shear stress on hairy roots, allowing elicitation and on-line product recovery by permeabilization as well as simultaneous separation of red and yellow pigments through an on-line adsorption column. The continuously changing rheology and floating nature of hairy roots have posed higher challenges for engineers to design advanced mathematical models for biomass estimation and product monitoring as well as smarter recovery methods, which are discussed in Chaps. 11, 13 and 14.

10.14.2 Model for Pigment Regulation Studies

Although many betacyanins and betaxanthins have been structurally known since 1960, their biosynthetic pathways were partially elucidated only recently (see Chap. 2). Such elucidations were possible because the biosynthesis of betalains can be fairly regulated by altering the nutrients and other culture conditions, and such responses are fairly consistent in cultured beet hairy roots. Qualitative and quantitative investigations of the betaxanthins, especially of vulgaxanthin II as well as amino acid analyses were tracked in paralleled analyses, with the intention of understanding the problem of how species-specific patterns of betaxanthin can occur (Steglich and Strack 1990) despite the absence of an enzymatic regulation of such a decisive biosynthetic condensation step. Feeding amino acids to hairy roots of the yellow beet (*Beta vulgaris* var. lutea) normally results in the formation of the respective betaxanthins. One exception is (*S*)-glutamate, whose feeding resulted in an increase in vulgaxanthin I (glutamine as amino acid moiety) instead of vulgaxanthin II (glutamate as amino acid moiety) (Hempel and Böhm 1997; Schliemann et al. 1999). To elucidate this phenomenon, hairy roots were cultivated in modified standard medium and (S)-glutamate was fed to the cultures. Under most nutrient conditions tested, glutamine and vulgaxanthin I dominated in the tissue over glutamate and vulgaxanthin II. Glutamate, as opposed to glutamine, was readily metabolized. Therefore, its concentration was lower than that of glutamine in the medium. The maximum glutamate accumulation was reached when the activity of glutamine synthetase was low. Even then, however, vulgaxanthin II remained at a low level. In contrast, the level of vulgaxanthin I increased with increasing concentrations of glutamine in the tissue. Also, 4-aminobutyric acid (GABA) was found as a major amino acid in the hairy roots. Its concentration reached maximum levels either when (S)-glutamate, a GABA precursor, was fed, or when the carbon source sucrose was replaced by glucose. The respective GABA–betaxanthin, however, was hardly detectable. When (S)-glutamate and glucose both were supplied, the GABA concentration exceeded that of all other amino acids to a level that allowed the GABA–betaxanthin characterization. Interestingly, levels of the main betaxanthin, miraxanthin V, consisting of betalamic acid and dopamine, were most markedly reduced by a replacement of sucrose with glucose. These observations infer that, in yellow beet hairy roots, the reaction of betalamic acid with glutamate and GABA remains considerably lower than with glutamine and dopamine, irrespective of the concentration of the amino acid in the tissue (Hempel and Böhm 1997; Böhm and Mack 2004).

10.14.3 Model for Sugar Signaling Studies

Cultured roots are most suitable for studying cellular responses to sugars, owing to their heterotrophic nature, free from the interference of sugar-synthesizing pathways of photosynthesis. Red beet hairy root cultures, in particular, appear more suitable than other systems because of their hormone-independent growth with high sensitivities to their environments, which are translated into easily distinguishable morphological changes. Their uniform growth and differential synthesis of yellow and red pigments are easy to monitor. Above all, they are free from the interference of photosynthesis unlike *Arabidopsis*, which has always served as a model for various physiological studies. In higher plants, the nutrients function as substrates for growth and also act as signals triggering cascades of both primary and secondary metabolic events (Koch 1996; Smeekens 1998, 2000; Sheen et al. 1999; Gibson and Graham 1999; Gibson 2000; Rolland et al. 2006). Many plant genes controlled by sugars are involved in a variety of processes such as photosynthesis, storage of protein/starch/lipid and production of homopolysaccharides and heteropolysaccharides (Nakamura et al. 1991; Karrer and Rodriguez 1992; Krapp et al. 1993; Mc Laughlin and Smith 1994; Sheen et al. 1999; Gibson 2000; Koch et al. 2000; Winter and Huber 2000; Rolland et al. 2006). Sugars are also known to interact with several growth regulators (Lazzeri et al. 1988), imparting changes at a morphological level (Kraemer et al. 2002). Sucrose metabolism is the first step in carbon assimilation in the majority of the carbon-importing cells (Koch et al. 2000) in autotrophic and sucrose-dependent heterotrophic cell cultures.

Sugar sensing pathways do not operate in isolation but are part of cellular regulatory networks (Gibson 2000; Rolland et al. 2006). Recent results clearly show cross-links between different signaling systems, especially those of sugars, phytohormones and light (Rolland et al. 2006). While sucrose is sensed via hexokinase (HK) systems acting at transcriptional and translational levels (Rook et al. 1998a, b), hexose sensing can occur via separate HK-independent/dependent systems (Smeekens 2000). Sucrose may readily be hydrolyzed at the cell surface into glucose and fructose (Street 1969; Last and Brettell 1990; Schripsema et al. 1991, 1996; Kraemer et al. 2002), and these hexoses in turn act via HK or other kinases, fructokinase/glucokinase, in a mediated step. The cell cultures that were able to hydrolyze sucrose to glucose and fructose were also able to utilize externally supplied glucose/fructose (Street 1969; Kino-Oka et al. 1992; Kraemer et al. 2002). In a sucrose sensing system, it is the influx of sucrose into the cell that is sensed rather than the actual cytosolic sucrose concentration (Gibson 2000; Winter and Huber 2000), although the latter effects on membrane osmotic sensitivity cannot be ignored.

In most plant cells, sugar sensing is specific for sucrose (Gibson and Graham 1999), while, in other cases, glucose or fructose can also serve as signaling molecules (Jang and Sheen 1994, 1997). When hairy root clone LMG-150 was grown in media containing different sugars where the root inoculum was derived from the stock cultures (grown in MS medium with 3% sucrose), highest biomass and betalain synthesis occurred in the presence of sucrose, followed by maltose, glucose and fructose, whereas the other sugars, such as lactose, galactose, xylose and glycerol, failed to support the growth or pigment formation (Bhagyalakshmi et al. 2004). This means that although this root clone efficiently uses sucrose, it can also partially utilize maltose (a glucose dimer) and other hexoses to limited extents. These cultures as well as red beet seedlings were unable to grow in glucose plus fructose medium, indicating that such hexoses are not utilized as a carbon source by red beets. However, other plant systems could hydrolyze sucrose to glucose–fructose by invertase at the cell surface (Kraemer et al. 2002; Schripsema et al. 1991) and maltose to glucose (Last and Brettell 1990) before being absorbed into the cell where either glucose or fructose may then be preferentially utilized (Street 1969). The fact that these roots were unable to habituate to respective growth conditions (with specific sugars) is indicative of their specific preference for sucrose.

The ratio of betacyanin to betaxanthin differed significantly, with a much higher level of betaxanthin in medium with maltose (1:4.4) than in sucrose (1:3) and glucose (1:12). Contrarily, in fructose-supplemented medium, betaxanthin content in hairy roots was much higher than betaxanthin (1:0.7) (Bhagyalakshmi et al. 2004). Thus sugar sensing is an extremely interesting aspect of study in red beet hairy roots. Such hexose insensitivity has also been reported in mutants of *Arabidopsis* (Wenzler et al. 1989; Yokoyama et al. 1994).

10.14.4 Influence of Sugars on Hairy Root Phenotype

Hexoses and di-hexoses significantly influenced red beet hairy root morphology. Their branching pattern differed significantly between treatments. While sucrose induced very elongated moderately branched luxuriant growth of roots, roots in media with other sugars, glucose, maltose, fructose and glucose plus fructose showed variable branching (Bhagyalakshmi et al. 2004). Sugars have been implicated in both the number and placement of lateral roots through signal pathways wherein particularly sucrose is known to retard lateral root initiation under low nitrogen levels (Malamy and Ryan 2001). The beet root cultures in fructose medium showed frequent branching with callusy cells in the root elongation region; whereas the roots grown in maltose and glucose media showed loosely arranged cells that appeared to slough off from the root elongation region; and in a mixture of glucose and fructose, roots mainly expressed effects similar to those seen with fructose medium. Maltose appeared to suppress hairs when compared with sucrose-supplemented medium, whilst hairs were greatly enhanced by glucose, where tufts of root hairs appeared to originate from a common point (Fig. 10.14). Fructose treatment showed densely arranged hairs of variable lengths. The hairs were sturdier in glucose medium than in sucrose medium. However, when both glucose and fructose were present, there were only rudimentary hairs. In maltose medium, the root hairs remained at formation level without any further elongation (Fig. 10.14). To ascertain whether these responses are akin to hairy roots, beet seedlings were treated similarly. For seedlings, fructose was generally root inhibitory; this was more pronounced in the presence of glucose, although glucose alone did not cause a similar retardation of root

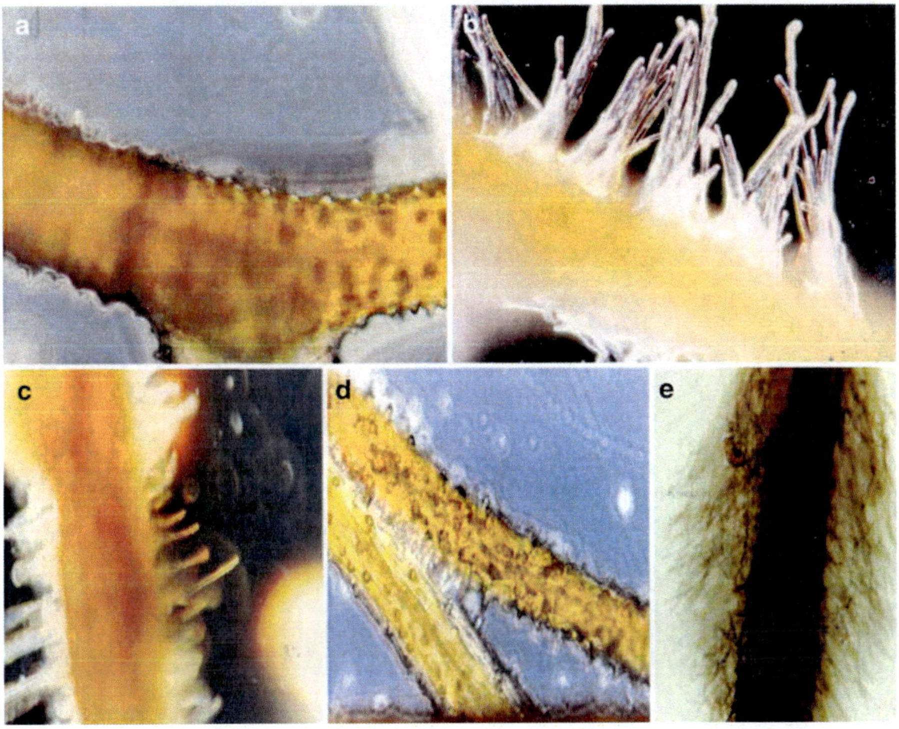

Fig. 10.14 Microscopic pictures showing highly rudimentary root hairs in maltose treatment (**a**), suppressed root hair elongation in glucose + fructose treatment (**d**), tufts of root hairs arising together from a common point in glucose treatment (**b**), sparsely arranged variable length root hairs in fructose treatment (**c**) associated with a complete absence of betacyanin pigment, and profuse and long root hairs with a high level of red pigment synthesis (betaxanthin + betacyanin) in sucrose (**e**) treatment

growth; especially lateral roots were seen with maltose, with the seedling weights nearly same as those in sucrose, suggesting that there is a good level of metabolic activity in seedlings. When these growth patterns were compared with hairy root growth, root lengths were more pronounced in glucose rather than in maltose, suggesting that there are certain differences between the genetically transformed hairy roots and the non-transformed seedling roots as far as sugar utilization patterns are concerned (Bhagyalakshmi et al. 2004).

The possibility of reversing the adverse effect caused by the hexoses and rejuvenation of sucrose sensing was tested. Roots grown in fructose medium were dark lignified and appeared dead, but upon transfer to sucrose medium were revived and grew with normal branching and dense root hairs (Fig. 10.15). This was also the case with other individual hexoses, with various levels of biomass. Contrarily, cultures from glucose and fructose media that had scanty hairs and stunted lateral branching continued to show the abnormality even after 2 weeks of incubation in sucrose (Fig. 10.15). The roots grown on maltose and glucose recovered quickly by

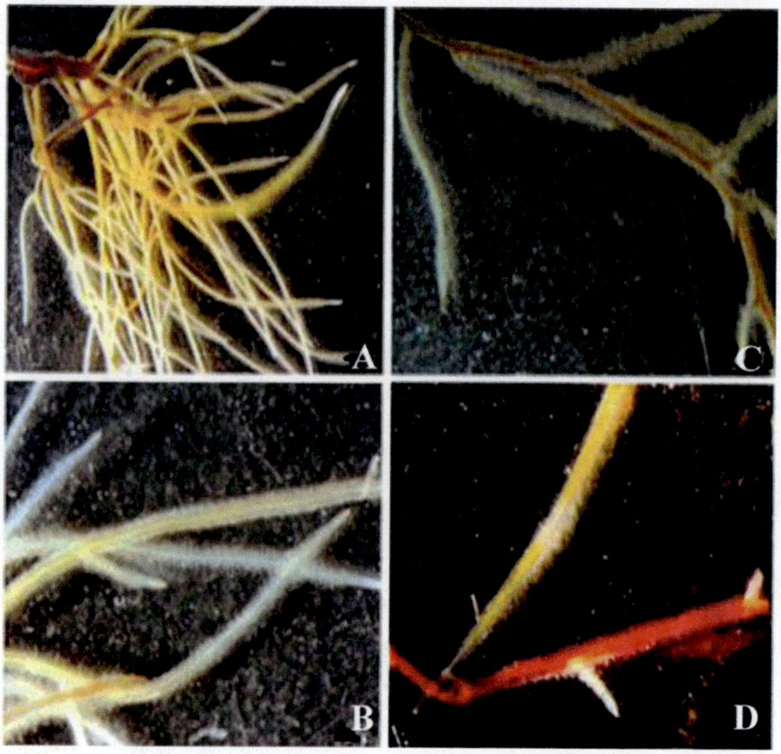

Fig. 10.15 Hairy roots 2 weeks after transfer to sucrose from maltose (**a**), glucose (**b**), fructose (**c**) and glucose plus fructose (**d**) showing variously types of branching and pigment synthesis

the end of the second week, indicating that maltose imparted less suppressive effects than hexoses, and the extent of rejuvenation directly correlated with the extent of repression caused by unfavorable hexoses. It was also suggested that these root cultures probably lacked sensing mechanisms needed for the phosphorylation through the hexokinase step, as has been well documented in other systems (Graham et al. 1994; Jang and Sheen 1997; Pego et al. 1999) or simply that the hexoses send repressive signals, although such speculations need to be authenticated.

10.14.4.1 Signaling Role of Sucrose

The failure of beet hairy roots to sense hexoses as the external carbon source has also been reported in other studies (Smeekens 2000; Wenzler et al. 1989). To know whether the presence of a small quantity of sucrose would support the uptake of other forms of carbon source, sucrose was partly supplemented with glucose plus fructose, or with other hexoses, making a total of 30 g/L (Bhagyalakshmi et al.

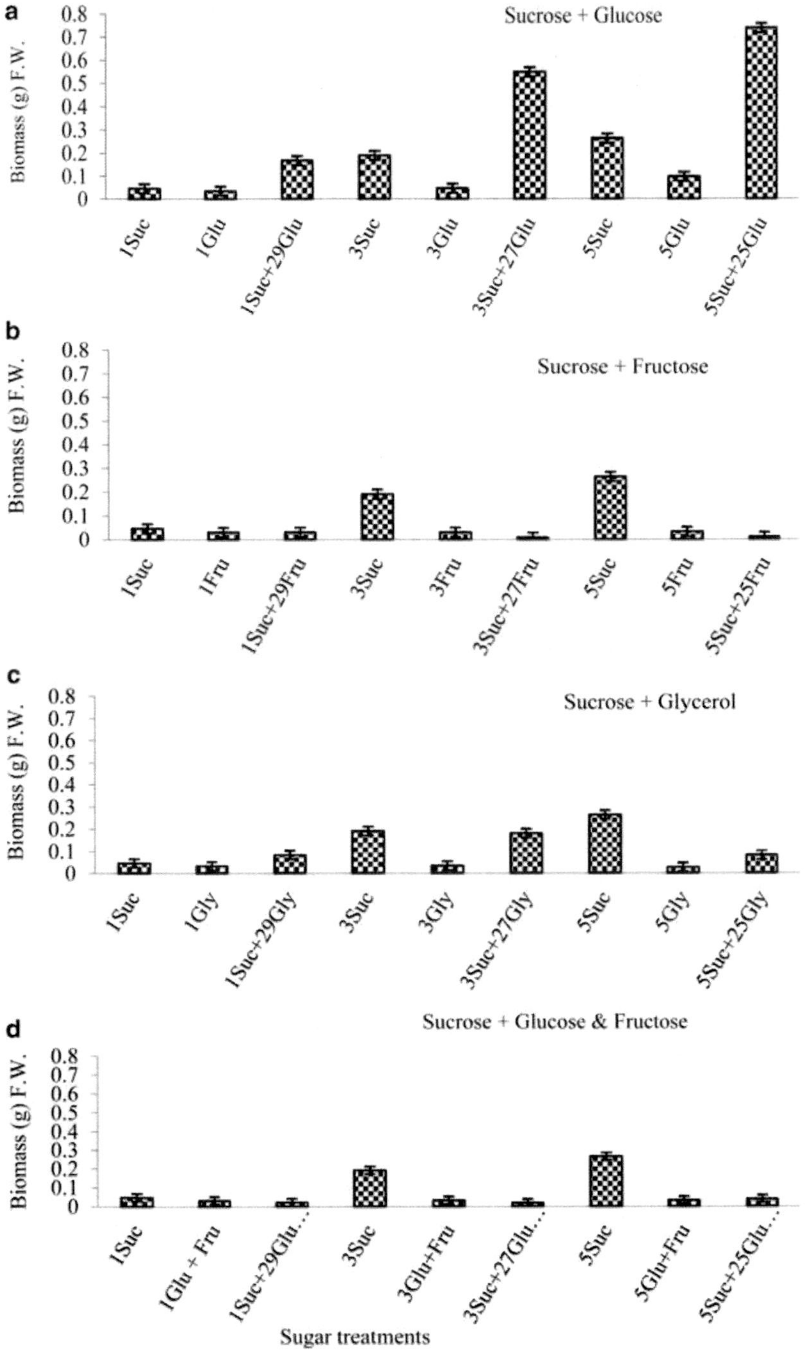

Fig. 10.16 Complementary effects of different levels of sucrose on its hexoses/glycerol acting as a signal and/or a nutrient. *Suc* sucrose, *Gluc* glucose, *Gly* glycerol, *Fru* fructose, *Glu + Fru* 1:1. Prefix numbers represent g/L of the respective carbohydrates in medium

2004). In this study, glycerol was also included since it is an intermediary compound in various cellular functions and a far-off intermediate in the glycolytic pathway. It was found that glycerol alone was not utilized by the hairy roots, but was taken up in the presence of a low level of sucrose (1 g/L). However, a steady decline in glycerol uptake in the presence of higher levels of sucrose was apparent, indicating the preference for sucrose. While sucrose was thought to help the uptake of glycerol, triggering the initial steps responsible for the uptake of the glycerol molecule, such glycerol uptake sparingly supported only root growth without supporting pigment synthesis (Fig. 10.16). The higher level of growth value observed in the presence of 1 g/L sucrose plus 29 g/L glucose was nearly 10-fold higher than the respective controls of either 1 g sucrose or only 29 g of glucose, clearly indicating the signaling role of sucrose, as negligible growth was observed in the presence of glucose alone. Higher levels of sucrose along with glucose were utilized in a dose-dependent manner for both hairy root growth and pigment synthesis, especially for betaxanthin; whereas, for the induction of betacyanin, higher levels of sucrose were essential. Such participations of sugar-sensing pathways for secondary metabolite synthesis have been observed in a few other systems (Gibson 2000; Mita et al. 1997; Muller et al. 2000; Rolland et al. 2006). Fructose, on the other hand, was inhibitory, imparting a strong growth repressive effect irrespective the level of sucrose. Even in the presence of glucose (i.e., glucose plus fructose), the positive effects of both glucose and sucrose were strongly repressed by fructose, indicating that fructose was highly inhibitory to this hairy root clone (LMG-150 of red beet).

10.15 Concluding Remarks

Red beet hairy roots have displayed diverse physiologies and extreme sensitivities on one hand and great genetic stabilities with consistencies in product formation on the other hand. Transformation efficiencies as high as 20% and large variations in the clones of hairy roots within the cultivar as well as different cells within the same explant herald that such transfection events, if properly characterized, would unravel a large number of interactions between the plant genome and the new gene inserts. While their growth patterns, nutritional requirements and product formation mostly remained similar for root clones extensively experimented in various laboratories, the sensitivities of root cultures to their environments were much different. Red beet hairy roots displayed highest morphological and metabolic differences in response to sugars, which has been demonstrated by the author's group in a preliminary study, inferring their suitability for elucidating sugar signaling networks. In support of this, large biochemical alterations occurred in response to sugars even in yellow beet hairy roots (Böhm and Mack 2004). The influence of *rol* genes on carbohydrate metabolism has seldom been studied even in other plant systems. When *rolB* and *rolC* gene transformed potato microtubers were analyzed, the structures of the starch granules were much different in each case from those of normal tubers (Aksenova et al. 2010), indicating the commercial importance of such gene interactions. Red beet is a rich source of folates and L-DOPA. Natural sources of these compounds are steadily gaining importance due to the fear of potential poisoning by higher doses of their synthetic forms. Red beet hairy roots can concomitantly synthesize an array of these compounds and such other products not explored so far.

The selective secretion of proteins such as peroxidase enzymes from red beet hairy roots has demonstrated the inherent secretory nature, which is yet another phenomenon worth characterizing for commercial exploitation in the context that hairy roots ably synthesize human proteins and other animal proteins. Cultured hairy roots have a much greater advantage for this purpose than field-grown plants, because the transformed genes are mostly confined to the aseptic roots grown in laboratory conditions, avoiding transgene or pharmacologically active protein dissemination into other organisms and to the environment. Hairy roots offer better technological edge over shoot cultures owing to their genetic stabilities and low requirements for energy inputs (no light requirement and hence low energy for temperature control), making them the ideal system for molecular farming, for which the high biomass-forming red beet hairy roots are perfect.

References

Ahn, J.C., B. Hwang, H. Tada, K. Ishimaru, K. Sasaki, and K. Shimomura. 1996. Polyacetylenes in hairy roots of *Platycodon grandiflorum*. *Phytochemistry* 42: 69–72.

Aird, E.L.H., J.D. Hamill, and M.J.C. Rhodes. 1988. Cytogenetic analysis of hair y root cultures from a number of plant species transformed by *Agrobacterium rhizogenes*. *Plant Cell, Tissue and Organ Culture* 15: 47–57.

Akita, T., Y. Hina, and T. Nishi. 2000. Production of betacyanins by a cell suspension culture of table beet (*Beta vulgaris* L.). *Bioscience, Biotechnology, and Biochemistry* 64: 1807–1812.

Akita, T., Y. Hina, and T. Nishi. 2002. New medium composition for high betacyanin production by a cell suspension culture of table beet (*Beta vulgaris* L.). *Bioscience, Biotechnology, and Biochemistry* 66: 902–905.

Aksenova, N.P., L.A. Wasserman, L.I. Sergeeva, T.N. Konstantinova, S.A. Golyanovskaya, A.V. Krivandin, I.G. Plashchina, W. Blaszczak, J. Fornal, and G.A. Romanov. 2010. Agrobacterial rol genes modify thermodynamic and structural properties of starch in microtubers of transgenic potato. *Russian Journal of Plant Physiology* 57: 656–663.

Allan, E.J., T. Stuchbury, and Mordue Luntz AJ. 1999. *Azadirachta indica* A. Juss. (neem tree): In vitro culture, micropropagation and the production of azadiractin and other secondary metabolites. In *Biotechnology in agriculture and forestry science series: Medicinal and aromatic plants*, vol. 43, ed. Y.P.S. Bajaj, 11–41. New York: Springer.

Allan, E.J., J.P. Eeswara, A.P. Jarvis, A.J. Mordue Luntz, E.D. Morgan, and T. Stuchbury. 2002. Induction of hairy root cultures of *Azadirachta indica* A. Juss. and their production of azadirachtin and other important insect bioactive metabolites. *Plant Cell Reports* 21: 374–379.

Angelova, Z., S. Georgiev, and W. Roos. 2006. Elicitation of plants. *Biotechnology & Biotechnological Equipment* 20: 72–83.

Bais, H.P., J. George, and G.A. Ravishankar. 1999. Influence of polyamines on growth and production of coumarins in hairy root cultures of *Cichorium intybus* L. cv Lucknow local (Witloof chicory). *Plant Growth Regal* 18: 33–37.

Bais, H.P., R. Madhusudhan, N. Bhagyalakshmi, T. Rajasekaran, B.S. Ramesh, and G.A. Ravishankar. 2000. Influence of polyamines on growth and formation of secondary metabolites in hairy root cultures of *Beta vulgaris* and *Tagetes patula*. *Acta Physiologiae Plantarum* 22: 151–158.

Baiza, A.M., A. Quiroz-Moreno, J.A. Ruiz, and V.M. Loyola-Vargas. 1999. Genetic stability of hairy root cultures of *Datura stramonium*. *Plant Cell, Tissue and Organ Culture* 59: 9–17.

Balen, B., D. Leljak-Levanić, S. Mihaljevič, S. Jelenić, and S. Jelaska. 2004. Formation of embryogenic callus in hairy roots of pumpkin (*Cucurbita pepo* L.). *In Vitro Cellular and Developmental Biology – Plant* 40: 182–187.

Barbier-Brygoo, H., C. Maurel, W.H. Shen, G. Ephritikhine, A. Delbarre, and J. Guern. 1990. Use of mutants and transformed plants to study the action of auxins. In *Hormone perception and signal transduction in animals and plants*, ed. J.A. Roberts, C. Kirk, and M. Venis, 66–77. Cambridge: Company of Biologists.

Barz, W., and J. Koster. 1990. Turnover and degradation of secondary (natural products). In *The biochemistry of plants*, ed. P.K. Stumpf and E.E. Conn, 35–84. New York: Academic.

Bhagyalakshmi, N., R. Thimmaraju, and M.S. Narayan. 2004. Various hexoses and di-hexoses differently influence growth, morphology and pigment synthesis in transformed root cultures of red beet (B*eta vulgaris*). *Plant Cell, Tissue and Organ Culture* 78: 183–195.

Böhm, H., and G. Mack. 2004. Betaxanthin formation and free amino acids in hairy roots of *Beta vulgaris* var. lutea depending on nutrient medium and glutamate or glutamine feeding. *Phytochemistry* 65: 1361–1368.

Bonhomme, V., Mattar-Laurain, D., and M.L. Fliniaux. 1999. Tropane alkaloid production by *Atropa belladona* roots transformed by the *rol*C or the *rolABC* genes. In: *Proceedings of IAPTC colloquium on "The plant biotechnology."* Wageningen, 8–10 Dec 1999.

Bonhomme, V., D. Laurain Mattar, and M.A. Fliniaux. 2000. Effects of the rolC gene on hairy root: induction development and tropane alkaloid production by *Atropa belladonna*. *Journal of Natural Products* 63: 1249–1252.

Bulgakov, V.P. 2008. Functions of *rol* genes in plant secondary metabolism. *Biotechnology Advances* 26: 318–324.

Bulgakov, V.P., M.V. Khodakovskaya, N.V. Labetskaya, G.K. Tchernoded, and Y.N. Zhuravlev. 1998. The impact of plant rolC oncogene on ginsenoside production by ginseng hairy root cultures. *Phytochemistry* 49: 1929–1934.

Bulgakov, V.P., G.K. Tchernoded, N.P. Mischenko, M.V. Khodakovskaya, V.P. Glazunov, S.V. Radchenko, E.V. Zvereva, S.A. Fedoreyev, and Y.N. Zhuravlev. 2002. Effect of salicylic acid, methyl jasmonate, ethephon and cantherdin on anthraquinone production by *Rubia cordifolia* callus cultures transformed with *rolB* and *rolC* genes. *Journal of Biotechnology* 97: 213–221.

Bulgakov, V.P., G.K. Tchernoded, N.P. Mischenko, Yu-N Shkryl, S.A. Fedoreyev, and Yu-N Zhuravlev. 2004. The *rolB* and *rolC* genes activate synthesis of anthraquinones in *Rubia cordifolia* cells by mechanism independent of octadecanoid signalling pathway. *Plant Science* 166: 1069–1075.

Cabrera-Ponce, J.L., A. Vegasgarcia, and L. Herreraestrella. 1996. Regeneration of transgenic papaya plants via somatic embryogenesis induced by *Agrobacterium rhizogenes*. *In Vitro Cellular and Developmental Biology – Plant* 32: 86–90.

Cai, D.G., M. Kleine, S. Kifle, H.J. Harloff, N.N. Sandal, K.A. Marcker, R.M. Klein-Lankhorst, E.M.J. Salentijn, W. Lange, W.J. Stiekema, U. Wyss, F.M.W. Grundler, and C. Jung. 1997. Positional cloning of a gene for nematode resistance in sugar beet. *Science* 275: 832–834.

Cai, Y., M. Sun, and H. Corke. 2003. Antioxidant activity of betalains from plants of the Amaranthaceae. *Journal of Agricultural and Food Chemistry* 51: 2288–2294.

Camilleri, C., and L. Jouanin. 1991. The TR-DNA region carrying the auxin synthesis genes of the *Agrobacterium rhizogenes* agropine-type plasmid pRiA4: nucleotide sequence analysis and introduction into tobacco plants. *Molecular Plant-Microbe Interactions* 4: 155–162.

Capone, I., M. Cardarelli, M. Trovato, and P. Costantino. 1989a. Upstream non-coding region, which confers polar expression to Ri plasmid root inducing gene *rolB*. *Molecular and General Genetics* 261: 239–244.

Capone, I., L. Spano, M. Cardarelli, D. Bellincampi, A. Petit, and P. Constantino. 1989b. Induction and growth properties of carrot roots with different complements of *Agrobacterium rhizogenes* T-DNA. *Plant Molecular Biology* 13: 43–52.

Capone, I., M. Cardarelli, D. Mariotti, M. Pomponi, A. De Paolis, and P. Costantino. 1991. Different promoter regions control level of and tissue specificity of expression of *Agrobacterium rhizogenes rolB* gene in plants. *Plant Molecular Biology* 16: 427–436.

Cardarelli, M., D. Mariotti, M. Pomponi, L. Spano, I. Capone, and P. Constantino. 1987. *Agrobacterium rhizogenes* T-DNA genes capable of inducing hairy root phenotype. *Molecular and General Genetics* 209: 475–480.

Christen, P., M.F. Roberts, J.D. Phillipson, and W.C. Evans. 1989. High yield production of tropane alkaloids by hairy root cultures of *Datura candida* hybrid. *Plant Cell Reports* 8: 75–77.

Clark, D.S., and H.W. Blanch. 1996. *Biochemical engineering – Bioreactor design and analysis*, 276–342. New York: Mercel Decker.

Claudia, G. 1991. The insertion of *rol* genes in cell-cycle synchronized carrot cell suspensions. In: The genetic dissection of plant cell process. *Journal of Cellular Biochemisty* 47: 1–157.

Clifford, M.N. 1999. Chlorogenic acids and other cinnamates - nature, occurrence and dietary burden. *Journal of the Science of Food and Agriculture* 79: 362–372.

De Paolis, A., M. Mauro, M. Pomponi, M. Cardarelli, L. Spano, and P. Constantino. 1985. Localization of agropine – Synthesizing functions in the TR-region of the root inducing plasmid of *Agrobacterium rhizogenes* 1855. *Plasmid* 13: 1–7.

Dehio, C., K. Grossmann, J. Schell, and T. Schmulling. 1993. Phenotype and hormonal status of transgenic tobacco plants over expressing the *rol* A gene of *Agrobacterium rhizogenes* T-DNA. *Plant Molecular Biology* 23: 1199–1210.

Delbarre, P., V. Muller, H. Imhoff, C. Barbier-Brygoo, N. Maurel, C. Lebanc, and G.J. Perrot-Rechenmann. 1994. The *rolB* gene of *Agrobacterium rhizogenes* does not increase the auxin sensitivity of tobacco protoplast by modifying the intracellular auxin concentration. *Plant Physiology* 105: 563–569.

Delgado-Vargas, F., A.R. Jiménez, and O. Paredes-López. 2000. Natural pigments: Carotenoids, anthocyanins, and betalains – Characteristics, biosynthesis, processing, and stability. *Critical Reviews in Food Science and Nutrition* 40: 173–289.

Dicola, A., A. Poma, and L. Spano. 1997. *rolB* expression pattern in the early stages of carrot somatic embryogenesis. *Cell Biology International* 21: 595–600.

Dong, D., S. Jiang, Y. Ni, and B. Jiang. 2001. Synthesis and properties of thermotropic copolyesters of *p*-hydroxybenzoic acid. *European Polymer Journal* 37: 611–617.

Doran, P.M. 2002. Properties and applications of hairy root cultures. In *Plant biotechnology and transgenic plants*, ed. K.M. Oksman-Caldentey and W.H. Barz, 143–162. New York: Mercel Dekker.

Doran, P.M. 2006. Foreign protein degradation and instability in plants and plant tissue cultures. *Trends in Biotechnology* 24: 426–432.

Escribano, J., M. Angeles Pedreño, F. Garcia-Carmona, and R. Muñoz. 1998. Characterization of the antiradical activity of betalains from *Beta vulgaris* L. roots. *Phytochemical Analysis* 9: 124–127.

Estruch, J.J., D. Chriqui, K. Grossman, J. Schell, and A. Spena. 1991a. The plant oncogene *rolC* is responsible for the release of cytokinins from glucoside conjugates. *EMBO Journal* 10: 2889–2895.

Estruch, J.J., J. Schell, and A. Spena. 1991b. The protein encoded by the *rolB* plant oncogene hydrolyses indole glucosides. *EMBO Journal* 10: 3125–3128.

Evron, T., B.C. Geyer, I. Cherni, M. Muralidharan, J. Kilbourne, S.H. FletcherSP, and T.S. Mor. 2007. Plant-derived human acetylcholinesterase-R provides protection from lethal organophosphate poisoning and its chronic aftermath. *The FASEB Journal* 21: 2961–2969.

Faiss, M., M. Strnad, P. Redig, K. Doležal, J. Hanuš, H. Van Onckelen, and T. Schmülling. 1996. Chemically induced expression of the rolC-encoded b-glucosidase in transgenic tobacco plants and analysis of cytokinin metabolism: rolC does not hydrolyze endogenous cytokinin glucosides in planta. *The Plant Journal* 10: 33–46.

Flores, H.E. 1986. Use of plant cell and organ culture in the production of biological chemicals. In *Application of biotechnology to agricultural chemistry*, Proceedings of American Chemical Society symposium series, vol. 190, ed. H. Leberon, R.O. Mumma, R.C. Honeycutt, and J.H. Duesing. Washington, DC: American Chemical Society.

Frank, M., H.M. Rupp, E. Prinsen, M. Motyka, H. Van Onckelen, and T. Schmulling. 2000. Hormone autotrophic growth and differentiation identifies mutant lines of *Arabidopsis thaliana* with altered cytokinin and auxin content or signaling. *Plant Physiology* 122: 721–729.

Fukui, H., A.F.M. Hasan, T. Ueoka, and M. Kyo. 1998. Formation and secretion of a new brown benzoquinone by hairy root cultures of *Lithospermum erythrorhizon*. *Phytochemistry* 47: 1037–1039.

Gaume, A., S. Komarnytsky, N. Borisjuk, and I. Raskin. 2003. Rhizosecretion of recombinant proteins from plant hairy roots. *Plant Cell Reports* 21: 1188–1193.

Gelvin, S.B. 2000. *Agrobacterium* and plant genes involved in T-DNA transfer and integration. *Annual Review of Plant Physiology and Plant Molecular Biology* 51: 223–256.

Georgiev, V., M. Ilieva, T. Bley, and A. Pavlov. 2008. Betalain production in plant in vitro systems. *Acta Physiologiae Plantarum* 30: 581–593.

Georgiev, V.G., J. Weber, E. Kneschke, P.N. Denev, T. Bley, and A.I. Pavlov. 2010. Antioxidant activity and phenolic content of betalain extracts from intact plants and hairy root cultures of the red beetroot *Beta vulgaris* cv. Detroit Dark Red. *Plant Foods for Human Nutrition* 65: 105–111.

Gibson, S.I. 2000. Plant sugar-response pathways: Part of a complex regulatory web. *Plant Physiology* 124: 1532–1539.

Gibson, S.I., and I.A. Graham. 1999. Another player joins the complex field of sugar-regulated gene expression in plants. *Proceedings of the National Academy of Sciences of the United States of America* 96: 4746–4748.

Gliszczy□ska-Świglo, H., P. Szymusiak, and P. Malinowska. 2006. Betanin, the main pigment of red beet: Molecular origin of its exceptionally high free radical-scavenging activity. *Food Additives and Contaminants* 23: 1079–1087.

Graham, I.A., K.J. Denby, and C.J. Leaver. 1994. Carbon catabolite repression regulates glyoxalate cycle gene-expression in cucumber. *The Plant Cell* 6: 761–772.

Granicher, F., P.P. Cristen, and I. Kaptanidis. 1995. Production of valepotriates by hairy root cultures of *Centranthus rubber*. *Plant Cell Reports* 14: 294–298.

Guillon, S., J. Trémouillaux-Guiller, P.K. Pati, M. Rideau, and P. Gantet. 2006. Harnessing the potential of hairy roots: Dawn of a new era. *Trends in Biotechnology* 24: 403–409.

Gutierrez-Pesce, P., K. Taylor, R. Muleo, and E. Rugini. 1998. Somatic embryogenesis, shoot regeneration from transgenic roots of the cherry rootstock Colt (*Prunus avium, P. pseudocerasus*) mediated by *pRi1855* T-DNA of *Agrobacterium rhizogenes*. *Plant Cell Reports* 17: 574–580.

Hamill, D., A.J. Parr, R.J. Robins, and M.J.C. Rhodes. 1986. Secondary product formation by cultures of *Beta Vulgairs* and *Nicotiana rustica* transformed with *Agrobacterium rhizogenes*. *Plant Cell Reports* 5: 111–114.

Hamill, J.D., A.J. Parr, M.J.C. Rhodes, R.J. Robins, and N.J. Walton. 1987. New routes to plant secondary products. *Biotechnology* 5: 800–804.

Hamill, J.D., R.J. Robins, and M.J.C. Rhodes. 1989. Alkaloid production by transformed root cultures of *Cinchona ledgeriana*. *Planta Medica* 55: 354–357.

Hansen, G., M. Larribe, D. Vaubert, J. Tempé, B.J. Biermann, A.L. Montoya, M.D. Chilton, and J. Brevet. 1991. *Agrobacterium rhizogenes pRi8196* T-DNA mapping and DNA sequence of functions involved in mannopine synthesis and hairy root differentiation. *Proceedings of the National Academy of Sciences of the United States of America* 88: 7763–7767.

Hempel, J., and H. Böhm. 1997. Betaxanthin pattern of hairy roots from *Beta vulgaris* var. lutea and its alteration by feeding of amino acids. *Phytochemistry* 44: 847–852.

Hildebrand, E.M. 1934. Life history of the hairy-root organism in relation to its pathogenesis on nursery apple trees. *Journal of Agricultural Research* 48: 857–885.

Hilton, M.G., and M.J.C. Rhodes. 1990. Growth and hyoscyamine production of "hairy root" cultures of *Datura stramonium* in a modified stirred tank reactor. *Applied Microbiology and Biotechnology* 33: 132–138.

Hitaka, Y., Y. Takahashi, M. Kino-oka, M. Taya, and S. Tone. 2000. Culture of red beet hairy roots by considering variation in sensitivity of tip meristems to hydraulic stress. *Biochemical Engineering Journal* 6: 1–6.

Hooykaas, P.J.J., and R.A. Schilperoort. 1992. Agrobacterium and plant genetic engineering. *Plant Molecular Biology* 19: 15–38.

Howles, P.A., V.J.H. Sewalt, N.L. Paiva, Y. Elkind, N.J. Bate, C. Lamb, and R.A. Dixon. 1996. Overexpression of L-phenylalanine ammonia-lyase in transgenic tobacco plants reveals control points for flux into phenylpropanoid biosynthesis. *Plant Physiology* 112: 1617–1624.

Huffman, G.A., F.F. White, M.P. Gordon, and E.W. Nester. 1984. Hairy-root-inducing plasmid: physical map and homology to tumor-inducing plasmids. *Journal of Bacteriology* 157: 269–276.

Inze, D., A. Follin, and M. van Lijsebettens. 1984. Genetic analysis of individual T-DNA genes of *Agrobacterium tumefaciens*; Further evidence that two genes are involved in indole-3-acetic acid synthesis. *Molecular and General Genetics* 194: 265–274.

Ishizaki, T., Y. Hoshino, K. Masuda, and K. Oosawa. 2002. Explants of Ri-transformed hairy roots of spinach can develop embryogenic calli in the absence of gibberellic acid, an essential growth regulator for induction of embryogenesis from non-transformed roots. *Plant Science* 163: 223–231.

Jang, J.C., and J. Sheen. 1994. Sugar sensing in higher plants. *The Plant Cell* 6: 1665–1679.

Jang, J.C., and J. Sheen. 1997. Sugar sensing in higher plants. *Trends in Plant Science* 2: 208–214.

Jounin, L. 1984. Restriction map of an agropine type Ri plasmid and its homologies with Ti plasmids. *Plasmid* 12: 91–102.

Kamada, H., N. Okamura, M. Satake, H. Harada, and K. Shimomura. 1986. Alkaloid production by hairy root cultures in *Atropa belladonna*. *Plant Cell Reports* 5: 239–242.

Kang, S., H. Ajjappala, H.H. Seo, J.S. Sim, S.H. Yoon, B.S. Koo, Y.H. Kim, and B.S. Hahn. 2011. Expression of the human tissue-plasminogen activator in hairy roots of oriental melon (*Cucumis melo*). *Plant Molecular Biology Reporter*. doi:10.1007/s11105-011-0303-8.

Karrer, E.E., and R.L. Rodriguez. 1992. Metabolic regulation of rice α-amylase and sucrose synthase genes *in planta*. *The Plant Journal* 5: 193–281.

Kifle, S., M. Shao, C. Jung, and D.G. Cai. 1999. An improved transformation protocol for studying gene expression in hairy roots of sugar beet (*Beta vulgaris* L.). *Plant Cell Reports* 18: 514–519.

Kino-Oka, M., Y. Hongo, M. Taya, and S. Tone. 1992. Culture of red beet hairy root in bioreactor and recovery of pigment released from cells by repeated treatment of oxygen starvation. *Journal of Chemical Engineering of Japan* 25: 490–495.

Kino-Oka, M., H. Nagatome, and M. Taya. 2001. Characterization and application of plant hairy roots endowed with photosynthetic functions. *Advances in Biochemical Engineering* 72: 186–218.

Koch, K.E. 1996. Carbohydrate-modulated gene expression in plants. *Annual Review of Plant Physiology and Plant Molecular Biology* 47: 509–540.

Koch, K.E., Z. Ying, Y. Wu, and W.T. Avigne. 2000. Multiple paths of sugar-sensing and a sugar/oxygen overlap for genes of sucrose and ethanol metabolism. *Journal of Experimental Botany* 51: 417–427.

Kraemer, K.H., E.P. Schenkel, and R. Verporte. 2002. *Ilex paraguariensis* cell suspension culture characterization and response against ethanol. *Plant Cell, Tissue and Organ Culture* 68: 257–263.

Krapp, A., B. Hofmann, C. Schafer, and M. Stitt. 1993. Regulation of the expression of *rbcS* and other photosynthetic genes by carbohydrates: A mechanism for the "sink regulation" of photosynthesis. *The Plant Journal* 3: 817–828.

Lacroix, B., T. Tzfira, A. Vainstein, and V. Citovsky. 2006. A case of promiscuity: Agrobacterium's endless hunt for new partners. *Trends in Genetics* 22: 29–37.

Last, D.I., and R.I.S. Brettell. 1990. Embryo yield in wheat anther culture is influenced by the choice of sugar in the culture medium. *Plant Cell Reports* 9: 14–16.

Lazzeri, P.A., D.F. Hildebrandt, J. Sunega, E.G. Williams, and G. Collins. 1988. Soybean embryogenesis: Interactions between sucrose and auxin. *Plant Cell Reports* 7: 517–520.

Leathers, R.R., C. Davin, and J.P. Zryd. 1992. Betalain producing cell cultures of *Beta vulgaris* L var Bikores Monogerm (red beet). *In Vitro Cellular and Developmental Biology – Plant* 28: 39–45.

Lemcke, K., and T. Schmulling. 1998a. Putative *rolB* gene homologue of *Agrobacterium rhizogenes* TL-DNA that alter plant morphogenesis or hormone sensitivity. *The Plant Journal* 15: 423–434.

Lemcke, K., and T. Schmulling. 1998b. A putative rolB gene homologue of the Agrobacterium rhizogenes TR-DNA has different morphogenetic activity in tobacco than rolB. *Plant Molecular Biology* 36: 803–808.

Lemcke, K., and T. Schmulling. 1998c. Gain of function assays identify non-rol genes from *Agrobacterium rhizogenes* TL-DNA that alter plant morphogenesis or hormone sensitivity. *The Plant Journal* 15: 423–433.

Lemcke, K., E. Prinsen, H. van Onckelen, and T. Schmülling. 2000. The ORF8 gene product of *Agrobacterium rhizogenes* TL-DNA has trypto-phan 2-monooxygenase activity. *Molecular Plant-Microbe Interactions* 13: 787–790.

Levesque, H., P. Delepelaire, P. Rouze, J. Slightom, and D. Tepfer. 1988. Common evolutionary origin of the central portions of the Ri TL-DNA of *Agrobacterium tumefaciens*. *Plant Molecular Biology* 11: 731–744.

Ley, S.V., J.C. Anderson, W.M. Blaney, P.M. Jones, Z. Lidert, E.D. Morgan, G. Nicholas, G. Robinson, D. Santafianos, M.S.J. Simmonds, and P.L. Toogood. 1989. Insect antifeedants from *Azadirachta indica* part-5. Chemical modification and structure-activity relationships of azadirachtin and related limonoids. *Tetrahedron* 45: 5175–5192.

Li, F.X., Z.P. Jin, D.X. Zhao, L.Q. Cheng, C.X. Fu, and F. Ma. 2006. Over-expression of the *Saussurea medusa* chalconeisomerase gene in *S. involucrata* hairy root cultures enhances their biosynthesis of apigenin. *Phytochemistry* 67: 553–560.

Liu, C., C. Guo, Y. Wang, and F. Ouyang. 2002. Effect of light irradiation on hairy root growth and artemisinin biosynthesis of *Artemisia annua* L. *Process Biochemistry* 38: 581–585.

Liu, C., M.J. Towler, G. Medrano, C.L. Cramer, and P.J. Weathers. 2009. Production of mouse interleukin-12 is greater in tobacco hairy roots grown in a mist reactor than in an airlift reactor. *Biotechnology and Bioengineering* 102: 1074–1108.

Malamy, J.E., and K.S. Ryan. 2001. Environmental regulation of lateral root initiation in *Arabidopsis*. *Plant Physiology* 127: 899–909.

Maldonado-Mendoza, I.E., and V.M. Loyola-Vargas. 1995. Establishment and characterization of photosynthetic hairy root cultures of *Datura stramonium*. *Plant Cell, Tissue and Organ Culture* 40: 197–208.

Mano, Y., H. Ohkawa, and Y. Yamada. 1989. Production of tropane alkaloids by hairy root cultures of *Duboisia leichhardtii* transformed by *Agrobacterium rhizogenes*. *Plant Science* 59: 191–201.

Mateus, L., S. Cherkoani, P. Christen, and K. Oksman-Caldenty. 2000. Simultaneous determination of scopolamine, hyocyamine and littorine in plants and different hairy root clones of *Hyocyamus muticus* by micellar electrokinetic chromatography. *Phytochemistry* 54: 517–523.

Maurel, S., H. Barbier-Bryqoo, A. Spena, G. Tempe, and G. Guern. 1991. Single rol genes from the *Agrobacterium rhizogenes* T(L)-DNA alter some of the cellular responses to auxin in *Nicotiana tobacum*. *Plant Physiology* 97: 212–216.

Mayer, M.J., A. Narbad, A.J. Parr, M.L. Parker, N.J. Walton, F.D. Mellon, and A.J. Michael. 2001. Rerouting the plant phenylpropanoid pathway by expression of a novel bacterial enoyl-CoA hydratase/lyase enzyme function. *The Plant Cell* 13: 1669–1682.

Mc Laughlin, J.C., and M. Smith. 1994. Metabolic regulation of glyoxalate-cycle enzyme synthesis in detached cucumber cotyledons and protoplasts. *Planta* 195: 22–28.

McQualter, R.B., B.F. Chong, K. Meyer, D.E.V. Dyk, M.G. O'Shea, N.J. Walton, P.V. Viitanen, and S.M. Brumbley. 2005. Initial evaluation of sugarcane as a production platform for *p*-hydroxybenzoic acid. *Plant Biotechnology Journal* 3: 29–41.

Medina-Bolivar, F., et al. 2003. A non-toxic lectin for antigen delivery of plant-based mucosal vaccines. *Vaccine* 21: 997–1005.

Memelink, J. 2005. The use of genetics to dissect plant secondary pathways. *Current Opinion in Plant Biology* 8: 230–235.

Merali, Z., M.J. Mayer, M.L. Parker, A.J. Michael, A.C. Smith, and K.W. Waldron. 2007. Metabolic diversion of the phenylpropanoid pathway causes cell wall and morphological changes in transgenic tobacco stems. *Planta* 225: 1165–1178.

Mischenko, N.P., S.A. Fedoreyev, V.P. Glazunov, G.K. Techernoded, V.P. Bulgakov, and Y.N. Zhuravlev. 1999. Anthraquinone production by callus cultures of *Rubia cardifolia*. *Fitoterapia* 70: 552–557.

Mita, S., N. Murano, M. Akaike, and K. Nakamura. 1997. Mutants of *Arabidopsis thaliana* with pleiotropic effects on the expression of the gene for β-amylase and on the accumulation of anthocyanin that are inducible by sugars. *The Plant Journal* 11: 841–851.

Mitra, A., Y. Kitamura, M.J. Gasson, A. Narbad, A.J. Parr, J. Payne, M.J.C. Rhodes, A. Mitra, M.J. Mayer, F.A. Mellon, A.J. Michael, A. Narbad, A.J. Parr, K.W. Waldron, R.J. Molyneux, Y.T. Pan, A. Goldmann, D.A. Tepfer, and D.A. Elbein. 1993. Calystegines: A novel class of alkaloid glycosidase inhibitors. *Archives of Biochemistry and Biophysics* 304: 81–88.

Mitra, A., M.J. Mayer, F.A. Mellon, A.J. Michael, A. Narbad, A.J. Parr, K.W. Waldron, and J. Walton. 2002. 4-Hydroxycinnamoyl-CoA hydratase/lyase, an enzyme of phenylpropanoid cleavage from *Pseudomonas*, causes formation of C6–C1 acid and alcohol glucose conjugates when expressed in hairy roots of *Datura stramonium* L. *Planta* 215: 79–89.

Morris, R.O. 1986. Genes specifying auxin and cytokinin biosynthesis in phytopathogens. *Annual Review of Plant Physiology* 37: 509–538.

Molyneux, R.J., Y.T. Pan, A. Goldmann, D.A. Tepfer, and A.D. Elbein. 1993. Calystegins, a novel class of alkaloid glycosidase inhibitors. *Archives of Biochemistry and Biophysics* 304: 81–88.

Mukundan, U., E.B. Carvalho, and W.R. Curtis. 1998. Growth and pigment production by hairy root cultures of *Beta vulgaris* L. in a bubble column reactor. *Biotechnology Letters* 20(5): 469–474.

Mukundan, U., V. Bhide, and H. Dawda. 1999. Production of betalains by hairy root cultures of *Beta vulgaris*. In *Plant cell and tissue culture for the production of food ingredients*, ed. T.J. Fu, G. Sing, and W. Curtis, 121–128. New York: Kluwer.

Muller, J., R.A. Aeschbacher, N. Sprenger, T. Boller, and A. Weimken. 2000. Disaccharide-mediated regulation of sucrose: Fructan-6-Fructosyltransferase, a key enzyme of fructose synthesis in barley leaves. *Plant Physiology* 123: 265–273.

Murashige, T., and F. Skoog. 1962. A revised medium for rapid growth and bioassays with tobacco cultures. *Physiologia Plantarum* 15: 473–497.

Nakamura, K., M.A. Ohto, N. Yoshida, and K. Nakamura. 1991. Sucrose-mediated accumulation of α-amylase occurs concomitant with the accumulation of starch and sporamin in leaf-petiole cuttings of sweet potato. *Plant Physiology* 96: 902–909.

Neelwarne, B., and R. Thimmaraju. 2009. Bioreactor for cultivation of red beet hairy roots and in situ recovery of primary and secondary metabolites. *Engineering in Life Sciences* 9: 227–238.

Nemoto, K., M. Hara, S. Goto, K. Kasai, H. Seki, M. Suzuki, et al. 2009a. The *aux1* gene of the Ri plasmid is sufficient to confer auxin autotrophy in tobacco BY-2 cells. *Journal of Plant Physiology* 166: 729–738.

Nemoto, K., M. Hara, M. Suzuki, H. Seki, A. Oka, T. Muranaka, and Y. Mano. 2009b. Function of the *aux* and *rol* genes of the Ri plasmid in plant cell division in vitro. *Plant Signaling & Behavior* 4: 1145–1147.

Nilsson, O., and O. Olsson. 1997. Getting to the root: the role of the *Agrobacterium rhizogenes rol*-genes in the formation of hairy roots. *Physiologia plantarum* 100: 463–473.

Ninković, S., T. Djordjević, B. Vinterhalter, B. Uzelac, A. Cingel, J. Savić, and S. Radovic. 2010. Embryogenic responses of *Beta vulgaris* L. callus induced from transgenic hairy roots. *Plant Cell, Tissue and Organ Culture* 103: 81–91.

Palazón, J., R.M. Cusido, J. Gonzalo, M. Bonfill, S. Morales, and M.T. Pinol. 1998a. Relation between the amount of *rolC* gene product and indole alkaloid accumulation in *Catheranthus roseus* transformed root cultures. *Journal of Plant Physiology* 153: 712–718.

Palazón, J., R.M. Cusido, C. Roig, and M.T. Pinol. 1998b. Expression of rolC gene and nicotine production in transgenic roots and their regenerated plants. *Plant Cell Reports* 17: 384–390.

Park, N.I., J.H. Park, and S.U. Park. 2011. Overexpression of cinnamate 4-hydroxylase gene enhances biosynthesis of decursinol angelate in *angelica gigas* hairy roots. *Molecular Biotechnology*. doi:10.1007/s12033-011-9420-8.

Parr, A.J. 1989. Production of secondary metabolites by plant cell cultures. *Journal of Biotechnology* 10: 1–25.

Pavlov, A., and T. Bley. 2005. Betalains biosynthesis by *Beta vulgaris* L. hairy root culture in different bioreactor systems. *Scientific Works, University of Food Technologies* 52: 299–304.

Pavlov, A., and T. Bley. 2006. Betalains biosynthesis by *Beta vulgaris* L. hairy root culture in a temporary immersion cultivation system. *Process Biochemistry* 41: 848–852.

Pavlov, A., P. Kovatcheva, V. Georgiev, I. Koleva, and M. Ilieva. 2002. Biosynthesis and radical scavenging activity of betalains during the cultivation of Red beet (*Beta vulgaris*) hairy root cultures. *Zeitschrift für Naturforschung. C* 57: 640–644.

Pavlov, A., V. Georgiev, and P. Kovatcheva. 2003. Relationship between type and age of inoculum and betalains biosynthesis by *B. vulgaris* hairy root culture. *Biotechnology Letters* 25: 307–309.

Pavlov, A., P. Kovatcheva, D. Tuneva, M. Ilieva, and T. Bley. 2005. Radical scavenging activity and stability of betalains from *Beta vulgaris* hairy root culture in simulated conditions of human gastrointestinal tract. *Plant Foods for Human Nutrition* 60: 43–47.

Pavoković, D., G. Rusak, V. Besendorfer, and M. Krsnik-Rasol. 2009. Light-dependent betanin production by transformed cells of sugar beet. *Food Technology and Biotechnology* 47: 153–158.

Pego, V.J., P.J. Weisbeek, and S.C.M. Smeekens. 1999. Mannose inhibits *Arabidopsis* germination via a hexokinase-mediated step. *Plant Physiology* 119: 1017–1023.

Pennica, D., W.E. Holmes, W.J. Kohr, R.N. Harkins, G.A. Vehar, C.A. Ward, et al. 1983. Cloning and expression of human tissue-type plasminogen activator cDNA in *E. coli. Nature* 301: 214–221.

Pradedova, E.V., O.D. Isheeva, and R.K. Salyaev. 2011. Antioxidant defense enzymes in cell vacuoles of red beet roots. *Russian Journal of Plant Physiology* 58: 36–44.

Prinsen, E., D. Chriqui, F. Viliane, M. Tepfer, and H. Van Onckelen. 1994. Endogenous phytohormones in tobacco plants transformed with *Agrobacterium rhizogenes* pRi TL-DNA genes. *Journal of Plant Physiology* 114: 80–85.

Putalun, W., et al. 2003. Anti-solasodine glycoside single-chain Fv antibody stimulates biosynthesis of solasodine glycoside in plants. *Plant Cell Reports* 22(3): 44–349.

Rahman, L., H. Kouno, Y. Hashiguchi, H. Yamamoto, A. Narbad, A. Parr, N. Walton, T. Ikenaga, and Y. Kitamura. 2009. HCHL expression in hairy roots of *Beta vulgaris* yields a high accumulation of p-hydroxybenzoic acid (pHBA) glucose ester, and linkage of pHBA into cell walls. *Bioresource Technology* 100: 4836–4842.

Rao, R., U. Tripathi, B. Suresh, and G.A. Ravishankar. 2001. Enhancement of secondary Metabolite production in hairy root cultures of *Beta vulgaris* and *Tagetes patula* under the influence of microalgal elicitors. *Food Biotechnology* 15(1): 35–46.

Riker, A.J. 1930. Studies on infectious hairy root of nursery apple trees. *Journal of Agricultural Research* 41: 507–540.

Rolland, F., E. Baena-Gonzalez, and J. Sheen. 2006. Sugar sensing and signaling in plants: Conserved and novel mechanisms. *Annual Review of Plant Biology* 57: 675–709.

Rook, F., N. Gerrits, A. Kortstee, M. van Kempe, and M. Borrias. 1998a. Sucrose specific signaling represses translation of the *Arabidopsis ATB2* bZIP transcription factor gene. *The Plant Journal* 15: 253–263.

Rook, F., P.J. Weisbeek, and S.C.M. Smeekens. 1998b. The light controlled *Arabidopsis* b*ZIP* transcription factor gene *ATB2* encodes a protein with an unusually long lucine zipper domain. *Plant Molecular Biology* 37: 171–178.

Samuelian, S., M. Kleine, C.P. Ruyter-Spira, R.M. Klein-Lankhorst, and C. Jung. 2004. Cloning and functional analyses of a gene from sugar beet up-regulated upon cyst nematode infection. *Plant Molecular Biology* 54: 147–156.

Santarem, E.R., H.N. Trick, J.S. Essig, and J.J. Finer. 1998. Sonication-assisted *Agrobacterium*-mediated transformation of soybean immature cotyledons: Optimization of transient expression transient expression. *Plant Cell Reports* 17: 752–759.

Savitha, B.C., R. Thimmaraju, N. Bhagyalakshmi, and G.A. Ravishankar. 2006. Different biotic and abiotic elicitors influence betalain production in hairy root cultures of *Beta vulgaris* in shake flask and bioreactor. *Process Biochemistry* 41: 50–60.

Schell, J., C. Koncz, A. Spena, K. Palme, and R. Walden. 1993. Genes involved in control of growth and differentiation in plants. *Gene* 135: 245–249.

Schliemann, W., N. Kobayshi, and D. Strack. 1999. The decisive step in betaxanthin biosynthesis is a spontaneous reaction. *Plant Physiology* 119: 1217–1232.

Schmulling, T., J. Schell, and A. Spena. 1988. Single genes from *Agrobacterium rhizogenes* influence plant development. *EMBO Journal* 7: 2621–2629.

Schmulling, T., J. Schell, and A. Spena. 1989. Promoters of the *rolA,* B and C genes of Agrobacterium rhizogenes are differentially regulated in transgenic plants. *The Plant Cell* 1: 665–670.

Schmulling, T., M. Fladung, K. Grossman, and J. Shell. 1993. Hormonal content and sensitivity of transgenic tobacco and potato plants expressing single rol genes of *A. rhizogenes* T-DNA. *The Plant Journal* 3: 371–382.

Schripsema, J., C. Erkelens, and R. Verpoorte. 1991. Intra and extracellular carbohydrates in plant cell cultures investigated by ^{1}H NMR. *Plant Cell Reports* 9: 527–530.

Schripsema, J., S.Y. Fung, and R. Verpoorte. 1996. Screening of plant cell cultures for new industrially interesting compounds. In *Plant cell culture secondary metabolism towards industrial application*, ed. F. Di Cosmo and M. Misawa, 1–10. New York: CRC press.

Sedira, M., E. Butler, T. Gallagherb, and M. Welandera. 2005. Verification of auxin-induced gene expression during adventitious rooting in *rolB*-transformed and untransformed apple. *Plant Science* 168: 1193–1198.

Seishiro, A., and S. Kunihiko. 2000. The roles of Ri-*rol* and *Ng-rol* genes in hairy root induction in *Nicotiana debney*. *Plant Science* 159: 183–189.

Sevon, N., B. Drager, R. Hiltunen, and K.M. Caldentey. 1997. Characterization of transgenic plants derived from hairy roots of *Hyocyamus muticus*. *Plant Cell Reports* 16: 605–611.

Sharp, J.M., and P.M. Doran. 2001. Strategies for enhancing monoclonal antibody accumulation in plant cell and organ cultures. *Biotechnology Progress* 17: 979–992.

Sheen, J., L. Zhou, and J.C. Jang. 1999. Sugars as signaling molecules. *Current Opinion in Plant Biology* 2: 410–418.

Sheikholeslam, S.N., and D.P. Weeks. 1987. Acetosyringone promotes high efficiency transformation of *Arabidopsis thaliana* explants by *Agrobacterium tumefaciens*. *Plant Molecular Biology* 8: 291–298.

Shilpa, K., K. Varun, and B.S. Lakshmi. 2010. An alternate method of natural drug production: eliciting plant secondary metabolite using plant cell culture. *Journal of Plant Sciences* 5: 222–247.

Shin, K.S., D. Chakrabarty, J.Y. Ko, S.S. Han, and K.Y. Paek. 2003. Sucrose utilization and mineral nutrient uptake during hairy root growth of red beet (*Beta vulgaris* L.) in liquid culture. *Plant Growth Regulation* 39: 187–193.

Shin, K.S., H.N. Murthy, J.Y. Ko, and K.Y. Paek. 2002. Growth and betacyanin production by hairy roots of *Beta vulgaris* in airlift bioreactors. *Biotechnology Letters* 24: 2067–2069.

Shkryl, Y.N., N.G. Veremeichik, P.V. Bulgakov, G.K. Tchernoded, N.P. Mischenko, S.A. Fedoreyev, and Y.N. Zhuravlev. 2008. Individual and combined effects of the rolA, B and C genes on anthraquinone production in *Rubia cordifolia* transformed calli. *Biotechnology and Bioengineering* 100: 118–125.

Singh, K., S. Kumar, A. Rani, A. Gulati, and P. Ahuja. 2009. Phenylalanine ammonia-lyase (PAL) and cinnamate 4-hydroxy-lase (C4H) and catechins (flavan-3-ols) accumulation in tea. *Functional & Integrative Genomics* 9: 125–134.

Sivakumar, G., K. Yu, J. Lee, J. Kang, L. Lee, W. Kim, and K. Paek. 2006. Tissue cultured mountain ginseng adventitious roots. *Engineering in Life Sciences* 6: 372–383.

Slightom, J.L., M. Durand-Tardiff, L. Jouanin, and D. Tepfer. 1986. Nucleotide sequence analysis of TL-DNA of *Agrobacterium rhizogenes* agropine type plasmid. *Journal of Biological Chemistry* 261: 108–121.

Smeekens, S. 1998. Sugar regulation of gene expression in plants. *Current Opinion in Plant Biology* 1: 230–234.

Smeekens, S. 2000. Sugar-induced signal transduction in plants. *Annual Review of Plant Physiology and Plant Molecular Biology* 51: 49–81.

Spena, A., T. Schmülling, C. Koncz, and J.S. Schell. 1987. Independent and synergistic activity of *rolA*, B and C loci in stimulating abnormal growth in plants. *EMBO Journal* 6: 3891–3899.

Srivastava, S., and A.K. Srivastava. 2007. Hairy root culture for mass-production of high-value secondary metabolites. *Critical Reviews in Biotechnology* 27: 29–43.

Steeghs, M., H.P. Bais, J. de Gouw, P. Goldan, W. Kuster, M. Northway, R. Fall, and J.M. Vivanco. 2004. Proton-transfer-reaction mass spectrometry as a new tool for real time analysis of root-secreted volatile organic compounds in *Arabidopsis*. *Plant Physiology* 135: 47–58.

Steglich, W., and D. Strack. 1990. Betalains. In *The alkaloids*, ed. A. Brossi, 1–62. London: Academic.

Street, H.E. 1969. Growth of organized and unorganized systems-knowledge gained by culture of organs and tissue explants. In *Plant Physiology – A treatise*, vol. VB, ed. F.C. Steward, 3–224. New York: Academic.

Sundaram, M.S. 1996. Azadirachtin biopestide: A review of studies conducted on its analytical chemistry, environmental behaviour and biological effects. *Journal of Environmental Science and Health* 31: 913–948.

Suresh, B., R. Thimmaraju, N. Bhagyalakshmi, and G.A. Ravishankar. 2004. Methyljasmonate and L-Dopa as elicitors of betalaine production in bubble column bioreactor. *Process Biochemistry* 39: 2091–2096.

Takahashi, Y., Y. Hitaka, M. Kino-Oka, M. Taya, and S. Tone. 2001. Evaluation of growth property of red beet hairy roots depending on condition of inocula and its application to culture control with fuzzy logic theory. *Biochemical Engineering Journal* 8: 121–127.

Tanaka, H., C. Hirao, H. Semba, and Y. Tozawin. 1985. Release of intracellularly stored 5′phosphodiesterase with preserved plant cell viability. *Biotechnology and Bioengineering* 27: 890–892.

Tanaka, N., M. Takao, and T. Matsumoto. 1994. *Agrobacterium rhizogenes* mediated transformation and regeneration of *Vinca minor* L. *Plant Tissue Culture Letters* 11: 191–198.

Tanaka, N., Y. Fuzikawa, M.A.M. Aly, H. Sanoeka, K. Fujita, and I. Yamashita. 2001. Proliferation and rol gene expression in hairy root lines of Egyptian clover. *Plant Cell, Tissue and Organ Culture* 66: 175–182.

Taya, M., K. Mine, M. Kino-Oka, S. Tone, and T. Ichi. 1992. Production and release of pigments by culture of transformed hairy root of red beet. *Journal of Fermentation and Bioengineering* 73: 31–36.

Taya, M., K. Yakura, M. Kino-Oka, and S. Tone. 1994. Influence of medium constituents on enhancement of pigment production by batch culture of red beet hairy roots. *Journal of Fermentation and Bioengineering* 77: 215–217.

Tepfer, D. 1984. Genetic transformation of special species of higher plants by *Agrobacterium rhizogenes* phenotypic consequences and sexual transmission of transformed genotype and phenotype. *Cell* 37: 459–467.

Tesoriere, L., M. Allegra, D. Butera, and M.A. Livrea. 2004. Absorption, excretion, and distribution of dietary antioxidant betalains in LDLs: Potential health effects of betalains in humans. *American Journal of Clinical Nutrition* 80: 941–945.

Thimmaraju, R., and N. Bhagyalakshmi. 2002. Application of rol genes for enhanced production of bio-active compounds from medicinal plants. In *Role of biotechnology in medicinal and aromatic plants*, vol. VII, ed. I.A. Khan and A. Khanum, 139–159. Hyderabad: UKAAZ Publications.

Thimmaraju, R., N. Bhagyalakshmi, M.S. Narayan, and G.A. Ravishankar. 2003a. Kinetics of pigment release from hairy root cultures of red beet under the influence of pH, sonication, temperature and oxygen stress. *Process Biochemistry* 38: 1067–1074.

Thimmaraju, R., N. Bhagyalaskshmi, M.S. Narayan, and G.A. Ravishankar. 2003b. Food grade chemical and biological agents permeabilize red beet hairy roots assisting the release of betalains. *Biotechnology Progress* 19: 1274–1282.

Thimmaraju, R., N. Bhagyalakshmi, and G.A. Ravishankar. 2004. In situ and ex situ adsorption and recovery of betalaines from hairy root cultures of *Beta vulgaris*. *Biotechnology Progress* 20: 777–785.

Thimmaraju, R., V. Kumar, L. Venkatachalam, N. Bhagyalakshmi, and G.A. Ravishankar. 2005. Peroxidase production from hairy root cultures of red beet (*Beta vulgaris*). *Electronic Journal of Biotechnology* 8: 185–196.

Thimmaraju, R., L. Venkatachalam, R.V. Sreedhar, N. Bhagyalakshmi, M.S. Narayan, and G.A. Ravishankar. 2006. Elicitation of peroxidase from hairy root cultures of red beet. *Electronic Journal of Biotechnology* 9: 512–521.

Thimmaraju, R., L. Venkatachalam, K. Roohie, M.S. Narayan, and N. Bhagyalakshmi. 2007. Purification and characterization of an intracellular peroxidase from genetically transformed roots of red beet (*Beta vulgaris* L.). *Food Chemistry* 105: 1312–1320.

Thimmaraju, R., L. Venkatachalam, and N. Bhagyalakshmi. 2008. Morphometric and biochemical characterization of red beet (*Beta vulgaris* L.) hairy roots obtained after single and double transformations. *Plant Cell Reports* 27: 1039–1052.

Thurau, T., S. Kifle, C. Jung, and D.G. Cai. 2003. The promoter of the nematode resistance gene *Hs1pro* – 1 activates a nematode-responsive and feeding site-specific gene expression in sugar beet (*Beta vulgaris* L.) and *Arabidopsis thaliana*. *Plant Molecular Biology* 52: 643–660.

Vasconsuelo, A., and R. Boland. 2007. Molecular aspects of the early stages of elicitation of secondary metabolites in plants. *Plant Science* 172: 861–875.

Vaughan, Martha M., Dorothea Tholl, and J.G. Tokuhisa. 2011. An aeroponic culture system for the study of root herbivory on *Arabidopsis thaliana*. *Plant Methods* 7: 5. 1–10.

Walden, R., I. Czaja, T. Schmulling, and J. Schell. 1993. Rol genes alter hormonal requirements for protoplast growth and modify the expression of an auxin-responsive promoter. *Plant Cell Reports* 12: 551–554.

Walton, N.J., M.J. Mayer, and A. Narbad. 2003. Vanillin. *Phytochemistry* 63: 505–515.

Wang, K., L. Herrera-Estrella, M.V. Montagu, and P. Zambryski. 1984. Right 25 bp terminus sequence of the nopaline T-DNA is essential for and determines direction of DNA transfer from *Agrobacterium* to the plant genome. *Cell* 38: 455–462.

Weathers, P.J., and R.W. Zobel. 1992. Aeroponics for the culture of organisms, tissues and cells. *Biotechnology Advances* 10: 93–115.

Weber, J., V. Georgiev, C. Haas, T. Bley, and A. Pavlov. 2010. Ploidy levels in *Beta vulgaris* (red beet) plant organs and in vitro systems. *Engineering in Life Sciences* 10: 139–147.

Weisshaar, B., and G.I. Jenkins. 1998. Phenylpropanoid bio-synthesis and its regulation. *Current Opinion in Plant Biology* 1: 251–257.

Wenzler, H.C., G. Mignery, L. Fischer, and W. Park. 1989. Sucrose-regulated expression of chimeric potato tuber gene in leaves of transgenic tobacco plants. *Plant Molecular Biology* 13: 347–354.

White, F.F., B.H. Taylor, G.A. Huffman, M.P. Gordon, and E.W. Nester. 1985. Molecular and genetic analysis of the transferred DNA regions of the root-inducing plasmid of *Agrobacterium rhizogenes*. *Journal of Bacteriology* 164: 33–44.

Winter, H., and S.C. Huber. 2000. Regulation of sucrose metabolism in higher plants: Localization and regulation of activity of key enzymes. *Critical Reviews in Biochemistry and Molecular Biology* 35: 253–289.

Yang, D.C., and Y.E. Choi. 2000. Production of transgenic plants via *Agrobacterium rhizogenes*-mediated transformation of *Panax ginseng*. *Plant Cell Reports* 19: 491–496.

Yokoyama, R., T. Hirose, N. Fujii, E.T. Aspuria, A. Kato, and H. Uchimiya. 1994. The *rolC* promoter of *Agrobacterium rhizogenes* Ri plasmid is activated by sucrose in transgenic tobacco plants. *Molecular and General Genetics* 244: 15–22.

Yoshikawa, T., and T. Furuya. 1987. Saponin production by cultures of *Panax ginseng* transformed with *Agrobacterium rhizogenes*. *Plant Cell Reports* 6: 449–453.

Zhang, L., R. Ding, Y. Chai, M. Bonfill, E. Moyano, K.M. Oksman-Caldentey, T. Xu, Y. Pi, Z. Wang, H. Zhang, G. Kai, Z. Liao, X. Sun, and K. Tang. 2004. Engineering tropane biosynthetic pathway in *Hyoscyamus niger* hairy root cultures. *Proceedings of the National Academy of Sciences of the United States of America* 101: 6786–6791.

Zhang, H.C., J.M. Liu, H.Y. Lu, and S.L. Gao. 2009. Enhanced flavonoid production in hairy root cultures of *Glycyrrhiza uralensis* Fisch by combining the over-expression of chalcone isomerase gene with the elicitation treatment. *Plant Cell Reports* 28: 1205–1213.

Chapter 11
Bioreactors for the Cultivation of Red Beet Hairy Roots

Vasil G. Georgiev, Thomas Bley, and Atanas I. Pavlov

Abstract Since hairy root cultures of red beet were obtained at the end of the twentieth century, they have been intensively investigated over the past decade as producers of biologically active compounds, including food colorants (betacyanin and betaxanthin pigments), phenolic compounds, enzymes and other proteins and for understanding basic cellular physiologies as well as for engineering their container designs, such as bioreactors. The analyses of available data concerning their growth potential, nutrient needs and secondary metabolite profiles outline these types of in vitro cultures as effective producers of biologically active colorants for the food and pharmaceutical industries, although commercial scales are yet to be realized. The main limitation for the development of such technologies is the cost-effectiveness of the process and the need for refinement of an appropriate bioreactor system suitable for further scale-up. This chapter summarizes the recent advances in bioreactor cultivation of beet hairy root cultures. The application of bioreactor systems with different designs, optimization of the cultivation conditions, monitoring and assessment of the cultivation process, methods for enhancement of pigment production and strategies for on-line

V.G. Georgiev
The Stephan Angeloff Institute of Microbiology, Bulgarian Academy of Sciences, Laboratory of Applied Biotechnologies, 139 Ruski blvd, Plovdiv 4000, Bulgaria
e-mail: vasgeorgiev@gmail.com; lbpmbas@yahoo.com

T. Bley
Institute of Food Technology and Bioprocess Engineering, Technische Universität Dresden, Dresden 01062, Germany
e-mail: thomas.bley@mailbox.tu-dresden.de

A.I. Pavlov (✉)
The Stephan Angeloff Institute of Microbiology, Bulgarian Academy of Sciences, Laboratory of Applied Biotechnologies, 139 Ruski blvd, Plovdiv 4000, Bulgaria

Department of Organic Chemistry and Microbiology, University of Food Technologies, 26 Maritza Blvd, Plovdiv 4002, Bulgaria
e-mail: at_pavlov@yahoo.com; lbpmbas@yahoo.com

B. Neelwarne (ed.), *Red Beet Biotechnology: Food and Pharmaceutical Applications*,
DOI 10.1007/978-1-4614-3458-0_11, © Springer Science+Business Media New York 2013

product recovery are reviewed, compared and discussed with the purpose of assessing their effectiveness for the development of an effective industrial process. The advantages of biotechnologically produced red beet colorants, their possible applications and the law frame for their regulation are discussed as well.

11.1 Introduction

Betalains constitute a class of water-soluble nitrogenous plant pigments, consisting of red–violet betacyanins and yellow–orange betaxanthins. They are produced by plants belonging to the families of *Caryophyllales*, with the exception of the *Caryophyllaceae* and the *Molluginaceae*. Information concerning their occurrence, biosynthesis, chemistry and physiological functions in plants is presented in many excellent research papers, reviews, and books (Mabry 2001; Strack et al. 2003; Stintzing and Carle 2004; Zryd and Christinet 2004; Gandia-Herrero et al. 2005a, b, c, d, 2010; Grotewold 2006; Stintzing and Carle 2007c; Tanaka et al. 2008; Moreno et al. 2008; Georgiev et al. 2008b; Han et al. 2009; Roberts et al. 2010). Although the name betalain was introduced in 1968 by Andre Dreiding and Tom Mabry, these pigments had been familiar to and well accepted by mankind for centuries. Dried or concentrated red beetroot extracts are widely used as a natural colorant in the processing of many foods, such as dairy products, beverages, candies, canine foods and cattle products (Martinez et al. 2006; Stintzing and Carle 2004, 2007a, b; Azeredo 2008; Moreno et al. 2008; Georgiev et al. 2008b; Georgiev et al. 2010). Recent discoveries concerning the molecular structure properties, stabilities and strong antioxidant activities of betaxanthins and betacyanins have regained scientific attention for this class of compounds (Kanner et al. 2001; Pavlov et al. 2002; Lila 2004; Pavlov et al. 2005b; Gliszczynska-Swiglo et al. 2006; Matkowski 2008; Gandía-Herrero et al. 2009, 2010; Tesoriere et al. 2009; Georgiev et al. 2010). As a result, the practical application of betalains has extended from food processing to pharmaceutical applications: treatment of osteoarthritis, acne, contact dermatitis and sinusitis (Pietrzkowski 2010); medicinal diagnostics: early diagnosis of hemochromatosis (Sotos 1999); dentistry: natural dye indicator for dental plaque (Gaffar 1984); chemistry: natural pH indicators (Soltan and Sirry 2002); to solar electricity: development of dye-sensitized solar cells (Zhang et al. 2008; Calogero et al. 2010).

Nowadays, storage roots of red beet (*Beta vulgaris* L.) form the main source for commercial production of betalains (Stintzing and Carle 2007b; Moreno et al. 2008; Georgiev et al. 2010). Red beet is the only allowed source of betalains approved as food additive in the European Union (Commission Directive 95/45/EC) and the USA (Title 21 CFR Part 73) and exempt from certification. In Europe, 20,000 t of beetroot are processed annually for juice and pigment extraction (Davies 2004). However, the market for betalains grows increasingly due to enhanced interest in natural food colorants. Following the latest tendencies, investigations of the biotechnological production of betalains have remarkably expanded during the last decade. Many cell suspension cultures derived from different *B. vulgaris* L. cultivars are investigated for betalain production, since they possess well-known advantages.

Most of the results are presented in Chap. 9 of this book and other review papers (Leathers and O'Riordain 1990; Jimenez-Aparicio and Gutierrez-Lopez 1999; Georgiev et al. 2008b; Moreno et al. 2008). In summary, the cultivation of *B. vulgaris* cell suspension cultures in different bioreactor types leads to a significant decrease or retention of the betalain yields during their submerged cultivations in flasks (Khlebnikov et al. 1995; Rodriguez-Monroy and Galindo 1999; Sanchez et al. 2002). This is not surprising, since primary *B. vulgaris* callus cultures possess significant somaclonal variability and seem to comprise a mixture of cell populations having different potentials to produce betacyanins, betaxanthins, or to be colorless (Girod and Zryd 1991; Akita et al. 2000; Pavokovic et al. 2009; Weber et al. 2010). Recent flow cytometry investigations of mechanically separated yellow and red lines from rhizogenic callus of *B. vulgaris* cv. Detroit Dark Red have demonstrated that they have different levels of polysomaty compared with the primary callus, and undergo two endoreduplication cycles less (Georgiev et al. 2009; Weber et al. 2010). Moreover, several studies have demonstrated that the addition of growth regulators into the culture medium can affect the plant cell endocycle in various, unpredictable ways (Mishiba et al. 2001; Campanoni and Nick 2005). The negative effects of extracellular auxins and cytokinins could be avoided by development of *Agrobacterium*-transformed cultures, which do not need additional phytohormones for their growth. Transformed by *A. tumefaciens* Conn 1942, callus and cell suspensions have been recently initiated for sugar beet *B. vulgaris* L., but the investigations of betalain production are at a very early stage (Pavokovic et al. 2009; Kriznik and Pavokovic 2010). However, the other "natural genetic engineer", *A. rhizogenes* Conn 1942 is widely used for induction of *B. vulgaris* hairy root cultures. The genetically transformed hairy root cultures are the result of the inter-kingdom horizontal transfer of a suite of oncogenes from the root-inducing (Ri) plasmids of *A. rhizogenes* into the plant cell genome (Britton et al. 2008). Detailed information about the techniques and the molecular mechanism of *A. rhizogenes* genetic transformation can be found in Chap. 10 and elsewhere (Gelvin 2000; Tzfira and Citovsky 2000, 2003; Georgiev et al. 2007, 2008a; Britton et al. 2008; Soltani et al. 2008; Young 2008). During the last two decades, hairy root cultures have attracted researchers' attention as potential in vitro systems for both studying plant biochemistry and for biotechnological production of secondary metabolites. The transformed roots are considered to be genetically stable, fast growing without necessity of exogenous plant hormones, and able to produce identical secondary metabolites like the intact plant roots at comparable levels. From a historical point of view, betalain production by hairy root culture of *B. vulgaris* var. Boltardy is one of the first reports to reveal the potential of *A. rhizogenes*-transformed root cultures for in vitro production of desirable plant secondary metabolites (Hamill et al. 1986; Flores et al. 1987). Up till now, hairy root cultures have been initiated from several beet cultivars, including: red beets *B. vulgaris* cv. Detroit Dark Red, *B. vulgaris* cv. Bordo, *B. vulgaris* cv. Detroit 2, *B. vulgaris* cv. Egyptian, *B. vulgaris* var. Ruby Queen and *B. vulgaris* var. Mahyco Red; yellow beets *B. vulgaris* var. Burpee's Golden, and *B. vulgaris* var. *lutea*; and table beets *B. vulgaris* var. *lutea*, *B. vulgaris* subsp. *vulgaris* var. Altamo, *B. vulgaris* subsp. *vulgaris* var. Fumona, and *B. vulgaris* subsp. *vulgaris* var. Brigadied. These in vitro cultures have been involved in numerous studies, including the discovery of the betalain biosynthesis pathway (Kobayashi et al. 2001), feeding experiments for controlled biosynthesis of desired betaxanthins (Hempel and Bohm 1997), production of peroxidase (EC 1.11.1.7) (Thimmaraju et al. 2006; Bhagyalakshmi and Thimmaraju 2009), and last but not least, for production of betacyanins and betaxanthins (Hamill et al. 1986; Taya et al. 1992; Weathers and Zobel 1992; Mukundan et al. 1998a; Pavlov et al. 2002; Shin et al. 2002; Thimmaraju et al. 2003b). Most of the investigations for pigment production have been performed at shake flask level (Thimmaraju et al. 2003b; Bhagyalakshmi et al. 2004; Pavlov et al. 2005a; Mukundan et al. 1998a; Taya et al. 1994; DiIorio et al. 1993; Taya et al. 1992), but several of them have reached the stage of laboratory bioreactor cultivation and are discussed in the current chapter.

11.2 Bioreactor Designs and Cultivation Conditions

Cultivation of hairy root cultures in different bioreactor systems presents a serious challenge to researchers. The proven bioreactors, including stirred tank and airlift types that are usually applied for large-scale cultivation in microbial biotechnologies, fail to support the cultivation of differentiated plant in vitro systems. Although laboratory experiments have demonstrated that hairy roots grow with sufficient productivities in these cultivation systems, the scale-up of the process is inconceivable without considerable changes in the bioreactor design and control algorithms supported with data of actual trials. In most of the cases, the introduced modifications promote a significant decrease in the system productivities and impair the abilities for the final root biomass harvesting during subsequent processing. Several characteristics that are the specific nature of hairy roots predetermine the need for establishing specialized bioreactor constructions to ensure their adequate growth, taking into consideration their nonpareil behavior during growth in submerged conditions of cultivation. Current information about the different constructions of bioreactors used for cultivation of hairy roots is available elsewhere (Sajc et al. 2000; Kim et al. 2002; Weathers et al. 1997; Wilson 1997; Georgiev et al. 2008a, b). Some of the specific characteristics of hairy roots in bioreactors are summarized below, with emphasis on different cultivars of *Beta vulgaris*.

11.2.1 Growth Rate and Culture Density

To evaluate the growth rate of hairy roots cultured in liquid systems, calculation of the doubling time (T_d) of the cultures is used most frequently.

$$T_d = \ln 2 * \frac{\Delta t}{\ln\left(\frac{X_f}{X_i}\right)},$$

where Δt is the culture interval, and X_i and X_f are the initial and final biomasses.

Table 11.1 Performances of different *Beta vulgaris* hairy root cultures during their cultivation in different bioreactor systems as compared with shake flasks

Cultivation systems	Vessel volume/ effective volume (L)	Doubling time (Td) (days)	Betalains (mg/g DW)	Reference
Stirred tank	5.0/3.0	5.11	10.06	Georgiev et al. (2006)
Bubble column	3.0/2.0	4.47	26.0	Pavlov et al. (2007)
Temporary immersion system RITA	0.5/0.2	6.22	18.8	Pavlov and Bley (2006)
Nutrient mist	1.8/1.29	2.9	3.3	Weathers and Zobel (1992)
Bubble column with support matrix	2.0/1.85	3.2	1.64	Mukundan et al. (1998b)
Cone-type airlift bioreactor	5.0/3.0	5.80	27.0	Shin et al. (2002)
Low-cost bottle bioreactor operating as bubble column	1.0/0.8	5.23	8.4	Unpublished data
Low-cost bottle bioreactor operating as bubble column	10.0/6.0	4.58	7.9	Unpublished data
Flasks	0.250/0.05	3.12	42.2	Pavlov et al. (2005a)
Flasks	–	2.9	2.9	Weathers and Zobel (1992)

Hairy root cultures show slower growth rates compared with microorganisms. Their effective T_d value is in the range of 1–29 days, depending on the root clone and culture conditions. This is a very large period in comparison with microorganisms (T_d is about 20 min for *Escherichia coli* and about 2 h for *Saccharomyces cerevisiae*). Moreover, hairy roots usually possess maximal growth at the shake flask level, whereas, when cultivated in bioreactors, their doubling times increase. In the case of *Beta vulgaris* hairy root cultures, the doubling time varies from 2.9 days in flasks to 6.28 days in a bubble column bioreactor (Table 11.1). Several examples for T_d values of *B. vulgaris* hairy roots, cultivated in different bioreactors and shake flasks, are presented in Table 11.1. The target requirement for successful choice of an effective bioreactor system is a T_d value comparable to (or lower than) the value reached at the shake flask level.

In contrast with microorganisms, where the growth rate depends on the increase in the cell number, hairy roots develop as branched filamentous organs, consisting of dividing, developed, specialized and senescing cells. The growth rate of hairy roots in submerged conditions depends on their linear extension, the formation of new meristems (growing points) and on the increase in the root diameter as a result of cell expansion and differentiation. Once transferred into the bioreactor, every single root distributed in the medium serves as inoculums from which branching begins, leading to compact clump formation. At this stage, the morphology of the tissue is such that the biomass reaches a stage when it becomes self-immobilizing, occupying either the surface of the medium or the bottom of the reactor, depending on the specific gravity of the cultures (Fig. 11.1). At the same time, the other part of the bioreactor stays filled with culture medium only, without any biomass. For these reasons, designing and operating hairy root bioreactors are much more challenging than with microbial or cell cultures.

Fig. 11.1 Self-immobilization of 14-days-old *B. vulgaris* cv. Detroit Dark Red hairy roots on the medium surface during cultivation in stirred tank bioreactor

The specific gravity (ρ) of *Beta vulgaris* cv. Detroit Dark Red hairy roots is determined to be 1.005×10^3 kg fresh weight (FW)/m^3 (Kino-oka et al. 1995; Hitaka et al. 2000; Takahashi et al. 2001), which makes the culture float at the top when cultured in a simple column bioreactor (Fig. 11.2). The formation of highly dense clumps severely limits the mass transfer within the culture and causes its growth limitation.

Because of the heterogeneous distribution of the biomass, growth limitation inside the clumps and the low content of dry matter (approximately 6–9%), the expected maximal density of submerged hairy root cultures is about 30–35 g dry weight (DW)/L. Our experience has revealed that a maximum culture density of 29.4 g DW/L is reached by cultivation of *B. vulgaris* cv. Detroit Dark Red hairy roots in shake flasks (Pavlov et al. 2005a), whereas, when cultivated in bioreactors, it decreases significantly (Table 11.1).

11.2.2 Mass Transfer Limitations

Clump formation during the cultivation of hairy roots in submerged systems leads to airflow channeling and impaired liquid mixing. As a consequence, several resistances to oxygen and nutrient transport appear and cause serious reduction in hairy roots growth. Resistance to oxygen transport is especially important because the oxygen requirement of the plant cells is relatively high. There are several kinds of resistance

Fig. 11.2 Floating 9-days-old *B. vulgaris* cv. Detroit Dark Red hairy roots cultivated in simple column bioreactors

to oxygen transport from the medium to the hairy root cells. The first barrier is in the liquid–solid boundary layer at the surface of hairy roots. The second is resistance to oxygen transfer inside the clump, and last but not least, is the internal oxygen transfer resistance from the epidermis to the stele of the root. The source of oxygen supply is important as well. Aeration with oxygen-enriched air enhances the hairy root growth, whereas the growth could be inhibited when pure oxygen is used (Weathers et al. 1997; Carvalho et al. 1997). The presence of carbon dioxide in aeration air used in cultured *B. vulgaris* hairy roots results in shortened lag phase of the culture and does not decrease its growth rate (Kim et al. 2002). Experiments with hairy root cultures of *Arabidopsis thaliana* and *Hyoscyamus muticus* have demonstrated that the root hairs that facilitate the nutrient uptake in static soil cultivation are detrimental to growth in a submerged environment (Shiao and Doran 2000; Bordonaro and Curtis 2000). When transferred into bioreactors, the root hairs represent the dominant resistance for oxygen transport and therefore contribute extensively to stagnation of fluid flow and limitation in oxygen availability. The elimination of root hairs by pyrene butyric acid boosts the bioreactor performance by improving the biomass accumulation, enhancing the mixing and increasing the respiratory activity of the culture (Bordonaro and Curtis 2000). Moreover, when the transformed roots are cultured in bioreactors, the increased surface area provided by the root hairs is not needed for nutrient uptake (Carvalho et al. 1997). Mass transfer resistance varies with plant species and the hairy root lines. Many factors influence nutrient transport in hairy roots, including the surface characteristics and size of the roots, the distance between the roots in the clumps, the nutrient concentration and the liquid properties. The specific surface area (S_V) of *B. vulgaris* hairy roots is determined to be 2.7×10^4 m^{-1} and the effective diameter of the hairy roots, considering the root hairs, is estimated to be 1.5×10^{-4} m (Hitaka et al. 1997, 2000). For effective cultivation of *B. vulgaris* hairy roots in large volumes, it is reasonable to search new approaches for decreasing their hairiness, while the reduction of the specific surface area has significant impact on scale-up by improving the mixing and mass transfers.

11.2.3 Sensitivity to Shear Stress

Hairy root cultures are sensitive to physical stress caused by shaking, mixing, flowing and stirring. When exposed to vigorous stirring, callus is induced on hairy roots, resulting in a decline in the formation of secondary metabolite (Wilson 1997). The sensitivity varies significantly between plant species and the hairy root lines. When exposed to high-shear stress, *B. vulgaris* hairy roots decrease the number of their growth points (GPs) by 33% compared with low-shear stress cultivation (Hitaka et al. 2000). The viability of growth points (A_{GP}) during shear stress loading on hairy root cultures is calculated with the following equation:

$$A_{GP} = \frac{n}{n_T},$$

where n is the number of GPs (m^{-3}), and n_T is the total number of tips examined (m^{-3}).

The tolerance of *B. vulgaris* cv. Detroit Dark Red hairy roots to shear stress has been investigated in a single column bioreactor operating in a convective flow mode (Hitaka et al. 2000). The value of shear stress (τ) is evaluated under conditions of laminar flow in the growth unit using the equation:

$$\tau = \frac{\left(\frac{\Delta P}{z}\right)}{(1-\varepsilon)^* S_V},$$

where ΔP is the pressure drop through the hairy root bed (Pa); z is the height of growth unit (m); S_V is the specific surface area of the hairy roots (m^{-1}), and ε is the void fraction in hairy root bed calculated as:

$$\varepsilon = 1 - \frac{4 * V_H}{\pi * d^2 * z} \text{ and } V_H = \frac{X * V}{\rho * (1-\omega_C)},$$

where d is the inner diameter of growth unit (m); V_H is the volume of hairy roots in a single column bioreactor (m^3); X is the concentration of hairy roots (kg DW/m^3);

V is the working volume in a single column bioreactor (m^3); ρ is the specific gravity of hairy roots (kg FW/m^3) and ω_C is the water content of hairy roots (Hitaka et al. 2000). The experiments have demonstrated that *B. vulgaris* hairy roots have tolerance to low-shear stress of $\tau=0.05\ N/m^2$, where the viability of growth points decrease with the increase of τ to $1.00\ N/m^2$. The authors have found that when acclimation treatment of 50 h exposure to low-shear stress ($\tau=0.05\ N/m^2$) is applied, the culture gains tolerance to shear stress, and the subsequent increases in τ up to $1.00\ N/m^2$ do not affect the culture growth. Therefore, a two-stage method of cultivation was suggested, in which the first stage includes 50 h of low-shear stress exposure and a second stage with 110 h exposure to high-shear stress (Hitaka et al. 2000). To reduce the resistance to mass transfer resistance in submerged systems, obtaining hairy root lines that have high tolerance to shear stress and improving culture efficiency is a critical step in large-scale cultivations.

11.2.4 Inoculation of Bioreactors

The specific nature of hairy roots presents serious handling problems for preparation of inoculums. For laboratory scale bioreactors, this is usually performed using forceps in the laminar flow chamber. This method is not appropriate for large-scale bioreactors. In contrast with plant cell suspensions, hairy roots do not require a critical minimum inoculation density. The inoculation of large volume vessels can be accomplished with homogenized root tissue transferred into the bioreactors as slurry (Ramakrishnan et al. 1994). A relatively small amount of inoculum is needed as long as there is sufficient tissue with viable meristems to provide the sufficient distribution throughout the bioreactor and to ensure the culture growth (Ramakrishnan et al. 1994). When bioreactors equipped with inner support matrix are used, usually a two-stage operation is applied; at the first stage they work as bubble columns to entrap the inoculum root fragments on the matrix, and at the second stage they are turned to trickle bed or convective flow mode to ensure the culture growth. Such types of bioreactors will be discussed in more detail later in the section. The other important factor influencing the culture growth and secondary product formation is the age and the acclimation of hairy roots used as inoculum. It has been found that inoculation with a 14-day-old culture of *B. vulgaris* hairy roots (stationary phase of growth) leads to a 12.5% increase in the biomass and a 14% increase in the betalain content (Pavlov et al. 2003).

11.2.5 Monitoring the Growth of Hairy Roots in Bioreactors

The accurate on-line estimation of hairy roots growth in bioreactors is essential to adequate monitoring and control of the cultivation process. Their unique structure precludes the direct and representative sampling. The direct measurement of the total hairy roots weight could be monitored in bioreactors by removing the medium into a sterile vessel and measuring the changes in the total weight of the bioreactor, including the cultured roots (Wilson 1997). This system has been successfully installed on a 500-L bioreactor during the cultivation of *Datura stramonium* hairy roots (Wilson 1997). At the current stage, this is the largest bioreactor applied for hairy root cultivation ever reported (discussed later). However, specialized equipment and additional apparatus as aseptic pumps and a storage reservoir are required for application of this method, which increases the risk of culture contamination.

Several methods for indirect monitoring of growth have been proposed, based on the correlations between tissue mass accumulation and liquid medium properties such as electric conductivity, sugar content and medium osmolality (Taya et al. 1989; Ramakrishnan et al. 1999; Mukundan et al. 1998b; Bhagyalakshmi et al. 2004; Pavlov and Bley 2005, 2006; Pavlov et al. 2007). The correlations based on conductivity and refractive index are commonly used for bioreactor cultures of *B. vulgaris* hairy roots. The decrease in the conductivity or refractive index values of the medium reflects the absorption of the ions or sugars by the hairy roots and, as such, they are proportional to the biomass increase (Ramakrishnan et al. 1999). These correlations are accurate and present a fast and effective way for on-line monitoring of roots growth. They can be calculated as simple proportionality between the media conductivity or refractive index decline (ΔC) and the biomass increase (ΔDW):

$$\Delta DW = K * (\Delta C),$$

where K is the proportionality constant that is obtained experimentally.

The above-mentioned relationship suffers from the inability to account for water loss due to sampling, evaporation and water uptake into the roots. To overcome this problem, a new technique based on mass balance has been introduced for hairy root cultures (Ramakrishnan et al. 1999; Mukundan et al. 1998b):

$$\Delta DW = K_b * \left[V_0 * C_0 - V_s * C_s + \sum (V_{add} * C_{add}) - \sum (V_{rem} * C_{rem}) \right],$$

where K_b is the constant for linear correlation; V is the liquid volume; C is the conductivity or refractive index; the subscript "*0*" refers to an initial time point, the subscript "*s*" refers to a subsequent sampling point and the subscripts "*add*" and "*rem*" represent liquid addition and removal (including evaporation), respectively.

The observation of the dependence of hairy root water content on medium (liquid) osmolality is critical for biomass estimation in large-scale bioreactors (Ramakrishnan et al. 1999). This relationship is based on the assumption that the tissue is at osmotic equilibrium with the surrounding medium. The ratio of fresh weight/dry weight (FW/DW) is calculated by the following equation (Mukundan et al. 1998b):

$$FW / DW = 1 + \frac{\beta}{S},$$

where β is the proportionality constant that is obtained experimentally and S is the medium osmolality (mM/kg). This equation allows the calculation of the corresponding fresh weight from the previously calculated dry weight value.

The number of lateral branches emerging from the inoculated hairy roots is used for identification of the current culture phases. By monitoring the variation in the number of growing points of *B. vulgaris* cv. Detroit Dark Red hairy roots, cultivated on a solid medium, fuzzy logic interpretation has been employed for the control of shear stress in parallel with similar culture in a single-column bioreactor system (Takahashi et al. 2001). Semiautomatic non-destructive image analysis is applied for on-line monitoring of morphological and secondary metabolite changes during the cultivation of red and yellow *B. vulgaris* hairy roots in a rotating drum bioreactor (Berzin et al. 1999, 2000). However, this method suffers from several disadvantages originating from the clump formation with the cultivation progression, which affect the correct estimations of the main root segmentation plane and measurements of the individual root tips length and diameter.

11.2.6 Stirred Tank Bioreactors

The classic stirred tank fermentors have been successfully applied for the cultivation of *B. vulgaris* hairy roots (Pavlov and Bley 2005; Georgiev et al. 2006; Kino-oka et al. 1992; Kino-oka and Tone 2001). However, to ensure good growth and quality of hairy roots as well as to avoid tissue disorganization induced by the impeller, a few modifications were required (Wilson 1997). To reduce the executed mechanical stress, it is convenient to use a propeller/impeller working at low speed (usually 50–100 rpm) rather than the widely used turbine blades (Fig. 11.3a). When a turbine blade impeller is used, roots need to be enclosed in a cage to protect from mechanical damage (Fig. 11.3b).

After the inoculation of bioreactors, the roots are uniformly dispersed in the medium by the impeller. When the culture starts growing, the clumps can easily form in the spaces between the electrodes (of pH and gas sensors), sampling tubes and the air sparger (Fig. 11.4a). This immobilization strongly increases the culture heterogeneity and decreases the system productivity, especially when small volume vessels are used. To avoid this undesirable effect, a fed-batch process is implemented, where culture volume increased step-wise after achieving a specific level of biomass (Georgiev et al. 2006). Feeding experiments with *B. vulgaris* cv. Detroit Dark Red hairy roots are conducted in a 5-L bioreactor with a 3-L working volume (Fig. 11.4b). The results show that when the culture is fed in five portions with additions of 0.2 L of MS medium, the obtained dry biomass (10 g/L) and betalain content (10.95 mg/g DW) are higher (13% and 11%, respectively) than when a single-feed procedure with 1.0 L of MS medium is performed (Georgiev et al. 2006). The results outline the advantages of fed-batch cultivation of hairy roots in stirred tank systems. These bioreactors are available in a wide range of volumes and, as such, they are of great interest for the future scale-up of red beet hairy root cultures.

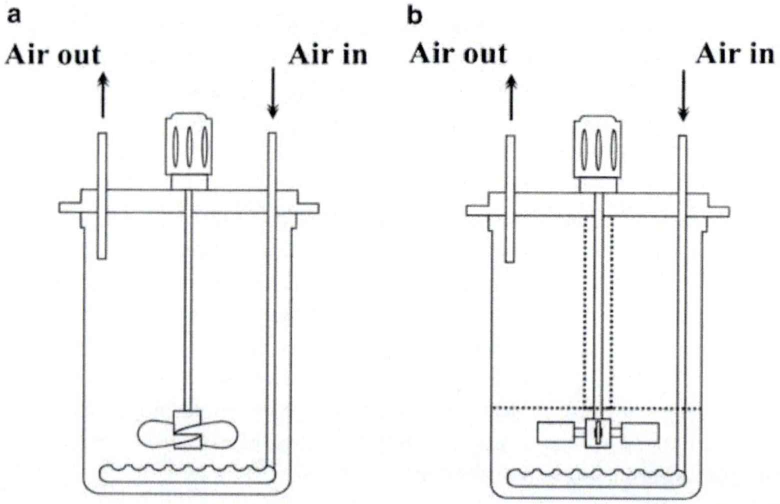

Fig. 11.3 Schematic diagram of stirred tank bioreactors with propeller impeller (**a**) and isolated turbine blade impeller (**b**)

Fig. 11.4 14-days-old *B. vulgaris* cv. Detroit Dark Red hairy roots cultivated in stirred tank bioreactors: clump formation between the electrodes in 3-L vessel (**a**); homogenous culture during fed-batch cultivation in 5-L vessel (**b**)

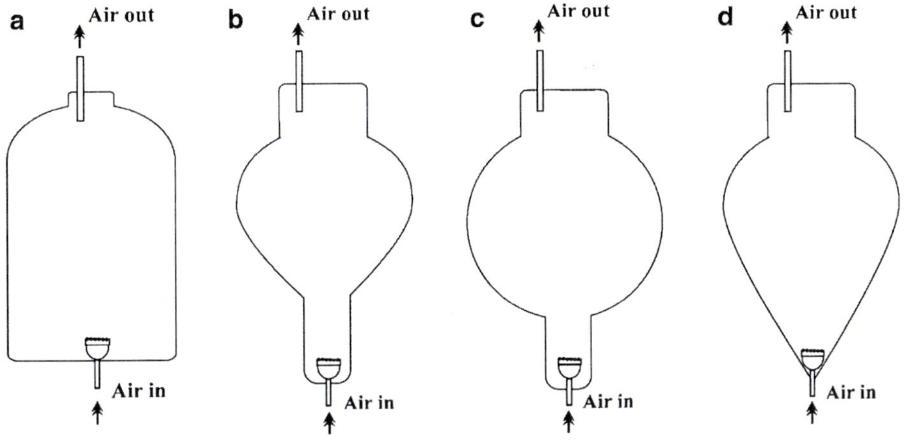

Fig. 11.5 Schematic diagram of airlift bioreactors: drum (**a**), balloon (**b**), bulb (**c**) and cone (**d**)

11.2.7 Airlift Bioreactors

The common bubble column bioreactors have very simple designs and are usually applied in large-scale cultivations where mechanical agitation is no longer practical. The typical construction consists of a tubular vessel with a sparger integrated at the bottom (Fig. 11.5a). The energy for biomass mixing and medium circulation is delivered by the air flows.

The bioreactor configuration recently used by the authors for the cultivation of red beet hairy roots is presented in Fig. 11.6. In this bubble column (3 L with 2-L working volume), hairy roots of *B. vulgaris* cv. Detroit Dark Red resulted in highest biomass and betalain yields ever reported (Table 11.1) (Pavlov et al. 2007). The accumulated biomass increased by 4.5% (up to 13.31 g/L) when a fed-batch process was performed by adding 1 L of MS medium on day 13 after the start of cultivation (Pavlov et al. 2007). When the feeding is performed with additions of 0.2 L of MS medium (from day 13 to day 17 of cultivation), a 10.7% increase in the betalain content (up to 29.1 mg/g DW) is achieved. Moreover, during the fed-batch cultivations, the betacyanins–betaxanthins ratio keeps changing, favoring the formation of betacyanins (Pavlov et al. 2007), the more desirable pigment for both food and pharmaceutical applications.

For improving the cultivation of hairy roots, the design of the bioreactor systems can be further improved by adding a support matrix to provide a more uniform distribution of roots in the medium (Fig. 11.5b). When red beet *B. vulgaris* var. Mahyco Red and yellow beet *B. vulgaris* var. Burpee's Golden hairy roots are cultured in a 2-L column bioreactor (1.85 L working volume) with support matrix, yields of 66.1 g FW/L (6.04 g DW/L) and 66.4 g FW/L (4.8 g DW/L), respectively, are

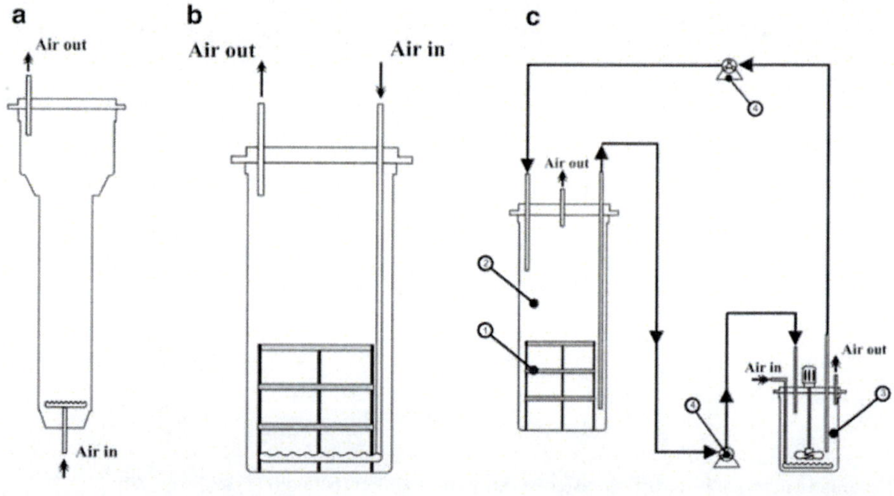

Fig. 11.6 Schematic diagram of column bioreactors: simple bubble column (a), bubble column with internal support matrix (b), bubble column with internal support matrix and external aeration vessel (c); 1 support matrix, 2 cultivation chamber, 3 aeration vessel, 4 peristaltic pumps

obtained (Mukundan et al. 1998b). Optimization of the support matrix construction can provide additional increase in the system productivity. The addition of one extra tier to the support basket in a 3-L bubble column bioreactor (1.75 L working volume) increases both the biomass and pigment production in *B. vulgaris* var. Ruby Queen hairy roots by 13% (Bhagyalakshmi and Thimmaraju 2009). However, in most cases, the supply of a support matrix slightly increases the system productivity in exchange for the significant decreases in the ability to recover biomass, especially when large volumes are involved.

A considerable decrease in the environmental stress in bubble column bioreactors could be achieved by employing an external aeration vessel (Fig. 11.5b). In this case, the oxygen enriches the medium circulating between the cultivation column and the external aeration vessel. As far as the cultivation column is free of air bubbling, the levels of the shear stress are vastly reduced, which improves culture growth. The application of an external aeration vessel during cultivation of *B. vulgaris* var. Ruby Queen hairy roots in a 3-L bubble column bioreactor (1.75-L working volume) leads to a 1.25-fold increase in both biomass and betalain content compared with systems operating as simple bubble columns with a support matrix (Bhagyalakshmi and Thimmaraju 2009). Moreover, the setup of an external aeration vessel provides a suitable opportunity for the development of an additional secondary circulation circuit, in which a column filled with adsorbent resins could be added and on-line recovery process for permeabilized as well as extracellular metabolites from the medium could be accomplished.

In contrast to bubble columns, airlift bioreactors are designed to provide both fluid–gas exchange and liquid circulation by creating directed narrow air flow.

Fig. 11.7 Cultivation of *B. vulgaris* cv. Detroit Dark Red hairy roots in 3-L simple bubble column bioreactor with 2-L working volume

When a homogenous cell suspension is cultured, the airlift bioreactor design can be implemented both with an internal draft tube (with the sparger situated either on inside or outside) or with an external circulation loop (Sanchez et al. 2002). However, when hairy roots are cultured, this design is not appropriate because the short space between the internal draft tube and the culture vessel, which is very suitable for immobilized hairy roots, strongly inhibits the medium circulation. In this case, the airlift effect is achieved by asymmetric position of the air sparger or by changing the geometric shape of the cultivation vessel. Depending on the geometric shape, different flows are formed in the medium by the sparging air. Shin and coauthors (Shin et al. 2002) compared four types of 5-L airlift bioreactors with working volumes of 3 L for cultivation of *B. vulgaris* cv. Detroit Dark Red hairy root cultures (Fig. 11.7). The authors found that the cone type (Fig. 11.5d) supported maximum biomass accumulation (6.0 g/L DW) and betacyanin content (27.0 mg/g DW). Further they supplied irradiation with 60 μM/m/s blue plus far-red light, which enhanced the accumulated biomass to 7.0 g/L and the betacyanin content to 34.0 mg/L (Shin et al. 2002, 2004). Recently, a 500-L airlift balloon type bioreactor has been developed and successfully applied for cultivation of ginseng adventitious roots; 74.8 kg of fresh roots are harvested after 7 weeks of cultivation (Murthy et al. 2008). This is a 150-fold increase in the growth, which defines this bioreactor system as potentially applicable for industrial process development.

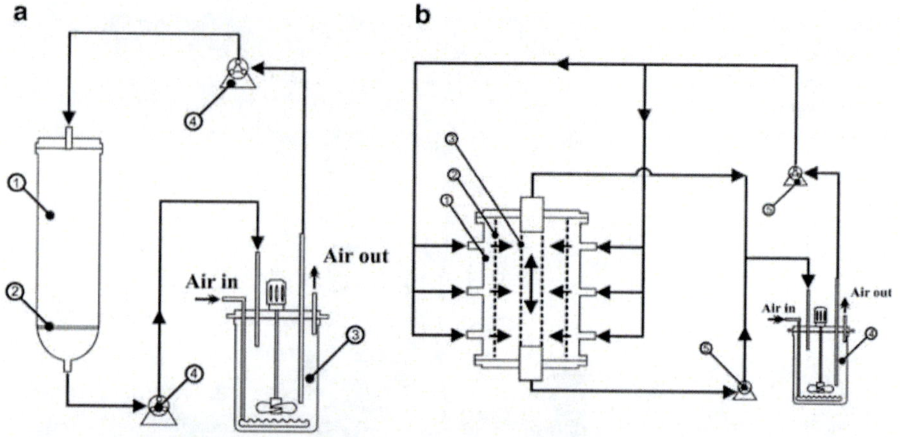

Fig. 11.8 Schematic diagram of convective flow bioreactor (**a**): *1* tubular cultivation chamber, *2* support mesh and *3* aeration vessel; and radial flow bioreactor (**b**): *1* cultivation chamber with radial side ports, *2* outer cylindrical mesh, *3* inner cylindrical mesh, *4* aeration vessel and *5* peristaltic pumps

11.2.8 Convective Flow Bioreactors

Several convective flow bioreactors are used for cultivation of *B. vulgaris* hairy roots. This cultivation system includes a tubular cultivation chamber with a support mesh attached on the bottom. During cultivation, the cultured hairy roots are anchored to the mesh and the mass transfer is performed by means of a convective medium flowing around them (Fig. 11.8a). Oxygen is supplied to the medium by an external aeration vessel.

The convective flow bioreactor has been used to investigate the effect of shear stress loaded on *B. vulgaris* cv. Detroit Dark Red hairy root culture considering the number of newly formed root tips and culture growth (Hitaka et al. 2000). By using fuzzy logic, an automated control system of shear stress in the culture has been developed, based on the appropriate recognition of the culture phases and compensating for the undesirable effects derived by the variation in inoculum states (Takahashi et al. 2001). A biomass density of 14.2 g DW/L is reached in 28 days of cultivation of *B. vulgaris* L. hairy roots in a 2-L convective flow bioreactor (0.7 L in the growth column and 1.3 L in the aeration vessel) (Kino-oka et al. 1995). In this system, the culture shows optimal growth at medium flow rate of 270 mL/min, whereas when the flow decreases to 130 mL/min, the culture is exposed to oxygen starvation. This is used for development of a bioreactor system for betalain release by controlled oxygen insufficiency (Kino-oka et al. 1995). To improve the oxygen supply between the flowing medium and the hairy roots, located at different distances from the bottom of the cultivation chamber, a radial flow bioreactor has been constructed (Fig. 11.8b) (Kino-oka and Tone 2001). This design allows the oxygen-rich medium to flow simultaneously from several ports to center through an external cylinder attached to a side of the cultivation chamber, from where the exhausted medium leaves the vessel. This design is constructed to operate with a high-density culture, and the 280 h of cultivation of beet hairy roots in it leads to a 2.2-fold higher dry biomass accumulation compared with shake flask cultivation (Kino-oka and Tone 2001). However, although the tubular and radial convective flow bioreactors provide adequate growth of *B. vulgaris* hairy roots at the laboratory scale, they are definitely not suitable for large-scale cultivation, due to various practical problems associated with their assembly, asepsis and operations.

11.2.9 Trickle Bed (Droplet) Bioreactors

Trickle bed or so-called droplet bioreactors are systems in which the cultured roots are not continuously submerged in the medium. In this bioreactor, the medium sprayed from inside top of the cultivation chamber trickles onto the anchored roots, flows down due to gravity and then is drained into a reservoir for the next recirculation by pumps. The design allows overcoming most of the mixing, oxygen and nutrient transfer limitations of the submerged root culture systems. As a result of the sustained availability of oxygen and improved transfer in the liquid–solid boundary layer, increased growth rates and higher tissue concentrations are expected. Up to now, there has been only one report of successful large-scale cultivation of hairy roots (500-L volume) in this bioreactor system (Wilson 1997). The bioreactor construction includes a cultivation chamber (500 L) with a support matrix consisting of a series of chains with barbs (70-cm diameter, 120-cm height, barb spacing 10 cm), a specially designed inoculation vessel (10-L volume) serving for both inoculum propagation and its transfer into the cultivation chamber, a holding vessel (75-L volume), feeding ports, pumps and a system for gravimetric monitoring of accumulated biomass (Fig. 11.9) (Wilson 1997). This setup, developed for the cultivation of *Datura stramonium* hairy roots operates as a submerged bubble column with consecutive alternations of the left and right aeration spargers, located at the bottom of the apparatus. This causes the appearance of two independent fluid flows, which provide uniform hairy roots distribution on the support matrix. Once the roots are immobilized (21 days after the start of cultivation), the medium is drained from the vessel and the bioreactor is switched to droplet mode (Wilson 1997). A total of 39.8 kg FW of *Datura stramonium* hairy roots are harvested after 40 days of cultivation in this system (Wilson 1997). One of the important parameters of this setup is the spacing between the immobilization barbs. This spacing is referred to as the "effective volume", which is the space that could be filled by roots from a single inoculation point largely without callus formation. Although the authors have projected this volume for *B. vulgaris* hairy roots to be $10 \times 10 \times 10$ cm, which means one barb per liter of working volume, the data for cultivation of red beet hairy roots in this bioreactor system is not presented (Wilson 1997).

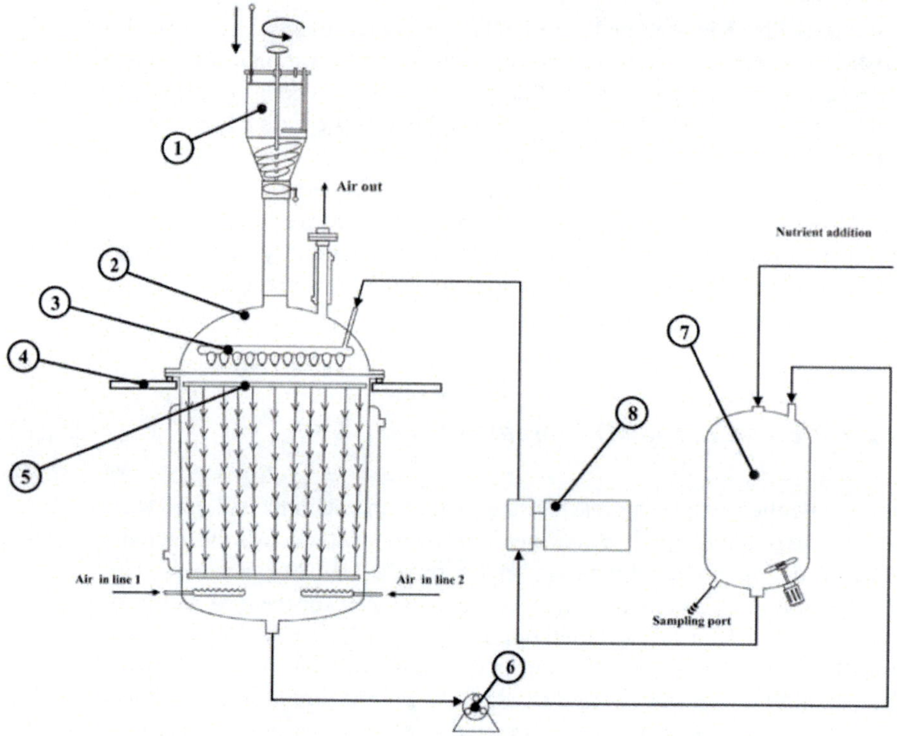

Fig. 11.9 Schematic diagram of trickle bed (droplet) bioreactor: *1* inoculation vessel, *2* cultivation chamber, *3* concentric rings with spray nozzles, *4* weight sensor, *5* support matrix, *6* peristaltic pump, *7* aeration vessel with feeding and sampling ports and *8* diaphragm pump

Trickle bed bioreactors are a prospective alternative, especially for continuous product recovery from the medium. However, the complicated harvesting of the accumulated biomass remains the main disadvantage for large-scale application of these cultivation systems.

11.2.10 Mist Bioreactors

Nutrient mist bioreactors offer a number of advantages for the growth of hairy roots. These systems have been designed to deliver the liquid phase as aerosol (Weathers and Zobel 1992). The aerosol particles are so fine that their penetration into the hairy root clumps is primarily dependent on the convective gas flow (Fig. 11.10). As a result of the improved nutrient availability, the better gas exchange and the reduced shear stress, a significant increase in growth rates can be achieved in a much lesser medium volume.

B. vulgaris cv. Detroit Dark Red hairy roots are cultured in a nutrient mist bioreactor with a 1.29-L growth chamber operating in a continuously culturing mode

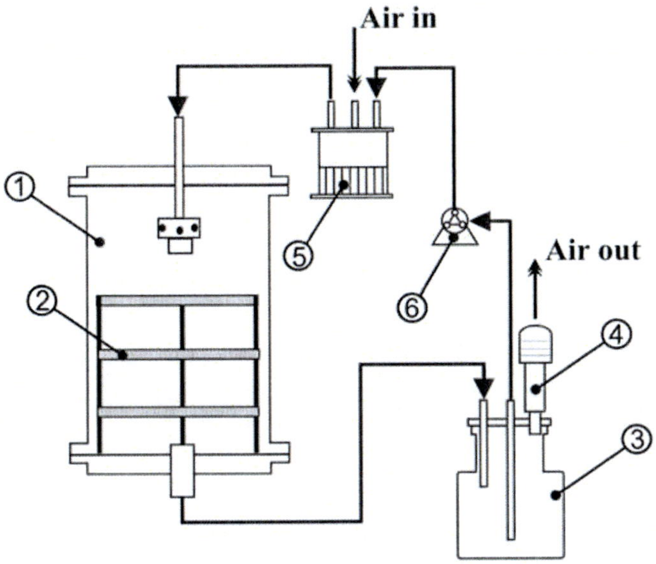

Fig. 11.10 Schematic diagram of nutrient mist bioreactor: *1* cultivation chamber, *2* support matrix, *3* medium reservoir, *4* mist coalescer, *5* mist generator and *6* peristaltic pump

(Weathers and Zobel 1992). The culture has shorter doubling time and higher betalain content (ranging from 2.9 mg/g DW to 3.3 mg/g DW) after 1-week cultivation in mist, compared with shake flask (2.9 mg/g DW) cultivation (Table 11.1). Further improvement of the system including a 1.8-L growth chamber operating in batch mode (recycle of 750 mL growth medium) with a 5/6 misting cycle (5 min misting ON/6 min misting OFF) leads to improved biomass accumulation (3.5-fold compared with 2.8-fold when a continuous culture mode is applied) (DiIorio et al. 1992). However, authors have found poor performance of hairy roots in the mist reactor because of the formation of mucilage coating on the root tips, which acts as a barrier inhibiting mass transfer, particularly when not washed away, as it is in the shake flasks (DiIorio et al. 1992).

The cultivation of *B. vulgaris* hairy roots in nutrient mist bioreactors has many advantages with economical effectiveness of the process due to increased the growth rates, shorter culture time and reduced nutrient medium volumes. However, these systems are counterproductive when betalain recovery from the recirculating medium is not appropriately matched with the production system.

11.2.11 Temporary Immersion Systems

Temporary immersion systems are designed to expose the hairy root tissue to cyclic submerged and non-submerged environments (Fig. 11.11). The main advantages of this apparatus are the reduction both of hyperhydricity and shear stress; improved gas exchange; overcoming the liquid–solid mass transfer limitations.

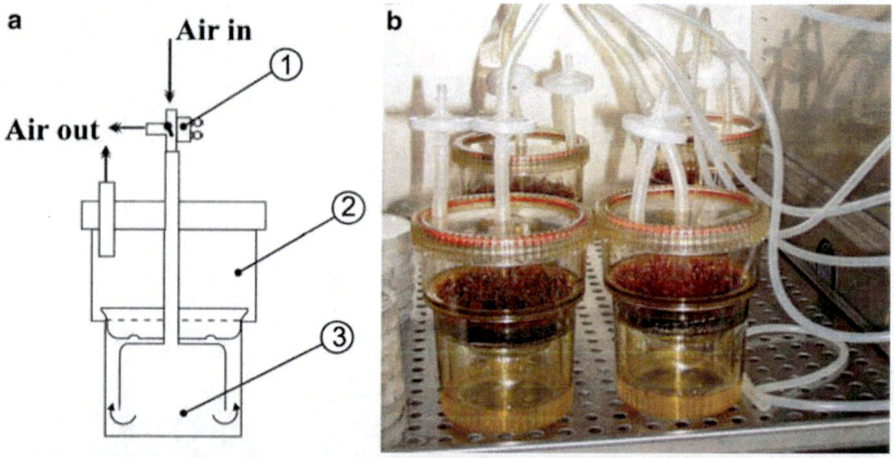

Fig. 11.11 Schematic diagram of a temporary immersion RITA® system (**a**): *1* three-way solenoid valve, *2* culture chamber and *3* medium compartment. 15-days-old *B. vulgaris* cv. Detroit Dark Red hairy roots cultivated in temporary immersion RITA® systems (**b**)

These cultivation systems are developed mainly for the needs of plant micropropagation, but recently we have found them very suitable for cultivating hairy roots by means of secondary metabolite production (Pavlov and Bley 2006; Georgiev et al. 2008c). *B. vulgaris* cv. Detroit Dark Red hairy roots cultured in a temporary immersion RITA® system at immersion frequency of 15-min flooding and 75-min stand-by periods accumulate maximal amounts of biomass (14.5 g DW/L) and 15.9 mg/g DW of betalains. The change of immersion frequency to 15-min flooding and 60-min stand-by periods decreases the accumulated amounts of biomass (12.5 g DW/L), but increases the betalain content up to 18.8 mg/g DW (Pavlov and Bley 2006).

The relatively high biomass and betalain yields, the simple design and the low price of the temporary immersion RITA® apparatus makes this system interesting for application on semi-industrial scale for production of betalains. The main disadvantages are the unpredictable behavior of the systems during future scale-up and the current lack of large volume apparatus systems in the market, although specific in-house designs are easy to fabricate.

11.2.12 Low-Cost Cultivation Systems

In cultivating *B. vulgaris* hairy roots in different bioreactor systems, in the authors' experience, it appears that the process is more economical when applied on small scale. In this way, the industrial process for betalain production should be based on the

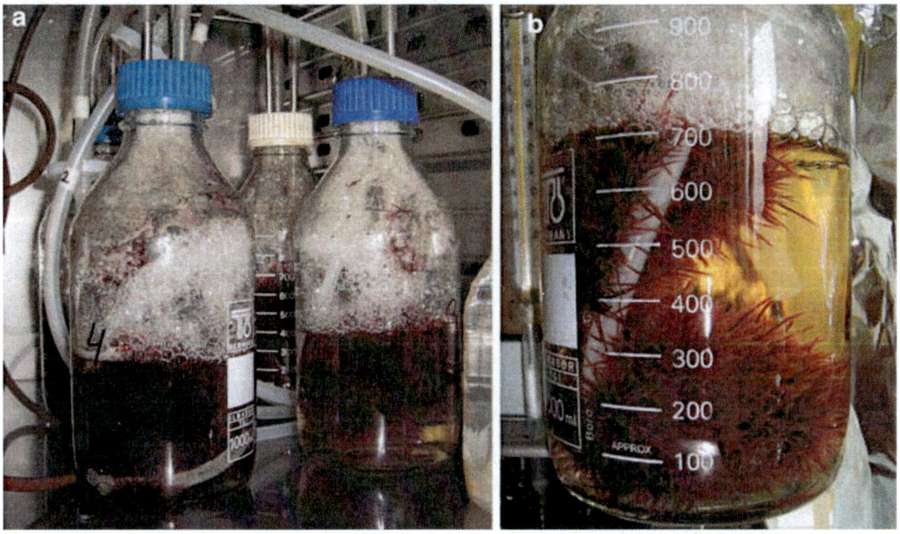

Fig. 11.12 21- (**a**) and 15- (**b**) days-old *B. vulgaris* cv. Detroit Dark Red hairy roots cultivated in low-cost bottle cultivation systems, operating as bubble columns

usage of several low-volume bioreactors, which are more efficient than the process in which one large vessel is involved. However, the capital costs for equipment with several bioreactors are always higher than those for one large volume vessel installation. In order to minimize investment expenses, the authors propose a simple alternative to traditional stainless steel bioreactors named, Low-cost cultivation systems (Fig. 11.12).

The design consists of simple autoclavable bottles, equipped with three-orifice caps connected with silicon tubes to the sparger, sampling port and air outlet. The sparger comprises a silicon closed end tube with 32 holes (1 mm in diameter) evenly distributed in four groups of eight at opposite sides on the tube. The inoculation is accomplished by forceps in a laminar flow chamber. Then the banks are cultured in a thermostat chamber at 26°C in darkness. The data for the accumulated biomass, betalain content and sugar uptake during the 21 days of cultivation of *B. vulgaris* cv. Detroit Dark Red hairy roots in 1.0 L (0.8 L working volume) low-cost cultivation systems under different aeration regimes are presented below (Figs. 11.13 and 11.14).

A maximum amount of accumulated dry biomass (16.2 g DW/L) is achieved at aeration rate of 26.25 L/L.h, whereas the maximum amount of betalains (9.6 mg/g DW) is reached at aeration rate of 35.00 L/L.h (Fig. 11.13). In both regimes, the culture consumes glucose and fructose in parallel, whereas at the lower and higher aeration rates, glucose is preferred (Fig. 11.14). The presented results demonstrate the potential of *B. vulgaris* hairy root cultures to grow with a relatively high growth

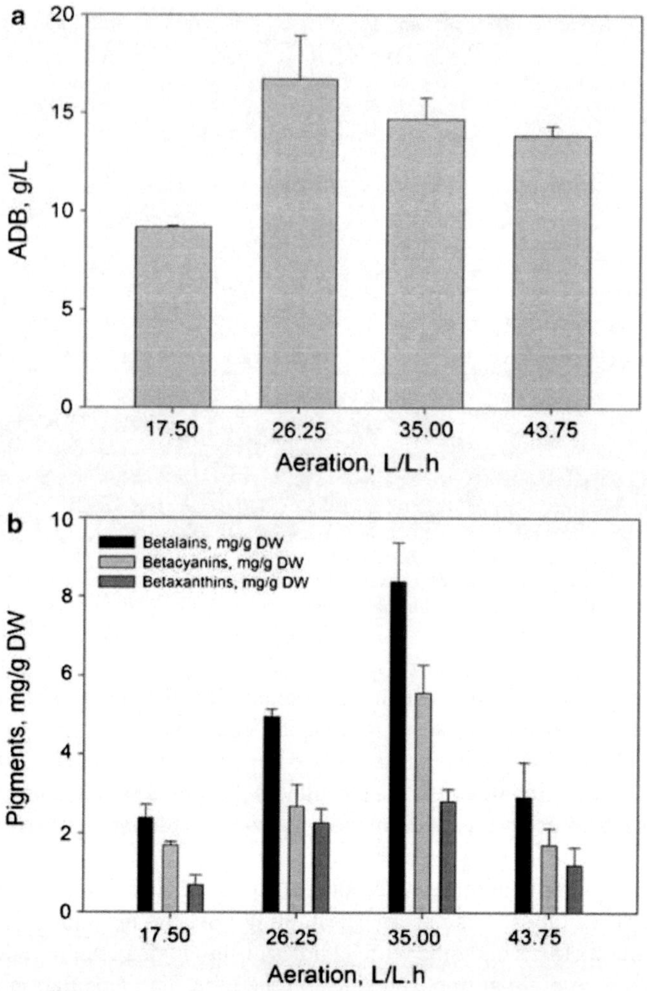

Fig. 11.13 Accumulated dry biomass (**a**) and betalains content (**b**) by 21-days-old *B. vulgaris* cv. Detroit Dark Red hairy roots cultivated in low-cost bottle cultivation systems at different aeration flow rates. *Error bars* represent standard deviations

rate and to reach good betalain yields even when cultured in less sophisticated low-cost vessels. Importantly, the applications of such low-cost cultivation systems reduce not only the cost of equipment, but also the costs for consumables, automated control systems and labor. Low-cost cultivation systems appear suitable even for commercial production, where, by using a great number of them with volumes up to 20 L in parallel, loss due to contamination can also be reduced. They offer dependable and flexible cost-effective systems from small-scale process development to medium-scale betalain production.

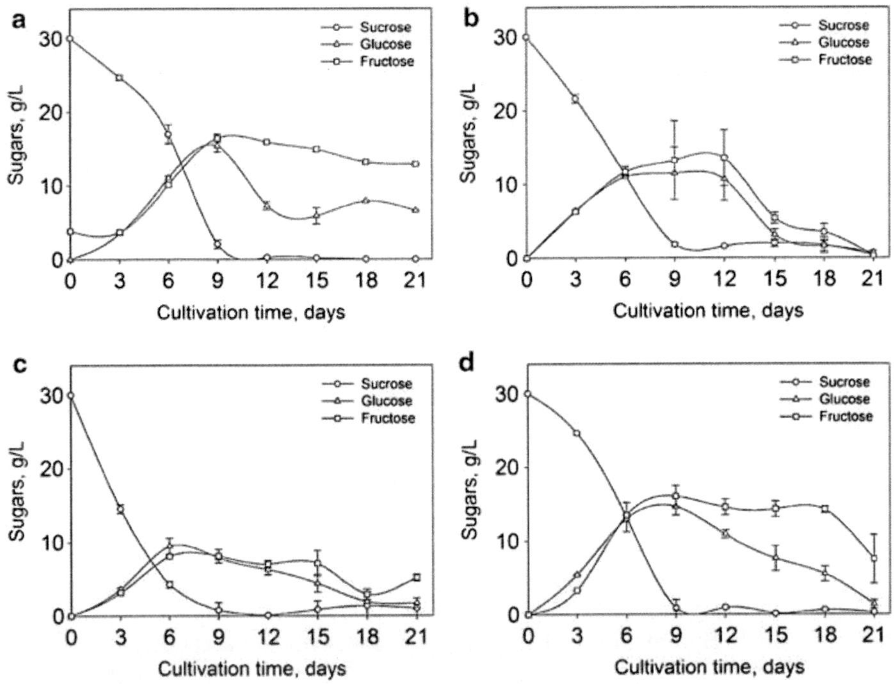

Fig. 11.14 Time courses of sugars utilization of *B. vulgaris* cv. Detroit Dark Red hairy roots during their cultivation in low-cost bottle cultivation systems at air flow rates of: 17.50 L/L.h (**a**), 26.25 L/L.h (**b**), 35.00 L/L.h (**c**) and 43.75 L/L.h (**d**). *Error bars* represent standard deviations

11.3 Cell Permeabilization and Product Recovery

As has been discussed, *B. vulgaris* hairy roots possess relatively slow growth rates and the culture durations for maximum biomass accumulation vary between 7 and 28 days depending on the root line and bioreactor system applied. Moreover, to provide higher biomass density, different internal support matrices are used, which complicate the harvest of biomass. On the other hand, the available high-density root biomass, immobilized on the support matrix, is a logical choice for development of a continuous metabolite secretion-recovery system. When betalains are the target secondary metabolites of interest in *B. vulgaris* hairy root cultures, the system productivity could be improved by involving their secretion and continuous recovery from the cultured medium. If this production strategy is applied, then the slow growth rate of hairy root cultures can have minimal impact on the overall process economics. Once the highest root biomass with high betalains is reached, the root cells may then be permeabilized to release pigments from their vacuoles while still keeping their viability. Several nondestructive approaches have been demonstrated for the permeabilization of *B. vulgaris* hairy roots for releasing betalain, including oxygen starvation, pH decrease, temperature, ionic stress, sonication and treatments with chemicals (dimethyl sulfoxide, Triton X-100, cetyltrimethylammonium bromide and Tween-80) or biological agents (*Lactobacillus helveticus*, *Saccharomyces cereviseae* and *Candida utilis*) (Taya et al. 1992; DiIorio et al. 1993; Kino-oka et al. 1995; Mukundan et al. 1998a; Kim et al. 2002; Thimmaraju et al. 2003a, b). These strategies are either fully or partially applicable, depending on the bioreactor designs. For example, oxygen starvation can be performed in bioreactor systems involving stirred tanks, column with external aeration vessel or convective flow types. This approach is practically inconceivable for trickle bed bioreactors or temporary immersion systems, where the roots are exposed to a gaseous atmosphere most of the time. The latter are more appropriate for sonication treatment, short-term temperature or pH shocks. The released betalains can be removed from the medium using various adsorbents such as silica/alumina (1:1), alumina/sand (2:1), and styrene-divinylbenzene copolymers (Sepabeads SP207), ensuring their further recovery with minimal losses (Kino-oka et al. 1995; Kino-oka and Tone 2001; Thimmaraju et al. 2004; Bhagyalakshmi and Thimmaraju 2009).

Several bioreactor designs exist that facilitate high biomass loadings and provide possibilities of attaching a second circulation cycle for on-line product recovery. Currently, only few bioreactors are applied for successful permeabilization of *B. vulgaris* hairy roots for continuous betalain recovery (Kino-oka et al. 1995; Kino-oka and Tone 2001; Thimmaraju et al. 2003a; Bhagyalakshmi and Thimmaraju 2009). A turbine blade reactor with medium circulation through an absorbent column packed with Sepabeads SP207 resin is used for pigment recovery from *B. vulgaris* hairy roots (Kino-oka and Tone 2001). The permeabilization is implemented 172 hours after the initiation of cultivation, by replacing the aerating gas with pure nitrogen for 16 hours. The cycles are repeated four times to result in the total recovery of betalains of 0.8–1.3 mg/L in each operation (Kino-oka and Tone 2001). Another earlier study by the same lab applied more sophisticated system controls for monitoring the level of dissolved oxygen, using a convective flow bioreactor (Kino-oka et al. 1995). In this setup, the medium is constantly aerated in an external aeration vessel, where oxygen starvation is induced by decreasing the flow rate of the medium from 270 mL/min to 130 mL/min. The culture medium, enriched with betalains, is then circulated through an external absorbent column packed with Sepabeads SP207 resin. At the end of the process (28 days), 8.0 mg of betalains are recovered from the column and 16.3 mg of betalains remain in the root biomass (Kino-oka et al. 1995). A column bioreactor with an internal support matrix has been used for permeabilization of *B. vulgaris* cv. Ruby Queen hairy roots (Thimmaraju et al. 2003a). The treatment of 20-day-old hairy roots with 0.002% cetyltrimethylammonium bromide and 10 mL of free lipid fraction from *L. helveticus* causes 80% (48 h after treatment) and 84% (12 h after treatment) betalain release, respectively, without significant losses of culture viability (Thimmaraju et al. 2003a). The above bioreactor system has been further improved by adding an external aeration vessel and simultaneous medium circulation through an external absorbent column packed with alumina/sand (2:1), and thus 97% of released betalains are recovered (Bhagyalakshmi and Thimmaraju 2009). Moreover, the authors propose a strategy for simultaneous recovery of primary and secondary metabolites (peroxidase EC 1.11.1.7 and betalains) by using aqueous two-phase extraction, which is an added value to cultured *B. vulgaris* hairy roots (Bhagyalakshmi and Thimmaraju 2009).

11.4 Ecological and Economic Considerations

The fast progress of the modern food industry has regained an interest in betalains due to their importance as natural food colorants with desired technological properties and valuable biological activities. Currently, red beet is the only food source commercially exploited for production of betanin juices, concentrates and powders, which are authorized for coloring purposes of foodstuffs in Europe and North America (Georgiev et al. 2008b). Breeding red beet is commercially satisfactory because of the high crop yields and pigment content, which have not yet been reached by any other betalain-producing crop (Stintzing and Carle 2007c). The annual yields vary between 18 and 70 t of fresh beetroots per hectare, with betanin content between 0.2 and 0.6 mg/g FW depending on the cultivars, soils, fertilizers and climate. Commercially available dehydrated red beetroot powder contains 1% betanin and color unit at $E^{1\%}/_{1\ cm}$, 535 nm between 10 and 50, which corresponds to final production ranging from 360 to 4,200 kg of red beet powder per hectare. The real market price of dehydrated red beetroot powder with the above-mentioned characteristics is between 2,000 and 2,500 USD per ton (valid for January–February 2011; from the international exchange site www.alibaba.com), which presumes very attractive profits from this crop. All these prospects have motivated farmers to use more factory areas to grow red beet, which they usually do by restricting other food crops. This raises serious problems with planning the agricultural lands, particularly when red beet prefers more sunlight for optimal growth, but the arable areas with solar illumination are limited. The problem seems to grow more with time because of the exponentially increasing demand for natural products, not only in food processing, but also in the pharmaceutical, cosmetic and biofuel industries. The shift of betalain production from agriculture to modern plant biotechnology offers a promising alternative for solving these problems in the near future. *B. vulgaris* hairy root cultures are the most promising candidates for realization of such technologies (Georgiev et al. 2008b). When these in vitro cultures result from "natural" transformation by wild *Agrobacterium rhizogenes* strains, they and their products are not classified as Genetically Modified Organisms (GMOs), as defined by the European Parliament Directive 2001/18/EC, and hence are not subject to EU regulations regarding GMOs (OJ, L 106, 17.4.2001, p.1–38). Unfortunately, the development of economically effective biotechnology for betalain production is more visionary than close to reality, and remains a big challenge for researchers. Even if the best reported biomass and betalain yields that are registered in shake flasks (ADB 28 g DW/L and 42.2 mg/g DW betalains; Table 11.1) are reached, a bioreactor with a 4,000-L working volume will need to operate at 10 cultivation cycles per year to achieve the maximal yield of dehydrated red beetroot powder from cultivated beetroot reported above. Moreover, the price of the end product will be approximately five times higher (40,000 L MS medium with 3% sucrose, energy for 1 year operation, equipment and labor costs, etc.). The higher price can be partially justified by the better qualities of the biotechnologically produced pigments compared with colorants extracted from agriculturally bred red beetroot. For example, when produced from hairy roots, several adverse properties, including the presence of an unpleasant smell due to pyrazine derivatives or geosmin and increased level of microbial contaminations can be eliminated from the final colorant. It has been found that the betalain extract, obtained from *B. vulgaris* cv. Detroit Dark Red hairy roots, has a sixfold higher antioxidant activity compared with the extract from non-transformed mature beetroots, which makes it more valuable (Georgiev et al. 2010). If the production process involves pigment recovery from culture medium, the highly purified betacyanin and betaxanthin extracts can be produced by selective gradient elution of pigments from the adsorbent columns. Moreover, by controlled alterations of simple cultivation parameters, such as the rate of aeration, extracts with desired color shades occurring as a consequence of a shift in the ratio between betaxanthins and betacyanins can be obtained. The color quality of red beet extracts is currently evaluated by color shade as the most important parameter and as such strongly affects consumer preference.

11.5 Conclusive Remarks

The recent data presented in this chapter clearly indicate that red beet hairy roots are amenable for scaled-up production, with performance levels almost matching commercially successful processes, although further extensive investigations are necessary for improvement of betalain yields. Available information indicates the need for concerted efforts towards an integrated optimization approach, including all elements of the optimization process, such as media optimization, line selection, bioreactor design optimization, environmental conditions optimization, process modeling, etc. Furthermore, the so-far proven techniques for increasing betalain pigment yields, such as elicitation, feeding, product release and recovery, are to be transferred from shake flasks into bioreactors. The investigated production processes should be scaled up to cultivation in bioreactor systems with bigger volumes; and their effectiveness should be proved at a semi-industrial scale. For economic realizations, it seems that the future belongs to searching for new alternatives and cheap bioreactor systems that should provide acceptable biomass and betalain yields. Various disposable bioreactor systems have recently been developed that provide economically effective cultivation of mammalian and plant cell cultures both at semi-industrial and industrial scales (Zhang et al. 2009; Ducos et al. 2009). The availability of such alternative cultivation systems indicates a completely new possibility for culturing *B. vulgaris* hairy roots. This could contribute to significant cost reductions with regard to equipment, consumables, labor and maintenance.

References

Akita, T., Y. Hina, and T. Nishi. 2000. Production of betacyanins by a cell suspension culture of table beet (*Beta vulgaris* L.). *Bioscience, Biotechnology, and Biochemistry* 64(9): 1807–1812.

Azeredo, H.M.C. 2008. Betalains: Properties, sources, applications, and stability – A review. *International Journal of Food Science and Technology* 44(12): 2365–2376.

Berzin, I., D. Mills, and J.C. Merchuk. 1999. A non-destructive method for secondary metabolite determination in hairy root cultures. *Journal of Chemical Engineering of Japan* 32(2): 229–234.

Berzin, I., B. Cohen, D. Mills, I. Dinstein, and J.C. Merchuk. 2000. RHIZOSCAN: A semiautomatic image processing system for characterization of the morphology and secondary metabolite concentration in hairy root cultures. *Biotechnology and Bioengineering* 70(1): 17–24.

Bhagyalakshmi, N., and R. Thimmaraju. 2009. Bioreactor for cultivation of red beet hairy roots and in situ recovery of primary and secondary metabolites. *Engineering in Life Sciences* 9(3): 227–238.

Bhagyalakshmi, N., R. Thimmaraju, and M.S. Narayan. 2004. Various hexoses and di-hexoses differently influence growth, morphology and pigment synthesis in transformed root cultures of red beet (*Beta vulgaris*). *Plant Cell, Tissue and Organ Culture* 78(2): 183–195.

Bordonaro, J.L., and W.R. Curtis. 2000. Inhibitory role of root hairs on transport within root culture bioreactors. *Biotechnology and Bioengineering* 70(2): 176–186.

Britton, M.T., M.A. Escobar, and A.M. Dandekar. 2008. The oncogenes of *Agrobacterium tumefaciens* and *Agrobacterium rhizogenes*. In *Agrobacterium: From biology to biotechnology*, ed. T. Tzfira and V. Citovsky, 523–563. New York: Springer.

Calogero, G., G. Di Marco, S. Cazzanti, S. Caramori, R. Argazzi, A. Di Carlo, and C.A. Bignozzi. 2010. Efficient dye-sensitized solar cells using red turnip and purple wild sicilian prickly pear fruits. *International Journal of Molecular Sciences* 11(1): 254–267.

Campanoni, P., and P. Nick. 2005. Auxin-dependent cell division and cell elongation. 1-Naphthaleneacetic acid and 2,4-dichlorophenoxyacetic acid activate different pathways. *Plant Physiology* 137(3): 939–948.

Carvalho, E.B., S. Holihan, B. Pearsall, and W.R. Curtis. 1997. Effect of root morphology on reactor design and operation for production of chemicals. In *Hairy roots: Culture and applications*, ed. P.M. Doran, 151–167. Amsterdam: Harwood Academic.

Davies, K. 2004. An introduction to plant pigments in biology and commerce. In *Plant pigments and their manipulation*, Annual Plant Reviews, vol. 14, ed. K. Davies, 1–22. Oxford, U.K.: CRC Press/Blackwell.

DiIorio, A.A., R.D. Cheetham, and P.J. Weathers. 1992. Growth of transformed roots in a nutrient mist bioreactor: Reactor performance and evaluation. *Applied Microbiology and Biotechnology* 37(4): 457–462.

DiIorio, A.A., P.J. Weathers, and R.D. Cheetham. 1993. Non-lethal secondary product release from transformed root cultures of *Beta vulgaris*. *Applied Microbiology and Biotechnology* 39(2): 174–180.

Ducos, J.-P., B. Terrier, and D. Courtois. 2009. Disposable bioreactors for plant micropropagation and mass plant cell culture. In *Disposable bioreactors*, Advances in biochemical engineering/biotechnology, vol. 115, ed. R. Eibl and D. Eibl, 89–115. Heidelberg: Springer.

Flores, H.E., M.W. Hoy, and J.J. Pickard. 1987. Secondary metabolites from root cultures. *Trends in Biotechnology* 5(3): 64–69.

Gaffar, M.C.S. 1984. Natural dye indicator for dental plaque. US Patent 4,431,628.

Gandia-Herrero, F., J. Escribano, and F. Garcia-Carmona. 2005a. Betaxanthins as pigments responsible for visible fluorescence in flowers. *Planta* 222(4): 586–593.

Gandia-Herrero, F., F. Garcia-Carmona, and J. Escribano. 2005b. Floral fluorescence effect. *Nature* 437(7057): 334.

Gandia-Herrero, F., F. Garcia-Carmona, and J. Escribano. 2005c. Fluorescent pigments: New perspectives in betalain research and applications. *Food Research International* 38(8–9): 879–884.

Gandia-Herrero, F., M. Jimenez-Atienzar, J. Cabanes, F. Garcia-Carmona, and J. Escribano. 2005d. A new role for tyrosinase in the biosynthesis of the plant pigments betalains. *The FEBS Journal* 272: 462.

Gandía-Herrero, F., J. Escribano, and F. García-Carmona. 2009. The role of phenolic hydroxy groups in the free radical scavenging activity of betalains. *Journal of Natural Products* 72(6): 1142–1146.

Gandía-Herrero, F., J. Escribano, and F. García-Carmona. 2010. Structural implications on color, fluorescence, and antiradical activity in betalains. *Planta* 232(2): 449–460.

Gelvin, S.B. 2000. Agrobacterium and plant genes involved in T-DNA transfer and integration. *Annual Review of Plant Biology* 51: 223–256.

Georgiev, M., A. Pavlov, and T. Bley. 2006. Betalains by transformed *Beta vulgaris* roots in stirred tank bioreactor: Batch and fed-batch processes. Paper presented at the FinMed 2006 – 2nd International conference on bioreactor technology in cell, tissue culture and biomedical applications, Saariselka.

Georgiev, M., A. Pavlov, and T. Bley. 2007. Hairy root type plant in vitro systems as sources of bioactive substances. *Applied Microbiology and Biotechnology* 74: 1175–1185.

Georgiev, M., V. Georgiev, J. Weber, T. Bley, M. Ilieva, and A. Pavlov. 2008a. *Agrobacterium rhizogenes*-mediated genetic transformations: A powerful tool for the production of metabolites. In *Genetically modified plants*, ed. T. Wolf and J. Koch, 99–126. Hauppauge: Nova.

Georgiev, V., M. Ilieva, T. Bley, and A. Pavlov. 2008b. Betalain production in plant in vitro systems. *Acta Physiologiae Plantarum* 30(5): 581–593.

Georgiev, V., A. Stukert, T. Bley, and A. Pavlov. 2008c. Hyosciamine biosynthesis by diploid and tetraploid *Datura stramonium* L. hairy root cultures in a temporary immersion cultivation system. *Advances in Bulgarian Science* 2(3): 42–47.

Georgiev, V., J. Weber, T. Bley, and A. Pavlov. 2009. Improved procedure for nucleus extraction for DNA measurements by flow cytometry of red beet (*Beta vulgaris* L.) hairy roots. *Journal of Bioscience and Bioengineering* 107: 439–441.

Georgiev, V., J. Weber, E. Kneschke, P. Denev, T. Bley, and A. Pavlov. 2010. Antioxidant activity and phenolic content of betalain extracts from intact plants and hairy root cultures of the red beetroot *Beta vulgaris* cv. Detroit Dark Red. *Plant Foods for Human Nutrition* 65: 105–111.

Girod, P.A., and J.P. Zryd. 1991. Secondary metabolism in cultured red beet (*Beta vulgaris* L) cells – Differential regulation of betaxanthin and betacyanin biosynthesis. *Plant Cell, Tissue and Organ Culture* 25(1): 1–12.

Gliszczynska-Swiglo, A., H. Szymusiak, and P. Malinowska. 2006. Betanin, the main pigment of red beet: Molecular origin of its exceptionally high free radical-scavenging activity. *Food Additives and Contaminants* 23(11): 1079–1087.

Grotewold, E. 2006. The genetics and biochemistry of floral pigments. *Annual Review of Plant Biology* 57: 761–780.

Hamill, J.D., A.J. Parr, R.J. Robins, and M.J.C. Rhodes. 1986. Secondary product formation by cultures of *Beta vulgaris* and *Nicotiana rustica* transformed with *Agrobacterium rhizogenes*. *Plant Cell Reports* 5(2): 111–114.

Han, X.H., Z.J. Gao, and X.G. Xiao. 2009. Enzymes and genes involved in the betalain biosynthesis in higher plants. *African Journal of Biotechnology* 8(24): 6735–6744.

Hempel, J., and H. Bohm. 1997. Betaxanthin pattern of hairy roots from *Beta vulgaris* var *lutea* and its alteration by feeding of amino acids. *Phytochemistry* 44(5): 847–852.

Hitaka, Y., M. Kino-oka, M. Taya, and S. Tone. 1997. Effect of liquid flow on culture of red beet hairy roots in single column reactor. *Journal of Chemical Engineering of Japan* 30(6): 1070–1075.

Hitaka, Y., Y. Takahashi, M. Kino-Oka, M. Taya, and S. Tone. 2000. Culture of red beet hairy roots by considering variation in sensitivity of tip meristems to hydraulic stress. *Biochemical Engineering Journal* 6(1): 1–6.

Jimenez-Aparicio, A., and G. Gutierrez-Lopez. 1999. Production of food related colorants by culture of plant cells – The case of betalains. *Advances in Experimental Medicine and Biology* 464: 195–210.

Kanner, J., S. Harel, and R. Granit. 2001. Betalains – A new class of dietary cationized antioxidants. *Journal of Agricultural and Food Chemistry* 49(11): 5178–5185.

Khlebnikov, A., B. Dubuis, O.M. Kut, and J.E. Prenosil. 1995. Growth and productivity of *Beta vulgaris* cell culture in fluidized bed reactors. *Bioprocess Engineering* 14(1): 51–56.

Kim, Y., B.E. Wyslouzil, and P.J. Weathers. 2002. Secondary metabolism of hairy root cultures in bioreactors. *In Vitro Cellular and Developmental Biology* 38(1): 1–10.

Kino-oka, M., and S. Tone. 2001. Transgenic beetroot (*Beta vulgaris*). In *Biotechnology in agriculture and forestry: Transgenic crops II*, vol. 47, ed. Y.P.S. Bajaj, 55–68. Heidelberg: Springer.

Kino-oka, M., Y. Hongo, M. Taya, and S. Tone. 1992. Culture of red beet hairy root in bioreactor and recovery of pigment released from the cells by repeated treatment of oxygen starvation. *Journal of Chemical Engineering of Japan* 25: 490–495.

Kino-oka, M., M. Taya, and S. Tone. 1995. Culture of red beet hairy roots in a column-type reactor associated with pigment release. *Plant Tissue Culture Letters* 12(2): 201–204.

Kobayashi, N., J. Schmidt, V. Wray, and W. Schliemann. 2001. Formation and occurrence of dopamine-derived betacyanins. *Phytochemistry* 56(5): 429–436.

Kriznik, M., and D. Pavokovic. 2010. Enhancement of betanin yield in transformed cells of sugar beet (*Beta vulgaris* L.). *Acta Botanica Croatica* 69(2): 173–182.

Leathers, R.R., and G. O'Riordain. 1990. Betalain biosynthesis in *Beta vulgaris* cell suspention cultures. *Experientia* 46w(563): A66.

Lila, M.A. 2004. Plant pigments and human health. In *Plant pigments and their manipulation*, Annual plant reviews, vol. 14, ed. K. Davies, 248–274. Oxford: CRC Press/Blackwell.

Mabry, T.J. 2001. Selected topics from forty years of natural products research: Betalains to flavonoids, antiviral proteins, and neurotoxic nonprotein amino acids. *Journal of Natural Products* 64(12): 1596–1604.

Martinez, L., I. Cilla, J.A. Beltran, and S. Roncales. 2006. Comparative effect of red yeast rice (*Monascus purpureus*), red beet root (*Beta vulgaris*) and betanin (E-162) on colour and consumer acceptability of fresh pork sausages packaged in a modified atmosphere. *Journal of the Science of Food and Agriculture* 86(4): 500–508.

Matkowski, A. 2008. Plant in vitro culture for the production of antioxidants – A review. *Biotechnology Advances* 26(6): 548–560.

Mishiba, K.I., T. Okamoto, and M. Mii. 2001. Increasing ploidy level in cell suspension cultures of Doritaenopsis by exogenous application of 2,4-dichlorophenoxyacetic acid. *Physiologia Plantarum* 112(1): 142–148.

Moreno, D.A., C. Garcia-Viguera, J.I. Gil, and A. Gil-Izquierdo. 2008. Betalains in the era of global agri-food science, technology and nutritional health. *Phytochemistry Reviews* 7(2): 261–280.

Mukundan, U., V. Bhide, G. Singh, and W.R. Curtis. 1998a. pH-mediated release of betalains from transformed root cultures of *Beta vulgaris* L. *Applied Microbiology and Biotechnology* 50(2): 241–245.

Mukundan, U., E.B. Carvalho, and W.R. Curtis. 1998b. Growth and pigment production by hairy root cultures of *Beta vulgaris* L. in a bubble column reactor. *Biotechnology Letters* 20(5): 469–474.

Murthy, H.N., E.J. Hahn, and K.Y. Paek. 2008. Adventitious roots and secondary metabolism. *Chinese Journal of Biotechnology* 24(5): 711–716.

Pavlov, A., and T. Bley. 2005. Betalains biosynthesis by *Beta vulgaris* L. hairy root culture in different bioreactor systems. Paper presented at the University of Food Technologies, Plovdiv.

Pavlov, A., and T. Bley. 2006. Betalains biosynthesis by *Beta vulgaris* L. hairy root culture in a temporary immersion cultivation system. *Process Biochemistry* 41(4): 848–852.

Pavlov, A., P. Kovatcheva, V. Georgiev, I. Koleva, and M. Ilieva. 2002. Biosynthesis and radical scavenging activity of betalains during the cultivation of red beet (*Beta vulgaris*) hairy root cultures. *Zeitschrift für Naturforschung. Section C* 57(7–8): 640–644.

Pavlov, A., V. Georgiev, and P. Kovatcheva. 2003. Relationship between type and age of the inoculum cultures and betalains biosynthesis by *Beta vulgaris* hairy root culture. *Biotechnology Letters* 25(4): 307–309.

Pavlov, A., V. Georgiev, and M. Ilieva. 2005a. Betalain biosynthesis by red beet (*Beta vulgaris* L.) hairy root culture. *Process Biochemistry* 40(5): 1531–1533.

Pavlov, A., P. Kovatcheva, D. Tuneva, M. Ilieva, and T. Bley. 2005b. Radical scavenging activity and stability of betalains from *Beta vulgaris* hairy root culture in simulated conditions of human gastrointestinal tract. *Plant Foods for Human Nutrition* 60(2): 43–47.

Pavlov, A., M. Georgiev, and T. Bley. 2007. Batch and fed-batch production of betalains by red beet (*Beta vulgaris*) hairy roots in a bubble column reactor. *Zeitschrift für Naturforschung. Section C* 62(5–6): 439–446.

Pavokovic, D., G. Rusak, V. Besendorfer, and M. K-R. 2009. Light-dependent betanin production by transformed cells of sugar beet. *Food Technology and Biotechnology* 47(2): 153–158.

Pietrzkowski, Z. 2010. Betalain compositions and uses thereof. Patent No WO 2010/014839A1.

Ramakrishnan, D., J. Salim, and W.R. Curtis. 1994. Inoculation and tissue distribution in pilot-scale plant root culture bioreactors. *Biotechnology Techniques* 8(9): 639–644.

Ramakrishnan, D., D. Luyk, and W.R. Curtis. 1999. Monitoring biomass in root culture systems. *Biotechnology and Bioengineering* 62(6): 711–721.

Roberts, M.F., D. Strack, and M. Wink. 2010. Biosynthesis of alkaloids and betalains. In *Biochemistry of plant secondary metabolism*, Annual plant reviews, vol. 40, 2nd ed, ed. M. Wink, 20–91. Oxford, UK: Blackwell.

Rodriguez-Monroy, M., and E. Galindo. 1999. Broth rheology, growth and metabolite production of *Beta vulgaris* suspension culture: A comparative study between cultures grown in shake flasks and in a stirred tank. *Enzyme and Microbial Technology* 24(10): 687–693.

Sajc, L., D. Grubisic, and G. Vunjak-Novakovic. 2000. Bioreactors for plant engineering: An outlook for further research. *Biochemical Engineering Journal* 4(2): 89–99.

Sanchez, M.J., A. Jimenez-Aparicio, G.G. Lopez, G.T. Tapia, and M. Rodriguez-Monroy. 2002. Broth rheology of *Beta vulgaris* cultures growing in an air lift bioreactor. *Biochemical Engineering Journal* 12(1): 37–41.

Shiao, T.L., and P.M. Doran. 2000. Root hairiness: Effect on fluid flow and oxygen transfer in hairy root cultures. *Journal of Biotechnology* 83(3): 199–210.

Shin, K.S., H.N. Murthy, J.Y. Ko, and K.Y. Paek. 2002. Growth and betacyanin production by hairy roots of *Beta vulgaris* in airlift bioreactors. *Biotechnology Letters* 24(24): 2067–2069.

Shin, K.S., H.N. Murthy, J.W. Heo, and K.Y. Paek. 2004. Induction of betalain pigmentation in hairy roots of red beet under different radiation sources. *Biologia Plantarum* 47(1): 149–152.

Soltan, M.E., and S.M. Sirry. 2002. Usefulness of some plant flowers as natural acid-base indicators. *Journal of the Chinese Chemical Society* 49(1): 63–68.

Soltani, J., G. Paul, H. van Heusden, and P.J.J. Hooykaas. 2008. *Agrobacterium*-mediated transformation of non-plant organisms. In *Agrobacterium: From biology to biotechnology*, ed. T. Tzfira and V. Citovsky, 649–675. New York: Springer.

Sotos, J.G. 1999. Beeturia and iron absorption. *Lancet* 354(9183): 1032.

Stintzing, F.C., and R. Carle. 2004. Functional properties of anthocyanins and betalains in plants, food, and in human nutrition. *Trends in Food Science and Technology* 15(1): 19–38.

Stintzing, F.C., and R. Carle. 2007a. Betalains – Emerging prospects for food scientists. *Trends in Food Science and Technology* 18(10): 514–525.

Stintzing, F.C., and R. Carle. 2007b. Betalains in food: Occurrence, stability, and postharvest modifications. In *Food colorants: Chemical and functional properties*, ed. C. Socaciu, 277–302. Boca Raton: CRC/Taylor & Francis.

Stintzing, F.C., and R. Carle. 2007c. N-heterocyclic pigments: Betalains. In *Food colorants: Chemical and functional properties*, ed. C. Socaciu, 87–100. Boca Raton: CRC/Taylor & Francis.

Strack, D., T. Vogt, and W. Schliemann. 2003. Recent advances in betalain research. *Phytochemistry* 62(3): 247–269.

Takahashi, Y., Y. Hitaka, M. Kino-oka, M. Taya, and S. Tone. 2001. Evaluation of growth property of red beet hairy roots depending on condition of inocula and its application to culture control with fuzzy logic theory. *Biochemical Engineering Journal* 8(2): 121–127.

Tanaka, Y., N. Sasaki, and A. Ohmiya. 2008. Biosynthesis of plant pigments: Anthocyanins, betalains and carotenoids. *The Plant Journal* 54: 733–749.

Taya, M., A. Yoyama, O. Kondo, T. Kobayashi, and C. Matsui. 1989. Growth characteristics of plant hairy roots and their cultures in bioreactors. *Journal of Chemical Engineering of Japan* 22(1): 84–89.

Taya, M., K. Mine, M. Kinooka, S. Tone, and T. Ichi. 1992. Production and release of pigments by culture of transformed hairy root of red beet. *Journal of Fermentation and Bioengineering* 73(1): 31–36.

Taya, M., K. Yakura, M. Kino-oka, and S. Tone. 1994. Influence of medium constituents on enhancement of pigment production by batch culture of red beet hairy roots. *Journal of Fermentation and Bioengineering* 77(2): 215–217.

Tesoriere, L., M. Allegra, C. Gentile, and M.A. Livrea. 2009. Betacyanins as phenol antioxidants Chemistry and mechanistic aspects of the lipoperoxyl radical-scavenging activity in solution and liposomes. *Free Radical Research* 43(8): 706–717.

Thimmaraju, R., N. Bhagyalakshmi, M.S. Narayan, and G.A. Ravishankar. 2003a. Food-grade chemical and biological agents permeabilize red beet hairy roots, assisting the release of betalaines. *Biotechnology Progress* 19(4): 1274–1282.

Thimmaraju, R., N. Bhagyalakshmi, M.S. Narayan, and G.A. Ravishankar. 2003b. Kinetics of pigment release from hairy root cultures of *Beta vulgaris* under the influence of pH, sonication, temperature and oxygen stress. *Process Biochemistry* 38(7): 1069–1076.

Thimmaraju, R., N. Bhagyalakshmi, and G.A. Ravishankar. 2004. In situ and ex situ adsorption and recovery of betalains from hairy root cultures of *Beta vulgaris*. *Biotechnology Progress* 20(3): 777–785.

Thimmaraju, R., N. Bhagyalakshmi, L. Venkatachalam, V. Sreedhar Reddampalli, and G.A. Ravishankar. 2006. Elicitation of peroxidase activity in genetically transformed root cultures of *Beta vulgaris* L. *Electronic Journal of Biotechnology* 9(5): 513–521.

Tzfira, T., and V. Citovsky. 2000. From host recognition to T-DNA integration: The function of bacterial and plant genes in the *Agrobacterium*-plant cell interaction. *Molecular Plant Pathology* 1(4): 201–212.

Tzfira, T., and V. Citovsky. 2003. The *Agrobacterium*-plant cell interaction. Taking biology lessons from a bug. *Plant Physiology* 133(3): 943–947.

Weathers, P.J., and R.W. Zobel. 1992. Aeroponics for the culture of organisms, tissues and cells. *Biotechnology Advances* 10(1): 93–115.

Weathers, P.J., B.E. Wyslouzil, and M. Whipple. 1997. Laboratory scale studies of nutrient mist reactors for culturing hairy roots. In *Hairy roots: Culture and applications*, ed. P.M. Doran, 191–199. Amsterdam: Harwood Academic.

Weber, J., V. Georgiev, C. Haas, T. Bley, and A. Pavlov. 2010. Ploidy levels in *Beta vulgaris* (red beet) plant organs and in vitro systems. *Engineering in Life Science* 10(2): 139–147.

Wilson, P.D.G. 1997. The pilot-scale cultivation of transformed roots. In *Hairy roots: Culture and applications*, ed. P.M. Doran, 179–190. Amsterdam: Harwood Academic.

Young, J.M. 2008. Agrobacterium - taxonomy of plant pathogenic rhizobium species. In *Agrobacterium: From biology to biotechnology*, ed. T. Tzfira and V. Citovsky, 183–220. New York: Springer.

Zhang, D., S.M. Lanier, J.A. Downing, J.L. Avent, J. Lum, and J.L. McHale. 2008. Betalain pigments for dye-sensitized solar cells. *Journal of Photochemistry and Photobiology A: Chemistry* 195(1): 72–80.

Zhang, X., M. Stettler, D. De Sanctis, M. Perrone, N. Parolini, M. Discacciati, M. De Jesus, D. Hacker, A. Quarteroni, and F. Wurm. 2009. Use of orbital shaken disposable bioreactors for mammalian cell cultures from the milliliter-scale to the 1,000-liter scale. In *Disposable bioreactors*, Advances in biochemical engineering/biotechnology, vol. 115, ed. R. Eibl and D. Eibl, 33–53. Heidelberg: Springer.

Zryd, J.P., and L. Christinet. 2004. Betalains. In *Plant pigments and their manipulation*, Annual Plant Reviews, vol. 14, ed. K. Davies, 185–213. Oxford, U.K.: CRC Press/Blackwell Publisher.

Chapter 12
Peroxidases and Other Enzymes from Red Beet Hairy Roots

Bhagyalakshmi Neelwarne and Thimmaraju Rudrappa

Abstract Enzymes are natural biocatalysts that initiate and accelerate a large number of biochemical reactions in living cells. The fact that an isolated enzyme could also accomplish specific reactions outside the cell has resulted in an enormous demand for industrial-scale production of different enzymes from microbes and cultured plant cells/organs. The soluble protein of *Beta vulgaris* was observed to display high activities of several enzymes of commercial importance, of which, type III peroxidases (POD) (EC 1.11.1.7), are interesting. These PODs are encoded by multigenic families in land plants, where they are involved in several important physiological, developmental and ecological processes. Because of their versatile functionalities, a wide range of chemicals can be modified by dual catalytic activity of POD (oxidoreductase) and, therefore, are useful for several novel applications. Presently, horseradish is sourced for high-quality POD for biochemical/clinical applications, whereas certain agricultural wastes/by-products have been suggested for commercial applications, although with limitations associated with difficulties in their purification. Alternatively, cultured plant cells and hairy roots have recently attracted profound attention. Cultured red beet hairy roots (RBHR) exhibit massive activities of PODs with some isomers released directly into the medium. The characteristics of RBHR POD were found comparable with those of horseradish peroxidase (HRP), and hence

B. Neelwarne (✉)
Plant Cell Biotechnology Department, Central Food Technological Research Institute
(Council of Scientific and Industrial Research), Mysore 570 020, India
e-mail: bneelwarne@cftri.res.in

T. Rudrappa
DABT, Dupont Knowledge Center, E.I. DuPont India Pvt. Ltd., ICICI Knowledge Park,
Genome Valley, Turkapally, Shamirpet (M), R. R. District, Hyderabad 500 078, India
e-mail: rajurt@yahoo.com

B. Neelwarne (ed.), *Red Beet Biotechnology: Food and Pharmaceutical Applications*,
DOI 10.1007/978-1-4614-3458-0_12,

can be similarly used in reagents for clinical diagnosis, for tracking various biochemical events in laboratory experiments and for pollution monitoring. This chapter provides an overview of recent advances made in the identification and characterization of different enzymes, particularly POD and their commercial applicability. Other presumed/hypothetical applications and research trends in this direction are also discussed.

Abbreviations

POD	Peroxidases
RBHR	Red Beet Hairy Roots
PPO	Polyphenoloxidases
MS	Murashige and Skoog's
HRP	Horseradish Peroxidase
SOD	Superoxide dismutase

12.1 Introduction

Higher plant cells and organs are cultured in vitro for two types of products: secondary metabolites and proteins. In addition to the native proteins, plant molecular techniques have made plant cells to express several other native and foreign proteins at levels that are feasible for commercial-scale ventures. Among such proteins are enzymes, which initiate and accelerate thousands of biochemical reactions in living cells. Enzymes process reactions that are otherwise not possible under chemical catalysis. In fact, almost all biochemical reactions require one or more enzymes for their completion, although a few reactions undergo non-enzymatic conversions. Since enzymes are proteins by nature, they can be extracted from living tissues, purified and even crystallized. Under controlled conditions of isolation and storage, they retain their original level of activity and, in some cases, even exhibit an increased activity. Thus, the purified enzyme can be used to accomplish a specific biochemical reaction outside the cell. The greatest advantage offered by enzymes for industrial processes is a combination of high specificity, mild treatment conditions and low energy requirements. This property of enzymes has been employed for laboratory experiments, as well as for commercial production of several important biochemical compounds, drugs, clinical reagents, diagnostics and industrial products. Thus some enzymes are high-value primary metabolites produced in large quantities by cultured cells/organs. When soluble protein extracts from hairy root cultures of red beet were screened for enzymes of commercial interest, high activities of mainly PODs were identified, although low activities of other enzymes, such as polyphenol oxidase (PPO) and superoxide dismutase (SOD), were also identified.

This chapter reviews the progress of research made in the production and recovery of enzymes from cultured red beet cells and hairy roots.

12.2 Peroxidase (EC 1. 11. 1. 7) (POD)

POD, in its widest sense, includes a group of specific enzymes involved in oxido-reductive functions such as NAD-peroxidase, NADP-peroxidase, fatty acid peroxidase, glutathione peroxidase, etc., as well as a group of very non-specific enzymes from different sources, which are simply known as peroxidases (H_2O_2 oxidoreductase). Thus POD is one of the leading molecules involved in cell defense, exhibits a high affinity for hydrogen peroxide and is enhanced during oxidative burst, where it is known to chiefly scavenge free radicals (peroxides) (Regalado et al. 2004). PODs also supply oxygen atoms to various anabolic processes during regular cellular functions. In fact, PODs seems to have equipped higher plants to evolve and diversify their colonization from water to land, offering them characteristic plasticity during evolutionary adaptations (Cosio and Dunand 2009). Different isoforms of this enzyme are present in almost all cellular compartments and some of these isoforms are membrane-bound (Gazaryan et al. 2006). PODs are mainly localized in the central vacuole, where they are represented by several of their isoforms; the role of POD in this compartment remains poorly characterized. Outside the plant system, POD catalyzes the dehydrogenation of a large number of phenolics, aromatic amines, hydroquinones, etc., thus having an important role in chemical synthesis. Several substrates are used to assay the activities of PODs. Generally, guaiacol is used as a substrate for the assay of POD, which is based on the following reaction:

$$\text{Guaiacol} + H_2O_2 \xrightarrow{\text{POD}} \text{oxidized guaiacol} + 2H_2O.$$

The resulting oxidized (dehydrogenated) guaiacol is more than one compound and depends on the reaction conditions. The rate of formation of dehydrogenated product is a measure of the POD activity in a spectrophotometric assay system. Since a wide range of chemicals can be modified by catalytic activity of POD, several novel applications have been suggested for this enzyme, including the treatment of wastewater containing phenolic compounds, synthesis of various aromatic chemicals and removal of peroxide from materials such as food stuffs and industrial wastes (discussed in a subsequent section). Owing to its dual oxido-reductive properties and the ease of conjugating POD protein with several reagents, plant POD has been widely used as an important component of reagents used in clinical diagnosis and in various laboratory experiments where high specificity in detection is required. A large number of applications are also found for PODs in food processing (Chung et al. 2004; Hilhorst et al. 2002; Takasaki et al. 2005; Matheis and Whitaker 1984; Buchert et al. 2010). Regarding the history and developments of PODs, research on HRP has contributed significantly to our understanding of the properties of POD (Saunders et al. 1964; Paul 1986; Veitch 2004).

Regarding its physiological roles, POD exists in plants for a number of functions it has to perform; some of the important aspects in which POD has been implicated are:

- POD is regarded as an enzyme chiefly involved in defense and host plant resistance reactions, which are ascertained through its over-production in the challenged tissue by either biotic or abiotic stress (Isheeva et al. 2009; Ferreres et al. 2011).
- POD is associated with the hypersensitive response, with a proven significant role in plant disease resistance (Levine et al. 1994; Bolton 2009).
- POD is induced systemically in response to herbivore attack in a number of plants (Davis 1987).
- POD is involved in the regulation of endogenous auxin and abscisic acid levels, thereby modulating plant growth processes (Biles et al. 1990; Medina et al. 1993)
- POD isozymes were found to be involved in the differentiation of tracheary elements and lignification in *Picea stichensis* (Richardson et al. 1997).
- In addition to a number of protective functions, POD has been reported to impart UV tolerance and lignin polymerization in plants (Richardson and Mc Dougall 1997).
- POD is involved in the turnover and degradation of some of the plant pigments, such as anthocyanins (Lim et al. 1989; Calderon et al. 1992).

Many of these physiological functions of POD in plants are due to its ability to generate hydroxyl radicals through various mechanisms (Fig. 12.1).

Based on the sequence of amino acids, PODs have been classified into three broad categories as class I, class II and class III. Class I PODs include yeast cytochrome-C POD, gene-duplicated bacterial PODs and ascorbate PODs. Class II PODs include most of the fungal PODs; Class III PODs include classic secretary plant PODs as those of the well-known horseradish POD (HRP) isoenzymes and a few other PODs from bacteria and fungi (Welinder 1992). This amino acid sequence-based classification is well supported by the three-dimensional structural data obtained for the PODs of each group (Veitch 2004; Watanabe et al. 2010). The reaction mechanisms of most of the PODs generally involve a reducing substrate with the formation of reaction intermediates, similar to its catalytic cycle with ferulate, as shown in Fig. 12.2.

12.3 Peroxidases in Red Beet and Hairy Roots

Over the years, horseradish has been the only commercial source of POD (Veitch 2004). However, other sources, particularly red beet (as described in Chap. 1) can also provide large amounts of PODs with similar or better substrate specificities, yield and economic feasibility. Seedling roots of red beet expressed the highest POD activity, of 9,700 U/g fresh tissue (FW); and in the mature tuber, the peel tissue showed the highest activity, of 6,116 U/g FW, whereas the pulp showed only 1,492 U/g FW. Other varieties of red beet showed much lower POD activity values

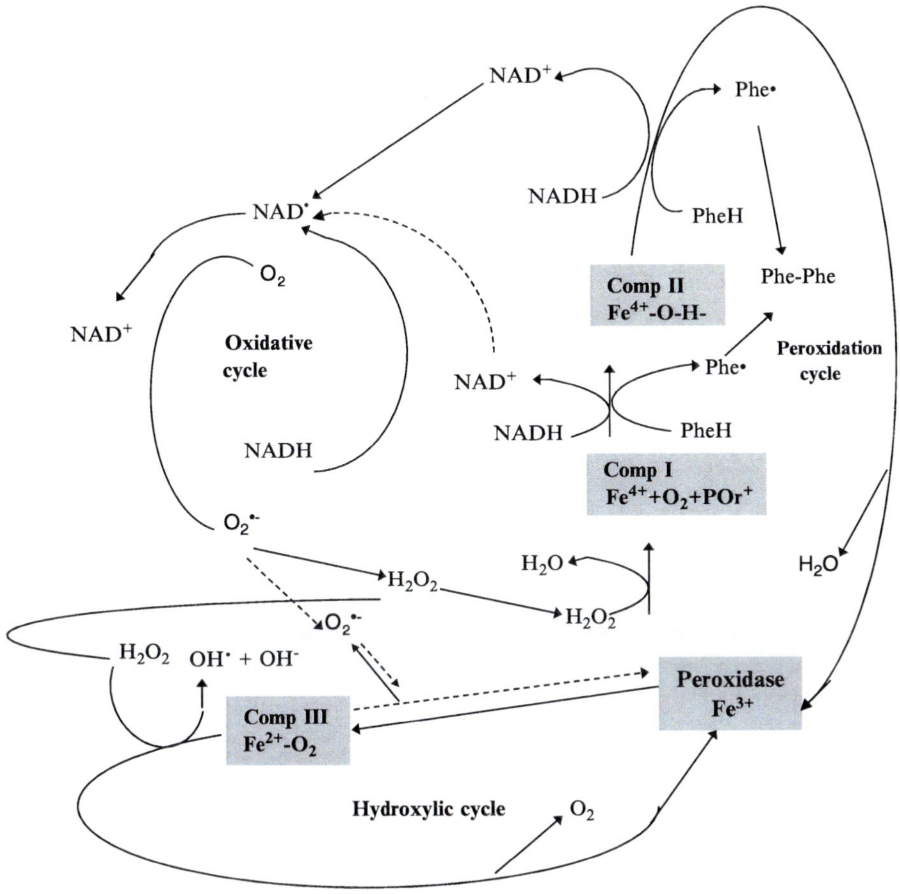

Fig. 12.1 Physiological role of peroxidases in plants. Plant cells generate hydroxyl radicals through different pathways such as oxidative, hydroxylic and peroxidative where POD mediates the oxidation of phenolic substrates (PheH) by H_2O_2 to phenoxy (Phe•) radicals that polymerize to generate molecules such as lignins and extensins. Reduction of substrates such as NADH (dihydroxyfumerate) initiates the oxidative cycle that reduces O_2 to O_2• and H_2O_2, which is utilized to produce compound III that reduces H_2O_2 to OH in the hydroxylic cycle (Adapted after modification from Chen and Schopfer (1999))

(Isheeva et al. 2009; Escribano et al. 2002). Hairy roots are obtained by infecting any part of the plant (preferably the aerial parts) with the soil bacterium *Agrobacterium rhizogenes*, in which case, a part of the DNA from a plasmid residing in the bacterium (Ri) is transferred to the plant cell and integrated into the nuclear DNA (Guillon et al. 2006). Thus, hairy roots are formed after genetic transformation where the newly inserted plasmid-borne DNA fragment (T-DNA) carries a number of genes that alter the hormone balance of the cell and instigate a signal network in such a way that the transformed cell starts growing in the form of root covered densely with root hairs. Such roots can then be grown continuously in vitro by supplying

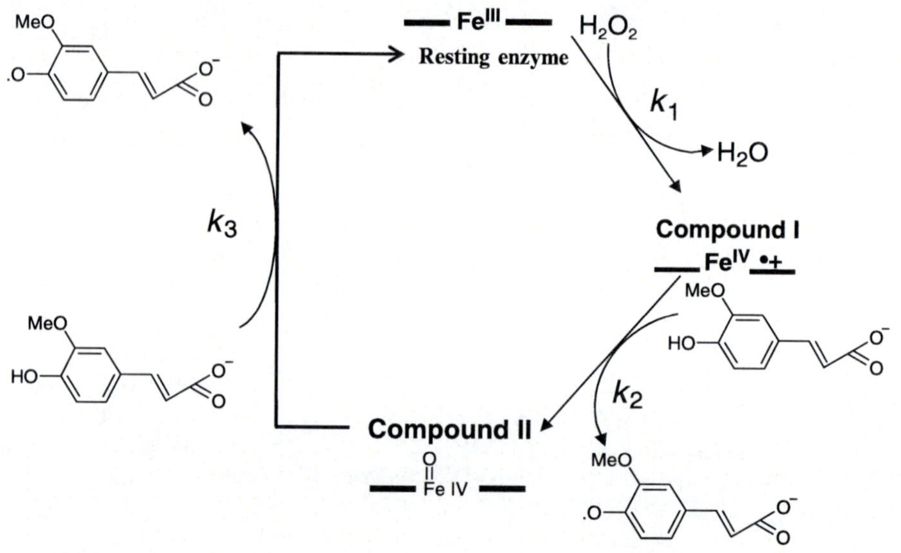

Fig. 12.2 Functioning of ferulate as the reducing substrate in the catalytic cycle of horseradish peroxidase where *k1*, *k2* and *k3* are the rate of formation of compound I and the formation and reduction of compound II, respectively (Adapted after modification from Veitch (2004))

appropriate nutrients and other conditions. In red beet, such genetically transformed hairy roots do not form tubers, instead they grow as a highly branched tangled mass, without tuberization (Bhagyalakshmi et al. 2004). Thimmaraju et al. (2005) established different clones of hairy roots, of which, one transformed red beet (hairy root) clone expressed very high POD activity, of over 9,000 U/g FW, with a specific activity of 600 U/mg protein (Thimmaraju et al. 2005). The transformed nature of RBHR harboring genes for root morphology i.e., *rol* genes (Lemcke and Schmulling 1998a, b) (see Chap. 10), is probably one reason for the over-expression of POD, as in the case of seedling roots rather than in the mature tubers. The higher expression of POD at the surface (skin) of the tuber may be attributed to defense-related activities elicited by surface-borne microorganisms, since several soil-borne microbes are known to elicit phenylpropanoid pathway-related enzymes (Gomez-Vasquez et al. 2004; Gundlach et al. 1992; Kino-Oka et al. 2001; Klibanov et al. 1980; Radman et al. 2003).

12.3.1 *Influence of A. rhizogenes Strains and Selection of Best Hairy Root Clone*

Since the hairy root clone resulting from each transformed cell is expected to be genetically different with respect to T-DNA integration and its copy number, and since such events are influenced by the physiological status of the host cell, each

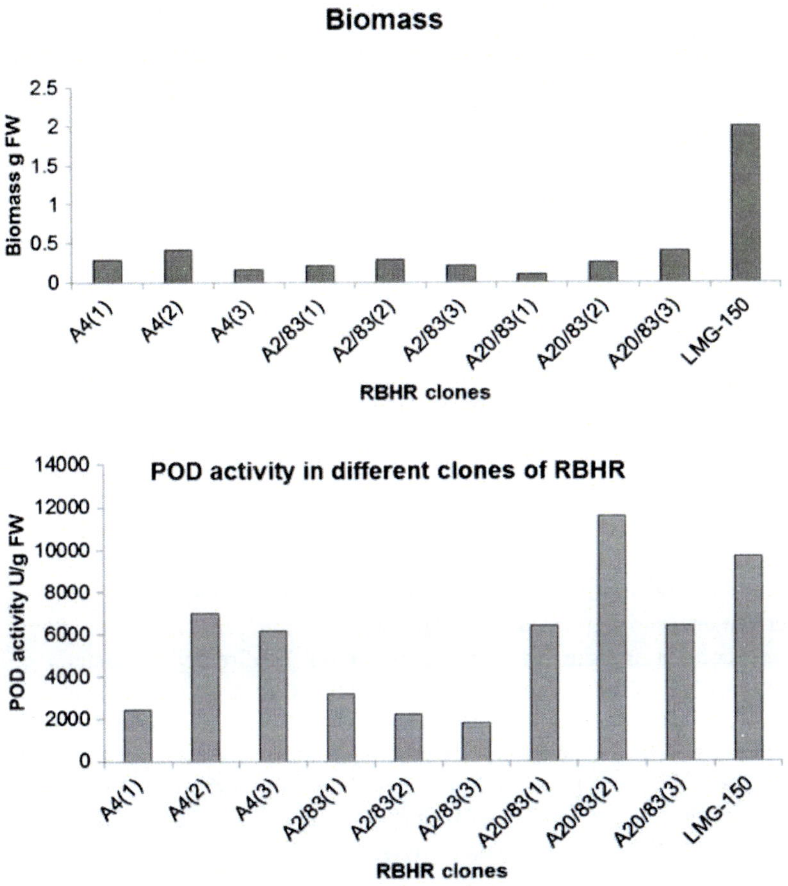

Fig. 12.3 Biomass and POD activities in the ten clones of RBHR during a period of 25 days (grown in 15-mL medium)

transformed root clone is expected to behave differently with respect to growth and product synthesis. Such differences are apparent due to the variation in the T-DNA insertion, copy number, size and location of integration of T-DNA of the Ri plasmid in the plant genome (see Chap. 10), observed in other such studies (Doran 1997) and among clones of horseradish hairy roots (Flocco et al. 1998). Therefore selection for the best-performing hairy root clone is an important strategy to obtain stable and high yields of enzymes. Infecting seedling nodes of red beet cv. Ruby Queen with different clones of *A. rhizogenes*, Thimmaraju et al. (2005) established ten clones of hairy roots, which displayed distinct differences in the rate of biomass accumulation as well as in POD activities (Fig. 12.3). While the root clone LMG-150 showed high fresh biomass productivity of 140 g/L on hormone-free medium, the other clones produced very low biomass. The high biomass-producing clone was also found having very high POD activity, of 9,000 U/g FW and 1.18×10^6 U/L medium, having a specific activity of 600 U/mg of total soluble protein. Some of the other clones also expressed high POD activities, however with very low biomass formation (Fig. 12.3). A nearly neutral-active (pH 6.0) POD was observed in almost all the hairy root clones; whereas the spent medium showed highest POD activity at pH 4.0 (Thimmaraju et al. 2005).

12.3.2 Kinetics of POD Formation in Hairy Roots

The kinetic studies in RBHR showed that the POD activity was high during the mid-exponential growth phase and declined further as the biomass reached its peak on the 25th day (Thimmaraju et al. 2005), similar to the root cultures of horseradish (Flocco and Guilietti 2003). In RBHR clone LMG-150, the total protein content was found to be highest on the 15th day (15 mg/g FW) and remained constant till the end of growth phase (Thimmaraju et al. 2005). The activity of POD in proteins extracted at pH 6.0 was found to coincide with the highest total protein content, i.e., on the 15th day, during which, the roots expressed POD of 9,000 U/g FW. This level declined to 8,000 U/g FW on the 20th day, with no concomitant decline in total protein. The fact that enzyme activity keeps fluctuating depending on kinetics of its milieu, the protein content and the enzyme activity cannot be directly correlated. Such fluctuations in POD activity could probably be linked to high metabolism, during which the pro-oxidants are formed, triggering signals for the release of POD to quench the oxidative radicals, as has also been suggested for *Ipomoea aquatica* (Kino-Oka et al. 2001). The low formation of PODs showing activity at pH other than 6.0 may also be due to the same reason that most of the cellular metabolic activities occur at pHs ranging from 5.5 to 6.0, which is well known. In the medium, only 50 U of POD was recorded on the 15th day of hairy root growth, of which about 13.5 U was that of acidic, and 6.0, 0.4 and 5.0 U of activity were recorded at pH 6.0, pH 7.0 and pH 9.0, respectively. One root clone, A20/83(2), expressed a much higher level of POD, of about 12,000 U/g FW than the other clones on the 20th day, however with lower biomass than in clone LMG-150 (Fig. 12.3), and hence accounted for lower productivity per liter culture medium (Thimmaraju et al. 2005).

12.3.3 Influence of Salts on Total Turnover and Secretion of POD

Cellular primary and secondary metabolic functions of cultured plant tissues are known to respond to the nutrients in the medium (Neumann et al. 2009), bringing changes in metabolic profile. The salts of Murashige and Skoog's (MS) medium (1962) were found adequate for RBHR to express POD activity of 1.18×10^6 U/L. However, the addition of extra (5 mM) calcium chloride or potassium chloride to culture medium enhanced POD activity up to 1.21×10^6 U/L. The cations Mg^{2+}, Ca^{2+} and K^+ caused slight increases in POD activity while other metal ions, such as SO_4^{2-}, CO_3^- and Cl^- appeared to impart antagonistic effects. Surprisingly, NO_3^- ions suppressed POD activity at lower concentrations and increased POD expression at higher concentrations (Thimmaraju et al. 2005). For horseradish hairy roots, the additional ion supplementation considerably enhanced the release of POD into the medium (Uozumi et al. 1992). Expecting similar effects, when Thimmaraju et al. (2005) treated RBHR with additional levels of cations and anions, PODs were enhanced in hairy roots. On the contrary, additional Na_2–EDTA and $FeCl_3$ completely suppressed the growth of RBHR. Nitrates, potassium nitrate in particular, fairly enhanced POD activity, and higher level resulted in higher POD activity.

In addition to the enhancement of POD turnover in the tissues, metal ions also caused release of POD into the medium. In cell cultures of other plants, higher than normal levels of calcium supplied to the medium enhanced the synthesis as well as the secretion of POD into the medium (Sticher et al. 1981). A 5% increase of Ca^{2+} content of the medium caused an increase in the activity of POD that excreted into the medium with *Arachis hypogaea* culture. Such enzyme activity increase could be abolished by the addition of the calcium chelator ethylene glycol-bis-(-aminoethyl ether)-*N*,*N*,*N*′,*N*′-tetra acetic acid (EGTA), which confirmed the participation of calcium in the enzyme release process. Na^+ and Mg^{2+} ions enhanced the secretion POD into the medium, thus acting as effluxing agents. The involvement of these ions in membrane permeabilization, acting as ion channels, has been well established (Karpen and Ruiz 2002). The other cations, such as K^+ and Ca^{2+}, were not found to cause any significant increase in POD secretion into extracellular medium in RBHR, although these ions were also found to be involved in the membrane activities (Karpen and Ruiz 2002) and reported to cause secretion of POD in horseradish hairy roots (Uozumi et al. 1992). Supplementation of culture medium with anions, such as SO_4^{2-}, CO_3^{2-} and NO_3^-, increased the extracellular POD activity, with NO_3^- causing the highest secretion, of about 50% of the total POD in RBHR. This was in contrast to the earlier reports where NO_3^- had very little effect on secretion of POD (Uozumi et al. 1992). Taken together, the release of POD from RBHR was highest in the presence of 5 mM KNO_3 (nearly 50%), followed be NaCl (43%), which was much higher than 29% of the total activity found in the control medium. Treatment with other metal ions at three different concentrations did not cause any effect on the secretion of POD into the medium (Thimmaraju et al. 2005).

12.3.4 *Influence of Auxins*

Involvement of auxins in growth and morphogenesis of various types, especially roots, is well established. The exogenous application of two auxins for cultured red beet hairy roots did not cause any significant improvement in biomass accumulation in most of the root clones, except in clones A2/83(2) and A4 (1). These clones showed significant improvements in biomass after the supplementation of auxins.

Such hormone dependency for growth has been a feature of other POD-producing hairy root cultures, such as *Cichorium intybus* (Bais et al. 1999) and horseradish hairy roots (Nakashimada et al. 1994). Contrarily, a very high variability in the expression of POD protein was observed in different RBHR clones as a response to auxins (Thimmaraju et al. 2005). The POD productivity in terms of activity, in units per liter, pH 6.0, in the best clone, LMG-150, was observed to be almost similar, with or without auxin treatment, accounting for about 1×10^6 U POD activity/L of medium (Fig. 12.4). However, in other clones, the productivity was enhanced significantly by auxins; the levels in some were on par with those of LMG-150, the best clone. While it was noticed that some RBHR clones resulted in enhanced levels (1×10^6 U/L) of POD upon supplementation with 1 ppm of naphthalene acetic acid (NAA), other clones expressed higher POD activity in the presence of lower indole-3-acetic acid (IAA; 0.5 ppm) supplementation, accounting for about 0.7×10^6 U/L. In other plant cell cultures, POD activity was found to increase in response to specific concentrations of auxins, which was attributed to the support rendered by auxin or a sort of stress created on the hairy root clones, in addition to the well-known fact of the significant implication of PODs in auxin metabolism (Liu et al. 1996; Thimmaraju et al. 2008). Even when root clones expressed high growth rates, the oxidants produced as a result of high anabolic activities are thought to demand the synthesis of this enzyme for scavenging O_2 (pro-oxidants) and H_2O_2 (peroxides) (Kino-Oka et al. 2001). In addition to the need for POD in the cell during the stress-related metabolism, auxins may also be involved in switching on the stress signals, creating situations for the production of POD, which could be the case in the other RBHR clones with suppressed biomass. However, RBHR LMG-150 did not show significant responses for auxin, where POD levels remained almost constant even after treatment (Fig. 12.4b), from which, the authors inferred that the clone LMG-150 neither utilizes exogenous auxin nor that the addition of hormones imparts any stressed physiological conditions. The suppression of biomass in response to exogenous auxins was also attributed to the already abundant presence of endogenous auxin (IAA) in the clone LMG-150, where additional supply would have imparted toxic effects. Even under these stressed circumstances, one can expect the expression of POD for performing the oxidation of auxins.

12.3.5 Thermal Stability of Native POD

While most of the enzymes, in their native forms, show their highest activities at biological temperatures, they display an almost similar performance at temperatures slightly or greatly deviating from their biological temperatures. Since the performance of POD at higher temperatures is preferred for its commercial applications and for its storage, the thermostability of POD from RBHR was assessed by Thimmaraju et al. (2005). The thermal inactivation of POD isolated at acidic and neutral pH was negligible up to 50°C, even after 40 min, whereas the POD extracted at basic pH was quite thermo-sensitive, with 50% loss at 50°C, and with a great loss

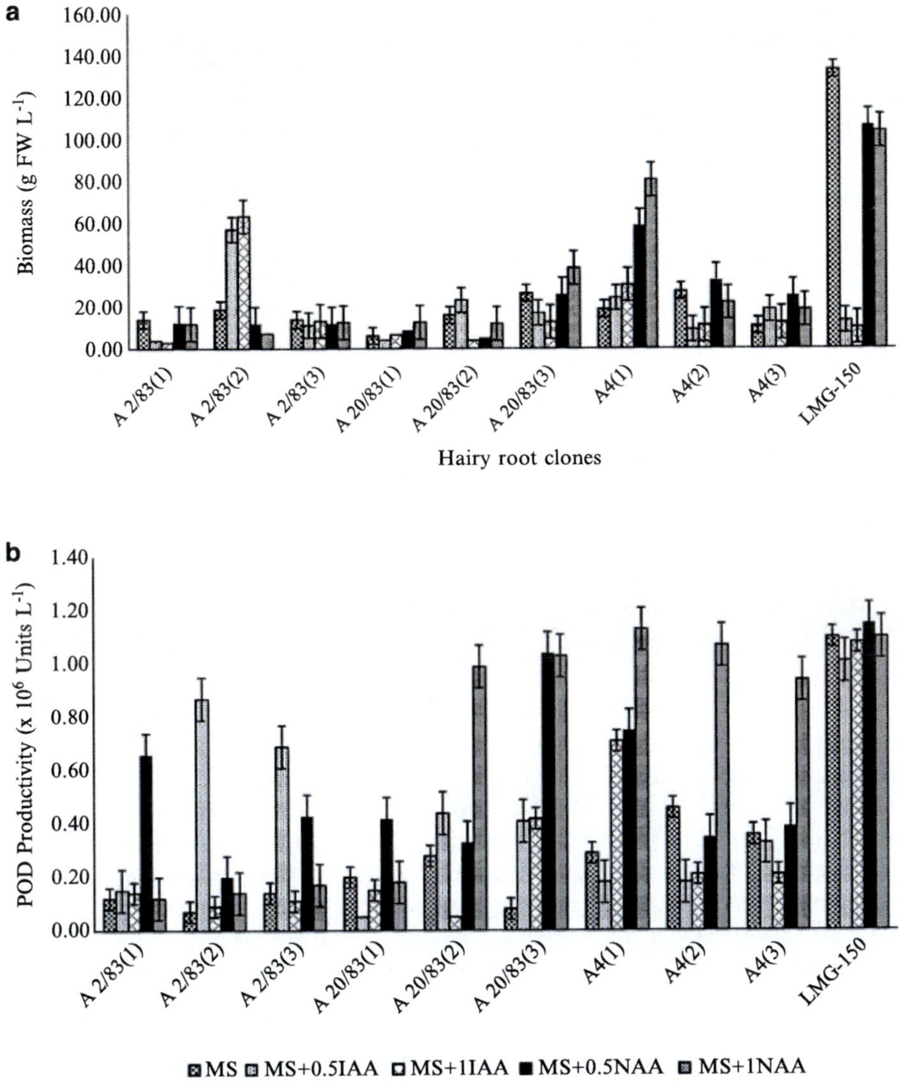

Fig. 12.4 Growth (**a**) and peroxidase productivity (**b**) in different hairy root clones grown in hormone-free MS liquid medium and in MS with either IAA or NAA at various levels observed on the 20th day of culture (Adapted with modification from Thimmaraju et al. (2005))

of activity at 60°C. At 60°C, the acidic and neutral PODs retained more than 70% of the activity up to 40 min, whereas they were completely inactivated at 70°C. Since PODs are useful for a number of commercial applications, its reasonably high stability at above-ambient temperatures, with only about 5% activity loss at 50°C even after 30 min, appears promising for a wide array of applications in various processes and for accomplishing biochemical reactions.

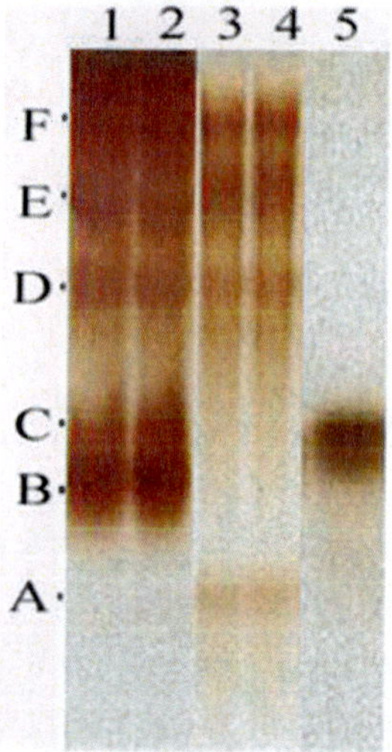

Fig. 12.5 Zymogram of isozymes of peroxidase found in soluble proteins from roots (*lanes 1* and 2) and from spent medium (*lanes 3* and *4*), where the gels were developed by activity staining of the crude enzyme separated on 7.2% native polyacrylamide gel and partially purified fraction (*lane 5*). R_m represents the relative mobility of the respective isoform (Adapted with permission and modification from Thimmaraju et al. (2007))

12.3.6 Isozymes of RBHR POD

All higher plant PODs have many isozymes, which are generally analyzed using polyacrylamide gel electrophoresis (PAGE) with activity staining. The spent medium of other plant cell cultures were found to contain remarkably high activities of oxidative enzymes, particularly POD (Barz and Koster 1990). In some cases, as that of horseradish (Uozumi et al. 1992; Parkinson et al. 1990) and turnip hairy roots (Agostini et al. 1997, 2000, 2003), continuous synthesis and release of such enzymes during growth phase was a regular feature. POD from RBHR, when analyzed for isozymes on PAGE with active staining, showed four distinct bands of R_m 0.06, 0.16, 0.25, 0.38 and 0.46 (R_m represents the relative mobility of the respective isoform). Spent culture medium was also observed to display three isozymes at R_m 0.06, 0.16, 0.25 and one extra band of R_m 0.575, which possibly is the isozyme that was completely secreted in to the medium. Isozymes of R_m 0.38 and 0.46 in the medium (Fig. 12.5) were found to be absent in this study, and the authors reported a total of about six POD isoforms, of which three were secreted partially into the medium and only one was completely secreted into the medium (Thimmaraju et al. 2005). Therefore, four of the six isozymes were of secretary-type proteins but the degree of secretion varied, allowing a good scope for improving their in situ secretion. The involvement of specific signals in assisting the leaching of a specific type of POD isozyme forms an interesting study for the future, which nudges the technological scope for probable continuous production and online recovery of the product. There was a clear distinction between any two isoforms differing by R_m value of at least 0.15, suggesting that the isoforms are amenable for easy separation and further purification (Fig. 12.5).

12.4 Elicitation

Enhancement of secondary metabolites and their related precursors, including the associated enzymes such as POD, has been accomplished using different strategies, such as selection of clone and changing growth conditions, particularly hormones and nutrients. Another interesting strategy is the exposure of cultures to changed environment. When heterotrophic hairy roots of *Ipomoea aquatica* were made autotrophic by exposing to light, the activity POD within the tissue nearly doubled, from 250 U/g to nearly 500 U/g FW, which the authors related to the requirement of POD for scavenging the high release of toxic oxidants, the peroxides during photosynthesis (Kino-Oka et al. 2001). Elicitation is an efficient and reliable way to enhance the synthesis of defense-related compounds in cultured plant cells and microbial cells, meaning that several intermediary compounds and related enzymes are also elicited. Being involved in several metabolic pathways in the plant, elicitation of POD has also been observed in a few other hairy root systems (Giri and Narasu 2000; Flocco et al. 2003; Lei et al. 2011). The compounds used to induce elicitation, commonly known as elicitors, are either of biological or non-biological origin, and that, upon contacting plant cells, trigger the increased production of pigments, flavones, phytoalexins and other defense-related compounds (Eilert et al. 1984; Graham and Graham 1999; Robins et al. 1985; Eilert et al. 1986; Flores and Curtis 1992; Sim et al. 1994; Bhagyalakshmi and Bopanna 1998; Buitelaar et al. 1992; Singh 1999). POD is known for its involvement in the utilization of many phenolic acids as substrates for its formation (Ferreres et al. 2011) as well as oxidation, reduction and inter-conversions of many intermediates of the phenyl propanoid pathway progressing towards lignin biosynthesis. Therefore, this enzyme is expected to respond positively to the treatment with elicitors (Gomez-Vasquez et al. 2004; Perera and Jones 2004). For example, treatment with different abiotic elicitors such as $AgNO_3$ and $CuSO_4$ caused about 100% increase in POD formation in transformed root cultures of horseradish. In these hairy roots, the metal ions caused synthesis with simultaneous secretion leading to an overall productivity of about 12-fold higher POD enzyme (Flocco et al. 1998).

A large number of signals perceived by plant cells are processed differently, and therefore, successful application of elicitation requires extensive screening. Elicitation of PODs *in planta* as a response to pathogenic organisms has been extensively documented (Perera and Jones 2004; Gomez-Vasquez et al. 2004). However, a few studies focused on the elicitation of POD activity in cultured cells and hairy roots (Uozumi et al. 1992; Agostini et al. 1997; Flocco et al. 1998; Flocco and Guilietti 2003; Funk et al. 1987; Gasper et al. 1983; Funk and Brodelius 1990; Xu et al. 2004) as well as in RBHR (Thimmaraju et al. 2006). In horseradish hairy roots, significant enhancement of POD activity (100%) occurred when cultures were treated with certain metal ions: $AgNO_3$, $CuSO_4$ and extracts of fungi, *Verticillium* sp., *Monodyctis cataneae* and *Aspergillus niger* (Uozumi et al. 1992). However in RBHR, which normally produce much higher levels of POD than other cultures, POD activity could be further enhanced by challenging the cultures with elicitors. Therefore, a large number of elicitors were systematically screened; their probable interactions with RBHR in eliciting PODs have been discussed (Thimmaraju et al. 2006). In this

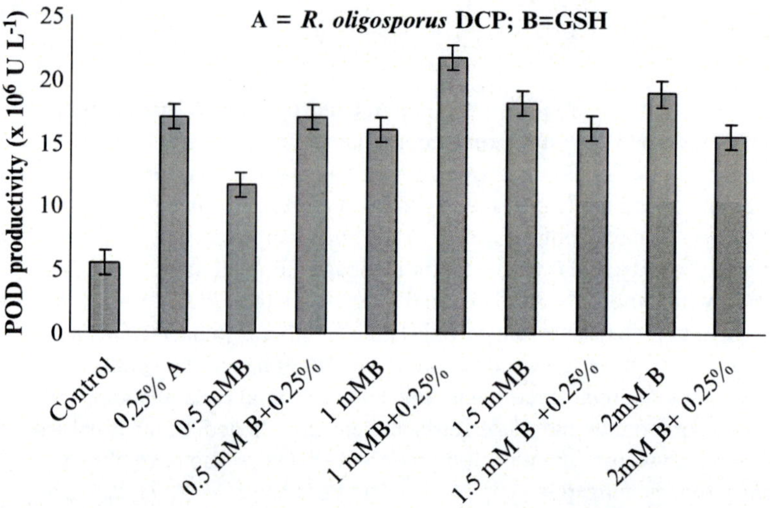

Fig. 12.6 Elicitation effect of a combination of glutathione (GSH) and 0.25% dry cell powder (DCP) of *Rhizopus oligosporus* (Adapted with modifications from Thimmaraju et al. (2006))

study, biotic elicitors, such as dried cell powders (DCP) of microbial cultures (0.1–0.5% w/v) and the respective culture filtrates (CF) (1–5% v/v) were periodically added to RBHR cultures. Other abiotic elicitors, such as metal ions (2- to 8-fold of that present in the nutrient medium), the plant hormone thidizuron (TDZ; 0.25–1 ppm) and bio-molecules such as glutathione (GSH; 0.5–10 mM) and methyl jasmonate (Mej; 20–100 mM), were also tested by monitoring POD activity continuously until the end of the growth phase. POD activity was most efficiently elicited in RBHR treated with DCP of *Candida versatalis* (3.52-fold higher than control), then by GSH (3.44-fold higher) and DCP of *Rhizophus oligosporus* (3.09-fold higher), and, among abiotic elicitors, TDZ, magnesium and calcium elicited 2.49-, 3.03- and 2.8-fold higher activities, respectively. Although Mej is known to elicit secondary metabolites in suspension cultures of various plant species (Gundlach et al. 1992), including betalains in RBHR clone LMG-150 (Suresh et al. 2004; Savitha et al. 2004), it was found ineffective in eliciting POD in the same root clone.

Whenever there were positive elicitation effects in RBHR, the time of elicitor addition was found to play a crucial role. Biotic elicitors were more effective when added on the 15th day of culture, whereas the abiotic elicitors were more effective when added on the 20th day (Thimmaraju et al. 2006). Among the selected good elicitors, a strategy of combining two elicitors for treatment indicated that the combination of GSH (1 mM) and 0.25% of DCP of *R. oligosporus* caused 4-fold enhancement, accounting for 21.8×10^6 U/L (Fig. 12.6) (Thimmaraju et al. 2006). Such elicitations find great applications for scaled-up production of POD from RBHR in bioreactors. Since most of the elicitors primarily influence the cell wall/membrane functions, a deeper understanding of the cascade of signal pathways would be of great use for a precise and desired level of elicitation, which may further be coupled with effluxing of POD enzyme and re-use of biomass.

For RBHR clone LMG-150, higher levels of calcium or magnesium in the medium correlated with the highest POD activity, 3.01-fold more than in control cultures. Elicitation effects of these metal ions could be attributed to their reported roles as both nutrient and secondary signaling molecules in causing responses that follow elicitation (Nishi 1994; Pitta-Alvarez et al. 2000). Since these metal ions are readily usable in their salt forms, they are more practical and cost-effective than biotic elicitors, which need several stringent processing steps.

Elicitation, which deviates cell's energy towards secondary metabolism, generally causes growth retardation. To overcome such losses during the elicitation of betalain pigments, the addition of elicitor at the late exponential phase was applied to enhance the overall productivity (Savitha et al. 2004). However, POD enhancement occurred when elicitors were added at either the early or middle exponential phase, between the 15th and 18th day, rather than at a later stage, on the 20th day (Thimmaraju et al. 2006). Therefore, by judiciously selecting and timing the addition of elicitor, there is a possibility of producing both POD and betalain in the same process. In the latter case, the process of online recovery of both pigment (Thimmaraju et al. 2003a, b, 2004) and POD is possible with the use of an additional on-line gel filtration column (Neelwarne and Thimmaraju 2009).

12.5 Purification of POD from RBHR

Despite the availability of hairy roots with high POD activity and the further elicitation of POD, the purification of the enzyme is a major hurdle for realizing commercial potential (Thimmaraju et al. 2007). Different strategies, such as ion exchange chromatography and fast protein liquid chromatography (FPLC) have been employed for POD purification in different systems (Aruna and Lali 2001; Brownleader et al. 1995; Christensen et al. 1998; Nair and Showalter 1996). Although the process of POD purification from RBHR is essentially similar to that applied for any plant, animal, or microbial source, there have been difficulties, where one needs to screen several protocols. Purification of PODs from a majority of plant sources involves precipitation with $(NH_4)_2SO_4$ followed by ion exchange chromatography with further partial characterization by SDS-PAGE and iso-electric focusing (Nair and Showalter 1996; Wititsuwannakul et al. 1997; Christensen et al. 1998; Aruna and Lali 2001). However, Srinivas et al. (1999) used a different strategy, which involved an aqueous two-phase extraction coupled with gel filtration for purification of a POD extracted from the leaves of *Ipomea palmata*. This method used PEG/ammonium sulphate/NaCl (24/7.5/2.0%, w/v) to obtain a purification factor of 2.18, a volume reduction of 57.5% and an approximate 49-fold purification using a Sephadex G-100 column with a recovery of about 75.3%. Later POD from the same source (*I. palmata*) was purified using a novel separation method of three-phase partitioning, where application of the first cycle involved crude extract and

Table 12.1 Different methods applied for the purification of RBHR POD showing the levels of purification after each separation

Sample	Protein (mg)	Activity	Specific activity (U/mg protein)	Fold (s) of activity
Crude extract	168.0	1.17×10^6 U	700.0	1.0
Ammonium sulphate precipitation	123.5	2.16×10^6 U	1,750.0	2.5
AEC* (DEAE cellulose)	9.5	0.99×10^6 U	10,500.0	15.0

*Anion exchange column with DEAE cellulose

t-butanol mixed in equal ratio with the third phase composed of 30% ammonium sulfate. Subsequent separation at 37°C resulted in about 160% activity recovery with a twofold purification, where POD migrated to the aqueous phase. Repeating the three-phase partition using the POD-rich aqueous phase, an 18-fold purification of POD (molecular weight ~21 kDa) was achieved with about 81% activity recovery (Narayan et al. 2008). Thimmaraju et al. (2007) purified RBHR POD by using a combination of $(NH_4)_2SO_4$ fractionation and ion exchange chromatography (DEAE anion exchange column [AEC]), which resulted in 15-fold enhancement of activity (Table 12.1). Aqueous two-phase extraction of red beet POD, seeking simultaneous separation of betalains, also resulted in a 2-fold purification of the enzyme (Neelwarne and Thimmaraju 2009).

12.5.1 *Molecular Weight and Purity*

Different molecular weight fractions, mainly of peak I and peak II, of red beet hairy roots obtained from a DEAE cellulose column were further separated by the SDS-PAGE method using Coomassie brilliant blue staining and native PAGE analysis, with activity staining in the latter (Fig. 12.7a). While some impurities persisted in peak I, peak II showed a single band (Fig. 12.7b), with a much higher purity than peak I, which does not qualify the band as a totally pure protein, since there is a chance that different types of PODs having similar properties could co-exist, and they are often difficult to separate by conventional biochemical methods (Thimmaraju et al. 2007). The specific activity of peak II POD in this study was 10,500 U/mg protein (Table 12.1), which is much higher than activities recorded in various similar sources (Duarte-Vazquez et al. 2001).

When the peak II fraction was subjected to SDS-PAGE analysis under reducing (ß-mercaptoethanol) conditions, a single band was obtained after staining with Coomassie brilliant blue, confirming the purity of the enzyme and also indicating that it is a single polypeptide chain with a molecular weight of 45 kDa (Fig. 12.7b), which is similar to that of HRP (40–46 kDa) (Paul and Stigbrand 1970), and sycamore maple POD (42 kDa) (Dean et al. 1994). Molecular weights of most PODs vary from 30 to 60 kDa (Srivastava and van Hyustee 1977). Such wide variation in molecular weights was attributed to post-translational modifications and also to the length and number of glycan chains in the polypeptide chain (van Huystee and Lobarzewski 1982).

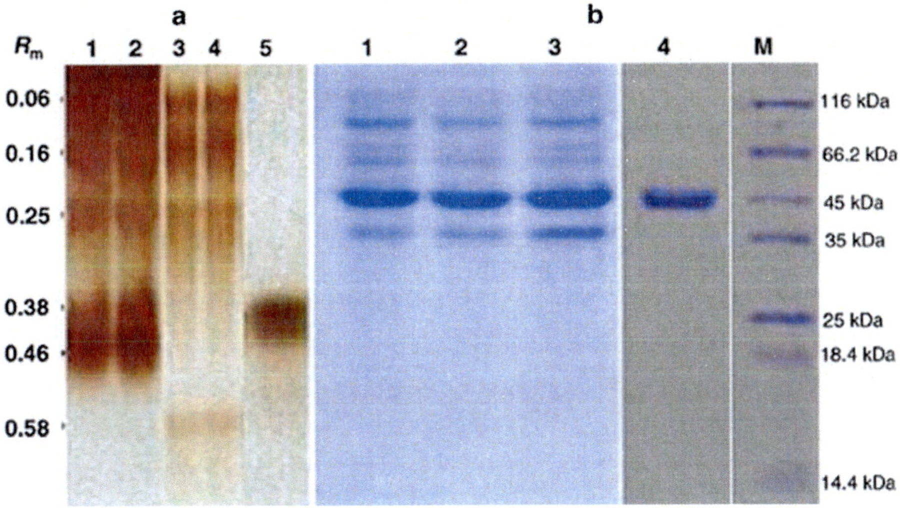

Fig. 12.7 (**a**) Zymograms of peroxidase isozymes from hairy roots (*lanes 1* and *2*) and from spent medium (*lanes 3* and *4*) developed by activity staining of the crude enzyme separated on 7.2% native polyacrylamide gel and partially purified fraction (*lane 5*). R_m represents the relative mobility of respective isoform. (**b**) Sodium dodecyl sulfate–polyacrylamide gel electrophoresis (SDS-PAGE) analysis of crude (*lanes 1–3*) and purified fraction from peak 2 (*lane 4*) of BHR-POD. *Lane M* represents molecular weight markers. Note the clear separation and dense presence of a 45-kDa protein, which is the same molecular weight observed for HRP (Adapted with permission from Thimmaraju et al. (2007))

12.6 Characteristics of Purified RBHR POD

12.6.1 Substrate Specificity

Similar to other PODs such as HRP, the red beet PODs showed a much higher affinity to the substrate hydrogen peroxide (H_2O_2), although there was also affinity to a number of ·H donors, such as orthodianisidine hydrochloride (Duarte-Vazquez et al. 2001). When the AEC-purified fraction of RBHR POD was assayed at various concentrations and the Lineweaver–Burk plots for substrates such as H_2O_2, orthodianisidine hydrochloride, 2,20-azino-bis(3-ethylbenz-thiazoline)-6-sulfonic acid (ABTS) and guaiacol were analyzed, the plot for H_2O_2 at 2 mM orthodianisidine showed the lowest K_m value of 0.1, indicating highest specificity of RBHR POD to H_2O_2. Among the H donors, the enzyme showed the highest affinity in the descending order of orthodianisidine (K_m value 2.134 mM), ABTS (K_m value 5.050 mM) and guaiacol (Km value 98.61 mM) (Fig. 12.8). The K_m value of RBHR POD for orthodianisidine was lower than the values found for guaiacol oxidation by POD from turnip roots (3.7 mM) (Duarte-Vazquez et al. 2001) and Korean radish roots (6.7–13.8 mM) (Lee and Kim 1994).

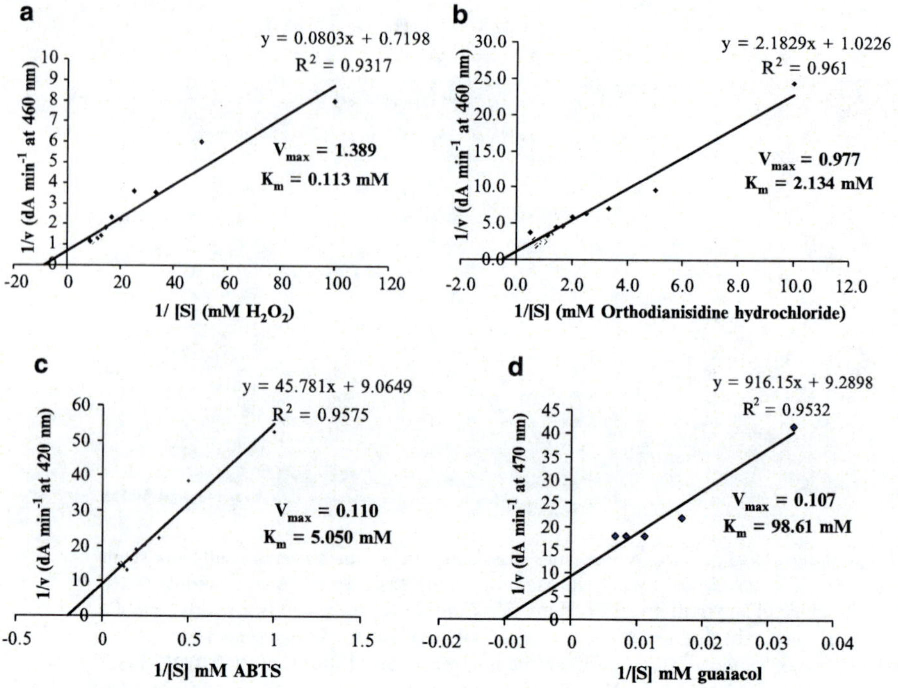

Fig. 12.8 Lineweaver–Burk plot for POD activity at various concentrations of H_2O_2 at 2 mM orthodianisidine hydrochloride (**a**), orthodianisidine hydrochloride at 0.5 mM H_2O_2 (**b**), ABTS at 0.5 mM H_2O_2 (**c**) and guaiacol at 0.5 mM H_2O_2 (**d**) assayed at 25°C in 0.2 M sodium phosphate buffer (pH 6.0). (Adapted with permission from Thimmaaju et al. (2007))

12.6.2 *Inhibition of RBHR POD*

When potent inhibitors of hemoprotein-catalyzed reactions, such as periodate and sodium azide, were tested (Duarte-Vazquez et al. 2001) for RBHR POD inhibition, data analysis by Dixon Plots (Segel 1993) indicated that there was competitive inhibition by periodate (Ki=0.2 mM), whereas sodium azide caused the enzyme inhibition in a non-competitive manner (Fig. 12.9). This inhibition of RBHR POD by sodium azide was in contrast to the property of turnip POD (Duarte-Vazquez et al. 2001). Turnip is also a tuber crop, but belongs to an altogether different family from that of red beet, and hence large biochemical differences may be expected.

12.6.3 *pH Optima for Activity and Stability of Crude and Purified POD*

pH optima is one of the very important criteria for efficient and wide applications of any enzyme. RBHR POD showed maximum activity at pH ranging from 5 to 6

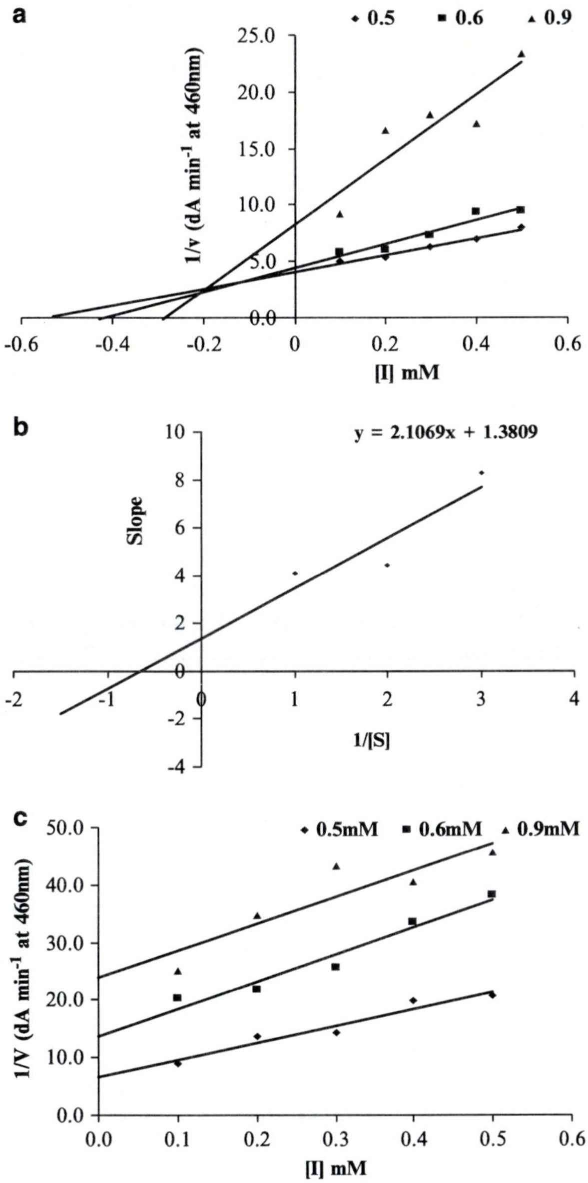

Fig. 12.9 Dixon plots for RBHR POD inhibition showing activity at different inhibitor concentrations (0–0.6 mM) of potassium periodate (**a**), slope re-plot (**b**) and sodium azide inhibition of hairy root POD (**C**). Plots infer that the periodate acts as a competitive inhibitor; whereas sodium azide acts in a non-competitive manner. (Adapted with permission from Thimmaraju et al. (2007))

whereas commercial HRP showed highest activity at pH ranging from 4 to 5. In addition, RBHR POD was stable over a wide range of pH, from pH 4 to 9, exhibiting the highest stability between pH 7 and 9. Purified BHR POD showed highest activity at pH 5, which was stable over a wide range, from pH 3 to pH 9, with the highest stability between 6 and 8 (Thimmaraju et al. 2007). The pH for optimum activity was very similar to that of strawberry fruit (pH 6.0) (Civello et al. 1995), tomato (pH 5.3–5.5) (Heidrich et al. 1983) and soybean (pH 5.4) (Sessa and Anderson

1981) PODs. Commercial HRP showed highest activity at slightly more acidic pH (pH 4–5) compared with crude and BHR POD. HRP was stable at a narrow range of pH between 4 and 6 and showed highest stability at pH 9, similar to crude beet hairy root POD (Fig. 12.5a). Therefore, the versatility of BHR POD is an added advantage, allowing for much wider applications than HRP.

12.6.4 *Thermostability of the Purified RBHR POD*

Stability of an enzyme at above ambient temperatures is one of the important requirements for its commercial feasibility. The thermostability of the crude enzyme as a function of three different temperatures during a period of 45 min showed either increase or decrease in activity. The PODs extracted at acidic and neutral pH showed negligible inactivation up to 50°C with almost 95% of the activity being retained even after 40 min. However, the POD of basic pH was very sensitive to temperature, with 50% loss at 50°C in 40 min, and total loss of activity at 60°C (Thimmaraju et al. 2005). At the latter temperature (60°C), the acidic and neutral PODs retained more than 70% of the activity for up to 40 min, with complete inactivation at 70°C. Nevertheless, the purified enzyme was more stable compared with its crude state, as it retained more than 70% activity at 70°C even after 20 min, whereas commercial HRP had lost most of its activity within 11 min at 70°C (Thimmaraju et al. 2007), as also observed in other studies (Duarte-Vazquez et al. 2001). Therefore, the purified intracellular RBHR POD reported in this communication showed properties either on par or better than commercial HRP.

12.7 Cultivation of Hairy Roots in Bioreactor for POD Production

Pre-requisites for considering the cultivation of different types of plant cell and organ cultures in bioreactor are based on data from shake-flask experiments, where growth and product formation occur concomitantly and continuously for certain periods. Several reports on cell and organ cultures, including roots, indicated that metabolites are suppressed in a bioreactor when compared with their performance in shake flasks (Bhagyalakshmi et al. 1998; Kim et al. 2002). After stringent screening of different clones of RBHR, Thimmaraju et al. (2005) selected clone LMG-150 for growing in a bioreactor, based on their earlier studies. This clone was found to perform well even in the bioreactor, producing POD at levels comparable to those in shake flasks (Table 12.2). Roots grown in a bubble column reactor for 10 days produced almost the same amount of total POD as observed in the shake flask, except that there was a slight reduction in the quantity of extracellular POD. Total POD of about 9,000 U/g fresh weight was produced in the bubble column reactor

Table 12.2 Activities (U/g FW) of different peroxidase fractions in hairy root clone LMG-150 grown in shake flask and in bubble column reactor, and in the spent medium of each system

	Total POD activity in biomass (U/g FW)		Total activity in the medium (% of the total)	
Peroxidase fraction	Shake flask (30 mL culture of 15 d)	Bioreactor (15 d)	Shake flask (30 mL culture of 15 d)	Bioreactor (15 d)
Acidic (pH 4.0)	48	13.8	20	13.4
pH 6.0	10,000	9,932.0	18	5.9
Neutral (pH 7.0)	1,000	3,495.1	43	0.4
Basic (pH 9.0)	450	4,754.3	24	4.9

(Table 12.2), although with slight suppression of biomass, leading to an insignificant lower productivity. Interestingly, the release of enzyme into the cell exterior was very limited in the bioreactor, except for at acidic pH. Contrarily, there are other organized systems grown in bubble column bioreactors, such as the somatic embryos of sandal wood, that release very high levels of PODs when grown in bioreactors (Pal et al. 2003).

The observation that nearly half of the POD is secreted into the medium by RBHR in shake flasks indicates the possibility of permeabilizing the roots for enhanced efflux of the enzyme, even in a bioreactor, as done for the pigments from the same roots (Thimmaraju et al. 2003a, b, 2004; Neelwarne and Thimmaraju 2009). Several other unit operations for scale-up of hairy root biomass have also been worked out and, therefore, RBHR appears very promising for the production of this expensive enzyme of high commercial utility.

12.8 Known and Presumed Applications of POD

Many of the potential applications suggested for commercial HRP may also be proposed for RBHR PODs. Some prospective applications for RBHR PODs are discussed in the following sections.

12.8.1 Reagent in Clinical Diagnostics

The POD enzyme is a key component of diagnostic kits used to evaluate numerous viral, bacterial and parasitic diseases, including AIDS and malaria. POD from cell cultures was tested for its application as a clinical diagnostic reagent. The crude extract of POD was used to measure uric acid for the reaction described by Nathan and Joan (1971):

$$\text{Uric acid} \xrightarrow{\text{Uricase}} \text{Alantoin} + CO_2 + H_2O_2$$

$$H_2O_2 + 3\text{ - methyl - }2 - 2\text{ - benzothiazoline} + \text{N}'\text{N}'\text{ dimethylamine} \xrightarrow{\text{POD}} \text{indamin pigments} + H_2O.$$

The increase in absorbance at 590 nm due to indamin could easily be quantified spectrophotometrically. POD derived from N5K6-S cells was found to give the same reaction curves as that of commercially available product at all the levels of activities examined, both with respect to the time taken to reach a plateau of absorbance at 590 nm and the maximum absorbance achieved (Yamada et al. 1987). Thus, despite their differences in isozyme patterns reported by several workers (van Huystee 1987), the POD from cultured cells and hairy roots could be used as an important reagent in clinical diagnostics, and it appears that cultured plant cells and organs serve as an excellent alternative source for the commercial production of consistently high-quality POD. The possibility of specific activity enhancement of isolated POD to several folds by ferulic acid, caffeic acid and *p*-coumaric acid (Prabha and Patwardhan 1986) indicate an additional advantage for the economic utilization of the enzyme in cell-free systems.

Lactose present in milk was traditionally estimated by various analytical methods requiring special technical skills and sophisticated instruments, such as spectrophotometer, biosensors, etc. A simple and economical biostrip for the detection/estimation of lactose was developed where β-galactosidase, galactose oxidase and HRP were immobilized onto a polymeric support. β-Galactosidase is the key enzyme, when the biostrip is dipped into milk, it develops color by oxidation of the added chromogen, with the ability to detect lactose at levels <20–100 g/L (Sharma et al. 2002). The ability of POD to oxidize various substrates to yield colored products at low concentrations and its relatively good stability makes it an enzyme of choice for the development of diagnostic kits. In human blood, cholesterol level varies in response to various disease conditions, necessitating the development of instant diagnostic kits. Ragland et al. (2000) developed an accurate, enzymatic and colorimetric kit containing cholesterol oxidase, cholesterol esterase and POD for the precise detection of blood cholesterol. But the enzyme stability and the high cost of ingredients restricted its use for routine applications. Employing the same enzymes and immobilizing them individually resulted in high selectivity and increased stability, and a kit was developed where the cholesterol ester was hydrolyzed first by cholesterol esterase to free fatty acid cholesterol that undergoes further oxidation of cholesterol oxidase to cholestenone and H_2O_2. The H_2O_2 could then be determined using HRP following the reaction shown below, where the resulting quinoneimine is measured at 520 nm, and forms the basis for quantifying cholesterol:

$$4\text{ - aminophenazone} + H_2O_2 + \text{Phenol} \xrightarrow{\text{Peroxidase}} \text{p - (benzoquinone)monominephenazone.}$$

Apart from these, HRP has been used in several diagnostic kits (Table 12.3), such as for the detection of the cystic fibrosis delta F508 mutation in blood based on enzyme immunoassay. That kit consists of an enzyme-linked HRP secondary antibody based on an antidouble-stranded DNA monoclonal antibody (Hopfer et al.

Table 12.3 Different types of POD-based diagnostic kits, their principles and applications

Analyte/toxin	Source/sample	Type of assay	Detection limit	References
Dengue virus	Human serum	Immunoblots	0.1 ng	Young (1989)
Streptomycin, dihydrostreptomycin	Milk	Double antibody solid phase enzyme immunoassay	0.4 and 0.6 ng/ml	Schnappinger et al. (1993)
Fumonisin B1	Corn-based foods	Competitive enzyme immunoassay	0.17 ng/ml	Usleber et al. (1994)
Zearalenone toxins	Cereals	Direct competitive ELISA	25 ng/g	Barna-Vetró et al. (1994)
Ochratoxin A	Barley/corn/soybean	Direct competitive ELISA	0.5 ng/g	Schnappinger et al. (1993)
	Corn/soybean/green coffee	ELISA	2–40 ppb	Zheng et al. (2005)
T2-toxin	Wheat	Flow through enzyme immunoassay	50 ng/g	De Saeger and Van Pethegem (1996)
Ochratoxin A	Wheat	Direct competitive ELISA	0.4 ng/ml	De Saeger and Van Pethegem. 1999
Alkaline phosphatase	Milk	Competitive indirect ELISA	0.5 μg/ml	Vega-Warner et al. (2000)
Gossypol	Cotton seed	Non-competitive ELISA	250 μg/ml	Wang and Plhak (2000)
T2-toxin	Maize, wheat, rye, barley	Direct enzyme immunoassay	0.1 ng/ml	Sibanda et al. (2000)
Hepatitis-E virus	Human serum	ELISA	–	Zhuang et al. (2001)
Gonyautoxins	Shellfish tissue/ dinoflagellates	Direct competitive ELISA	0.15–22.3 ng/ml	Kawatsu et al. (2002)

1995). Another assay kit based on a bi-specific antibody to hTNF-alpha and an HRP solid-phase enzyme immunoassay was developed to measure human tumor necrosis factor alpha (hTNF-alpha) for use in clinical research, and has a detection sensitivity of 1 ng/mL (Berkova et al. 1996). HRP and indole acetic acid (a plant hormone) have been used in gene-directed therapy, which represents an efficient system for enzyme/pro-drug-based anticancer approach (Greco et al. 2001). A HRP-based diagnostic kit has been regularly used for the detection of 8-hydroxydeoxyguanosine and its analogs in the urine to identify prostate and bladder cancer risks (Chiou et al. 2003).

12.8.2 POD as a Cell Marker

Cell labeling is one of the important techniques used for the study of the nervous system by neurologists. Traditionally a Golgi stain has been used for such studies. Recently, the utility of POD has been gaining importance mainly because of the formation of dense precipitate by the POD reaction product inside the neurons, which can be seen in both light and electron microscopy (Mesulam 1982). Since Kristensson and Olsson (1971) first used HRP to label neurons, the method has found large applications in retrograde tracing of neuronal terminal connections in the peripheral nervous system. Later, La vail and La vail (1972) demonstrated the efficacy of the use of HRP to label neurons in the central nervous system and worked out a further course of elucidation of the relevant cell biology. In general, the enzyme POD is used in three ways by neurobiologists. The most common use is as a retrograde label; in this approach, POD is placed on or near the injured axons, retrograde axonal transport then carries the enzyme molecules to cell bodies, which are made visible by catalysis of an appropriate substrate, such as diaminobenzidine. This simple but powerful approach shows the location of nerve cells with their ramifications within a tissue region, allowing easy tracking of neuronal projections. The other variant of this approach is the uptake of the enzyme by intact axon terminals, which can be augmented by electrical stimulation and then distinguish the active terminals (Holtzman et al. 1971). The enzyme can also be introduced near an injured nerve cell, and is transported distally by anterograde axon transport after it is taken up by the cell, probably by endocytosis. This approach can reveal cellular details and extents of projections as well as the anatomy of the terminal branching. POD can also be introduced by micro-injection, permitting the intracellular labeling. Thus, POD has been a very promising biomolecule that offers a number of elegant applications not only in neurobiological studies but also for other developmental studies in medicine.

12.8.3 Chemiluminescent POD Substrates

Similar to ELISA, it is possible to track many in vivo and in vitro reactions by POD-linked substrates, which allow linking luminescent markers. Accordingly, many in vitro microplate-based chemiluminescence ELISA methods are becoming popular due to their increased sensitivity over conventional colorimetric methods, reaction speed and simplified procedures, where no reagent is required for stopping the reaction (see Table 12.3). Several new chemiluminescent HRP substrates for ELISA are available, with extended linear detection ranges and long glow times; some with detection levels up to the picogram level, and some enable femtogram-level sensitivity. Along similar lines, reagent substrates are continuously being explored and marketed for fast in situ hybridization (FISH) analyses for in vivo tracking of biochemical reactions.

12.8.4 POD-Based Biosensors

A biosensor is an analytical device comprising a biological recognition element directly interfaced to a signal transducer, which collectively relates the concentration of a single analyte or a group of anylates to a measurable response, such as change in voltage potential, which is further measured as quantum of change in current. Since Clark Jr (1962) proposed the concept of an enzyme-based biosensor, significant progress has been made in the field of enzyme-based biosensors involving POD, particularly for the development of POD-based electrodes targeting the detection of hydro peroxides and hydrogen peroxide (H_2O_2) as well as for use in conjunction with H_2O_2-producing oxidases for measuring the oxidase substrate, such as glucose, alcohols, glutamate and choline (Ruzgas et al. 1996; Jia et al. 2002). As explained earlier, the first catalytic cycle of POD involves reaction of the active site with H_2O_2 as shown in the following equation:

$$Fe^{3+} + H_2O_2 \longrightarrow ((Fe^4 = O) - R^{+\bullet})\text{Compound I}^+. \qquad (12.1)$$

Compound I further oxidizes to produce a substrate radical and compound II, which is reduced by a second substrate molecule and regenerates the native enzyme molecule (ferric). When an electrode substitutes for the electron donating substrate, the process is called direct electron transfer (Freire et al. 2003). Enzyme immobilized on an electrode is oxidized by H_2O_2 and then gets reduced by the electrons donated by the electrode:

$$((Fe^4 = O) - R^{+\bullet})\text{Compound I} + 2e^- + 2H^+ \longrightarrow Fe^3 + H_2O. \qquad (12.2)$$

When an electron donor (A) is included in a POD-electrode system, the direct process occurs simultaneously, with the reduction of the oxidized donor A□ (Liu and Ju 2002) by the electrode:

$$A^\bullet + e^- + H^+ \longrightarrow AH. \qquad (12.3)$$

In this process, electrons act as a second substrate for the enzyme reaction causing a changed electrode potential with the measured current proportionate to H_2O_2 concentration (Everse et al. 1991). This process of communication of POD with

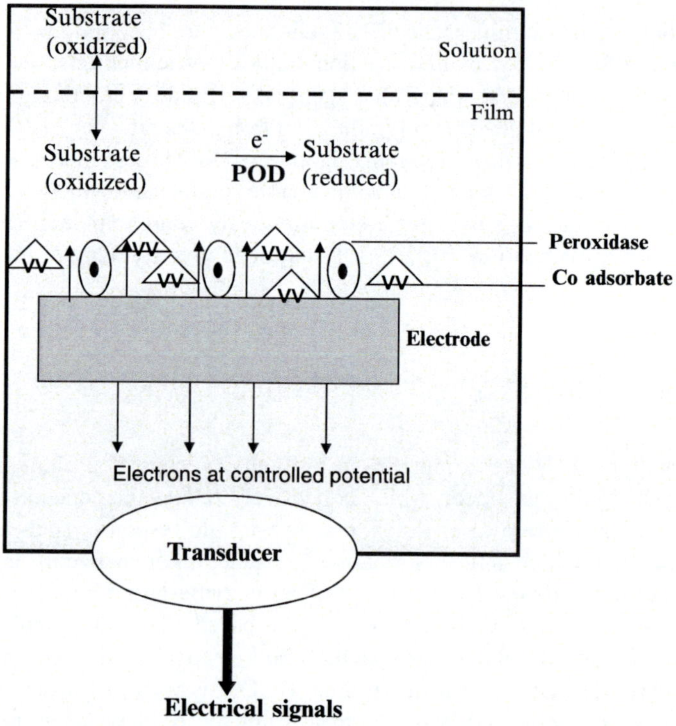

Fig. 12.10 Proposed scheme for a POD-based biosensor, based on the principle of direct electron transfer showing continuous generation of regulated electron potential. The latter, in turn, is detected after converting into electrical signal displayed digitally

electrode in a biosensor is known as exchange of electrons with the electrode. A hypothetical model of a POD biosensor based on this principle using protein film voltammetry is shown in Fig. 12.10. An alternative approach where a mediator (an electron donor, A) transfers the electrons between the enzyme and electrode (Ruzgas et al. 1996), is known as mediated electron transfer. In this system, the enzymatically oxidized donor (A□) is reduced electrochemically by the electrode. However, when H_2O_2 and an aromatic electron, for example, p-cresol or hydroquinone on catechol, is present, both direct and mediated electron transfer occur simultaneously (Lindgren et al. 2000b). This phenomenon has been successfully utilized for the development of a POD biosensor for the determination of phenols and aromatic amines at levels as low as the nanomolar range (Ruzgas et al. 1995; Lindgren et al. 1997; Munteanu et al. 1998). Three plant PODs, horseradish (HRP), peanut (PNP) and sweet potato (SPP), were compared for electrochemically determined rate constants. The biochemical characteristics for the above native PODs that were adsorbed on graphite electrodes in this test showed that PNP behaved almost the same as HRP and possessed superior electrochemical properties on the graphite electrode (Lindgren et al. 1997). This observation offers great promise for studying other

PODs for many applications. Likewise, POD can also be successfully employed for the development of optical biosensors where the principle involves the detection of light absorption between the reactants and products of a reaction by using rather low technology colorimetric techniques, such as disposable single-use cellulose strips impregnated with reagents. The well-known example for this technology is the assay for blood monitoring in diabetics, which involves a strip containing glucose oxidase, HRP and a POD substrate (e.g., *O*-toluidine or 3,3′5,5′-tetramethylbenzidine). In such a strip, the H_2O_2 produced by the oxidation of glucose further oxidizes the weakly colored chromogenic dye to colored dye as shown in the following reaction mechanism:

$$\text{Chromogenic substrate} + H_2O_2 \xrightarrow{\text{Peroxidase}} \text{Dye} + H_2O.$$

Based on the above-mentioned principles, a number of POD-based biosensors have been developed. Table 12.4 summarizes different POD biosensors, their principles and applications. Although a number of studies have been done to develop an HRP-based biosensor, the main limitation is the sluggish electron transfer rate of HRP; therefore biomimetic films have been first developed to achieve the goal. Quasi-reversible electron transfer was observed when HRP was incorporated in DDAB, DMPC, DHP ionomer poly (ester sulfonic acid), Eastman AQ29, etc. (Zhuang et al. 2001; Huang and Hu 2001). Inorganic clay, such as kieselguhr clay (Fan et al. 2001), has also been shown to be useful for fabricating HRP-based biosensors. In addition, gold nanoparticles and carbon nanotubes are other promising materials (Liu and Ju 2002; Jia et al. 2002). However, site-directed mutagenesis of HRP and exploring other electrochemically efficient PODs from plant sources may efficiently address this issue.

12.8.5 Reagent in Enzymatic Synthesis of Organic Compounds

POD is one of the few enzymes used for the catalysis of organic reactions; the enzyme brings about a number of chemical transformations by virtue of the free radicals produced during the reaction. Creation of new functional polymers was investigated by using oxidoreductases in oxidative polymerization of aromatic compounds, where a new set of phenolic resins was produced with good chemo selectivity (Hutterman et al. 2001), which otherwise requires the use of toxic formaldehyde for chemical polymerization. High yields of DMF soluble polyphenols (See Fig. 12.11 for illustration) were obtained by performing POD-catalyzed oxidative polymerization of phenols in the presence of H_2O_2 in an aqueous alcohol solvent system (Oguchi et al. 2000). Highly regio- and chemo-selective polymerization was observed, leading to the synthesis of polyphenols with mixed structures of phenelene and oxyphenelenes upon subjecting various phenols to oxidative polymerization by isolated enzymes such as POD under mild reaction conditions (Uyama and Kobayashi 2003). These useful and functional polymers are often difficult to synthesize by conventional methodologies. Further enzymatic curing of urushiol

Table 12.4 Different types of POD-based biosensors, their principles and applications

Enzyme used	Type of Electrode	Type of Biosensor	Application	Reference
HRP immobilized on TiO_2	Pyrolytic graphite	–	Electrochemistry and bioelectrocatalysis	Zhang et al. (2004)
Heme of HRP	Direct electro transfer	–	Electron transfer kinetics	Lindgren et al. (2001)
Peroxidase	Graphite–Teflon–peroxidase composite electrode	Amperometric	Detection of 18 phenolic compounds	Serra et al. (2001)
Cellobiose dehydrogenase and POD	–	–	Quantitation of phenolics and diphenolics	Lindgren et al. (2000a)
Sweet potato peroxidase	Rotating disc electrode	–	Detection of H_2O_2	Lindgren et al. (2000a)
HRP	Ferrocene-conjugated m-phenylenediamine-conducting polymer	Amperometric	H_2O_2 and organic peroxides	Mulchandani and Pan (1999)
HRP	*N*-methyl phenazine mediated	–	Detection of H_2O_2	Wang and Dordick (1998)
HRP–glucose oxidase–biotin	Hydrogel electrode	–	Glucose	Vreeke and Rocca (1996)
HRP	Carbon paste electrode	–	Quantitative and qualitative detection of oxidizable drugs	Erdem et al. (2000)
HRP–glucose oxidase	Ferrocene-embedded carbon paste electrode	Amperometric	Glucose	Matsumoto et al. (2002)
HRP	Glassy carbon electrode	Amperometric	H_2O_2	Mulchandani and Pan (1999)
HRP	SnO_2 electrode	Amperometric	H_2O_2	Thanachasai et al. (2002)

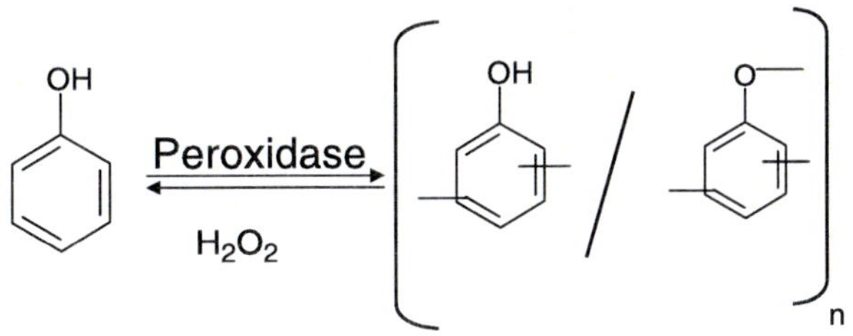

Fig. 12.11 Reaction scheme showing POD-catalyzed synthesis of polyphenols

analogs produced cross-linked polymeric films (artificial urushi), which possessed good elasticity and hardness (Uyama and Kobayashi 2003).

Cardanol used as a raw material in the production of resins and friction linings is obtained by thermal distillation process from cashew nut shell liquid and is a phenol derivative with a C15 unsaturated alkyl chain with one to three double bonds at its meta position (Ikeda et al. 2000). It is obtained by a soybean POD-mediated oxidative polymerization process in ethanol, 2-propanol, t-butyl alcohol, or 1–4 dioxane. However, a higher yield of about 62% was obtained only when 2-propanol was used as a solvent (Kim et al. 2003).

The area of conducting polymers is highly demanding and has attracted the interest of many researchers because of their wide range of applications, including applications in anticorrosive protection, optical display, light emitting diodes, etc. (Raitman et al. 2002). Polyaniline is one of the most extensively studied polymers because of its high environmental stability and highly promising electronic properties. Conventionally, polyaniline is synthesized by the oxidation of the monomer aniline under highly acidic condition at low temperature using ammonium persulphate as the initiator of radical polymerization (Rannau et al. 1998). This chemical method of synthesis of polyanilines has a number of disadvantages, such as that the reaction is radical catalyzed, hence it is not kinetically regulated and the reaction is carried out at acidic pH, thus it is not environmentally safe. Therefore enzymatic synthesis offers an attractive strategy for the synthesis of polyanilines. Synthesis of polyelectrolyte complex was achieved by using horseradish POD-mediated process (Lui et al. 1999). However, the HRP showed low catalytic activity towards aniline and was not stable at pH below 4.5 (Chottopadhyay and Mazumdar 2000). This disadvantage of HRP can be overcome by exploring the vast plant kingdom for acid tolerant PODs. An anionic POD was isolated and purified from African oil palm tree was reported to be stable at acidic pH. This enzyme was further used for the enzymatic polymerization of polyaniline–sulfonated polystyrene complex. The polymerization reaction was carried out in aqueous buffer of pH 3.5 and the synthesis of electro active polyaniline and polystyrene complex was confirmed by visible spectroscopy and electron paramagnetic resonance (EPR). The effect of aniline concentration on the reaction was studied further; it was found that even at higher concentration of aniline, the enzyme was not inhibited; whereas at a higher concentration of H_2O_2 (20 mM), the aniline polymerization was stopped (Sakharov et al. 2003).

Lignin is the second most abundant biopolymer on the earth after cellulose and has potential applications, particularly in the use as a template in the synthesis of phenolic polymers. Some of the processes, such as syntheses of polymeric dispersants, soil conditioning agents, adhesives or phenolic resins and laminates could use lignin as template. Several methods have been employed for the production of lignin-containing phenolic resins (Tock et al. 1987; Matt and Doucet 1988). However most of these methods use formaldehyde for hydroxymethylation of lignin and/or phenol to cross-link the polymer. Alternatively, the ability of HRP to catalyze the polymerization of phenols in presence of H_2O_2 (Nicell and Wright 1997) could be conveniently employed for the production of lignin-containing phenolic resins. Co-polymerization of lignin with cresol to produce a copolymer of mean molecular weight of about 1,890 KDa was successfully achieved by using a POD-mediated reaction in a reverse micellar system. Here the mean molecular weight of the product can be manipulated by using the surfactant as a design variable. The resultant copolymer had properties quite different from those of the native lignin, including a lower glass transition temperature (Liu et al. 1999).

A novel phenolic polymer containing thymidine pendent group was produced in a coupled enzymatic reaction. The regioselective acylation of thymidine was achieved in nearly anhydrous CH_3CN, using the lipase from *Candida antarctica*, which was followed by polymerization of the phenolic nucleoside derivative catalyzed by soybean POD (SBP). This is a novel reaction where polynucleosides with unnatural polymeric backbones were produced, which may provide for highly selective and stable materials for therapeutic, diagnostic and materials applications (Wang and Dordick 1998).

Enzymatic cross-linking of biopolymers, such as proteins, including collagens, yields semisolid gels (Prochaska et al. 2003) that have application as biocompatible wound sealants, as delivery vehicles or as binding agents in food product applications. A 3,4-dihydroxyphenylalanine (DOPA)-PEG hydrogel formation was studied by using HRP/H_2O_2 as the oxidizing system, this resulted in the production of fast-drying hydrogels (Prochaska et al. 2003). Electroenzymatic reactions used for enzymatic synthesis utilize electrons as the cheap reagents. The advantages of enzymes as enantioselective catalysts are combined here in the reaction cascades with waste-free electrochemical in situ generation or regeneration of reagents. Such a reaction can be combined and electroenzymatic reactors can be developed; a model is shown in Fig. 12.12. The best example, as shown in Fig. 12.12, for an electroenzymatic reaction is the oxidation of thioanisol, where the oxidant hydrogen peroxide is produced in situ by cathodic reduction of oxygen. The enantioselective oxidation is carried out by a POD such as chloroperoxidase (CPO). In this type of a set up, the stability of the enzyme can be increased significantly and the reactor can be further scaled up.

The group transfer catalytic property of the peroxidase enzyme can be used for synthetic applications. Some of the group transfer reactions catalyzed by peroxidases include heteroatom oxidations, oxidation of C–H bonds in allylic/benzylic compounds, alcohols and indoles and epoxidation. The reactions are mediated by peroxigenase rather than a peroxidase mechanism. HRP has been proved ineffective as a catalyst for the simple epoxidation of some compounds. Although the peroxidases have one-electron oxidation activities, their oxygen transfer activity is low, which reflects the limited substrate access to the ferryl oxygen. The inaccessibility

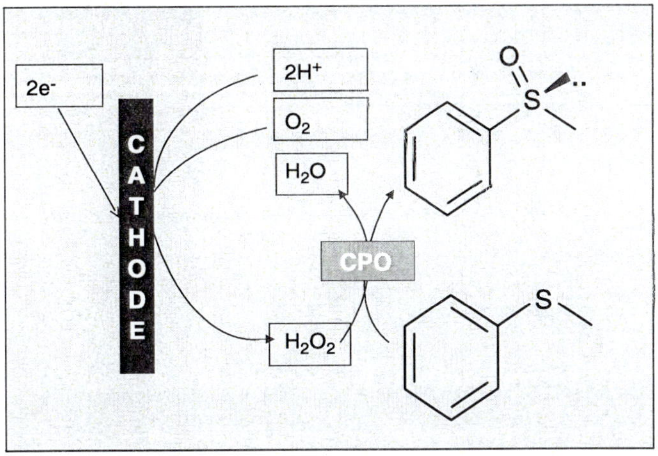

Fig. 12.12 Schematic diagram showing an electroenzymatic reactor with chloroperoxidase (CPO) as an example where the oxidation of thioanisol is coupled with in situ production of the oxidant hydrogen peroxide by cathodic reduction of oxygen

of the ferryl species in the enzyme was overcome by engineering the HRP by site-directed mutagenesis to replace Phe41 by smaller amino acids (Ozaki and Ortiz de Montellano 1995), such as leucine and threonine. When studied for catalytic properties, both mutants showed epoxidation properties and catalyzed the epoxidation of styrene and *cis*-β-methylstyrene, yielding *cis* and *trans* epoxides (Ozaki and Ortiz de Montellano 1995).

Using enzymes such as POD in the synthesis of pharmaceutically important organic compounds and intermediaries is an attractive solution to overcome the chirality problems of chemical synthesis. The unique properties, such as substrate specificity, stereospecificity, regiospecificity, low energy consumption, lack of by-products, lack of toxicity, and reduction of steps in synthesis make enzymatic catalysis an appealing method to the pharmaceutical industry. Coupling of catharanthine and vindoline has been successfully achieved to yield 3′,4′-anhydrovinblastine, which is a metabolic precursor of the anticancer compounds vinblastine and vincristine, where the synthesis of latter compounds involves a peroxidase-mediated reaction (Sottomayor et al. 1998). Further characterization of the reaction revealed the involvement of the H_2O_2-dependent enzyme, presumably a POD of *Catharanthus roseus*. This characterization has unraveled the possibility of a semi-synthetic step in the production of such important anticancer compounds.

12.8.6 Biotransformation and Coupled Enzyme Assays

With enzyme availability through commercially feasible routes such as cell cultures, the opportunities to perform biotransformation studies and subsequent scale up to industrial level are immense. The POD secreted into the culture media by cells

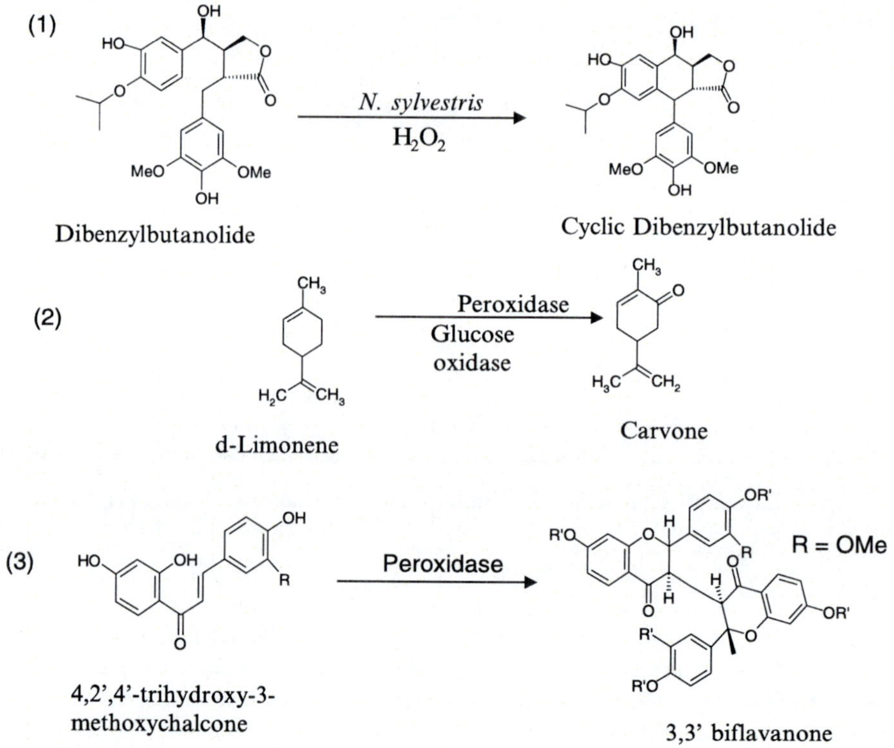

Fig. 12.13 Peroxidase enzyme-mediated biotransformation of dibenzylbutanolide (*1*), D-limonene (*2*) and 4,2′,4′-trihydroxy-3-methoxychalcone (*3*)

of *N. sylvestris* has been used for the biotransformation of dibenzylbutanolide precursors (Fig. 12.13 (*1*)) to the desired cyclic compounds, where about 87% yield of the desired compound has been recorded (Kutney 1998). Similarly, the dithiane derivative of piperonal was transformed to 4′-dimethyl epi-podophyllotoxin and its C-1 epimer. These and other such examples clearly show the possibility of altogether new routes for biotransformations, where PODs from plant cell cultures could be conveniently employed in gram-scale production of desired products, such as podophyllotoxins (Kutney 1998). One of the best examples of coupled enzyme assays has been the oxidation of aromatic hydrocarbons by HRP and H_2O_2 complex (Fang and Barcelona 2003). A novel method for the biotransformation of D-limonene to carvone (Fig. 12.13 (*2*)) using POD and glucose oxidase was developed; maximum yield occurred when 1.5% substrate at 50°C at pH 7.0 was used (Trytek and Fiedurek 2002). Hruby et al. (1997) successfully used HRP to transform benfluron, a potentially cytotoxic drug, to *N*-demethylated benfluron (demB). The study proved that benfluron, the cytostatic drug, can be metabolized extra-hepatically by a *N*-demethylation reaction by PODs, proving helpful in reversing hepatotoxicity. Peroxidase isozymes from the cell cultures of *Cassia didimobotrya* were used for biotransformation of chalcones such as 4,2′,4′-trihydroxy-3-methoxy chalcone in the presence of H_2O_2 as a cofactor; this study revealed that the compound was biotransformed to biflavanone when analyzed using TLC. Further analysis with NMR revealed the crude product mixture to be a mixture of two compounds (Vitali et al. 1998) (Fig. 12.13 (*3*)). These examples clearly indicate enormous applications for red beet POD for cost-effective synthesis of useful compounds.

12.8.7 POD in Bioremediation

Chemical contamination of soil and ground water is alarmingly increasing and bioremedial treatment approaches are difficult because of the large volumes. Alternative approaches such as phytoremediation are preferred, however, have disadvantages of slow remediation (Madsen et al. 1991). Making plant PODs available at cheaper costs through cell and organ cultures could be helpful. POD-based oxidation of organic compounds such as diesel, fuels and polycyclic aromatic hydrocarbons have been well documented (Torres et al. 1997; Ayala et al. 2000). Recently Fang and Barcelona (2003) reported the oxidation of *o*-xylene-d_{10} and naphthalene-d_8 by a POD–H_2O_2-coupled complex. Similarly efforts have been made to develop a process for the degradation of TNT (2,4,6-trinitrotoluene), which is toxic to all life forms, and presents a serious environmental hazard generated enormously due to chemical war near and around military establishments. Mineralization of TNT could be achieved by PODs; various stages involved in TNT degradation and the possible stages at which POD can be employed in the bioremediation process are shown in Fig. 12.14. A novel method for enzymatic biotransformation of limonene to carvone was developed, which involves addition of glucose oxidase and POD to the biotransformation medium; the maximum yield of carvone occurred in the medium containing 1.5% substrate, at 50°C and pH 7.0. Several lignin-degrading fungi were found to release POD (Barr and Aust 1994), along with other enzymes, allowing for the degradation of phenyl compounds and other polycyclic aromatic compounds found in soil (Lamar and Dietrich 1990; Davis et al. 1993). It was further demonstrated that these fungi were also able to degrade a number of xenobiotic toxicants, such as aromatic hydrocarbons, alkyl halide insecticides, nitrogen aromatics such as 2,4-dinitrotoluene and TNT, and several other compounds. It was demonstrated that lignin POD and manganese POD from *P. chrysosporium* were able to decolorize olive mill wastewater (Sami 1995; Novotny et al. 2000). HRP was shown to effectively degrade azo dyes and help in their precipitation, which appears useful in the treatment of paper industry effluent (Regalado et al. 2004). Although fungal systems are mostly studied as model systems for bioremediation, POD from cultured plant cells and organs could also be a commercially viable alternative, owing to the ease with which POD can be produced in large quantities. All these studies offer a great promise for the application of POD for bioremediation of the contaminated

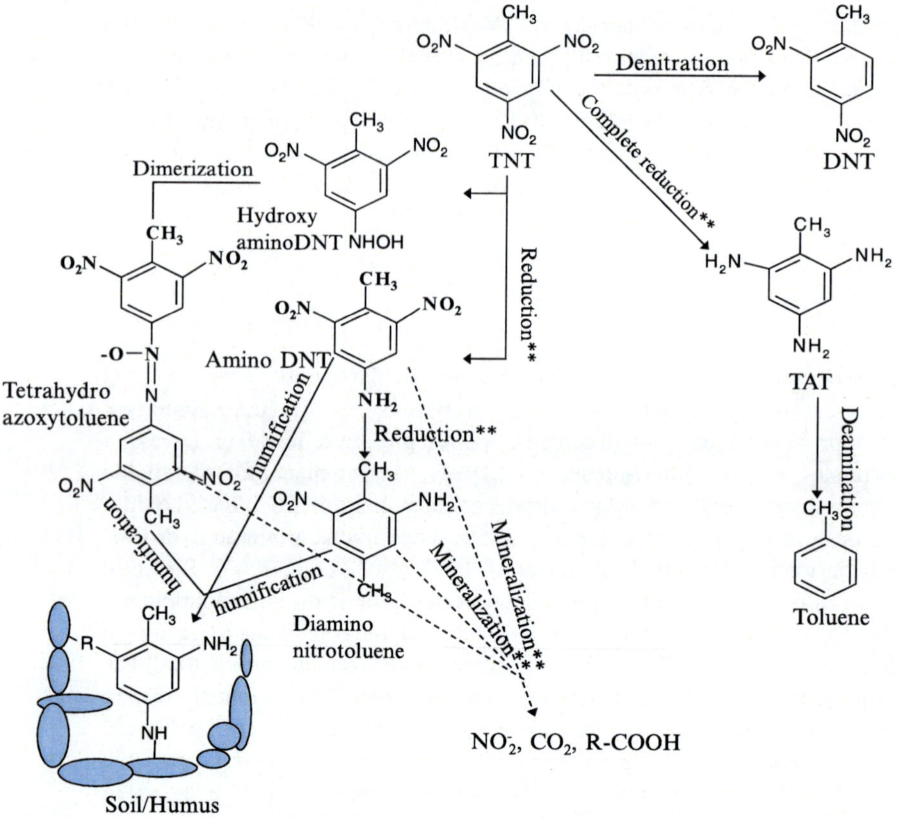

Fig. 12.14 TNT degradation and mineralization showing the possible steps of (**) involvement of POD

sites and aquifers, such as treating chlorinated soils and removal of aviation fuel hydrocarbons (mostly aromatics) and pesticides.

12.8.8 Biobleaching and Biopulping

Pulping is a process where the lignin in the wood is hydrolyzed, which results in about 10% of lignin modification in the wood, imparting a characteristic brown color to the wood. Modified lignin is further degraded in a process called bleaching. The processes may be chemical, microbial or enzymatic; if they are either microbial or enzymatic, they are called biobleaching and biopulping. Traditionally the white rot fungi are known to attack lignin and simultaneously convert the components of wood into carbon dioxide and water (Arana et al. 2002). Some of them selectively and efficiently degrade lignin rather than cellulose or hemicellulose (Li 2003). However there are several drawbacks in the use of microorganisms for breaking down lignocellulosic materials and cellulose fibers (Jimenez et al. 1997), where the inherent slower rate of degradation is a hindrance (Katagiri et al. 1995). Alternatively, application of lignin-degrading enzymes for the process of biobleaching and biopulping would not only hasten the process but also improve the quality of bleaching. Major lignin-degrading enzymes such as MnP (manganese peroxidase), laccase and LiP (Lignin peroxidase) have been reported in the fungi of Basidiomycetes (Hatakka et al. 2002). MnP, in the presence of Mn(II) chelated with an organic acid Tween-80, and a H_2O_2-generating system, was capable of depolymerizing milled pinewood (Hatakka et al. 2003; Maciel et al. 2010). The key element identified in the oxidation is Mn (III), a strong oxidizing agent, which is generated by MnP (Feijoo et al. 2002). Enzymatic delignification of wood pulp was effective and good results were obtained when LiP was used in the absence of peroxide, especially when the enzyme was chemically modified to avoid its binding to the pulp. The enzymatic process of wood pulping significantly reduced the amount of NaOH used in the alkaline pulping process (Poonpairoj et al. 2001). All these studies have clearly indicated that the use of lignin-degrading enzymes in the process of pulping and bleaching offer a number of advantages. Identification of such enzymes from plant sources and their in vitro production in large quantities is expected to further economize the process.

12.8.9 Treatment of Wastewaters

Traditionally, industrial wastewaters are treated to remove the harmful chemical contaminants in two ways: chemically and biologically. Recent research focuses on the application of pure enzymes isolated from the parent organisms for such a treatment. The newer enzymatic system falls between the two traditional categories because it involves chemical reactions based on biological catalysts. Isolated enzymes are preferred over parent organisms because of their greater specificity, of their amenability for better standardization, they are easier to handle and store and their concentration is not dependent on growth of the parental organism. Therefore, enzyme-based methods are being developed as an alternative strategy for industrial wastewater treatment where isolated enzymes, either in free form in solution or in immobilized form, are used (Yu et al. 1994). In some cases, enzymatic methods of analysis are developed to complement the spectroscopic methods that are used in process development and modeling. Enzymatic treatment of a strong oil refinery wastewater was investigated using hydrogen peroxide. Phenolic compounds in the refinery wastewater were enzymatically converted to colored polymeric products, which were subsequently removed by coagulation with alum. Although the chemical and biological oxygen demands were reduced in the wastewater by the enzymatic treatment and subsequent coagulation, the dissolved organic materials in the wastewater were apparently not affected by either process and tended to remain in the treated wastewater (Ikehata et al. 2003). HRP has been successfully used for the phenol conversion and dimerization in industrial wastewater (Yu et al. 1994; Lai and Lin 2005).

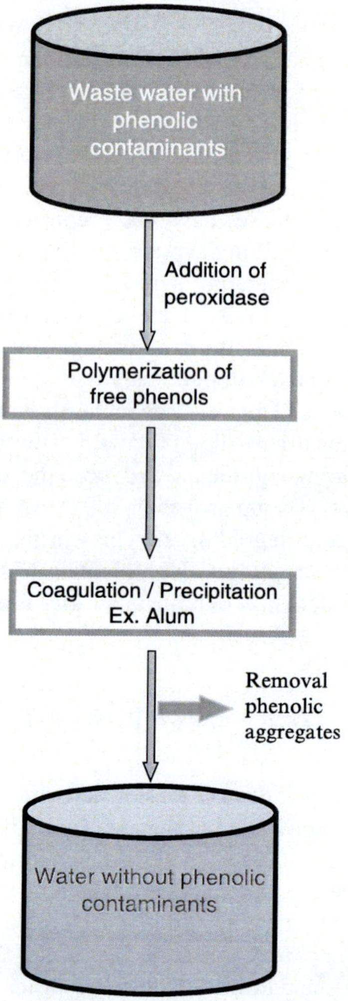

Fig. 12.15 Process for the removal of phenolics from contaminated wastewaters

Nicell et al. (1993) developed a reactor for POD-catalyzed polymerization and precipitation of phenol from wastewater. A schematic representation of the entire process of a POD-based polymerization and removal of phenolic contaminants from water is shown in Fig. 12.15. A process for the removal of 2,4-dichlorophenol (2,4-DCP) by oxidation in the presence of H_2O_2 and soybean peroxidase enzyme has been developed. It was found that the optimum pH of 8.2 was necessary for the removal of 2,4-DCP without the addition of PEG. However, addition of PEG increased the effectiveness of SBP by 10 and 50 times for PEG-3350 and PEG-8000 respectively. In this process, a highest 2,4-DCP removal of about 83.5% was achieved (Kennedy et al. 2002).

Although direct use of red beet hairy root cultures for 2,4-DCP has not been demonstrated, such accomplishments are imperative based on the observations made in other hairy root systems. For example, many attempts through inoffensive technologies have been made to remove 2,4-DCP, a harmful compound for aquatic life and human health (Petroutsos et al. 2008). Tomato hairy roots (González et al. 2008) and a selected clone of tobacco hairy roots was used for 2,4-DCP phytoremediation assays (Talano et al. 2010). These cultures removed 2,4-DCP in a short time and with high respective efficiencies of 98%, 88% and 83%, from solutions containing 250, 500 and 1,000 mg/L, respectively, where the removal process was associated mainly with POD activity. The process had an added advantage of re-using the hairy roots for almost three consecutive cycles. These findings might also have implications in the decontamination process of polluted waters.

A POD-mediated advanced oxidation strategy for the removal of nonylphenol (NP) was studied in which the optimum conditions for enzyme activity including enzyme concentration were studied and about 95% removal of NP was achieved (Wagner et al. 2002). Apart from wastewater treatment, POD has applications in the removal of specific chemicals from dilute mixtures that are hard to remove through conventional methods.

Red beet seedling roots with their high POD activities were able to convert the flavoring precursors of green vanilla bean to the respective volatile compounds, which reduced the length of curing period (Sreedhar et al. 2009). Beans thus cured showed a better flavor sensory profile with high flavor notes of vanilla, sweet, floral and low woody, beany and smoky notes, as these feature mark the good quality characteristics of natural vanilla (Hariom et al. 2006).

12.9 Other Enzymes from RBHR

12.9.1 Polyphenol Oxidase (PPO)

The common presence of polyphenol oxidases (PPO), which oxidize plant phenolic compounds, resulting in the formation of undesirable browning of fruit and vegetable products and causing off-flavor, is reported frequently in red beet (Escribano et al. 2002). As generally accepted, PPOs are described as catechol oxidase (EC 1.10.3.1) (which oxidize *o*-diphenols to quinones, the catecholase activity) and hydroxylates (monophenols to *o*-diphenols, the cresolase activity), the laccase (EC 1.10.3.2), which oxidizes both *o*- and *p*-diphenols and tyrosinase (EC 1.14.18.1), which catalyzes the formation of L-DOPA and DOPA-quinone, as shown in the following reaction:

$$\text{L - Tyrosine} + \text{L - DOPA} + O_2 = \text{L - DOPA} + \text{DOPA - qinone} + H_2O.$$

PPO is often used as an indicator of process completion, while processing fruit and vegetables, for organic synthesis, ascorbic acid determination, sugar beet pectin gelation and as a biosensor (Polaina and MacCabe 2007).

Similar to PODs, PPOs have several types that catalyze the oxidation of phenolic and such aromatic compounds using molecular oxygen (Mason and Wasserman 1987), and hence are highly useful biocatalysts for various biotechnological applications, particularly in the food industry, the pulp and paper industry, the textile industry and medicine and environmental technology, in a manner almost similar to those of POD. Although the browning effects caused by PPOs present in raw food materials is considered an undesirable feature, isolated PPOs are largely useful for the removal of phenolics in wines, and removal of haze formation and turbidity in beer and fruit juices. In baking industries, PPO (particularly the laccase) is used to enhance cross-linked biopolymers in the dough (Couto and Herrera 2006). PPOs are useful for the biosynthesis of antioxidants and food colorants particularly in the color formation and flavor enhancement of tea, cocoa and coffee (Simsek and Yemenicioglu 2007). In the paper industry, PPO replaces chlorine used for bleaching, and enhances the delignification process (Couto and Herrera 2006).

Although PODs have been suggested for the removal of hazardous phenolic compounds and their derivatives from wastewaters, because of their ability to function over wide ranges of pH and temperature, their stoichiometric requirements of hydrogen peroxide have often been difficult to meet. In this respect, PPOs appear more advantageous because of their requirement of only molecular oxygen to function (Edwards et al. 1999), which is easier to meet.

Another area of importance for the application of PPOs is in the production of L-DOPA, which is used in the treatment of Parkinson's disease (PD). PD is a common age-related neuronal disorder, particularly of central nervous system, caused by a deficiency of the neurotransmitter compound dopamine. L-3,4-Dihydroxyphenylalanine (L-DOPA) is a precursor of dopamine that is known to pass across the blood–brain barrier, and hence it has been used as a drug for the treatment for PD. Many hundred tons of L-DOPA is synthetically produced each year. This chemical process requires an expensive metal catalyst, with low conversion rates and low enantioselectivity.

By the action of PPO (specifically tyrosinase), L-tyrosine is converted to L-DOPA (Fig. 12.16) that is used to supplement the insufficient amount of natural dopamine for PD treatment (Asanuma et al. 2003; Xu et al. 1998). A novel technology for the production of L-DOPA through an electroenzymatic synthesis was demonstrated using a tyrosinase-immobilized cathode of –530 mV, having the potential to reduce DOPA-quinone. When compared with many previously known methods for L-DOPA synthesis, the new electroenzymatic approach showed the highest conversion rate (95.9%), resulting in a highly enhanced productivity of 47.27 mg/L/h (Min et al. 2010). Since tyrosinase is involved in betalain biosynthesis, several clones of RBHR rich in betalains could serve as sources for this enzyme. While using a tumor-suppressing agent during the treatment of vitiligo, an autoimmune disease, PPOs are used as markers for a pro-drug therapy, which aids in deciding the dose and effectiveness of the drug (Seo et al. 2003).

PPOs find additional applications in other fields of food industry. They can be used in the development of biosensors for immunoassays, for the detection of phenols and phenolic compounds in wastewaters, food and beverage (Duran and

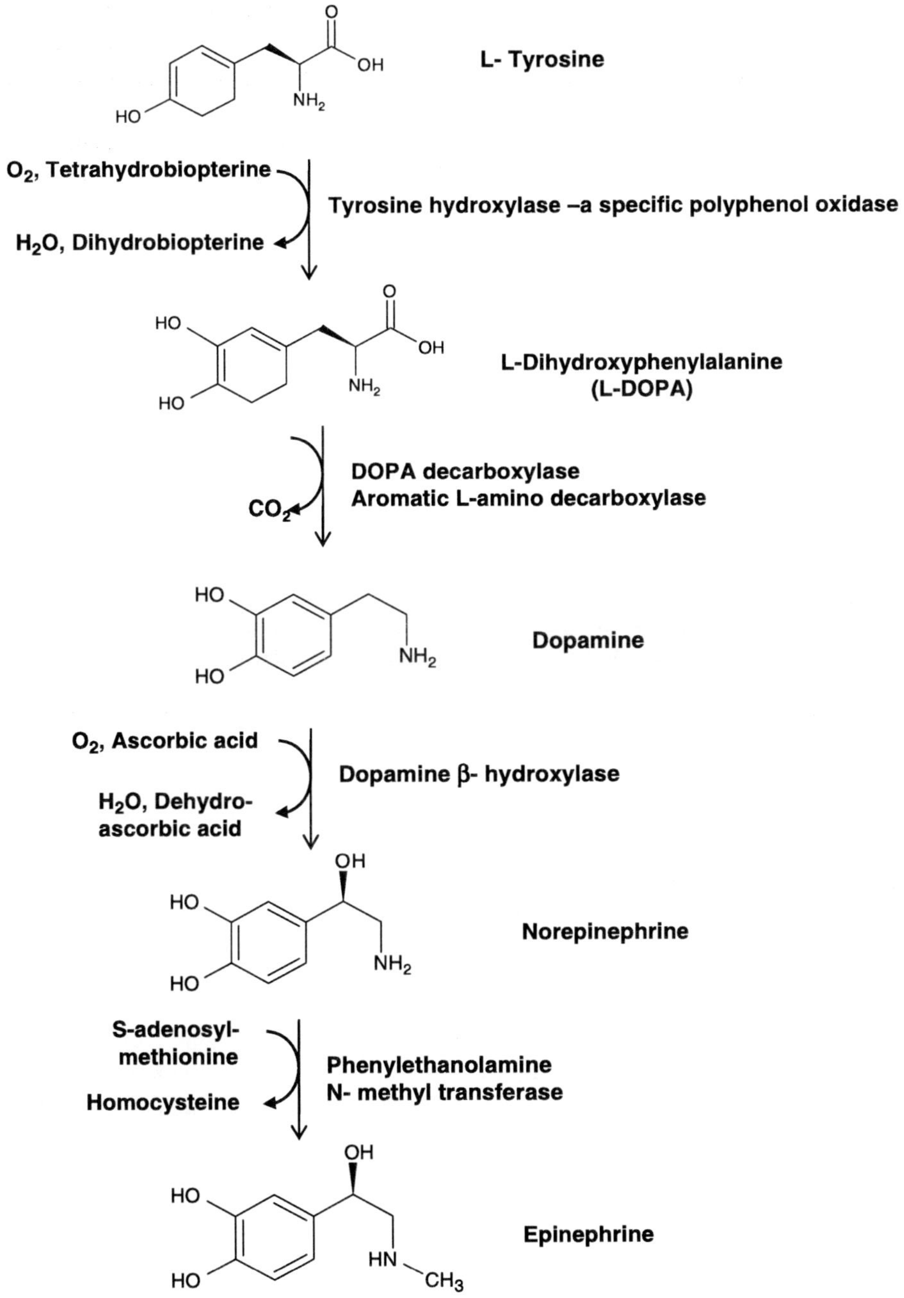

Fig. 12.16 Synthesis of L-DOPA, which is further converted into epineprine

Esposito 2000), for the detection of morphine, codeine and catecholamines. In cosmetics, some hair dyes and dermatological skin lightning preparations are based on laccase. In the textile industry, PPOs are used for denim bleaching and dye decolorization (Couto and Herrera 2006).

Glucosyltransferases (EC 2.4.1) isolated from cultured cells of red beet were found to perform flavonoid-specific glucose transfers. Therefore, these enzymes were of great importance in the glycosylation of low molecular weight compounds, thus increasing their solubility. Such enzymes find use in solubilization of xenobiotics, assisting their removal from the cells, and are proposed as regioselective biocatalysts (Lim et al. 2002; Isayenkova et al. 2006).

12.9.2 Antioxidant Enzymes from RBHR

Among the widespread types of cellular biochemical stress, the oxidative burst, induced by various factors of biotic and abiotic origin, is considered to cause major cellular damages. Although the reactive oxygen species (ROS) generated during oxidative burst fulfill the task of triggering the defense-signaling network, they are also self-destructive, damaging cellular structures. Therefore, cells have evolved systems for protection from excess oxygen and other ROS by the antioxidant system, of which many are enzymes. Apart from POD, vacuoles isolated from the tap root of red beet were found to possess good activities of superoxide dismutase (SOD; EC 1.15.1.1) and catalase (CAT; EC 1.11.1.6). SOD (Cu and Zn-SOD) was in three isoforms, located in the sap of the vacuole, without having any association with the membrane (Pradedova et al. 2009, 2011). Activity was higher during the initiation and termination of dormancy than in deeply dormant tubers, indicating that physiologically active cells of red beet display more of this protein, which could be true even in hairy roots. However, SOD or CAT enzymes need to be characterized in RBHR for their commercial applications. Since SOD catalyzes the dismutation of superoxide into oxygen and hydrogen peroxide, it has application in cosmetic products, where it is expected to reduce free radical damage to skin and suppress fibrosis occurring in breast after radiation treatment for cancer (Vozenin-Brotons et al. 2001). SOD has also been proven to be highly effective in treatment of colonic inflammation in experimental colitis. Treatment with SOD was found to decrease the generation of ROS and oxidative stress, thus inhibiting endothelial activation, which indicates its effectiveness in the modulation of leukocyte–endothelial interactions. These properties and its strong antioxidant potential are important in developing new therapies for the treatment of inflammatory bowel diseases (Segui et al. 2004, 2005).

Pavlov et al. (2002, 2005) found strong antioxidant activities in extracts of RBHR of four different cultivars, where Detroit and cv. Detroit Dark Red were found to possess higher potential than cv. Bado and Egyptian (Pavlov et al. 2002). Such activities were linked to betalains, rather than the enzymes, as was confirmed in a later study (Lee et al. 2005).

12.10 Concluding Remarks

The present review has provided an overview of the uniqueness of the POD enzyme and its vast array of applications for various purposes. The great potential this enzyme holds is limited by the high cost of production from various other reported sources. Apart from very high level of activity, the stable productivity of POD realized in cultured RBHR appears very promising, creating an altogether new platform for the enzyme production. Several novel applications suggested for this enzyme, such as the synthesis of various aromatic chemicals and removal of peroxide from materials such as food stuffs and industrial wastes and treatment of waste water containing phenolic compounds, coupled with its key role in designing biosensors, make POD from cultured higher plants an important molecule for environment monitoring as well as for several biotechnological processes at industrial scale. However, before the realization of full potential of PODs, some of the issues such as development of low-cost sources such as that of red beet and RBHR and improving the efficacies of such enzymes under their application conditions need to be addressed for economic realization of such processes. Although large similarities between the characteristics of RBHR POD and HRP are found through recent studies, more characterization is required for elevating RBHR POD to the status of HRP. Many RBHR clones expressed high POD activities, as did the seedling roots of red beet, indicating that the rooty morphology is mainly responsible for the expression of the genes for POD, calling for genetic characterization of POD proteins. Such characterizations would allow the expression of relevant genes in heterologous systems for further characterization and in microbial systems for the large-scale production of such proteins.

References

Agostini, E., S. Milrad de Forchetti, and H.A. Tigier. 1997. Production of peroxidases by hairy roots of *Brassica napus*. *Plant Cell, Tissue and Organ Culture* 47: 177–182.

Agostini, E., J. Hernández-Ruiz, M.B. Arnao, S.R. Milrad, H.A. Tigier, and M. Acosta. 2000. A peroxidase isoenzyme secreted by turnip (*Brassica napus*) hairy-root cultures: inactivation by hydrogen peroxide and application in diagnostic kits. *Biotechnology and Applied Biochemistry* 35: 1–7.

Agostini, E., M.S. Coniglio, S.R. Milrad, H.A. Tigier, and A.M. Giulietti. 2003. Phytoremediation of 2,4-dichlorophenol by *Brassica napus* hairy root cultures. *Biotechnology and Applied Biochemistry* 37: 139–144.

Arana, A., A. Téllez, T. González, and A. González. 2002. Aspectos generales de la biodegradación de la madera: Aplicaciones industriales de las laccasas. *Biotechnologia* 7: 40–55.

Aruna, N., and A. Lali. 2001. Purification of a plant peroxidase using reversibly soluble ion exchange polymer. *Process Biochemistry* 37: 431–437.

Asanuma, M., I. Miyazaki, and N. Ogawa. 2003. Dopamine- or L-DOPAinduced neurotoxicity: the role of dopamine quinone formation and tyrosinase in a model of Parkinson's disease. *Neurotoxicity Research* 5: 165–176.

Ayala, M., N.R. Robledo, A. Lopez-Munguia, and R. Vazquez-Duhalt. 2000. Substrate specificity and ionization potential in chloroperoxidase-catalyzed oxidation of diesel fuel. *Environmental Science and Technology* 34: 2804–2809.

Bais, H.P., J. George, and G.A. Ravishankar. 1999. Influence of polyamines on growth and production of coumarins in hairy root cultures of *Cichorium intybus* L. cv Lucknow local (Witloof chicory). *Journal of Plant Growth Regulation* 18: 33–37.

Barr, D.P., and S.D. Aust. 1994. Mechanisms the white rot fungi use to degrade pollutants. *Environmental Science and Technology* 28: 79A–87A.

Barz, W., and J. Koster. 1990. Turnover and degradation of secondary (natural products). In *The Biochemistry of Plants*, vol. 7, ed. P.K. Stumpf and E.E. Conn, 35–84. New York: Academic.

Berkova, N., L. Karawajew, V. Korobko, O. Behrsing, B. Micheel, O. Shamorant, E. Stukatcheva, and L. Shingarova. 1996. Development of an enzyme immunoassay for the measurement of human tumor necrosis factor-alpha (hTNF-alpha) using bispecific antibodies to hTNF-alpha and horseradish peroxidase. *Biotechnology and Applied Biochemistry* 23: 163–171.

Bhagyalakshmi, N., and K. Bopanna. 1998. Elicitation and immobilization of plant cell cultures for the enhanced synthesis of natural compounds. In *Role of biotechnology in medicinal and aromatic plants*, ed. I.A. Khan and A. Khanum, 305–325. Hyderabad: Ukaaz Publication.

Bhagyalakshmi, N., L.V. Venkataraman, and G.A. Ravishankar. 1998. Bioprocess parameters for the scaled-up production of hairy root cultures. In *Tissue culture and molecular biology-problems and prospects*, ed. P.S. Srivastava, 332–358. New Delhi: Narosa Publishing House.

Bhagyalakshmi, N., R. Thimmaraju, and M.S. Narayan. 2004. Various hexoses and di-hexoses differently influence growth, morphology and pigment synthesis in transformed root cultures of red beet (*Beta vulgaris*). *Plant Cell, Tissue and Organ Culture* 78: 183–195.

Biles, C.L., F.B. Abeles, and C.L. Wilson. 1990. The role of ethylene in anthracnose of cucumber (*Cucumis sativus*) caused by *Colletotrichum lagernarum*. *Phytopathology* 80: 726–732.

Bolton, M.D. 2009. Primary metabolism and plant defense – Fuel for the fire. *Molecular Plant-Microbe Interactions* 22: 487–497.

Brownleader, M.D., N. Ahmed, M. Trevan, M.F. Chaplin, and P.M. Dey. 1995. Purification and partial characterization of tomato extensin peroxidase. *Plant Physiology* 109: 1115–1123.

Buchert, J., D. Ercili Cura, H. Ma, C. Gasparetti, E. Monogioudi, G. Faccio, M. Mattinen, H. Boer, R. Partanen, E. Selinheimo, R. Lantto, and K. Kruus. 2010. Crosslinking food proteins for improved functionality. *Annual Review of Food Science and Technology* 1: 113–138.

Buitelaar, R.M., M.T. Cesario, and J. Tramper. 1992. Elicitation of thiophene production by hairy roots of *Tagetes patula*. *Enzyme and Microbial Technology* 14: 2–7.

Calderon, A.A., E. Garcia-Florenciano, R. Munoz, and A. Ros-Barcelo. 1992. Gamay grapewine peroxidase: Its role in vacuolar anthocyani(di)n degradation. *Vitis* 31: 139–147.

Chen, S.X., and P. Schopfer. 1999. Hydroxyl-radical production in physiological reactions: A novel function of peroxidase. *European Journal of Biochemistry* 260: 726–735.

Chiou, C.C., P.Y. Chang, E.C. Chan, T.L. Wu, K.C. Tsao, and J.T. Wu. 2003. Urinary 8-dydroxyguanosine and its analogs as DNA marker of oxidative stress: Development of an ELISA and measurement in both bladder and prostate cancers. *Clinica Chimica Acta* 334: 87–94.

Chottopadhyay, K., and S. Mazumdar. 2000. Structural and conformational stability of horseradish peroxidase: Effect of temperature and pH. *Biochemistry* 9: 125–132.

Christensen, J.H., G. Bauw, K.G. Welinder, M.V. Montagu, and W. Bioerjan. 1998. Purification and characterization of peroxidases correlated with lignification in poplar xylem. *Plant Physiology* 118: 125–135.

Chung, S.Y., S.J. Maleki, and E.T. Champagne. 2004. Allergenic properties of roasted peanut allergens may be reduced by peroxidase. *Journal of Agricultural and Food Chemistry* 52: 4541–4545.

Civello, P.M., G.A. Martinez, A.R. Chaves, and M.C. Anon. 1995. Peroxidase from strawberry fruit (*Fragaria ananassa* Duch): Partial purification and determination of properties. *Journal of Agricultural and Food Chemistry* 43: 2596–2601.

Clark, L.C., and C. Lyons. 1962. Electrode systems for continuous monitoring in vascular surgery. *Annals of the New York Academy of Sciences* 102: 29–45.

Cosio, C., and C. Dunand. 2009. Specific functions of individual class III peroxidase genes. *Journal of Experimental Botany* 60: 391–408.

Couto, R.S., and T.J.L. Herrera. 2006. Industrial and biotechnological applications of laccases: A review. *Biotechnology Advances* 24: 500–513.

Davis, E. 1987. Plant responses to wounding. In *The biochemistry of plants*, vol. 12, ed. D.D. Davis, 243–264. San Diego: Academic.

Davis, M.W., J.A. Glaser, J.W. Evans, and R.T. Lamar. 1993. Field evaluation of lignin degrading fungus *Phanerochaete sordida* to treat creosote contaminated soil. *Environmental Science and Technology* 27: 2572–2576.

Dean, J.F.D., R. Sterjiades, and K.-E.L. Eriksson. 1994. Purification and characterization of an anionic peroxidase from sycamore maple (*Acer pseudoplatanus*) cell suspension cultures. *Physiologia Plantarum* 92: 233–240.

De Saeger, S., and C. Van Pethegem. 1996. Dipstick enzyme immunoassay to detect *Fusarium* T-2 toxin in wheat. *Applied and Environmental Microbiology* 62: 1880–1884.

De Saeger, S., and C. Van Pethegem. 1999. Flow-through membrane based enzyme-immunoassay for rapid detetion of ochratoxin A in wheat. *Journal of Food Protection* 62: 65–69.

Doran, P.M. 1997. *Hairy roots: Culture and applications*, 1–235. Amsterdam: Harwood Academic.

Duarte-Vazquez, M.A., B.E. Garcia-Almendarez, C. Regalado, and J.R. Whitaker. 2001. Purification and properties of a neutral peroxidase isozyme from turnip (*Brassica napus* L. Var. purple top white globe) roots. *Journal of Agricultural and Food Chemistry* 49: 4450–4455.

Duran, N., and E. Esposito. 2000. Potential applications of oxidative enzymes and phenoloxidase-like compounds in wastewater and soil treatment: A review. *Applied Catalysis B: Environmental* 28: 83–99.

Edwards, W., R. Bownes, W.D. Leukes, E.P. Jacobs, R. Sanderson, P.D. Rose, and S.G. Burton. 1999. A capillary membrane bioreactor using immobilized polyphenol oxidase for the removal of phenols from industrial effluents. *Enzyme and Microbial Technology* 24: 209–217.

Eilert, U., A. Ehmke, and B. Wolters. 1984. Elicitor induced accumulation of acridone alkaloid epoxides in *Ruta graveolens* suspension cultures. *Planta Medica* 50: 508–512.

Eilert, U., F. Constabel, and W.G.W. Kurz. 1986. Elicitor-stimulation of monoterpene indole alkaloid, in suspension cultures of *Catharanthus roseus*. *Journal of Plant Physiology* 126: 11–22.

Erdem, A., A. Pabuccuoglu, B. Meric, K. Kerman, and M. Ozsoz. 2000. Electrochemical biosensor based on horseradish peroxidase for the determination of oxidizable drugs. *Turkish Journal of Medical Sciences* 30: 349–354.

Escribano, J., F. Gandia-Herrero, N. Cabellero, and M.A. Pedreno. 2002. Subcellular localization and isoenzyme pattern of peroxidase and polyphenol oxidase in beet root (*Beta vulgaris* L.). *Journal of Agricultural and Food Chemistry* 50: 6123–6129.

Everse, S.L., M.B. Everse, and M.B. Grisham. 1991. *Peroxidases in chemistry and biology*, vol. 2, 1. Boca Raton: CRC Press.

Fan, C., H. Wang, D. Zhu, G. Wagner, and G. Li. 2001. Incorporation of Horseradish peroxidase in a Kieselghur membrane and the application to a mediator-free hydrogen peroxide sensor. *Analytical Sciences* 17: 273–276.

Fang, J., and M.J. Barcelona. 2003. Coupled oxidation of aromatic hydrocarbons by horseradish peroxidase and hydrogen peroxide. *Chemosphere* 50: 105–109.

Feijoo, G., M.T. Moreira, and J.M. Lerna. 2002. Effect of Ca^{2+} in the delignification of kraft paste by *Phanerochaete chrysosporium*. *Afinidad* 59: 586–591.

Ferreres, F., R. Figueiredo, S. Bettencourt, I. Carqueijeiro, J. Oliveira, A. Gil-Izquierdo, D.M. Pereira, P. Valenta, P.B. Andrade, P. Durte, A.R. Barcelo, and M. Sottomayor. 2011. Identification of phenolic compounds in isolated vacuoles of the medicinal plant *Catharanthus roseus* and their interaction with vacuolar class III peroxidase: An H_2O_2 affair? *Journal of Experimental Botany*. doi:10.1093/jxb/erq458 P1-14.

Flocco, C.G., and A.M. Guilietti. 2003. Effect of chitosan on peroxidase activity and isozyme profile in hairy root cultures of *Armoracia lapathifolia*. *Applied Biochemistry and Biotechnology* 110: 175–183.

Flocco, C.G., M.A. Alvarez, and A.M. Guilietti. 1998. Peroxidase production in vitro by *Armoracia lapathifolia* (horseradish)-transformed root cultures: effect of elicitation on level and profile of isoenzymes. *Biotechnology and Applied Biochemistry* 28: 33–38.

Flores, H.E., and W.R. Curtis. 1992. Approaches to understanding and manipulating the biosynthetic potential. In *Biochemical engineering*, Cellular and reaction engineering, vol. VII, ed. H. Pederson, R. Mutharasan, and D. Di Biasio, 188–209. New York: New York Academy of Sciences.

Freire, R.S., C. Pessoa, L.D. Mello, and L.T. Kubota. 2003. Direct electron transfer: An approach for electrochemical biosensors with higher selectivity and sensitivity. *Journal of the Brazilian Chemical Society* 14: 230–243.

Funk, C., K. Gugler, and P. Brodelius. 1987. Increased secondary product formation in plant cell suspension cultures after treatment with a yeast carbohydrate preparation (Elicitor). *Phytochemistry* 26: 401–405.

Gasper, T., C. Kevers, C. Penel, and H. Greppin. 1983. Auxin control of calcium-mediated peroxidase secretion of auxin-independent sugar beet cells. *Phytochemistry* 22: 2657–2660.

Gazaryan, I.G., D.M. Khushpul'yan, and V.I. Tishkov. 2006. *Uspekhi Biologicheskoĭ Khimii* 46: 303–323.

Giri, A., and L. Narasu. 2000. Transformed hairy roots: Recent trends and applications. *Biotechnology Advances* 18: 1–22.

Gomez-Vasquez, R., R. Day, H. Buschmann, S. Randles, J.R. Beeching, and R.M. Cooper. 2004. Phenylpropanoids, phenylalanine ammonia lyase and peroxidases in elicitor – Challenged cassava (*Manihot esculenta*) suspension cells and leaves. *Annals of Botany* 94: 87–97.

González, P.S., E. Agostini, and S.R. Milrad. 2008. Comparison of the removal of 2,4-dichlorophenol and phenol from polluted water, by peroxidases from tomato hairy roots, and protective effect of polyethylene glycol. *Chemosphere* 70: 982–989.

Graham, T.L., and M.Y. Graham. 1999. Role of hypersensitive cell death in conditioning elicitation competency and defense potentiation. *Physiological and Molecular Plant Pathology* 55: 13–20.

Greco, O., S. Rossiter, C. Kanthou, L.K. Folkes, P. Wardman, G.M. Tozer, and G.U. Dachs. 2001. Horseradish peroxidase-mediated gene therapy: choice of pro-drugs in oxic and anoxic tumor conditions. *Molecular Cancer Therapeutics* 1: 151–160.

Guillon, S., J. Tremouillaux-Guiller, P.K. Pati, M. Rideau, and P. Gantet. 2006. Hairy root research: Recent scenario and exciting prospects. *Current Opinion in Plant Biology* 9: 341–346.

Gundlach, H., M.J. Muller, T.M. Kutchan, and M.H. Zenk. 1992. Jasmonic acid as a signal transducer in elicitor-induced plant cell cultures. *Proceedings of the National Academy of Sciences of the United States of America* 89: 2389–2393.

Hariom, B.N., M.P. Shyamala, and S.S. Bhat. 2006. Vanilla flavour evaluation by sensory and electronic nose techniques. *Journal of Sensory Studies* 21: 228–239.

Hatakka A, Hakala T, Lundell T, Horichter M, Maijala P (2002) Manganese peroxidase – The key enzyme in lignin biodegradation and biopulping by white-rot fungi. In *Abstracts of papers, 223rd ACS national meeting* (*CELL-028*). Orlando, April 7–11, Washington.

Hatakka, A., T. Lundell, M. Hofrichter, and P. Maijala. 2003. Manganese peroxidase and its role in the degradation of wood lignin. In *Applications of enzymes to lignocellulosics*, ACS symposium series, vol. 855, ed. S.D. Mansfield and J.N. Saddler, 230–243. Washington, DC: American Chemical Society.

Heidrich, E., G. Lorenz, and P. Schreier. 1983. Ultrathin – layer isoelectric focusing of partially purified peroxidase from tomato fruit. *Food Chemistry* 10: 285–296.

Hilhorst, R., H. Gruppen, R. Orsel, C. Laane, H.A. Schols, and A.G.J. Voragen. 2002. Effects of xylanase and peroxidase on soluble and insoluble arabinoxylans in wheat bread dough. *Journal of Food Science* 67(2): 497–506.

Holtzman, E., A.R. Freeman, and L.A. Kashner. 1971. Stimulation dependent alterations in peroxidase uptake at lobster neuromuscular junctions. *Science* 173: 733–736.

Hopfer, S.M., G.S. Makowsky, E.L. Davis, and J. Aslanzadeh. 1995. Detection of cystic fibrosis delta F508 mutation by anti-doublestranded DNA antibody. *Annals of Clinical and Laboratory Science* 25: 475–484.

Hruby, K., E. Anzenbacherova, P. Anzenbacher, and M. Nobilis. 1997. Potential cancerostatic benfluron is metabolized by peroxidase: In vitro biotransformation of benfluron by horseradish peroxidase. *General Physiology and Biophysics* 16: 321–327.

Huang, R., and N. Hu. 2001. Direct electrochemistry and electro-catalysis with horseradish peroxidase in Eastman AQ films. *Bioelectrochemistry* 54: 75–81.

Hutterman, A., C. Mai, and A. Kharazipour. 2001. Modification of lignin for the production of new compound materials. *Applied Microbiology and Biotechnology* 55: 387–394.
Ikeda, R., H. Tanaka, H. Uyama, and S. Kobayashi. 2000. A new crosslink-able polyphenol from a renewable resource. *Macromolecular Rapid Communication* 21: 496–499.
Ikehata, k, I.D. Buchanan, and D.W. Smith. 2003. Treatment of oil refinery wastewater using crude *Coprinus cinereus* peroxidase and hydrogen peroxide. *Journal of Environmental Engineering and Science* 2: 463–470.
Isayenkova, J., V. Wray, M. Nimtz, D. Strack, and T. Vogt. 2006. Cloning and functional characterisation of two regioselective flavonoid glucosyltransferases from *Beta vulgaris*. *Phytochemistry* 67: 1598–1612.
Isheeva, O.D., E.V. Pradedova, and R.K. Salyaev. 2009. Multiplicity of vacuolar peroxidase isozymes in higher plants by the example of red beet (*Beta vulgaris* L.) roots. *Doklady Biochemistry and Biophysics* 424: 53–55.
Jia, J., B. Wang, A. Wu, G. Cheng, Z. Li, and S. Dong. 2002. A method to construct a third-generation horseradish peroxidase biosensor: Self-assembling gold nanoparticles to three-dimensional sol-gel network. *Analytical Chemistry* 74: 2217–2223.
Jimenez, L., C. Martinez, E. Perez, and F. Lopez. 1997. Biobleaching procedures for pulp from agricultural residues using *Phanerochaete chrysosporium* and enzymes. *Process Biochemistry* 32: 297–304.
Karpen, J.W., and M.L. Ruiz. 2002. Ion channels: Does each subunit do something on its own? *Trends in Biochemical Sciences* 27: 402–409.
Katagiri, N., Y. Tsutsumi, and T. Nishida. 1995. Correlation of brightening with cumulative enzyme activity related to lignin biodegradation during biobleaching of kraft pulp by white rot fungi in the solid–state fermentation system. *Applied and Environmental Microbiology* 61: 617–622.
Kawatsu, K., Y. Hamano, A. Sugiyama, K. Hashizume, and T. Noguchi. 2002. Development and application of an enzyme immunoassay based on a monoclonal antibody against gonyautoxin components of paralytic shellfish poisoning toxins. *Journal of Food Protection* 65: 1304–1308.
Kennedy, K., K. Alemany, and M. Warith. 2002. Optimisation of soybean peroxidase treatment of 2,4-dichlorophenol. *Water SA* 28: 149–158.
Kim, Y., B.E. Wyslouzil, and P.J. Weathers. 2002. Secondary metabolism of hairy root cultures in bioreactors. *In Vitro Cellular and Development Biology – Plant* 38: 1–10.
Kim, Y.H., A.S. An, B.K. Song, D.S. Kim, and R. Chelikani. 2003. Polymerization of cardanol using soybean peroxidase and its potential application as anti-biofilm coating material. *Biotechnology Letters* 25: 1521–1524.
Kino-Oka, M., H. Nagatmoto, and M. Taya. 2001. Characterization and application of plant hairy roots endowed with photosynthetic functions. *Advances in Biochemical Engineering* 72: 186–218.
Klibanov, A.M., B.N. Alberti, E.D. Morris, and L.M. Felshin. 1980. Enzymatic removal of toxic phenols and anilines from waste waters. *Journal of Applied Biochemistry* 2: 414–421.
Kristensson, K., and Y. Olsson. 1971. Uptake and retrograde axonal transport of peroxidase in hypoglossal neurons. *Acta Neuropathologica* 19: 1–9.
Kutney, J.P. 1998. Biotechnology and synthetic chemistry - Routes to clinically important compounds. *Pure and Applied Chemistry* 70: 2093–2100.
La Vail, J.H., and M.M. La Vail. 1972. Retrograde axonal transport in the central nervous system. *Science* 176: 1416–1417.
Lai, Y.C., and S.C. Lin. 2005. Application of immobilized horseradish peroxidase for the removal of *p*-chlorophenol from aqueous solution. *Process Biochemistry* 40: 1167–1174.
Lamar, R.T., and D.M. Dietrich. 1990. In situ depletion of pentachlorophenol from contaminated soil by *Phanerochaete* spp. *Applied and Environmental Microbiology* 56: 3093–3100.
Lee, M.Y., and S.S. Kim. 1994. Characteristics of six isoperoxidases from Korean-radish root. *Phytochemistry* 35: 287–290.
Lee, C.-H., M. Wettasinghe, B.W. Bolling, L.-L. Ji, and K.L. Parkin. 2005. Betalains, phase II enzyme-inducing components from red beetroot (*Beta vulgaris* L.) extracts. *Nutrition and Cancer* 53: 91–103.

Lemcke, K., and T. Schmulling. 1998a. Putative *rolB* gene homologue of *Agrobacterium rhizogenes* TL-DNA that alter plant morphogenesis or hormone sensitivity. *The Plant Journal* 15: 423–434.

Lemcke, K., and T. Schmulling. 1998b. A putative *rolB* gene homologue of the *Agrobacterium rhizogenes* TR-DNA has different morphogenetic activity in tobacco than *rolB*. *Plant Molecular Biology* 36: 803–808.

Levine, A., R. Tenhaken, R. Dixon, and C. Lamb. 1994. H_2O_2 from the oxidative burst orchestrates the plant hypersensitive disease resistance response. *Cell* 79: 583–593.

Li, K. 2003. The role of enzymes and mediators in white-rot fungal degradation of lignocellulose. In *Wood deterioration and preservation*, ACS symposium series, vol. 45, ed. B. Goodell, D.D. Nicholas, and T.P. Schultz, 196–209. Washington, BC: American Chemical Society.

Lei, W., X. Shui, Y. Zhou, S. Tang, and M. Sun. 2011. Effects of Praseodymium on flavonoids production and its biochemical mechanism of *Scutellaria viscidula* hairy roots in vitro. *Pakistan Journal of Botany* 43: 2387–2390.

Lim, M.H., P.J. Velasco, R.M. Pangborn, and J.R. Whitaker. 1989. Enzymes involved in off-aroma formation in broccoli. In *Quality factors of fruits and vegetables chemistry and technology*, ed. J.J. Jen, 72–83. Washington, DC: American Chemical Society.

Lim, E.K., C.J. Doucet, Y. Li, L. Elias, D. Worrall, S.P. Spencer, J. Ross, and D.J. Bowles. 2002. Activity of the group I glycosyltransferases of *Arabidopsis* towards salicylic acid, parahydroxybenzoic acid and other benzoates. *Journal of Biological Chemistry* 277: 586–592.

Lindgren, A., J. Emnéus, T. Ruzgas, L. Gorton, and G. Marko-Varga. 1997. Amperometric detection of phenols using peroxidasemodified graphite electrodes. *Analytical Chimica Acta* 347: 51–62.

Lindgren, A., T. Ruzgas, L. Stoica, F. Munteanu, and L. Gorton. 2000a. Cellobiose dehydrogenase and peroxidasae biosensors for determination of phenolic compounds. In *Chemical and biological sensors for environmental monitoring*, ACS symposium series, vol. 762, ed. A. Mulchandani and O.A. Sadik, 113–124. Washington, DC: American Chemical Society.

Lindgren, A., T. Ruzgas, L. Gorton, E. Csoregi, A.G. Bautista, I.Y. Sakharov, and I.G. Azaryan. 2000b. Biosensors based on novel peroxidases with improved properties in direct and mediated electron transfer. *Biosensors and Bioelectronics* 15: 491–497.

Lindgren, A., T. Ruzgas, and L. Gorton. 2001. Direct electron transfer of native and modified peroxidases. *Current Topics in Analytical Chemistry* 2: 71–94.

Liu, S.Q., and H.X. Ju. 2002. Renewable reagentless hydrogen peroxide sensor-based on direct electron transfer horseradish peroxidase immobilized on colloidal gold-modified electrode. *Analytical Biochemistry* 307: 110–116.

Liu, J., Y. Weiping, and T. Lo. 1999. Copolymerization of lignin with cresol catalyzed by peroxidase in reversed micellar systems. *Electronic Journal of Biotechnology* 2: 82–87.

Lui, W., J. Kumar, S. Tripathy, K.J. Senecal, and L. Samuelson. 1999. Enzymatically synthesized conducting polyaniline. *Journal of the American Chemical Society* 121: 71–78.

Liu, Z.-H., H.Y. Liu, and H.Y. Wang. 1996. Effect of light on endogenous indole-3-acetic acid, peroxidase and indole-3-acetic-acid oxidase in soybean hypocotyls. *Botanica Bulletin Academica Sincia* 37: 113–119.

Maciel, M.J.M., A.C. Silva, and H.C.T. Ribeiro. 2010. Industrial and biotechnological applications of ligninolytic enzymes of the basidiomycota: A review. *Electronic Journal of Biotechnology* 13. doi:10.2225/vol13-issue6-fulltext-2.

Madsen, E.L., J.L. Sinclair, and W.C. Ghiorse. 1991. Determining in situ biodegradation. *Science* 252: 830–833.

Mason, T.L., and B.P. Wasserman. 1987. Inactivation of red beet root β-glucan synthase by native and oxidized phenolic compounds. *Phytochemistry* 26: 2197–2202.

Matheis, G., and J.R. Whitaker. 1984. Peroxidase-catalyzed cross linking of proteins. *Journal of Protein Chemistry* 3: 35–48.

Matsumoto, R., M. Mochizuki, K. Kano, and T. Ikeda. 2002. Unusual response in mediated biosensors with an oxidase/peroxidase bi-enzyme system. *Analytical Chemistry* 74: 3297–3303.

Matt, J.-F., and J. Doucet. 1988. Production of phenolic resins using lignin. *Cellulose Chemistry and Technology* 22: 71–78.

Mesulam, M.-M. (ed.). 1982. *Tracing neural connections with horseradish peroxidase*. New York: Wiley.

Min, K., D.H. Park, and Y.J. Yoo. 2010. Electroenzymatic synthesis of l-DOPA. *Journal of Biotechnology* 146: 40–44.

Mulchandani, A., and S. Pan. 1999. Ferrocene-conjugated *m*-phenylenediamine conducting polymer-incorporated peroxidase biosensors. *Analytical Biochemistry* 267: 141–147.

Munteanu, F.D., A. Lindgren, J. Emneus, L. Gorton, T. Ruzgas, E. Csoregi, A. Ciucu, R.B. van Hyustee, I.G. Gazaryan, and L.M. Lagrimini. 1998. *Analytical Chemistry* 70: 2596–2600.

Murashige, T., and F. Skoog. 1962. A revised medium for rapid growth and bioassays with tobacco tissue culture. *Physiologia Plantarum* 15: 473–497.

Nair, A.R., and A.M. Showalter. 1996. Purification and characterization of wound inducible cell wall cationic peroxidases from carrot roots. *Biochemical and Biophysical Research Communications* 226: 254–260.

Nakashimada, Y., N. Uozumi, and T. Kobayashi. 1994. Stimulation of emergence of root apical meristems in horseradish hairy root by auxin supplementation and its kinetic model. *Journal of Fermentation and Bioengineering* 77: 178–182.

Narayan, A.V., M.C. Madhusudhan, and K.S.M.S. Raghavarao. 2008. Extraction and purification of Ipomoea peroxidase employing three-phase partitioning. *Applied Biochemistry and Biotechnology* 151: 263–272.

Nathan, G., and S.M. Joan. 1971. Automated determination of uric acid with use of a uricase-peroxidase system. *Clinical Chemistry* 17: 1154–1159.

Neelwarne, B., and R. Thimmaraju. 2009. Bioreactor for cultivation of red beet hairy roots and in situ recovery of primary and secondary metabolites. *Engineering in Life Sciences* 9: 227–238.

Neumann, K.H., A. Kumar, and J. Imani. (2009) *Plant Cell and Tissue Culture – A Tool in Biotechnology and Tissue Culture: Basics and Application*. Berlin: Springer.

Nicell, J.A., and H. Wright. 1997. A model of peroxidase activity with inhibition by hydrogen peroxide. *Enzyme and Microbial Technology* 21: 302–310.

Nicell, J.A., J.K. Bewtra, N. Biswas, C.C. St. Pierre, and K.E. Taylor. 1993. Enzyme catalyzed polymerization and precipitation of aromatic compounds from aqueous solution. *Canadian Journal of Civil Engineering* 20: 725–735.

Nishi, A. 1994. Effect of elicitors on the production of secondary metabolites. In *Advances in plant biotechnology*, ed. D.D.Y. Ryu and S. Furasaki, 135–151. Amsterdam: Elsevier Science.

Novotny, C., P. Erbanova, T. Cajthaml, N. Rothschild, C. Dosoretz, and V. Sasek. 2000. *Irpex lacteus*, a white rot fungus applicable to water and soil bioremediation. *Applied Microbiology and Biotechnology* 54: 850–853.

Oguchi, T., S.-I. Tawaki, H. Uyama, and S. Kobayashi. 2000. Enzymatic synthesis of soluble polyphenol. *Bulletin of the Chemical Society of Japan* 73: 1389–1396.

Ozaki, S., and P.R. Ortiz de Montellano. 1995. Molecular engineering of horseradish peroxidase: Thioether sulfoxidation and styrene epoxidation by Phe-41 Leucine and threonine mutants. *Journal of the American Chemical Society* 117: 7056–7064.

Pal, S., S. Das, and S. Dey. 2003. Peroxidase and arabinogalactan protein as by-products during somatic embryo cultivation in airlift bioreactor. *Process Biochemistry* 38: 1471–1477.

Parkinson, M., T. Cotter, and P.J. Dix. 1990. Peroxidase production by cell suspension and hairy root cultures of horseradish (*Armoracia rusticana*). *Plant Science* 66: 271–277.

Paul, K.G. 1986. Peroxidases: Historical background. In *Molecular and physiological aspects of plant peroxidases*, ed. H. Greppin, C. Penel, and T. Gaspar, 1–14. Geneva: University of Geneva.

Paul, K.G., and T. Stigbrand. 1970. Four isoperoxidases from horseradish root. *Acta Chemica Scandinavica* 24: 3607–3617.

Pavlov, A., P. Kovatcheva, V. Georgiev, I. Koleva, and M. Ilieva. 2002. Biosynthesis and radical scavenging activity of betalains during the cultivation of red beet (*Beta vulgaris*) hairy root cultures. *Zeitschrift für Naturforschung* 57C: 640–644.

Pavlov, A., P. Kovatcheva, D. Tuneva, M. Ilieva, and T. Bley. 2005. Radical scavenging activity and stability of betalains from *Beta vulgaris* hairy root culture in simulated conditions of human gastrointestinal tract. *Plant Foods for Human Nutrition* 60: 43–47.

Perera, R.M., and M.G.K. Jones. 2004. Expression of the peroxidase gene promoter (*Shpx6b*) from *Stylosanthes humilis* in transgenic plants during insect attack. *Entomologia Experimentalis et Applicata* 111: 165–171.

Petroutsos, D., P. Katapodis, M. Samiotaki, G. Panayotou, and D. Kekos. 2008. Detoxification of 2,4-dichlorophenol by the marine microalga *Tetraselmis marina*. *Phytochemistry* 69: 707–714.

Pitta-Alvarez, S.I., T.C. Spollansky, and A.M. Giulietti. 2000. The influence of different biotic and abiotic elicitors on the production and profile of tropane alkaloids in hairy root cultures of *Brugmansia candida*. *Enzyme and Microbial Technology* 26: 252–258.

Polaina, J., and A.P. MacCabe. 2007. *Industrial enzymes: structure, function and applications*. Dordrecht: Springer.

Poonpairoj, P., C. Peerapatsakul, and L. Chitadron. 2001. Trend in using fungal enzymes, lignin -and pectin-degrading enzymes, an improvement of the paper mulberry pulping process. *Biotechnology Sustainable Utilization Biological Research Tropicales* 15: 56–65.

Prabha, T.N., and M.V. Patwardhan. 1986. In vitro effect of polyphenols on some enzyme systems. *Acta Alimentaria* 15: 129–135.

Pradedova, E.V., O.D. Isheeva, and R.K. Salyaev. 2009. Superoxide dismutase of plant cell vacuoles. *Membrane and Cell Biology* 3: 24–32.

Pradedova, E.V., O.D. Isheeva, and R.K. Salyaev. 2011. Antioxidant defense enzymes in cell vacuoles of red beet roots. *Russian Journal of Plant Physiology* 58: 36–44.

Prochaska, J.F., J.T.Keeton., I.R.Tizard., and D.R.Miller. 2003. System for polymerizing collagen and collagen composites in situ for a tissue compatible wound sealant, delivery vehicle, binding agent and/or chemically modifiable matrix. US Patent No 6,509,031

Radman, R., T. Saez, C. Bucke, and T. Keshavaraz. 2003. Elicitation of plants and microbial cell systems. *Biotechnology and Applied Biochemistry* 37: 91–102.

Raitman, O.A., E. Katz, A.F. Buckmann, and I. Willner. 2002. Integration of polyaniline/poly (acryl acid) films and redox enzymes on electrode supports: An in situ electrochemical/surface plasmon resonance study of the bioelectrocatalyzed oxidation of glucose or lactate in the integrated bioelectrocatalytic systems. *Journal of the American Chemical Society* 124: 6487–6496.

Rannou, P., A. Gawlicka, D. Berner, A. Pron, and M. Nechtsschein. 1998. Spectroscopic, structural and transport properties of conducting polyaniline processed from fluorinated alcohols. *Macromolecules* 31: 1307–1315.

Regalado, C., B.E. Garcia Almandarez, and M.A. Duarte-Vazquez. 2004. Biotechnological applications of peroxidases. *Phytochemistry Reviews* 3: 243–256.

Ragland, B.D., R.J. Konrad, C. Chaffin, C.A. Robinson, and R.W. Hardy. 2000. Evaluation of a homogeneous direct LDL-cholesterol assay in diabetic patients: Effect of glycemic control. *Clinical Chemistry* 46: 1848–1851.

Richardson, A., and G.J. Mc Dougall. 1997. A laccase-like polyphenol oxidase from lignifying tobacco xylem. *Phytochemistry* 44: 229–235.

Richardson, A., D. Stewart, and G.J. Mc Dougall. 1997. Identification and partial characterization of a coniferyl alcohol oxidase from lignifying xylem of Sitka spruce (*Picea sitchensis*). *Planta* 203: 35–43.

Robins, M.P., G.P. Bollwell, and R.A. Dixon. 1985. Metabolic changes in elicitor-treated bean cells. Selectivity of enzyme induction in relation to phytoalexin accumulation. *European Journal of Biochemistry* 148: 563–569.

Ruzgas, T., J. Emneus, L. Gorton, and G. Marko-Verga. 1995. The development of a peroxidase biosensor for monitoring phenol and related aromatic compounds. *Analytica Chimica Acta* 311: 245–253.

Ruzgas, T., E. Csoregi, I. Katakis, G. Kenansis, and L. Gorton. 1996. Preliminary investigations of amperometric oligosaccharide dehydrogenase based electrode for the detection of glucose and some other low molecular weight sacharides. *Journal of Molecular Recognition* 9: 480–484.

Sakharov, I.Y., A.C. Vorobiev, and J.J. Castillo Leon. 2003. Synthesis of polyelectrolyte complexes of polyaniline and sulfonated polystyrene by palm tree peroxidase. *Enzyme and Microbial Technology* 33: 661–667.

Sami, R. 1995. Role of lignin peroxidase and manganese peroxidase from *Phanerochaete chrysosporium* in the decolorization of olive mill wastewater. *Applied and Environmental Microbiology* 61: 1098–1103.

Saunders, B.C., A.G. Holmes-Siedle, and B.P. Stark. 1964. *Peroxidase: The properties and uses of a versatile enzyme and some related catalysts*. London: Butterworths.

Savitha, B.C., R. Thimmaraju, N. Bhagyalakshmi, and G.A. Ravishankar. 2004. Different biotic and abiotic elicitors influence betalain production in hairy root cultures of *Beta vulgaris* in shake-flask and bioreactor. *Process Biochemistry* 41: 50–60.

Schnappinger, P., E. Usleber, E. Märtlbauer, and G. Terplan. 1993. Enzyme immunoassay for the detection of streptomycin and dihydrostreptomycin in milk. *Food and Agricultural Immunology* 5: 67–73.

Segel, I.H. (ed.). 1993. *Enzyme kinetics – Simple inhibition systems*, 100–160. New York: A Wiley Inter-Science Publication. ISBN 0-471-30309-7.

Seguí, J., M. Gironella, M. Sans, S. Granell, F. Gil, M. Gimeno, P. Coronel, J.M. Piqué, and J. Panés. 2004. Superoxide dismutase ameliorates TNBS-induced colitis by reducing oxidative stress, adhesion molecule expression, and leukocyte recruitment into the inflamed intestine. *Journal of Leukocyte Biology* 76: 537–544.

Segui, J., F. Gil, M. Gironella, M. Alvarez, M. Gimeno, P. Coronel, D. Closa, J.M. Pique, and J. Panes. 2005. Down-regulation of endothelial adhesion molecules and leukocyte adhesion by treatment with superoxide dismutase is beneficial in chronic immune experimental colitis. *Inflammatory Bowel Diseases* 11: 872–882.

Seo, S.Y., V.K. Sharma, and N. Sharma. 2003. Mushroom tyrosinase: Recent prospects. *Journal of Agricultural and Food Chemistry* 51: 2837–2853.

Serra, B., B. Benito, L. Agui, A.J. Reviejo, and J. Pingarron. 2001. Graphite-teflon-peroxidase composite electrochemical biosensors, a tool for the wide detection of phenolic compounds. *Electroanalysis* 13: 693–700.

Sessa, D.J., and R.L. Anderson. 1981. Soybean peroxidase purification and some properties. *Journal of Agricultural and Food Chemistry* 29: 960–965.

Sharma, A.K., N. Sehgal, and A. Kumar. 2002. A quick and simple biostrip technique for detection of lactose. *Biotechnology Letters* 24: 1737–1739.

Sibanda, L., S. De Saeger, and C. Van Pethegem. 2000. Detection of T-2 toxin in different cereals by flow-through enzyme immunoassay with a simultaneous internal reference. *Journal of Agricultural and Food Chemistry* 48: 5864–5867.

Sim, S.J., H.N. Chang, J.R. Liu, and K.H. Jung. 1994. Production and secretion of indole alkaloids in hairy root cultures of *Catharanthus roseus*: Effects of in situ adsorption, fungal elicitation and permeabilization. *Journal of Fermentation and Bioengineering* 3: 229–234.

Simsek, S., and A. Yemenicioglu. 2007. Partial purification and kinetic characterization of mushroom stem polyphenoloxidase and determination of its storage stability in different lyophilized forms. *Process Biochemistry* 42: 943–950.

Singh, G. 1999. Elicitation-Manipulating and enhancing secondary metabolite production. In *Plant cell and tissue culture for the production of food ingredients*, ed. T.J. Fu, G. Sing, and W. Curtis, 121–128. New York: Kluwer.

Sottomayor, M., M. López-Serrano, F. Di Cosmo, and A. Ros Barceló. 1998. Purification and characterization of á-3, 4-anhydrovinblastine synthase (peroxidase-like) from *Catharanthus roseus* (L.) G. Don. *FEBS Letters* 428: 299–303.

Sreedhar, R.V., K. Roohie, P. Maya, L. Venkatachalam, and N. Bhagyalakshmi. 2009. Biotic elicitors enhance flavour compounds during accelerated curing of vanilla beans. *Food Chemistry* 112: 461–468.

Srinivas, N.D., K.R. Rashmi, and K.S.M.S. Raghavarao. 1999. Extraction and purification of a plant peroxidase by aqueous two-phase extraction coupled with gel filtration. *Process Biochemistry* 35: 43–48.

Srivastava, O.M.P., and R.B. van Hyustee. 1977. IAA oxidase and polyphenol oxidase activities of peanut peroxidase isozymes. *Phytochemistry* 16: 1527–1530.

Sticher, L., C. Penel, and H. Greppin. 1981. Calcium requirement for secretion of peroxidase by plant cell suspensions. *Journal of Cell Science* 48: 345–353.

Suresh, B., R. Thimmaraju, N. Bhagyalakshmi, and G.A. Ravishankar. 2004. Polyamine and methyljasmonate influenced enhancement of betalain production in hairy root cultures of *Beta vulgaris* in a bubble column reactor and studies on efflux of pigments. *Process Biochemistry* 39: 2091–2096.

Takasaki, S., Y. Kato, M. Murata, S. Homma, and S. Kawakishi. 2005. Effects of peroxidase and hydrogen peroxide on the dityrosine formation and the mixing characteristics of wheat-flour dough. *Bioscience, Biotechnology, and Biochemistry* 69: 1686–1692.

Talano, M.A., S. Frontera, P. González, M.I. Medina, and E. Agostini. 2010. Removal of 2,4-diclorophenol from aqueous solutions using tobacco hairy root cultures. *Journal of Hazardous Materials* 176: 784–791.

Thanachasai, S., S. Rokutanzono, S. Yoshida, and T. Watanabe. 2002. Novel hydrogen peroxide sensors based on peroxidase-carrying poly{pyrrole-co-[4-(3-pyrrolyl)butane sulfonate]} copolymer films. *Analytical Sciences* 18: 773–777.

Thimmaraju, R., N. Bhagyalakshmi, M.S. Narayan, and G.A. Ravishankar. 2003a. Kinetics of pigment release from hairy root cultures of red beet under the influence of pH, sonication, temperature and oxygen stress. *Process Biochemistry* 38: 1067–1074.

Thimmaraju, R., N. Bhagyalaskshmi, M.S. Narayan, and G.A. Ravishankar. 2003b. Food grade chemical and biological agents permeabilize red beet hairy roots assisting the release of betalains. *Biotechnology Progress* 19: 1274–1282.

Thimmaraju, R., N. Bhagyalakshmi, and G.A. Ravishankar. 2004. In situ and ex situ adsorption and recovery of betalaines from hairy root cultures of *Beta vulgaris*. *Biotechnology Progress* 20: 777–785.

Thimmaraju, R., V. Kumar, L. Venkatachalam, N. Bhagyalakshmi, and G.A. Ravishankar. 2005. Peroxidase production from hairy root cultures of red beet (*Beta vulgaris*). *Electronic Journal of Biotechnology* 8: 185–196.

Thimmaraju, R., L. Venkatachalam, R.V. Sreedhar, N. Bhagyalakshmi, M.S. Narayan, and G.A. Ravishankar. 2006. Elicitation of peroxidase from hairy root cultures of red beet. *Electronic Journal of Biotechnology* 9: 512–521.

Thimmaraju, R., L. Venkatachalam, K. Roohie, M.S. Narayan, and N. Bhagyalakshmi. 2007. Purification and characterization of an intracellular peroxidase from genetically transformed roots of red beet (*Beta vulgaris* L.). *Food Chemistry* 105: 1312–1320.

Thimmaraju, R., L. Venkatachalam, and N. Bhagyalakshmi. 2008. Morphometric and biochemical characterization of red beet (*Beta vulgaris* L.) hairy roots obtained after single and double transformations. *Plant Cell Reports* 27: 1039–1052.

Tock, R.W., R.S.J. Chen, and C.R. Richardson. 1987. Reaction of lignin catalyzed by peroxidases. *Chemical Engineering Communications* 56: 229–251.

Torres, F., R. Tinoco, and R. Vazquez-Duhalt. 1997. Biocatalytic oxidation of polycyclic aromatic hydrocarbons in media containing organic solvents. *Water Science and Technology* 36: 37–44.

Trytek, M., and J. Fiedurek. 2002. Biotransformation of d-limonene to carvone by means of glucose oxidase and peroxidase. *Acta Microbiologica Polonica* 51: 57–62.

Uozumi, N., Y. Kato, Y. Nakashimada, and T. Kobayashi. 1992. Excretion of peroxidase from horseradish hairy root in combination with ion supplementation. *Applied Microbiology and Biotechnology* 37: 560–565.

Usleber, E., M. Straka, and G. Terplan. 1994. Enzyme immunoassay for fumonisin B1 applied to corn-based food. *Journal of Agricultural and Food Chemistry* 42: 1392–1396.

Uyama, H., and S. Kobayashi. 2003. Enzymatic synthesis of polyphenols. *Current Organic Chemistry* 7: 1387–1397.

van Huystee, R.B. 1987. Plant peroxidase, isoenzymes. *Isozymes: Current Topics in Biological Medical Research* 16: 241–242.

van Huystee, R.B., and J. Lobarzewski. 1982. An immunological study of peroxidase release by cultured peanut cells. *Plant Science Letters* 27: 59–67.

Vega-Warner, A.V., H. Gandhi, D.M. Smith, and Z. Ustonol. 2000. Polyclonal-antibody-based ELISA to detect milk alkaline phosphatase. *Journal of Agricultural and Food Chemistry* 48: 2087–2091.

Veitch, N.C. 2004. Horseradish peroxidase: A modern view of a classic enzyme. *Phytochemistry* 65: 249–259.

Vitali, A., B. Botta, G. Delle Monache, S. Zappitelli, P. Ricciardi, and S. Melino. 1998. Purification and partial characterization of a peroxidase from plant cell cultures of *Cassia didymobotrya* and biotransformation studies. *Biochemical Journal* 331: 513–519.

Vozenin-Brotons, M.C., V. Sivan, N. Gault, C. Renard, C. Geffrotin, S. Delanian, J.L. Lefaix, and M. Martin. 2001. Antifibrotic action of Cu/Zn SOD is mediated by TGF-beta 1 repression and phenotypic reversion of myofibroblasts. *Free Radical Biology & Medicine* 30: 30–42.

Vreeke, M.S., and P. Rocca. 1996. Biosensors based on crosslinking of biotinylated glucose oxidase by avidin. *Electroanalysis* 8: 55–60.

Wagner, J., A. Bhandari, H. Singh, F. Macritchie (2002) An evaluation of advanced oxidation processes for the removal of 4-nonyl phenols from water and wastewater. In The proceedings of waste research technology, 20–28. Kansas City.

Wang, P., and J.S. Dordick. 1998. Enzymatic synthesis of unique thymidine containing polyphenols. *Macromolecules* 31: 941–943.

Wang, X., and L.C. Plhak. 2000. Production, characterization, and application of anti-gossypol polyclonal antibodies. *Journal of Agricultural Food Chemistry* 48: 5109–5116.

Watanabe, L., P.R. de Moura, B. Lucas, A.S. Nascimento, L.S. Zamorano, J.J. Calvete, L. Sanz, A. Pérez, S. Bursakov, M.G. Roig, V.L. Shnyrov, and I. Polikarpov. 2010. Crystal structure and statistical coupling analysis of highly glycosylated peroxidase from royal palm tree (*Roystonea regia*). *Journal of Structural Biology* 169: 226–242.

Welinder, K.G. 1992. Superfamily of plant, fungal and bacterial peroxidase. *Current Opinion in Structural Biology* 2: 388–393.

Wititsuwannakul, R., D. Wititsuwannakul, B. Sattaysevana, and P. Pasitkul. 1997. Peroxidase from *Hevea brasiliensis* bark: Purification and properties. *Phytochemistry* 44: 237–241.

Xu, Y.M., A.H. Stokes, R. Roskoski, and K.E. Vrana. 1998. Dopamine, in the presence of tyrosinase, covalently modifies and inactivates tyrosine hydroxylase. *Journal of Neuroscience Research* 54: 691–697.

Xu, T., L. Zhang, X. Sun, H. Zhang, and K. Tang. 2004. Production and analysis of organic acids in hairy root cultures of *Isais indigotica* fort. (Indigo woad) biotechnol. *Applied Biochemistry* 39: 123–128.

Yamada, Y., S. Kobayashi, K. Watanabe, and U. Hayashi. 1987. Production of horseradish peroxidase by plant cell culture. *Journal of Chemical Technology and Biotechnology* 38: 31–39.

Young, P.R. 1989. An improved method for the detection of peroxidase-conjugated antibodies on immunoblots. *Journal of Virological Methods* 24: 227–236.

Yu, J., K.E. Taylor, H. Zou, N. Biswas, and J.K. Bewtra. 1994. Phenol conversion and dimeric intermediates in horseradish peroxidase-catalyzed phenol removal from water. *Environmental Science and Technology* 28: 2154–2160.

Zhang, Y., P. He, and N. Hu. 2004. Horseradish peroxidase immobilized in TiO2 nanoparticle films on pyrolytic graphite electrodes: Direct electrochemistry and bioelectrocatalysis. *Electrochimica Acta* 49: 1981–1988.

Zheng, Z., J. Hanneken, D. Houchins, R.S. King, L. Peter, and J.L. Richard. 2005. Validation of an ELISA test kit for the detection of ochratoxin A in several food commodities by comparison with HPLC. *Mycopathologia* 159: 265–272.

Zhuang, H., Y. Cui, W. Zhu, Y. Zhu, and G. Xu. 2001. Development of a diagnostic kit of enzyme-linked immunoassay for detecting serum anti-hepatitis E virus IgG. *Zhonghua yu fang yi xue za zhi* 35: 315–317.

Chapter 13
Downstream Processing of Red Beet Hairy Roots

Bhagyalakshmi Neelwarne and Thimmaraju Rudrappa

Abstract Triggering cultured plant cells and organs to synthesize the highest possible extent of a particular metabolite is an advantage for its industrial-scale production. However, product recovery by smart techniques is equally important for the commercial success of the process. Considering the possibilities of in situ and ex situ product extraction methods, a large number of studies have focused on different modes of recovery of metabolites where intactness of molecules and their further stabilities have posed technological challenges. In red beet hairy roots, where chiefly betalains are of significance, reverse sequestration of vacuolar pigments into the cell exterior was accomplished by driving forces of both physical and chemical nature. Subsequent coupling technologies, such as pigment concentration by ex situ adsorption and simultaneous separation into red and yellow compounds during desorption, have also been successfully applied. Experiments conducted at bench-scale to standardize several unit operations were found adoptable for continuous process when such units were integrated. This chapter brings together research attempts made towards the recovery of betalains as well as the progress made towards developing strategic processes for the simultaneous recovery of betalains and peroxidase enzymes from red beet hairy roots grown in specialized bioreactors.

B. Neelwarne (✉)
Plant Cell Biotechnology Department, Central Food Technological Research Institute
(Council of Scientific and Industrial Research), Mysore 570 020, India
e-mail: bneelwarne@cftri.res.in

T. Rudrappa
DABT, Dupont Knowledge Center, E.I. DuPont India Pvt. Ltd., ICICI Knowledge Park,
Genome Valley, Turkapally, Shamirpet (M), R. R. District, Hyderabad 500 078, India
e-mail: rajurt@yahoo.com

B. Neelwarne (ed.), *Red Beet Biotechnology: Food and Pharmaceutical Applications*,
DOI 10.1007/978-1-4614-3458-0_13,

13.1 Introduction

Plant cell and organ cultures are gifted with an enormous potential for the production of natural bioactive molecules, which are diverse and often much different from those of in vivo plants. The compounds synthesized by higher plant cells are structurally more complicated than those synthesized by microbes, and not easy to synthesize by chemical methods. Among organ cultures, hairy root cultures display a set of characteristics such as rapid growth rates comparable to those of cell suspension cultures, amazing genetic stability and reliable functional metabolism. Therefore, hairy roots have attracted the attention of technologists to opt for such culture systems for secondary metabolite production in a practically feasible manner. Owing to such a great set of technological and commercial benefits, the industrial-scale root bioreactors have been a commercial success (Sivakumar et al. 2011). In plant cells, both synthesis and accumulation of different secondary metabolites have respective maxima, beyond which no further synthesis and accumulation occurs—a mechanism inherent in all living cells. One limitation for their high productivities is their self-toxic effects. However, if cells are made to quickly sequester such compounds into their vacuoles, cellular toxicities can be minimized. Thus, removing the metabolite from the site of its synthesis has been demonstrated to eliminate the metabolic feedback inhibition, shifting the metabolic flux towards product biosynthesis (Wong et al. 2004). The other way is by removing the metabolite from the accumulation site within the cell so that chances for further product accumulation are enhanced. In red beet hairy roots (RBHR), betalain pigments and many enzymes are synthesized in cytoplasm and sequestered into vacuoles within the cells (Wink 1997; Pradedova et al. 2009, 2011). Without causing great damage to the cell, vacuolar secondary metabolites may be reverse-sequestered to the cell exterior by various cell permeabilization techniques. Such processes offer an advantage for the repeated utilization of the valuable biomass over an extended period, resulting in an overall yield enhancement. Since the total cost of several unit operations of product recovery is often more than the total of biomass/product formation, both up-stream (biomass production) and downstream (product recovery) operations need efficient methods for realizing commercial success. The following sections describe different methods used for the recovery of betalains from cultured red beet hairy roots (RBHR) and their further separation.

13.2 Product Recovery by Permeabilization

The process of cell permeabilization fundamentally involves alteration of membrane properties (depolarization) by applying external stimuli, which are translated into intracellular signals causing opening of plasma membrane as well as vacuolar membrane either by way of opening ion channels, or by activating membrane located enzymes or receptors. Higher plant cells, like any other eukaryote, are equipped with multiple signalling pathways to sense and respond to changes in their

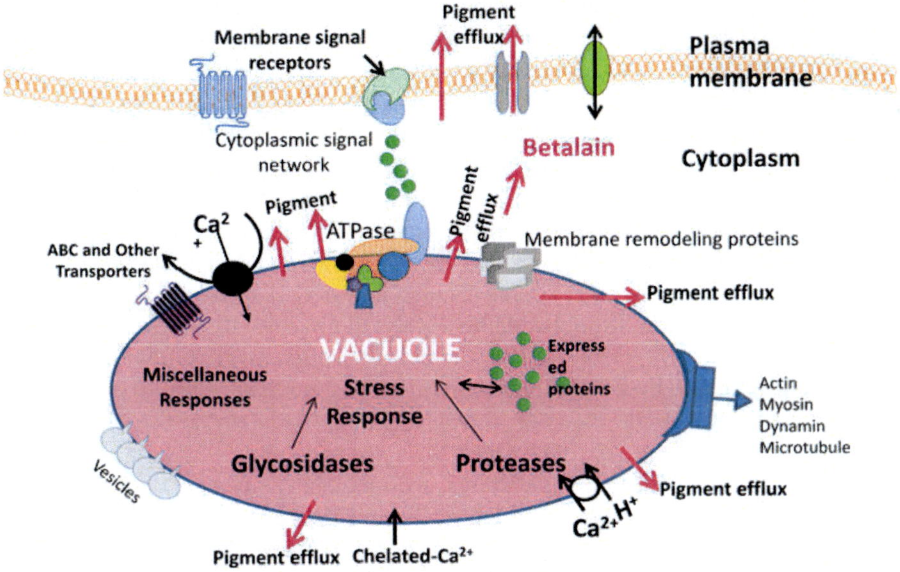

Fig. 13.1 Various signals involved in the release of betalains from the vacuole

exterior. Therefore, materials causing cell permeabilization are expected to act on different membrane-bound proteins and lipids, as well as cytosolic messenger molecules, which in turn exert their actions on intracellular vacuolar/tonoplast membranes. All such events result in a precisely orchestrated influx of information and efflux of vacuolar compounds (Fig. 13.1). Cell permeabilization is employed mostly for the uptake of macromolecules like DNA/protein etc., and employed extensively for the identification of drug targets. In addition, permeabilization is also used to gain access to intracellular sites to study intracellular processes by controlled manipulation of the cell interior. Any discrete treatment aimed at permeabilization may sometimes have severe limitations. Ideally, the method of choice should allow incorporation and/or secretion of substances without causing severe damage to the cell and retain cellular integrity to a maximum possible extent. Therefore, it comprises striking the right balance among various circumstances to allow perfect membrane permeabilization, allowing functional cellular integrity and viability on one hand and releasing the desirable set of metabolites on the other hand.

In certain initial experiments, the leached product caused medium toxicity, in these cases, introducing a foreign material with high affinity to the released product was useful. Such materials may either be an immiscible solvent or a solid resin that partitioned the cell culture medium so that the released metabolite accumulated in the partition material, allowing the repeated recovery of the metabolite (Deno et al. 1987; Byun et al. 1990; Shimomura et al. 1991; Byun and Pedersen 1994; Sim and Chang 1997; Peterson 1999; Zhang et al. 2002; Wong et al. 2004), sometimes increasing as high as fivefold productivity (Sim and Chang 1997). In case of Shikonin, for an anti-inflammatory dye produced by *Litheospermum erythrorhizon* cell cultures, the productivity was enhanced by introducing a partition into the medium in the form of adsorbents where the release of color and adsorption were spontaneous (Shimomura et al. 1991). The elicitation-enhanced terpene indole alkaloids were spontaneously secreted into the medium from hairy roots of *Catharanthus roseus* (Ruiz-May et al. 2009). However, cultured cells/hairy roots of red beet do not spontaneously release betalains to the cell exterior. Thus for reverse-sequestering vacuole-located betalains to the cell exterior, a well-programmed non-lethal permeabilizing method with the advantage of repetitive use of the biomass over an extended period needs to be developed. A few earlier trials made in this direction for other similar compounds in cell culture systems resulted in different degrees of success (Parr et al. 1984; Berlin et al. 1988; Brodelius 1988; Dilorio et al. 1993; Zhong et al. 1997; Pedersen et al. 1999; Bais et al. 2001), and were often associated with loss of culture viability caused by the treatment conditions. To facilitate leaching of rosmarinic acid accumulated in the vacuoles of *Coleus blumeii*, dimethylsulfoxide (DMSO) was used (Park and Martinez 1992, 1994; Martinez and Park 1993). In this case, 0.1% DMSO was continuously fed to cultures, acclimatizing cells to the treatment, because DMSO treatment adversely affected the viability of the cells after permeabilization (Parr et al. 1984; Park and Martinez 1992; Reuffer 1985). Such pre-treated cells of *C. blumeii*, upon exposure to a higher level of 0.5% DMSO, released nearly 65% of the total rosmarinic acid into the medium without loss of viability (Park and Martinez 1992, 1994). Nevertheless, the released rosmarinic acid was quickly destroyed by active peroxidases, which also leached into the medium (Peterson 1999). Therefore, balancing all vital parameters among several factors of permeabilization demands a stringent process characterization.

13.2.1 Permeabilization by Physical Parameters

In many permeabilization studies, *Beta vulgaris* hairy root cultures were selected as a model system because betalains are stored in the vacuole within the cell and secondary metabolites are not normally released into the medium. Hence, this represents a complicated case wherein the product has to traverse through two membranes to reach the cell exterior. Phospholipid bilayers of cells are formed by their energy balance, having high sensitivity to physical factors (may also be chemically induced) that cause disturbances in their electric potential. Various physical factors such as temperature, altered pH, ultrasonication, osmotic stress, oxygen stress and electric fields have resulted in considerable release of secondary product from live cells to their exterior milieu.

13.2.1.1 pH-Mediated Product Release

pH-mediated effluxing of secondary metabolites has been tried in a number of root cultures such as *Datura stramonium*, *Catharanthus roseus*, *Tagetes patula* and *Beta vulgaris* (Saenz-carbonell et al. 1993; Mukundan et al. 1998). When red beet hairy root cultures of cv. Mahyco red were incubated for a period of 6 h in culture medium ranging in pH from 3.0 to 7.0, no pigment release was observed (Mukundan et al. 1998). However, about 2.1% of the total pigment was released when subjected to a medium pH of 2.5 and lower, and substantial release of betalains occurred at pH 2.0. The kinetics of pigment release revealed that 22.7% of the total pigments were released in the first 30 min, with no further release (Mukundan et al. 1998). In another set of experiments, when RBHR grown for 12 days were exposed to pH 2.0 for different time periods, such as 10, 20 and 30 min, and then the medium was replaced with fresh B_5 medium (0.3 mM phosphate) of pH 5.5, there was an incremental release of pigments after 90 min, 240 min and 24 h, which resulted in overall 50%, 53.2% and 77.8% pigment recovery. The RBHR that received low pH treatment for 10 min was viable and re-grew further, accumulating additional quantities of the pigment. These initial trials indicated that permeabilized cultures could accumulate more than twice the amount of pigment than the control cultures during the same culture period, resulting in an increase in the specific as well as volumetric productivity (Mukundan et al. 1998). These authors inferred that the specific application of pH 2.0-mediated pigment release may be due to the existence of betalains under zwitterionic state at pH 2.0. In contrast to these observations, in RBHR from cv. Ruby Queen, over 70% of the pigment was released in less than 30 min of exposure to the medium of pH 2.0 with no further release later (Thimmaraju et al. 2003a). Only about 10% of the pigment was released at pH 3.0 and lesser at pH 4.0 in 30 min, with a very slight improvement up to 240 min. However, upon exposure of RBHR to different pH of 2.0, 3.0 and 4.0, the released color gradually degraded with time, which was more in pH 2.0. Addition of fresh medium with or without calcium enrichment did not support further growth of hairy roots in pH 2.0 treatments for 40 min and longer. However, cultures treated with pH 3.0 and 4.0 showed viability and normal growth pattern comparable to non-pH-treated controls (Thimmaraju et al. 2003a). At pH 2.0, even the viable cultures largely had disrupted cells away from the tips, which resulted in medium fouling from dead suspended particles. Thus continuation of growth and pigment synthesis may only be expected from the intact tips, while the presence of whole and ruptured cells contributes towards medium browning, which in turn results in low quality of subsequently extracted pigments. Most of the pH-mediated secondary metabolite release was attributed to the hydrophilic nature of metabolites. Because of their hydrophilic nature, they cannot be transported by passive diffusion across the membrane (e.g., betalains). Hence these hydrophilic molecules may be transported by specific carriers or by hijacking existing carriers. For example, the pH-dependent ionization of betalains provided the rationale for the release of betalains; they are cations below pH 2.0, zwitterions at pH 2.0, monoions between pH 2.0 and 3.5 and *bis* ions above pH 3.5 and 7.5 (Reznik 1980). In typical vacuolar pH conditions, betalains exist as ions, incapable of diffusing out of the vacuole. At pH 2.0, the zwitterionic state might permit diffusion and release from the vacuole. This behavior is similar to ion trapping where metabolites such as alkaloids are retained within the vacuole as a result of acquiring a net charge at the acidic vacuolar pH (Renaudin 1981).

Additional information on the mechanism of pigment release comes from the measurement of cellular respiration by cytochemical assay. The decrease of respiration in permeabilized cells exposed low pH suggested the possibility of a severe damage to the cells when exposed to acidic treatment and killing the cells (Mukundan et al. 1998; Kino-Oka et al. 1992).

Product stability is an important parameter during the recovery of pigments. Separate observations made for both the release and the degradation pattern of pigments in the medium and/or water of pH 2.0 showed a significant difference between the two; i.e., the release was faster in water than in medium. Similarly, there was a faster degradation of pigments in water of pH 2.0. This difference may be due to the interaction of ions present in the nutrient medium, which act as osmoticum and also may act as ionic support to a certain extent, as metal ion chelation of natural pigments is known (Deveoglu et al. 2009). Apart from protecting the pigments, the media components were also helpful in rendering membrane protection to the cells of RBHR (Thimmaraju et al. 2003a). To support this view, these authors tested the effect of calcium on the viability of hairy roots, since calcium is known to particularly support the membrane integrity of cells (Weathers et al. 1990). As in the case of an earlier experiment (Kino-Oka et al. 1992), there was a total loss of viability in RBHR exposed for 10 min to pH 2.0. The addition of fresh medium with normal levels or additional calcium was found to support the re-growth in 10% of the total cultures after pH 2.0-mediated pigment efflux conducted for more than 30 min (Thimmaraju et al. 2003a). Therefore, normal levels of calcium in MS medium is probably sufficient to support hairy root tips for further growth, particularly those tips (10%) that were resistant to the low pH treatment. It is quite possible that the exposure of RBHR to low pH water would have severely damaged the cells, where without actual permeabilization, the pigment release would have occurred from the dead cells, which was evident from the loss of their viability. A similar observation was made by some of the earlier workers (Mukundan et al. 1998; Kino-Oka et al. 1992). Reducing the treatment period from 30 to 5, 10, 15 and 20 min released 4.5%, 9.2%, 14.9% and 20.7% of pigments from RBHR with a slight improvement in the viability when root tips were grown on solid medium.

13.2.1.2 Sonication

The application of ultrasonic waves of specific frequencies is known to release intracellular products while retaining considerable level of cellular viability. Subjecting the cells to ultrasonic waves to harvest intracellular, vacuole-located secondary products is almost a conventional process, which is accomplished at the expense of cellular viability. When ultrasound waves travel in a matter, they create expansion and compression cycles causing negative and positive pressures within the cell and alterations in the membranes (Wang and Weller 2006). The advantage with this technique is that neither any chemical is introduced into the culture medium nor is there a requirement of medium replacement, as in the case of pH treatment or other chemical treatments, hence this is of great practical advantage. Above all, the sonication process can easily be integrated with the process line, for continuous product harvest.

Ultrasound treatment of red beet cell suspension stimulated the release of about 5–10% of the total pigment at 1.02 MHz frequency, with no adverse effect on cell viability, where the frequency of the ultrasonic sound was very important for both pigment stability and culture viability (Kilby and Hunter 1990, 1991). Thimmaraju et al. (2003a) found that 1.02 MHz was too high for permeabilizing the RBHR, since it resulted in complete maceration of the cultures. Therefore, a lower level of 0.02 MHz was adopted, which was also capable of releasing the same amount of pigment from hairy roots (Thimmaraju et al. 2003a) as reported for red beet suspension cultures (Kilby and Hunter 1990, 1991). The extent of release of betalains under the influence of sonication for different periods resulted in cell rupture to various extents, with the partial release of pigment into the medium. Sonication of RBHR for 15 s released about 8% of the pigment, which remained constant at higher levels of sonication, for 30 and 60 s, indicating the resistance of hairy roots to the treatment for up to 60 s, after which, the release of pigments increased to 12% in cultures treated for 120 s (Thimmaraju et al. 2003a). Post-sonication incubation of cultures, treated for different periods, in normal MS medium and calcium-enriched medium indicated that in all the cases the viability was preserved, although the growth rate declined by 50%, resulting in the decline of overall betalains recovery. The quantity of betalains effluxed during further sonication for a given span of time remained almost similar to that of the first treatment. Thus, by means of sonication, a maximum of only 12% of the total betalains could be recovered from RBHR without the loss of culture viability during first cycle, whereas after the second harvest viability was totally lost. Bubble formation during sonication is an indication of the involvement of cavitational events. Since such events are coupled with the release of free radicals (Kilby and Hunter 1991), sonic effluxing may not be suitable for pigments, particularly betalains, which are known to be highly sensitive to oxidation. The sonicated cultures were observed to contain a lot of suspended ruptured cells, making the culture medium unfit for re-use and calcium treatment did not have any beneficial effect on the viability of hairy roots (Thimmaraju et al. 2003a). Contrary to this, red beet cell suspensions were amenable to repetitive sonication (Kilby and Hunter 1990, 1991).

13.2.1.3 Temperature

Mild fluctuations in temperature around cultured cells and organs result in reverse sequestration of metabolites stored in their vacuoles, probably as a defence response. The release of betalains from RBHR was directly proportional to the increase in temperature in a given time span. Upon 30 min exposure to a temperature of 50°C, 45°C or 40°C, there were respectively 45%, 35% and 5% of the total content of betalains released into the medium (Thimmaraju et al. 2003a; Dilorio et al. 1993). The pigment release reached a plateau after 30 min and was about 50% at 50°C in 60 min. Though there was an increasing trend of pigment release at 40°C and 45°C, the released pigment quality was poor, with a brown tinge, indicative of degradation.

As explained in Chap. 3, betalains are highly sensitive to temperature, where they undergo rapid oxidative degradation.

A high temperature of 50°C also affected viability, as all the treated cultures lost their viability, despite additional calcium treatment (Thimmaraju et al. 2003a). Subjecting RBHR to 42°C for 10 min was found to release only 5% of the total betalains, without affecting the viability (Dilorio et al. 1993), whereas a slight increase in treatment temperature to 45°C for 2 min reduced the growth index to 80%. Even in these cultures, exposure to 50°C for 2 min completely killed roots. An increment of 10% of total betalains was recorded when permeabilized at 3-day intervals at 42° C for 1 h, followed by treatment with $CaCl_2$ (Dilorio et al. 1993). A separate study observed that $CaCl_2$ applied after heat treatment was observed to improve the viability of beetroot disks and hairy roots (Weathers et al. 1990), where CO_2 production from growing tissue was used as an index for the measurement of viability of biomass. In the study of Dilorio et al. (1993), the viability improved further when RBHR were exposed to $CaCl_2$ for 10, 20 and 30 min, with 10 and 20 min being the most effective. Benson and Hamill (1991) found that viability varied by as much as 50% of the experimentally determined value for post-thaw hairy root cultures, which remained mostly leaky. Therefore, the best method for viability determination was to study their re-growth and biomass accumulation, which is a factor of key importance in order to develop a practically feasible extraction process (Benson and Hamill 1991). Since the post-permeabilized RBHR of cv. Ruby Queen were not responsive to treatment with additional levels of calcium (Thimmaraju et al. 2003a), as in other root clones, such differences in their responses would be of great interest to study the molecular interactions with *rol* genes. Although temperature treatment resulted in the efflux of over 50% of pigments in some clones of RBHR, it appears that heat treatment may not be a suitable effluxing method considering its lethal effect on hairy root cultures as well as its adverse effects on the released pigment. Above all, uniform heat transfer to every part of the tangled root biomass is yet another practical problem for adopting this strategy as an online step for product recovery.

13.2.1.4 Oxygen Stress

Starving cells from their regular supply of oxygen imparts respiratory stress on them; plant cells respond by making various regulations in their physiological processes, the major one being to trigger the defence response. Hairy root cultures immersed in medium are normally aerated by incubating on a rotary shaker. When shaking was stopped, there was a gradual release of betalains up to 45 h, leading to about 5% efflux of total betalains. However, there was a steep increase in the pigment release after 45 h, resulting in the release of about 25% of the betalains in 50 h, particularly in the presence of light (2,000 Lux), where the viability was lost. There was no release of pigments under dark conditions even after 50 h (Thimmaraju et al. 2003a). By increasing the ionic strength of the medium, the cells of *Catharanthus roseus* were permeabilized to release certain cytosolic contents, including secondary metabolites, with no loss of viability (Tanaka et al. 1985). RBHR cultures of late exponential phase, under the influence of oxygen stress, released over 20% of the pigments into the medium (Kino-Oka et al. 1992), where the hairy root biomass was used repeatedly for three cycles that resulted in an overall production enhancement of betalains from unit biomass. In all of these permeabilization studies, product recovery was either very poor or the viability of hairy roots was lost; otherwise the process involved two to three changes of nutrient medium, adding to the process cost. Although oxygen-starved hairy root cultures released pigments, in contrast to other reports (Taya et al. 1992), it resulted in complete death of hairy root cultures and browning of released pigments almost immediately, indicating their poor stability.

13.2.1.5 Osmotic Stress

Change in osmotic potential of cell environment triggers specific signals in plant cells to which each cell system responds differently. One of the responses is the efflux of cellular secondary metabolites. Depending on the osmotic potential of the cell exterior, compounds can either flow into or flow out of the cell. Such cellular properties may be exploited to extract pigments from cultured red beet tissues. Thimmaraju et al. (2003a) tested the effect of osmotic stress on the release of any betalains using different levels of soribitol in culture medium. Even oxygen stress, induced by stopping the medium mobility, released about 25% of pigment, only in the presence of light, over a period of 50 h. The osmotic stress on the other hand did not result in the release of pigment even at as high as 5 M sorbitol levels in the medium for 24 h.

13.2.1.6 Electropermeabilization

Since plant cell membranes are formed by electrically balanced alignment of phospholipids, exposure of cells to electric fields brings about changes in their orientation, inducing both influx and efflux of materials, where the membrane reassumes normal structure once the voltage is stopped. This process of electropermeabilization is a valuable method for the recovery of metabolites from living cells, is similar to electroporation used in genetic engineering (Zimmermann 1986; Bates 1990) and is a very popular routinely used protocol for engineering *E. coli*. Yang et al. (2003) electropermeabilized cultured red beet cells, and found that a steady increase in the power level from 5 to 20 V produced high detection of pigment when measured through an online electrophoresis setup. A power level of 10 V was found adequate for both preserving the cell viability and for the reduction of foaming due to cell death. However, when two cycles of oscillatory variations in voltage from 5 to 20 V were imposed on cells during a span of 6 h, with foaming at the end of the cycle, pigment release occurred only after 5 h, with lesser quantity than in the continuous run (Yang et al. 2003). When viability is not under consideration, other field-assisted processes conventionally applied for other horticultural commodities are applied for the extraction of betalains from field-grown red beets (Chalermchat et al. 2004; Fincan et al. 2004; Loginova et al. 2011). See Chap. 14 for more information.

13.2.2 Permeabilization by Chemical Methods

Many chemicals have the ability to alter the balance of the electric potential of lipid bilayers, causing intermittent disruption and loss of their integrity, which causes leakage of soluble cellular components. Based on this principle, several transient permeabilizing chemicals, such as detergents, have been identified and used safely and effectively for the recovery of specific secondary metabolites. Table 13.1 lists the most effective chemical permeabilizing agents used for RBHR clone LMG-150, based on the reports of Thimmaraju et al. (2003b, 2004). When the detergent Tween-80 was used, the pigment release was observed to occur after 48 h, where 0.05% and 0.1% detergent in the medium imparted almost similar pigment release but greatly affected culture viability and stability. Increasing the detergent concentration to 0.15% caused a large increase in the release of betalains in static condition with no further increase in pigment efflux, darkening root cultures. In all chemical permeabilization treatments, one common factor that caused high pigment release was when treatment was combined with an additional stress parameter, oxygen starvation. Stopping medium rotation, which makes roots to sink into the medium, caused oxygen starvation. Thus, a ten-fold higher pigment release could be achieved without changing the concentration of Tween-80 when cultures were kept in a static condition. Use of another nonionic surfactant, Triton X-100, (0.2%) caused substantial release of pigments; over 70% of the pigment was found in the medium within 2 h and pigment was released for up to 4 h, with a subsequent steady degradation; by 8 h there was a nearly 12% loss of the released pigment. Cetyltrimethylammonium bromide (CTAB) is a food-grade cationic detergent extensively used for permeabilizing yeast cells (Joshi et al. 1987). This detergent, when used at similar levels as that of Triton X-100, caused immediate release of pigment, probably due to the dissolution of the phospholipid bilayer, as also observed by previous workers (Chao and Lee 2000). However, a lower concentration range, from 0.0005% to 0.05% was found more suitable. At the highest concentration (0.05%), over 90% of the total pigment was released within 1 h; the pigment degradation commenced after 4 h in static medium, although retaining nearly 80% of the released pigment even after 24 h (Thimmaraju et al. 2004). In all chemical treatments, the released pigment concentration was directly proportional to the treatment concentration with higher and faster release in static cultures. However, the stability of pigment and the viability of roots were drastically affected by high treatment levels (Table 13.1). As in the case of other chemical treatments, CTAB also caused a loss of pigment in the medium beyond 24 h. Treatment with appropriate levels of chemical permeabilizing agents was found to mostly retain the integrity of root tips (Fig. 13.2), unlike in physical treatments.

13.2.3 Repeated Recovery by Chemical Methods

For increasing the overall productivity, the re-use of biomass after each permeabilization and recurrent recovery of metabolites from cells is a desirable strategy, where

Table 13.1 Release of betalains from red beet hairy root clone LMG-150 by chemical permeabilizing agents

Chemical	Concentration (wt/vol) % in the medium	Additional parameter	% Pigment effluxed (in hours)	Viability after treatment	Pigment stability
Triton X-100	0.1	–	66 (1 h)	Greatly affected	Fairly stable
	0.2	–	70 (2 h)	Severely affected	Low stability
	0.2	O_2 starvation	77	Severely affected	Low stability
Tween-80	0.15	O_2 starvation	40	Severely affected	Fairly stable
	0.15	–	04	Severely affected	Fairly stable
CTAB	0.002	–	72 (2 h)	Mildly affected	Stable
	0.01	–	50 (1 h)	Mildly affected	Stable
			80 (4 h)	Mildly affected	Fairly stable
			70 (1 h)	Mildly affected	Fairly stable
	0.05	O_2 starvation	90 (1 h)	Mildly affected	Fairly stable

CTAB etyl-trimethylammonium bromide
Based on the report of Thimmaraju et al. 2003b, 2004

treatments ought to be precisely monitored after each recovery cycle. In case of RBHR, the biomass needs to be precisely monitored with specific post-permeabilization treatments. Normally, the repetition of permeabilization resulted in the yield of 4-, 3.6- and 2-fold higher betalains in RBHR effluxed thrice, twice and once with CTAB, respectively. Calcium is one of the key regulators of membrane integrity, and is an important component in integrating pectic polysaccharides of plant cell walls (Hepler et al. 2005). The role of calcium in the repair of damaged membranes is well known. Calcium treatment was beneficial to post-sonicated cell suspension cultures of carrot subjected for repeated efflux of metabolites (Kilby and Hunter 1990, 1991; Dilorio et al. 1993). The use of additional level calcium (1%) in the medium for the culture of RBHR after each permeabilization was found to result in a modest improvement in viability, leading to 10-, 5- and 3-fold higher pigment yield in thrice-, twice- and once-effluxed cultures, respectively (Fig. 13.3).

13.2.4 Biological Permeabilizing Agents

Apart from methods involving effects of physical and chemical agents, certain biological materials, such as live microbial cells, are traditionally used for food applications. These microbes either chew up the cell walls/membranes of higher plant cells by releasing hydrolytic enzymes or have the capacity to change the electrical potential of the cell membrane, which in turn causes transient permeabilization.

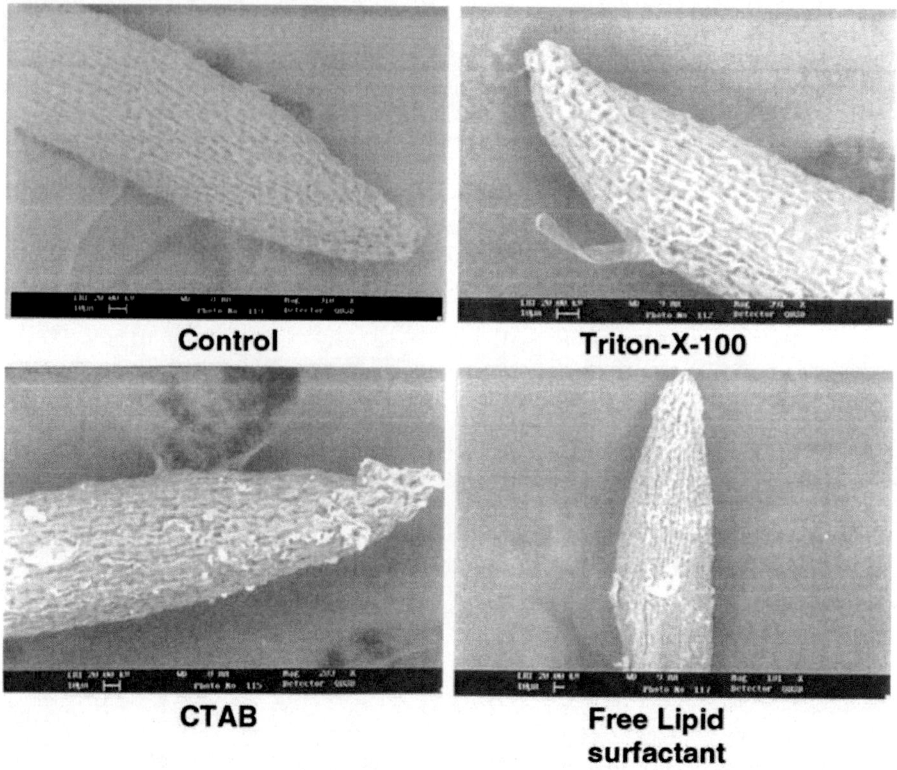

Fig. 13.2 Scanning electron photomicrographs of post-permeabilized hairy root tips of *red* beet treated with different permeabilizing agents: Triton-X-100 (0.05%), CTAB (0.0005%) and free lipid fraction of *L. helveticus*

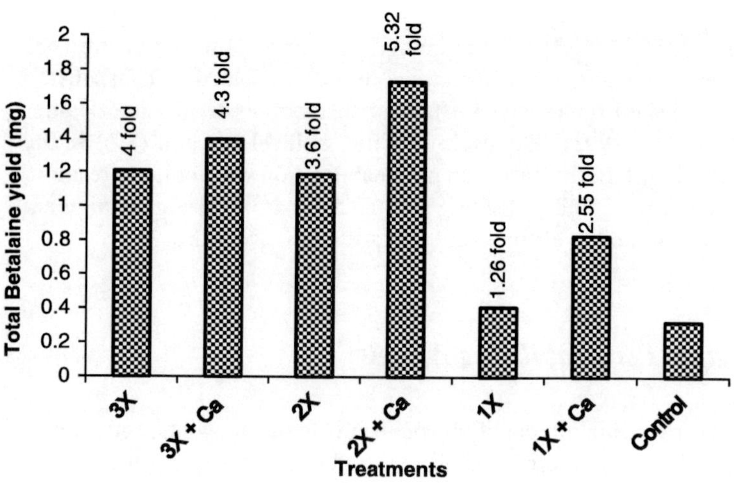

Fig. 13.3 Total yield of betalains from the hairy root cultures permeabilized with CTAB (0.005% w/v) where one set was treated with extra calcium. 1X, 2X and 3X represent yields after once, twice and three times efflux

Microbes capable of permeabilizing higher plant cells accomplished the same through alterations of cell surface activity by producing bio-surfactants (Joshi et al. 2008; Darvishi et al. 2011; Singh 1994) and hence are attractive to develop strategies for product recovery. Most of the natural pigments are stored in the cells as glycosides, and decoupling the pigment from its respective glycoside is known to improve the hue value of the pigment (von-Elbe 1979). Because the microbes used for permeabilization can also act on the glycoside of betanin (von-Elbe 1979), the use of biological agents may bring about a color intensification effect, in addition to cell permeabilization followed by pigment efflux. The food-grade microbes, particularly the species of *Lactobacillus*, synthesize and release organic acids, mainly lactic acid, which is proven to be a nutraceutical and an anti-microbial in food formulations (Barefoot and Netless 1993). Lactic acid bacteria utilize carbohydrates for the production of lactic acid and other organic acids anaerobically, which not only increase the color strength of betalains but may also protect pigments from oxidation. In view of such beneficial properties, several food-grade microorganisms and their cell fractions were analyzed for pigment recovery from cultured RBHR (Thimmaraju et al. 2004). When *Candida utilis*, *L. helveticus* and *Saccharomyces cereviseae* culture broth (0.5 OD at 660 ηm) were separately added at the rate of 1 ml to 5 ml in 40 ml culture medium containing about 6 g (fresh biomass) of fully grown RBHR, the release of 60%, 85% and 54% betalains occurred, respectively, in 24 h, while the addition of lower amounts of respective culture broths needed longer times, of 48 h, to release the same quantity of betalains. Apart from culture broth, whole cell powder obtained after drying *L. helveticus*, its insoluble carbohydrate fraction and isolated free and bound lipid fractions were also capable of releasing 10%, 0%, 85% and 10% pigment, respectively. The free lipid fraction was most efficient in pigment release (Fig. 13.4), without causing cellular death or any disruption of root tips (Fig. 13.2). Since live microbial cells also multiply in the hairy root culture medium, and, therefore, their application is restricted to batch cultivation. However, the released pigment and the by-products of the process of partial fermentation are expected to impart improved sensory/nutraceutical effects to the recovered pigment and hence may add value to products that incorporate the red beet pigment thus produced. For example, in minimally processed fresh vegetable products such as pickled vegetables, where lactic acid is a component, one can impart a natural red color effluxed using *L. helveticus*. In most of these permeabilization processes, as found in the case of chemical permeabilization, agitated cultures were found to result in a much lower level of pigment release. Thus, the RBHR has served as one of the earliest model systems where the application of biological permeabilizing agents has been demonstrated for pigment recovery, with a good level of success (Thimmaraju et al. 2003b, 2004). The food-grade microorganisms were efficient in pigment release probably because of their capacity to synthesize and release bio-surfactants (Darvishi et al. 2011; Singh 1994), which is evident from the highest pigment release when their lipid fractions were used. The use of isolated cellular lipid fraction of *L. helveticus* added an advantage that the pigments were highly stable, showing no signs of degradation even after 24 h after release (Thimmaraju et al. 2003b).

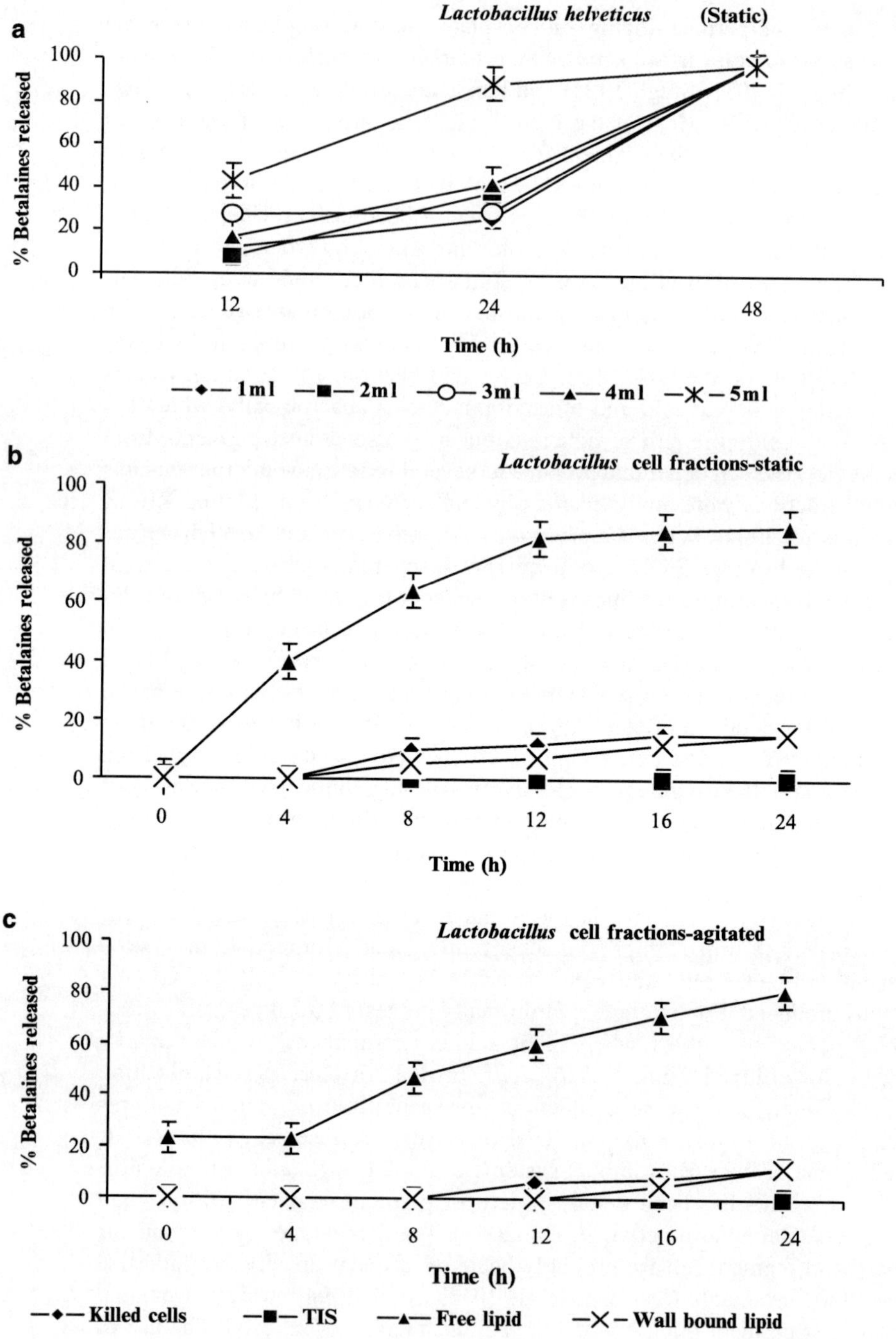

Fig. 13.4 Pigment release from RBHR cultures treated with *Lactobacillus helveticus* live cells by adding different volumes of culture broth (**a**), cell fractions (**b** and **c**) such as killed cells, total insoluble sugars (*TIS*), free lipids and cell wall-bound lipids under static (**b**) and agitated (**c**) conditions

13.2.5 Pigment Release in Bioreactor

It is now established that the synthesis and accumulation of betalains are growth-associated in field-grown storage roots as well as RBHR and that betalains are exclusively sequestered into vacuoles. In cultured RBHR, experiments aimed at product recovery are chiefly focused on the recovery of pigments (Chethana et al. 2007; Thimmaraju et al. 2004; Suresh et al. 2004). For establishing a continuous process and enhancing productivities, online recovery techniques are important for commercial feasibility of the product. The first step in this direction would be the use of commercial bioreactors where, when sufficient biomass quantity is attained, permeabilizing agent(s) could be added for in situ release of pigments, and the medium containing released pigment may then be pumped for their separation, and pigment-removed medium could be replenished with nutrients and re-used for hairy root cultivation. Here the most crucial factor is the viability of the biomass, their further performance in terms of metabolite synthesis and their ability to cope with further permeabilization cycles. Nevertheless, there are some successful experiments demonstrating that in situ product removal enhanced total metabolite production. The products that were selectively released from cells were either held by a second phase composed of solvents or held by the adsorbents. For example, a simultaneous in situ extraction and elicitation was feasible for suspension cultures of *Escholtzia californica* for the production of benzophenadrine alkaloids (Byun et al. 1990). For the release of pigments from RBHR grown in bioreactor, when the free lipid (non-cell wall) fraction of *L. helveticus* was added into the bubble column reactor, there was a time-bound pigment release. A highest amount of pigment release of 84% was observed after 12 h of treatment, preceded by about 50 and 19% pigment release after 8 and 4 h of treatment, respectively. Neither any loss of viability nor any degradation of pigment was observed up to 24 h after the highest release in 12 h (Thimmaraju et al. 2003b), indicating that unit operations standardized using shake flasks were helpful for developing bioreactor processes. Similarly, the best chemical permeabilizing agent for RBHR is CTAB; and the low level application of 0.002% CTAB to hairy roots in a bubble column reactor was adequate to release 80% of betalains, where the time requirement was 48 h, with only 30% and 44% pigment release after 24 h and 36 h treatment, respectively. In this study, an appreciable re-growth of biomass was observed in the bioreactor (Thimmaraju et al. 2003b).

13.2.6 Involvement of Calcium-Channel Modulators in Pigment Release

Mechanisms involved in pigment release are not always those resulting from changes involved in membrane charges. Various membrane channels and ion gates do exist in the plasmalemma and vacuolar membranes that are regulated by a vast array of signals of physico-chemical and biological nature (see Fig. 13.1). A large body of information is available, in classic physiology, on the various signal transduction mechanisms involved in both perceiving signals and executing the release of secondary metabolites, where calcium (Sticher et al. 1981) and calcium-channel modulators play an important role (Sudha and Ravishankar 2002). For characterizing the mechanism involved in the release of pigments as a response to permeabilizing agents, a study was conducted by the authors using calcium-channel modulators. In this study, the calcium-channel modulators were administered to 20-day-old cultures by replacing spent medium with fresh medium containing calcium-channel modulators such as ethylene glycol tetraacetic acid (EGTA, 3 mM), a calcium chelator known to inhibit calcium uptake; ionophore A23187 (calcimycin, 0.5 μM), a calcium-channel enhancer; chlorpromazine (10 μM), a calmodulin inhibitor; and verapamil hydrochloride (10 μM), a calcium-channel blocker. The treated cultures were further subjected to permeabilization using various effluxing agents such as CTAB (0.005%), Triton-X-100 (0.005%) and free lipid surfactant from *Lactobacillus helveticus*. Here, a lower concentration each of CTAB and Triton-X-100 were selected to achieve a regulated pigment release for easy monitoring and subjecting roots for pigment efflux at different time intervals.

Different calcium-channel modulators were found to modulate the pattern of release considerably. As compared with respective controls (treated with TritonX-100 or CTAB or lipid), about 80% suppression occurred with EGTA, a calcium chelator known to inhibit calcium uptake, inferring that calcium is needed for the release of pigments. Enhancement of 40% with ionophore and about 70% suppression of pigment release with chlorpromazine treatment occurred, which also suggested the respective enhancement of channels and channel blockers. Further, cultures treated with chlorpromazine in combination with ionophore or verapamil resulted in the suppression of pigment release to about 80% and 70%, respectively. Thus it is evident that Ca^{2+} channels as well as calmodulin are involved in the reverse-sequestering of betalains, in addition to the possible passive escape of pigment molecules when voltage-gated calcium channels are opened by unknown ionic interference.

Figure 13.5 shows that the most of the Ca^{2+}-channel modulators were effective when RBHR were permeabilized after 24 h. The trend was similar for all of the three selected permeabilizing agents. Compared with Triton-X-100 and CTAB, the calcium-channel modulators were more responsive to free lipid-mediated pigment release. EGTA caused more suppression of pigment release, causing about 90%, 52% and 41% decrease in pigment release compared with control cultures (not treated with EGTA). The ionophore A23187, similar to EGTA, caused highest enhancement of pigment release compared with its effect on Triton-X-100 and CTAB, causing 45%, 36.4% and 9.4% increases of pigment release (Fig. 13.5). The other calcium and calmodulin antagonists had similar effects to Triton-X-100 and CTAB-mediated pigment release. All in all, the results demonstrate various levels of involvement of calcium–calmodulin-mediated signalling processes in chemical and free lipid-mediated pigment release from RBHR. In cultured plant cells, such studies have been very well documented in different processes, especially in defence signalling during host pathogen interactions. Strong evidence for the role of Ca^{2+} has been demonstrated in cowpea epidermal cell death after challenging with the fungus *Uromyces*, which was further hypothesized because of the diffusion of

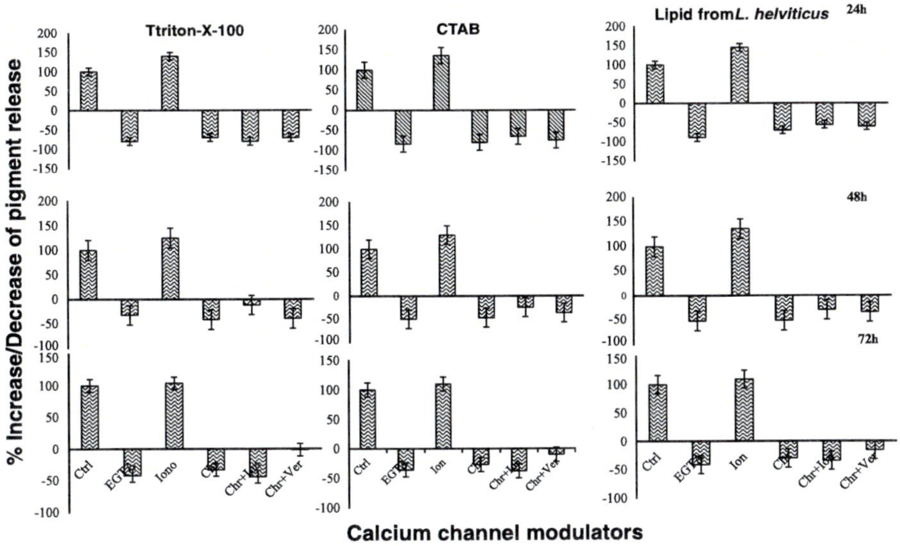

Fig. 13.5 Influence of calcium-channel modulators on pigment release as compared with control (100%) after different time intervals. *Ctrl* control, *EGTA* ethylene glycol tetraacetic acid, *Ion* ionophore A23187, *Chr* chlorpromazine and *Ver* verapamil

peptide elicitors from the invading fungi (Gelli et al. 1997). Ca^{2+} involvement was observed in triggering the defence responses when plant cell suspensions were treated with microbes or elicitors (Levine et al. 1996; Zimmerman et al. 1997).

13.3 Adsorption and Recovery of Betalains

After permeabilization, when betalains are outside the cell, they are highly vulnerable to degradation (Von-Elbe 1979; Stintzing and Carle 2007; Herbach and Stintzing 2006), therefore, ought to be recovered quickly. Employing either a lipophilic second phase (Becker et al. 1984) or a polar second phase has been reported to be beneficial for the accumulation and extraction of secondary metabolites (Sim and Chang 1997). Brodelius and Nilsson (1983) showed that some solvents were useful for extracting products from immobilized plant cells without affecting cell viability. In situ extraction of shikonin pigment with hydrocarbons was performed for suspension cultures of *L. erythrorhizon* (Deno et al. 1987). Typically the released product, which is in dilute solution is recovered and then concentrated using either a non-selective separation method such as ultrafiltration, precipitation, liquid–liquid extraction and adsorption or a selective method based on affinity. Further recovery and purifications involve a series of steps to specifically capture the product of interest and eliminate the contaminants. Various types of general chromatographic methods based on molecular size, surface charge, net charge, hydrophobicity and molecular recognition are then used to simultaneously achieve high purity and retain chemical conformity with biological activity, with the whole process being amenable for scale-up to a desirable extent. In certain bioreactor prototypes, extra-cellular products released into the medium were separated by passing the spent medium through a resin column (Amberlite XAD-7) attached as an external loop (Peterson 1999) or by directly adding the resin (XAD-2; XAD-7) into the medium (Sim and Chang 1997; Peterson 1999). Scopolamine, an alkaloid from *Duboisia leichhardtii*, was recovered from the medium using amberlite XAD-2 columns and the compound was later eluted with a mixture of methanol and ammonium hydroxide (Muranaka et al. 1992), leading to the recovery of 97% of the released compound. By repeated recovery of scopolamine, the production was improved fivefold when compared with that without in situ adsorption. Shikonin, a pigment derived from shikimic acid in the cell cultures of *L. erythrorhizon*, was enhanced by immobilization coupled with in situ extraction (Kim and Chang 1990), which was later adopted for pigment enhancement in hairy roots of the same species (Shimomura et al. 1991; Sim and Chang 1993). All these reports clearly indicate that selection of a suitable adsorbent depends on the characteristics of the target compound, which is specific to each culture system. All in all, the product removal by adsorption is mainly accomplished by two ways: the first, by in situ adsorption, where an adsorbent/affinity material is directly be added to the medium containing the released product and the second, by ex situ method, where the released product is passed through an affinity/adsorption surface. A defined temperature range is selected as adsorption isotherms while designing separation methods employing fixed-bed adsorption (Chern and Chein 2001). However for betalains, one temperature, i.e., 25°C and pH 5.5 could be conveniently used to keep betalains fairly stable (Thimmaraju et al. 2003b, 2004).

13.3.1 Selection of Adsorbents

For the recovery of red beet pigments from the medium, Thimmaraju et al. (2004) selected two sets of adsorbents, one set for in situ (Table 13.2 and 13.3) and the other for ex situ (in the column; Table 13.4). Assuming that the adsorbents do not bring about any chemical change to betalains, the temperature and the pH in the adsorbent bags and in the column remain constant throughout the operating time and the adsorption mass transfer occurs mainly by surface actions/convection (Snyder 1968).

13.3.2 In Situ Adsorption

Both the pigments (betaxanthin and betacyanin) are derivatives of betalamic acid, and exhibit similar properties except for their adsorption maxima and the absence of the glycosidic group in betaxanthin (Von-Elbe 1979). For in situ adsorption, the adsorbent bags may be directly inserted into the pigmented solution of a known

Table 13.2 Different adsorbents for in situ adsorption and recovery of betalains[a]

Adsorbent[b]	% Adsorption	% Desorption	% Recovery of betalains
Silica gel	22.60	77.20	17.45
Silica:alumina (1:1)	97.00	73.60	71.39
Alumina	68.20	21.80	15.00
Wheat starch	20.50	75.01	15.38
Corn starch	36.40	35.20	12.80
Maltodextrin	30.30	6.80	2.06
Cyclodextrin	17.90	28.60	5.12
XAD-2	0.00	0.00	0.00
XAD-4	6.70	100.00	6.70
Dextrin white	35.40	43.50	15.40

[a] In 25 ml medium containing equilibrium concentration of 0.13 mg ml^{-1} betalain
[b] Total weight of adsorbent in each case is 3 g

Table 13.3 Methanol pre-treated adsorbents for in situ adsorption and recovery of betalains[a]

Adsorbent[b]	% Adsorption	% Desorption	% Recovery of betalains
Silica gel	26.36	47.10	12.42
Silica:alumina (1:1)	79.00	27.50	21.73
Alumina	97.20	0.60	0.58
Wheat starch	32.72	44.40	14.53
Corn starch	21.80	58.30	12.71
Maltodextrin	31.80	4.14	1.32
Cyclodextrin	33.60	14.90	5.01
XAD-2	45.90	0.77	0.35
XAD-4	42.20	0.73	0.31
Dextrin white	26.80	47.40	12.70

[a] In 25 mL of medium containing an equilibrium concentration of betalain of 0.13 mg/mL^{-1}.
[b] Total weight of adsorbent in each case = 3 g

equilibrium concentration, and gyrated for a specific period, where the quantity adsorbed of adsorbed betalains may be calculated using the following formula (Payne and Schuler 1988):

$$q = \frac{(C_i - C)V}{A} \tag{13.1}$$

where C and C_i are the measured (un-adsorbed) and initial betalains concentrations, respectively (mg ml^{-1}); V is the liquid volume (ml); A is the amount of adsorbent in the nylon bag (g); and q is the pigment loaded onto the adsorbent (mg g^{-1} adsorbent). To find the adsorption isotherm constant, the amount of betalains adsorbed per unit mass of adsorbent (x/m) may be plotted against different equilibrium concentrations of betalains at pH 5.5 at 25°C and used for further calculations.

Among several adsorbents, alumina showed a maximum adsorption of 68.2%, of which only about 22% could be desorbed from the adsorbent, ultimately resulting in the recovery of about 15% of the total pigment that was loaded/effluxed. Silica gel showed 22.6% adsorption, of which, although 77% desorbed into acidified water, the overall pigment recovery of the unit operation of in situ adsorption was only 17.45%. When alumina and silica were mixed in equal ratio, the adsorption percentage of betalains increased significantly, showing 97% adsorption, and resulting in a recovery of 71.4% although desorption was only 73.6% in this case. Amberlite XAD-2 and XAD-4 were poor adsorbents, although found beneficial for a vast number of plant secondary metabolites (Payne et al. 1988; Payne and Schuler 1988). Both corn starch and dextrin white showed almost similar adsorption capacities (about 36%) with dextrin white being more suitable for desorption, allowing a recovery of 43.5%. In general, the materials that had poor adsorption capacity invariably showed high desorption properties, indicative of the low surface energy/ adsorption isotherms for betalains. Maltodextrin, which is extensively used for encapsulation of natural pigments (Choi et al. 2001), was less efficient than corn starch and dextrin white. Cyclodextrin, with its cyclic arrangement of dextrin molecules, generally capable of holding a vast number of natural flavor and pigment molecules (Furuya and Yoshi 2000), was a poor adsorbent of betalains, capable of holding only 17.1% of the pigment from the medium. These observations clearly indicate the involvement of specific interactions between adsorbent and adsorbate, where the net energy plays an important role rather than the contour of adsorbent particles, as has also been recorded by other such studies (Payne and Schuler 1988; Komaraiah et al. 2002).

The adsorbents are sometimes pre-treated with methanol before use (Sim and Chang 1993, 1997; Payne et al. 1988; Payne and Schuler 1988), as it is well known that such a treatment cause changes in adsorption energy (Snyder 1968). When the adsorbents (listed in Table 13.3) were pre-treated with methanol, there was a significant improvement in adsorption in the cases of alumina and silica, where alumina showed a very high adsorption of about 97%. Similarly, the nonionic resins XAD-2 and XAD-4 that adsorbed negligible quantities of pigment before pre-treatment adsorbed substantial levels, of about 46% and 42%, respectively, after methanol pre-treatment (Table 13.3). However, with the increase in adsorption capacities of these materials, there were concomitant reductions in pigment recovery during the desorption process, leading to a poor yield of betalains after desorption (Thimmaraju et al. 2004). Silica gel, wheat starch and corn starch, which were capable of moderate adsorption and desorption, did not change much even after methanol pre-treatment. Clearly, alumina, silica and the mixture of both without any pre-treatment were useful for in situ recovery of betalains released by RBHR. These three adsorbents were further evaluated to establish the effect of the time '*t*', because the knowledge of this value is important to accomplish complete adsorption of a particular concentration of solute from a solution/medium. Irrespective of the adsorbent, maximum adsorption of betalains occurred within the first 10 min and was complete within 20 min, with no net adsorption later (Thimmaraju et al. 2004). As observed for silica and alumina in other studies (Snyder 1968); only activated carbon was observed to exhibit higher '*t*' values, of 75 min (Shekinah et al. 2002) to 800 h (Nilsson 1970).

Since the solute concentration is one of the major parameters in adsorption process, using the '*t*' value, the effect of solute concentration on the adsorption process may

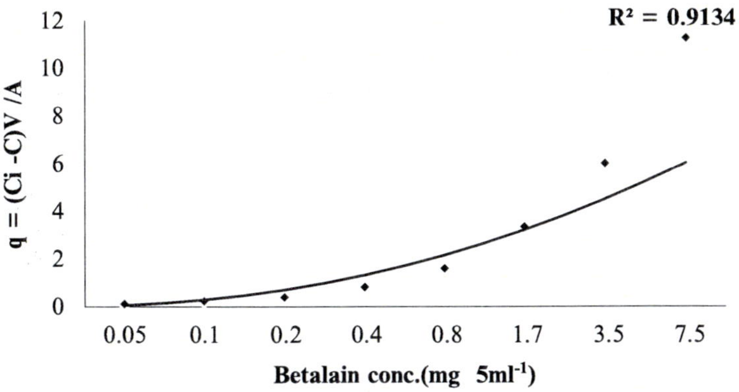

Fig. 13.6 Quantity of adsorbent (*q*) required for complete removal of different equilibrium concentrations of betalains. *Curve* shows the predicted value and *dots* represent experimental value

be discovered using the mixture of silica and alumina in a similar experimental model. Such trials conducted by Thimmaraju et al. (2004) showed that maximum adsorption occurred at the least solute concentration, with about 60% getting adsorbed at a very high solute level. Assuming that complete in situ adsorption occurs at the least pigment concentration, the adsorbent required to remove different equilibrium concentrations of pigment may then be calculated and presented as the '*q*' value using Eq. 13.1. As shown in Fig. 13.6, the predicted values (*curve*) matched well with the experimental values (*dots*) with a constant '*q*' value of 0.17 mg g^{-1} solute. By applying the same formula (Eq. 13.1), one can also predict the effect of volume, where the adsorbent required linearly increases with the increase in dilution of pigment in the medium. However, to check whether the adsorption process recorded here follows the standard adsorption phenomenon, such as the Langmuir isotherm, the following equation can be applied and the adsorption equilibrium can thus be calculated:

$$\frac{C_e}{q_e} = \frac{1}{Q_0 b} + \frac{C_e}{Q_0} \tag{13.2}$$

where C_e is the equilibrium concentration (mg ml^{-1}) and q_e is the amount adsorbed (mg g^{-1}) at equilibrium. Q_o and b are Langmuir's constants related to adsorption capacity and energy of adsorption respectively (obtained by plotting x/m and log x/m against different equilibrium concentrations). The plots of C_e/q_eVsC$_e$ are initially linear for the selected adsorbent, and the overall adsorption follows the Langmuir isotherm model for betalain adsorption (Fig. 13.7). The values of Qo and b were found to be 0.174 mg g^{-1} and 0.9, respectively, for RBHR pigments, which were calculated from the slope and intercept of Langmuir plot.

Based on the above properties, the quantity of adsorbent required to recover different levels of betalains using a silica:alumna mixture may be calculated by plotting (C_i–C) V/q versus the equilibrium concentration of betalains. Figure 13.7 shows

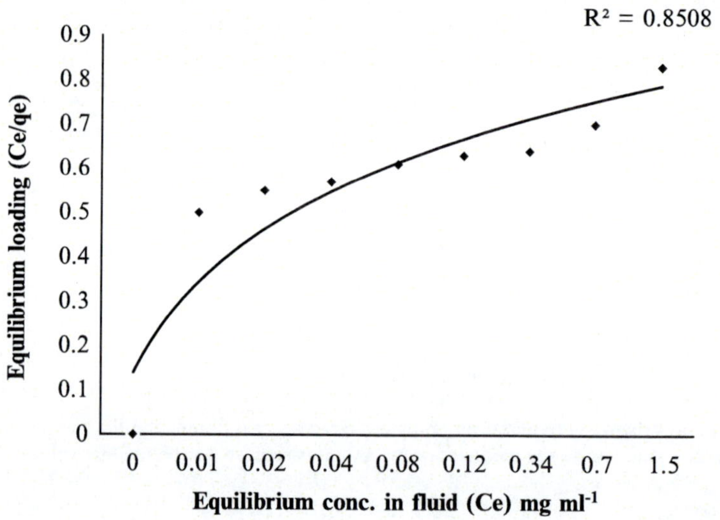

Fig. 13.7 Langmuir plot of betalains adsorption, showing predicted (*line*) and experimental (*dots*) values

the linearity of this relationship with the calculated values (*line*) correlating with the experimental values (*dots*). All in all, these types of preliminary observations made for in situ recovery of betalains are assumed to be very useful for rapid recovery of pigments permeabilized into the medium.

13.3.3 Ex situ Adsorption and Recovery

Since ex situ adsorption is normally conducted by passing the pigment solution through an adsorption column, it becomes essential to determine the best geometry of the column for both adsorption and further recovery of the pigment by desorption.

Extensive studies have been conducted on the recovery of pigments by passing it through adsorption columns, where the pigment passes through a series of adsorption plates leading to an array of interactions taking place between the adsorption material, pigment and the solvent (Snyder 1968; Altenau 1966; Cserháti 2006). In such cases, column geometry plays a significant role. After various permutations and combinations of column geometry and the adsorption material iterations, these authors found that a glass column of 2-cm diameter filled with adsorbent to a height of 6 cm (Fig. 13.8) was ideal. Keeping the column geometry constant, an initial screening showed that all adsorbent materials were capable of adsorption to various extents, where alumina alone showed 96% adsorption with no desorption occurring when eluted with 2% HCl water (Table 13.4). However, when used in combination

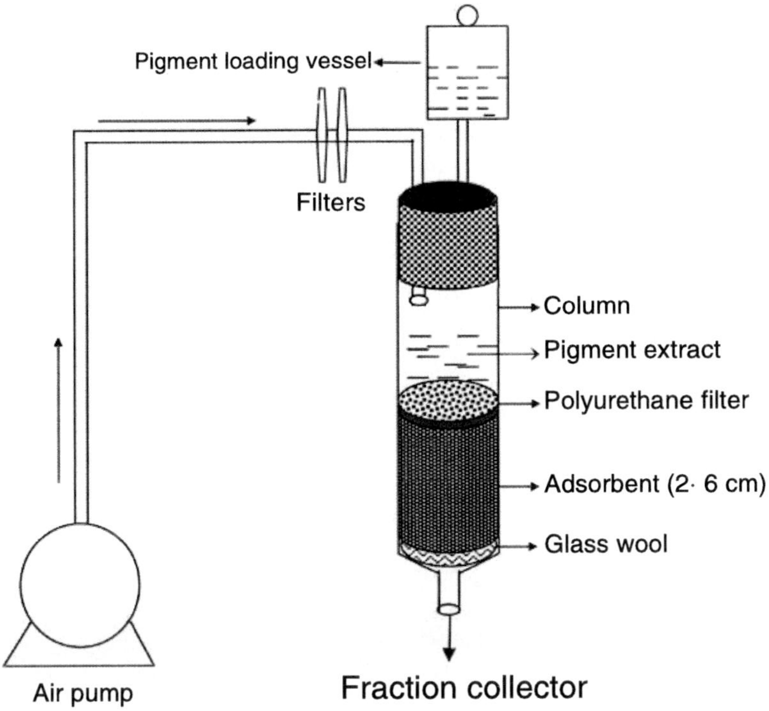

Fig. 13.8 Ex situ adsorption of betalains using a column for both normal and pressurized flow of pigment extract

Table 13.4 Adsorption and recovery of betalains[a] in a column[b] (ex situ) containing different adsorbents having the same geometry

Column material	% Adsorption	% Recovery[c]
Amberlite	0	0
Alumina	96	0
Alumina:silica (2:1)	94	47
Alumina:silica (1:1)	100	62
Alumina:silica (3:2)	100	55
Alumina:silica (7:3)	100	54
Silica gel	100	70
Alumina:sand (2:1)	97	84
XAD-2	0	0
XAD-4	0	0

[a] The content of pigment in the loaded solution was 24.4 mg in 15 ml
[b] The column containing the adsorbent was 3.5 cm in height and 2.0 cm in internal diameter
[c] Recovery by eluting with 2% (v/v) aqueous HCl, where immediately after elution the pH needs to be adjusted to 5.5 with 1% NaOH before quantifying the pigment by spectrophotometry

with other widely used adsorbents such as silica and activated sand, the adsorption property of alumina changed. A low level of 94% adsorption occurred when using alumina:silica at a ratio of 2:1, respectively. A similar ratio of alumina with processed sand showed 97% adsorption and a highest level of 84% desorption, and although the other alumina:silica combinations showed complete adsorption (100%), desorption was low (Table 13.4) (Thimmaraju et al. 2004).

In addition to fixed parameters, it also becomes necessary to predict the column performance under changed process conditions. A comparison of theoretical values with the available experimental values, followed by simulations, makes such predictions possible, where the column capacity for highest pigment adsorption may be calculated in advance. Generally, the column is said to have reached a state of adsorption equilibrium, a situation where no any adsorption energy remains at the surface of the adsorbent. This condition starts happening sequentially from top, progressing towards the bottom of the column, ultimately resulting in "adsorption breakthrough" where the concentration of pigment at the inlet and the outlet remains constant. At the time point when the effluent optical density starts steeply raising, the column is said to have reached a saturation point or the break point. In a model experiment for calculating the breakthrough points, an equilibrium concentration of pigment ($C_e = C_i = 0.18$ mg ml^{-1}; obtained as shown in Fig. 13.9) was loaded onto the fixed-bed column, where several variables, such as liquid velocity, pigment concentration at the mobile phase (C_m) and bed length, i.e., the distance travelled by the pigment extract, were considered at different feed flow rates, and the solute concentration at the outlet was recorded at fixed time intervals. Figure 13.10 shows that the flow patterns follow sigmoid curves and reach plateau immediately after the breakthrough point. At any constant equilibrium concentration, the very slow flow rate (0.02 ml s^{-1}) revealed breakthrough at 100 min (Fig. 13.10a), whereas a high flow rate resulted in early breakthrough. To properly design and operate a fixed-bed adsorption processes, the adsorption capacity, i.e., the breakthrough curves, must be known so that about 50% of the breakthrough value can be considered as the column adsorption capacity. A knowledge of the equilibrium relationship that determines the extent to which material can be adsorbed onto a particular surface will help quantify the adsorbent required for a specific unit operation of separation in an industrial-scale process. Several adsorption equilibrium data for various inorganic/organic compounds are available as adsorption isotherms (Choi et al. 2001). However, no such data was available for betalains, until the authors' report (Thimmaraju et al. 2004). In this study, when alumina-sand column material was used as the stationary phase, the isotherm of only alumina was considered because sand showed zero adsorption, functioning only as an improver of flow rate. The constant relationship 'k' between the quantum of pigment adsorbed onto alumina that had been calculated earlier for the in situ case ($r^2 = 0.9$) was almost similar for the ex situ column adsorption of different flow rates ($r^2 = 0.89$) (Fig. 13.10), indicating the small experiments are adequate to standardize unit operations. Here, increase in flow rate resulted in reduction of the x/m value. This behavior of liquid–solid adsorption is different from gas–solid adsorption, where, in the latter, increase in pressure (flow rate) generally causes an increase of x/m value. For normal flow rate

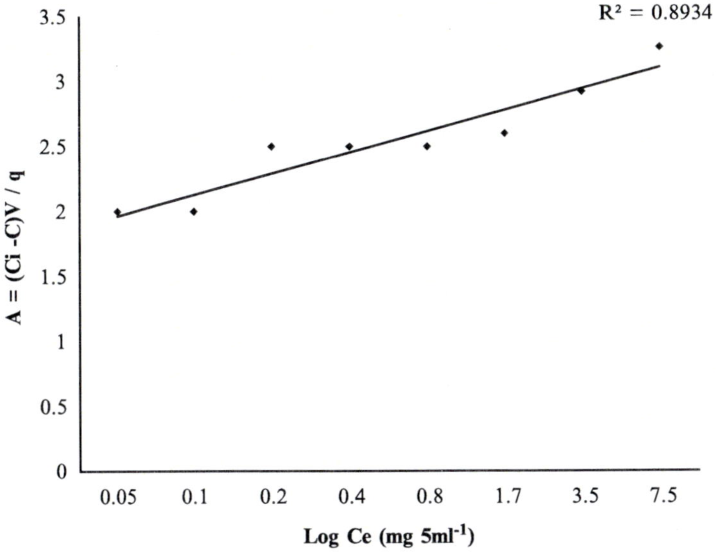

Fig. 13.9 Experimental and calculated values of the amount of adsorbent required to remove different equilibrium concentrations of betalains. The *line* indicates predicted values and the *dots* represent experimental data

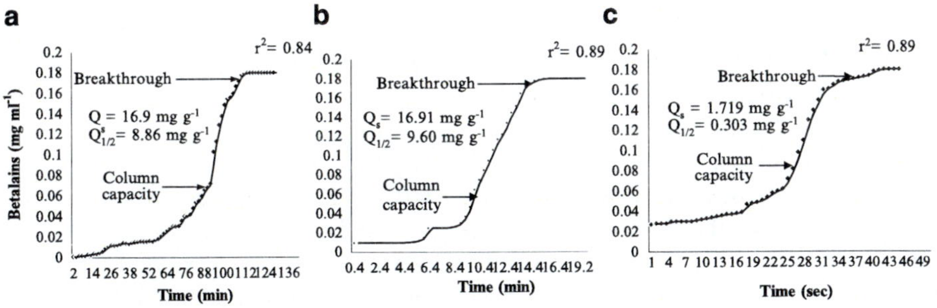

Fig. 13.10 Breakthrough curve for adsorption of betalains at normal flow rate (0.02 ml s^{-1}) (**a**), moderate flow rate (0.3 ml s^{-1}) (**b**), and high flow rate (3.1 ml s^{-1}) (**c**), where *dots* represent experimental values and the *line* shows calculated values

(e.g., 0.02 ml s^{-1}), using Eq. 13.2, Q_S and $Q_{1/2}$ were calculated, for which the flow details and elution pattern are plotted in Fig. 13.10. $Q_S = C_i - C_s/X$ and $Q_{1/2} = C_{i/2} - C_{1/2}/X$, which are 16.9 mg g^{-1} and 8.86 mg g^{-1} respectively. Although the higher flow rates of 0.3 ml s^{-1} resulted in early breakthrough, the net adsorptions at saturation ($Q_S = C_i - C_S/X$) and $Q_{1/2}$ were 16.91 mg g^{-1} and 9.60 mg g^{-1}, respectively (Fig. 13.10b), indicating that this is the ideal flow rate. A further higher flow rate of 3.1 ml s^{-1} (Fig. 13.10c) resulted in Q_S and $Q_{1/2}$ values of 1.72 mg g^{-1} and 0.30 mg g^{-1}, respectively. The low net adsorption values in the higher flow rates are due to a shorter contact period, where the 't' value is lower than the ideal 't' value of 20 min. A similar observation has been made for benzoic acid adsorption using activated charcoal (Chern and Chein 2001).

The pigment adsorbed by the column material (mg g^{-1}) at a given flow rate can be calculated by

$$Q_S = \frac{C_i - C_S}{X} \text{ and } Q_{1/2} = \frac{C_{i/2} - C_{1/2}}{X}$$

where C_i and $C_{i/2}$ are pigment loaded (mg) at saturation and breakthrough points, respectively; Q_S and $Q_{1/2}$ are the total quantity of pigment (mg g^{-1}) adsorbed at saturation and breakthrough point; C_S and $C_{1/2}$ are the quantity of pigment (equilibrium concentration, mg ml^{-1}) present in the effluent (un-adsorbed) at saturation and breakthrough points; and X is the quantity of adsorbent (g).

13.4 Desorption

It has been traditionally established that the adsorption phenomenon involves pH as one of the main functional parameters, where progressive decrease in pH results in linear desorption of natural pigments. For testing desorption/recovery of in situ adsorbed pigments, the adsorbent bags from in situ experiments may be removed and placed in a Buchner funnel and the un-adsorbed pigments may be removed by air suction. The adsorbed solute may further be desorbed by inserting the pigment-laden adsorbent bags in an appropriate volume of water with 2% hydrochloric acid (HCl) kept in a centrifuge tube, followed by centrifugation at 4°C for 5 min at 10,000 × *g*, where the desorbed pigments migrate to the supernatant, which needs to be immediately decanted and the pH brought back to 5.5, because long exposure of pigments to low pH may cause degradation of betalains. If the adsorbent bag appears to contain more pigment, the desorption step needs to be repeated. The pooled supernatant is adjusted to pH 5.5 before quantifying the pigments spectrophotometrically. Similarly, the pigments in the extract before and after the addition of adsorbent bags may also be quantified to ascertain the loss of pigment, which must be the same as that recovered from the adsorbent, and the difference in the value will explain the deviation. The ex situ adsorbed pigment could be eluted with different elution solvents and automatically collected as specific fractions until the pigment level in the effluent reaches a plateau. The ethanol and hydrochloric acid gradient (as shown in Fig. 13.11, where dilutions to obtain the respective concentration of each solvent were done using distilled water), was useful for the elution as well as separation. A fixed aliquot of each gradient mixture may then be allowed to flow through the column material where the required gradient mixture and the flow rates may be achieved by using HPLC pumps.

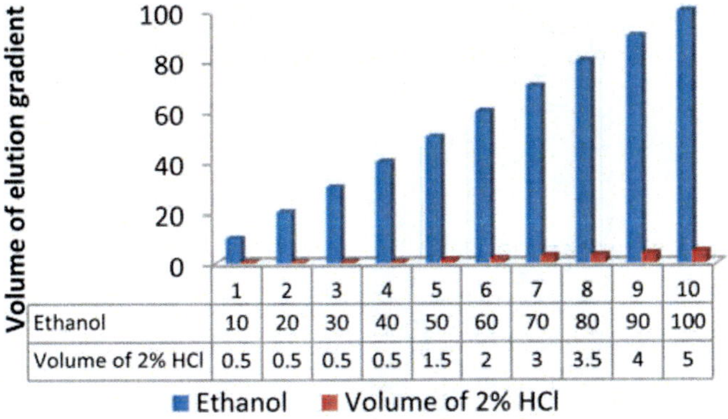

Fig. 13.11 Solvent gradient for elution and simultaneous separation of betacyanin and betaxanthin from silica–sand column

13.4.1 Separation of Betalains into Betaxanthin and Betacyanin

Development of an ex situ column for the separation of the two components of betalains, i.e., betaxanthin and betacyanin, is an important step in the production of an economically viable online production system for continuous production. Various adsorbent column materials and the elution solvents were employed to separate betalains into betaxanthin and betacyanin (Thimmaraju et al. 2004). Since the initial screening of various column materials indicated that a 3.5 × 2-cm column containing alumina and processed sand in the ratio of 2:1 was ideal for both maximum pigment adsorption and desorption, the efficacy of this column was further checked for further separation of betalains, where the elution parameter was changed using various elution solvents. While elution with an ethanol (EtOH) gradient in water with 2% HCl affected a small separation (Fig. 13.11), changing the acid strength from 2% to a gradient of 0.5–5% was found to cause an almost similar separation, inferring that yellow and purple pigments are not easily separable by these simple chromatographic methods. However, a series of experiments by the authors led to the best separation when the adsorbed pigment was eluted with the gradient composed of steadily increasing acid strength (Fig. 13.12), where initial fractions from 1 to 8 had only betaxanthin, followed by betacyanin (fractions 9–17) and a mixture of both (fractions 18–20). The initial fractions appeared bright yellow, followed by purplish red and a yellowish red (Fig. 13.12a). When the separated fractions were pooled into three groups, such as pool-1 comprising of all the yellow fractions (assumed as betaxanthins), followed by pool-2 comprising all the purplish red fractions and pool-3 comprising of all the mixed fractions (Fig. 13.12b), HPLC analysis of the pooled fractions showed a single peak each in pooled first and second group indicating the dominant presence either betaxanthin and betacyanin. The third set, which eluted last, showed three peaks, of which, two had different retention times than either betaxanthin or betacyanin, indicating the presence of other structurally different xanthine and cyanine pigments. Aronoff and Aronoff (1948) separated the beet

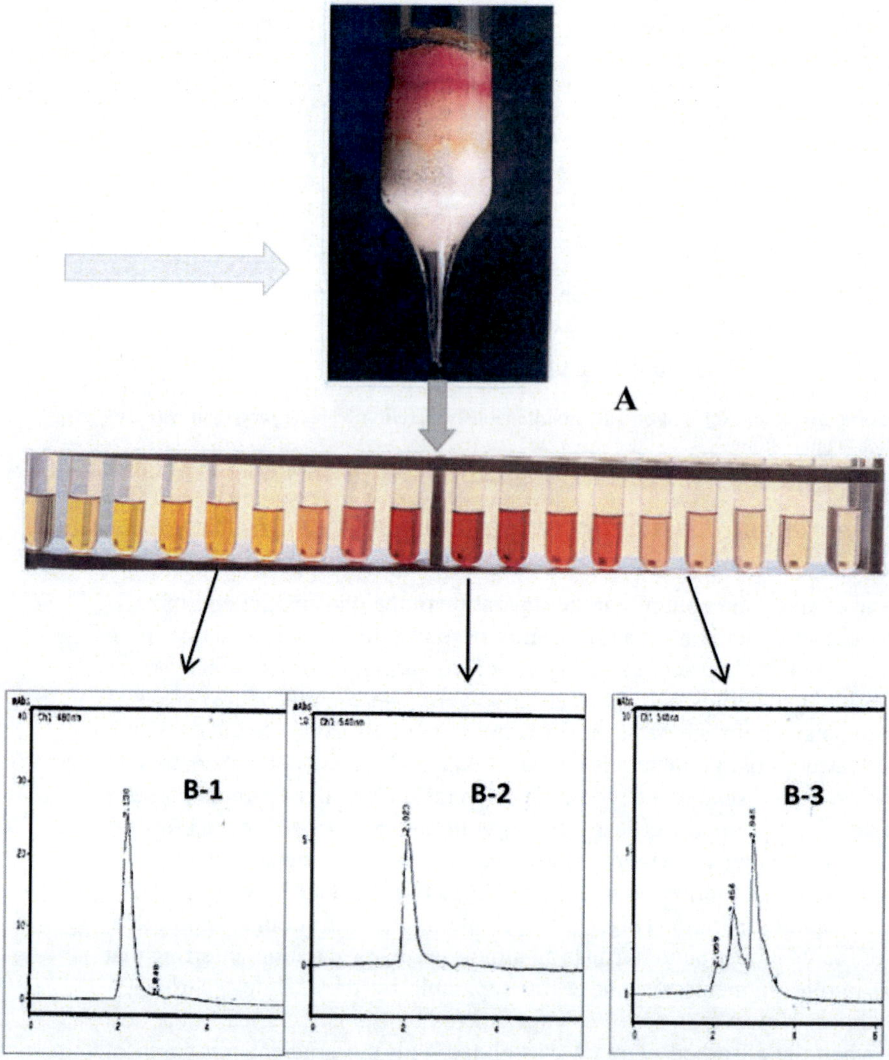

Fig. 13.12 Alumina:processed sand column (**a**) with the fractions separated in to betaxanthin, betacyanin and mixture of both, and the HPLC pattern of the eluted pigment (**B-1, B-2, B-3**) using a solvent gradient as in Fig. 13.10

pigments in an extract on a talc–siliceous earth column into at least 11 colored bands. However, several investigators have reported the separation of beet pigments to several fractions by electrophoresis (Lindstedt 1956; Wyler and Dreiding 1957; Peterson and Joslyn 1960).

Technologies that are more advanced replace the harsh treatment of molecules with milder treatments in a smart manner. While some of such extraction and separation methods have been compiled in Chap. 14, it is worth mentioning here that, in a

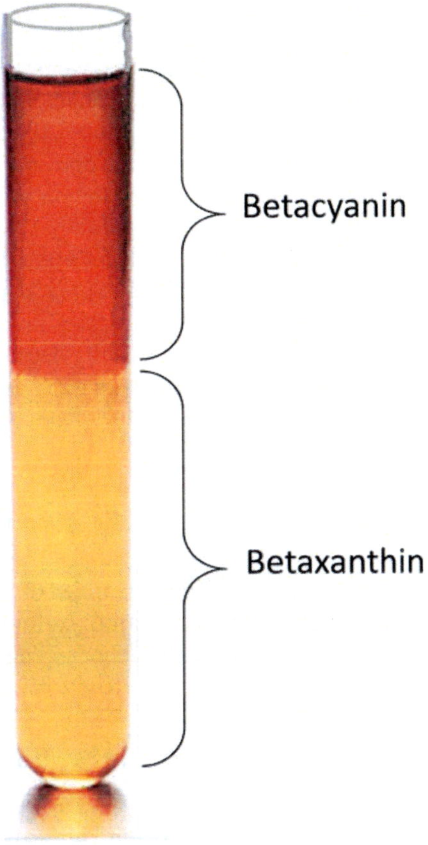

Fig. 13.13 Aqueous two-phase extraction and separation of betacyanin and betaxanthin from red beet hairy root extract

preliminary trial conducted by the authors, the use of the aqueous two-phase (ATP) system efficiently separated betacyanin and betaxanthin from RBHR. The use of polyethylene glycol (PEG) and ammonium sulfate (discussed in a subsequent section) were found efficient in separating the red and yellow components of beet color, although with some level of cross-contamination. Such differential partitioning may be linked to the structure of each pigment, where betacyanin with an aromatic ring preferentially partitioned to the top phase, leaving the betaxanthin in the bottom phase (Fig. 13.13). More details on ATP are available in Chap. 15.

13.5 Simultaneous Recovery of Betalains and Peroxidase by Aqueous Two-Phase Extraction (ATPE)

In addition to betalains, RBHRs synthesize high amount of peroxidases comprising various isoforms, of which, one isoform is spontaneously secreted (Thimmaraju et al. 2005, 2006, 2007). Since enzymes are large molecular weight proteins and betalains are very small molecules vulnerable to degradation, the simultaneous recovery of these two products from cultured RBHR is a greatly challenging proposition. Nevertheless, several smart engineering techniques, such as the liquid–liquid extraction process, can be safely executed. The important criterion for this type of extraction is the phase formation by two liquids. Aqueous two-phase systems (ATPS) are often used for large-scale continuous separation of proteins, natural pigments and removal of contaminants from fermentation broths, because they can rapidly produce an initial purification (Hart and Bailey 1991). ATPS can be used as a preliminary step in the purification of proteins and has been applied for the commercial purification of several products, including recombinant products. Thus the selection of the right combination of liquid mixtures involves judicious screening. Red beet has often been chosen as a model system, because of the delightful red color easily detectable in each liquid phase. Although Chap. 15 is dedicated to this purpose, some key features of hairy root-derived pigment separation are provided below.

13.5.1 Preparation of Aqueous Two-Phase System (ATPS)

ATPS prepared by mixing known quantities of polyethylene glycol (PEG) (of various molecular weights) and ammonium sulfate containing RBHR were screened (Neelwarne and Thimmaraju 2009). The length of separated phases, and hence their volumes in specific containers, indicate whether or not there is a distinct separation. Since enzymes are not visible to naked eyes, aliquots of the phases need to be assayed for enzyme activities, as well as for protein concentration. The partition coefficient (k) of the enzyme and pigment can be determined from the equation $k=(C_t/C_b)$, where C_t and C_b are the equilibrium concentrations in the top phase and bottom phase, respectively. Similarly, the recovery of the pigment and the enzyme can be calculated using the formula:

$$R = 100(Q_p / Q_t),$$

where R is the recovery in percentage, Q_p is the quantity in a particular phase, and Q_t is the total quantity taken for the experiment.

The preliminary set of experiments conducted using different phase compositions such as various molecular weight polyethylene glycols (PEG) and a constant ammonium sulfate concentration indicated that increasing the molecular weight of the PEG resulted in better phase formation, whereas low molecular weight PEG did not form any phases (Table 13.5). In this study it was observed that there was a tendency for peroxidases to move towards the aqueous and salt-rich bottom phase, as is known for many proteins, which salt out. However, it was bothersome that the distribution of the pigment was more or less even in both the phases, with a degradative loss of pigment ranging from 30% to 60%. Therefore, it was possible only to partition the maximum quantity of the enzyme to the bottom phase and the pigment to the top phase. From the table it is clear that, of the various mixtures, the ATPE system J appeared best (with a better trade-off) for the portioning and recovery of

Table 13.5 Effect of aqueous two-phase compositions on the partitioning and recovery of betalains and peroxidase

System			Recovery of pigment (%)		Recovery of peroxidase (%)		Partition coefficient	
PEG (%W/V)		Ammonium Sulfate (%W/V)	Top phase	Bottom phase	Top phase	Bottom phase	Top phase	Bottom phase
PEG 1,500								
A	4	13	Did not form distinct phases					
B	6	13						
C	8	13						
D	10	13						
PEG 4,000								
E	4	13	12.7	26.9	30.1	187.4	0.5	0.2
F	6	13	12.1	26.4	33.0	226.5	0.5	0.1
G	8	13	19.7	22.7	49.7	206.9	0.9	0.2
H	10	13	18.8	18.8	57.2	179.5	0.9	0.3
PEG 6,000								
I	4	13	25.7	41.1	21.0	124.9	0.6	0.2
J	6	13	25.5	38.1	42.4	137.7	0.7	0.3
K	8	13	10.8	30.0	27.7	124.4	0.4	0.2
L	*10*	*13*	*10.3*	*68.0*	*33.1*	*120.6*	*0.2*	*0.3*

PEG polyethylene glycol

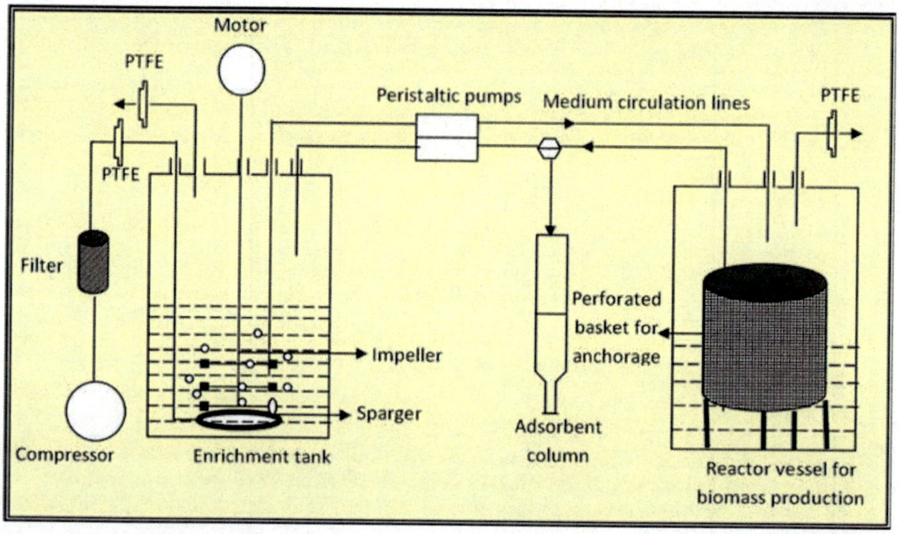

Fig. 13.14 Bioreactor model with an attached air enrichment tank where, the biomass can be permeabilized in situ and the released pigment can be passed through an adsorbent column for simultaneous separation and recovery of betaxanthin and betacyanin

betalains and peroxidase; the recovery of betalains in the top phase was only about 25%, whereas about 137% peroxidase activity was found in the bottom phase. The increase in recovery of the peroxidase above 100% may be because of the enhancement of the activity of the enzyme because of simultaneous purification in the salt phase (Miranda et al. 1998; Srinivas et al. 1999). When the partitioning behavior is considered, the partitioning of pigment to the top phase appeared to increase with increase in the molecular weight of PEG although there was a considerable amount of degradation of the pigment. On the other hand, for peroxidase, no definite trend was observed with respect to molecular weight of the PEG. However, the enzyme appeared to concentrate in the salt-rich bottom phase with enhancement in the total activity. The retention of considerable amount of pigment in the bottom phase and the enzyme in the top phase may however be improved further by studying various parameters such as different concentrations of NaCl, pH and temperature, as reported elsewhere (Srinivas et al. 1999; Miranda et al. 1998). Although there was a high partition coefficient for the pigment (0.9) in certain phase systems, there was a drastic reduction in the recovery of the pigment, possibly because of the degradation under such set of conditions. Probably because of these difficulties, most of the other studies with ATPE were focused only on one product, i.e., either enzyme or the pigment (Srinivas et al. 1999). Since betalains are very hydrophilic, one needs to conduct extensive experiments, taking into account the various kinetic parameters, to arrive at a congenial strategy for their simultaneous separation.

Integrating all the unit operations discussed above, the authors conducted a pilot-scale bioreactor study (Fig. 13.14), in which hairy roots were grown in a chamber supported with a basket and were bathed intermittently with aerated medium supplied from an aeration tank. In this system, there are ample opportunities for adding a permeabilizing agent and recovering the product through an online adsorption column, which may then be detached and the product may be separated into red and yellow pigment (Neelwarne and Thimmaraju 2009). The spent medium may then be stoichiometrically enriched.

13.6 Concluding Remarks

Product recovery and further concentration are of crucial importance, as downstream processing often accounts for 50–90% of the total production costs (Dwyer 1984; Sahai 1994). Although barely a handful of experiments have been conducted on the online product recovery from RBHR cultures, such studies have significantly contributed to our knowledge and form a basis for adopting such strategies in the future for red beet as well as other secondary metabolites. Permeabilizing higher plant cells in a desirable manner while retaining good extent of their viability has always remained a great challenge. Compared with normal beet tissues, hairy roots are easier to permeabilize because of their large surface area rendered by their hairs and delicate texture and they probably also carry inherent secretory signals since they exude opines when they naturally exist on their host plants. The fact that certain peroxidase isozymes are selectively released into the medium (Thimmaraju et al. 2005; see also Chap. 12) also supports their inherent secretory nature. The selective permeabilization of either betalains or peroxidases need a thorough characterization of each compound as well as of the cellular behavior in response to a particular permeabilizing agent. A few attempts by different research groups have also been made for the application of ATPE for the simultaneous recovery of the betalains and peroxidase from normal red beet as well as from cultured RBHR. Normal beet has served as a model system for several advanced separation techniques. A prototype reactor model for continuous biomass production and product recovery by permeabilization with options for online recovery of betalains and peroxidase enzyme suggested in this review forms the basis for other such biotechnological processes.

References

Altenau, A.G. 1966. Chromatography as an analytical tool. *Annals of the New York Academy of Sciences* 137: 335–359.

Aronoff, S., and E.M. Aronoff. 1948. Thermal degradation of dehydrated beets. II. Chromatographic separation of red beet-root pigments. *Food Research* 13: 59.

Bais, H.P., G. Sudha, B. Suresh, and G.A. Ravishankar. 2001. Permeabilization and *in situ* adsorption studies during growth and coumarin production in hairy root cultures of *Cichorium intybus*. *Indian Journal of Experimental Biology* 14: 231–239.

Barefoot, S.F., and C.G. Netless. 1993. Antibiosis revisited: bacteriocins produced by dairy starter cultures. *Journal of Dairy Science* 76: 2366–2379.

Bates, G.W. 1990. Transfer of tobacco mosaic virus resistance by asymmetric protoplast fusion. In *Progress in plant cellular and molecular biology*, ed. H.J.J. Nijkamp, L.H.W. van der Plas, and J. van Aartrijk, 293–298. Dordrecht/The Netherlands: Kluwer Academic.

Becker, H., J. H. Reichling, W. Bisson, S. Herald. 1984. Two phase culture-a new method to yield liphophilic secondary products from plant suspension cultures. Proceedings of the 3rd European Congress of Biotechnology. Munisch. Vol 1, 209–213.

Benson, E., and J.D. Hamill. 1991. Cryopreservation and post-freeze molecular and biosynthetic stability in transformed roots of *Beta vulgaris* and *Nicotiana rustica. Plant Cell, Tissue and Organ Culture* 24: 163–172.

Berlin, J., C. Mollenschot, and C. Wray. 1988. Triggered efflux of protoberberine alkaloids from cell suspension cultures of *Thalictrum rugusum. Biotechnology Letters* 10: 193–198.

Brodelius, P.E., and K. Nilsson. 1983. Permeabilization of immobilized plant cells resulting in release of intracellular stored products with preserved cell viability. *European Journal of Applied Microbiology and Biotechnology* 17: 275–280.

Brodelius, P. 1988. Permeabilization of plant cells for release of intracellularly stored products: viability studies. *Applied Microbiology and Biotechnology* 27: 561–566.

Byun, S.K., and H. Pedersen. 1994. Two – phase airlift fermentor operation with elicitation for the enhanced production of benzophenanthridine alkaloids in cell suspensions of *Escherichia californica. Biotechnology and Bioengineering* 44: 14–20.

Byun, S.K., H. Pedersen, and C. Chin. 1990. Two-phase culture for the enhanced production of benzophenandrine alkaloids in cell suspensions of *Escholtzia californica. Photochemistry* 29: 3135–3139.

Chalermchat, Y., M. Fincan, and P. Dejmek. 2004. Pulsed electric field treatment for solid–liquid extraction of red beetroot pigment: mathematical modelling of mass transfer. *Journal of Food Engineering* 64: 229–236.

Chao, H.P., and W.C. Lee. 2000. A bioelectrode for penicillin detection based on gluten-membrane-entrapped microbial cells. *Biotechnology and Applied Biochemistry* 32: 9–14.

Chern, J.M., and Y.W. Chein. 2001. Adsorption isotherms of benzoic acid onto activated carbon and breakthrough curves in fixed-bed columns. *Industrial Engineering and Chemical Research* 40: 3775–3780.

Chethana, S., C.A. Nayak, and K.S.M.S. Raghavarao. 2007. Aqueous two phase extraction for purification and concentration of Betalains. *Journal of Food Engineering* 81: 679–687.

Choi, J.W., G.H. Cho, S.Y. Byun, and D.I. Kim. 2001. Integrated bioprocessing for plant cell cultures. In *Advances in biochemical engineering principles: Plant cells*, ed. T. Scheper and J.J. Zhong, 63–103. Berlin: Springer.

Cserháti, T. 2006. *Liquid chromatography of natural pigments and synthetic dyes*. The Netherlands: Elsevier.

Darvishi, P., S. Ayatollahi, D. Mowla, and A. Niazi. 2011. Biosurfactant production under extreme environmental conditions by an efficient microbial consortium, ERCPPI-2. *Colloids and Surfaces B: Biointerfaces*. doi:10.1016/j.colsurfb.2011.01.011.

Deno, H., C. Suga, T. Morimoto, and Y. Fujita. 1987. Production of shikonin derivatives by cell suspension cultures of *Lithospermum erythrorhizon*. VI. Production of shikonin derivatives by a two layer culture containing an organic solvent. *Plant Cell Reports* 6: 197–199.

Deveoglu, O., R. Karadag, and T. Yurdunc. 2009. Preparation and HPLC analysis of the natural pigments obtained from Buckthorn (*Rhamnus petiolaris* Boiss) Dye Plants. *Jordan Journal of Chemistry* 4: 377–385.

Dilorio, A.A., P.T. Weathers, and R.D. Cheetham. 1993. Non-lethal secondary product release from transformed root cultures of *Beta vulgaris. Applied Microbiology and Biotechnology* 39: 174–180.

Dwyer, J.L. 1984. Scaling up of bioproduct separation with HPLC. *Biotechnology* 2: 975–984.

Fincan, M., F. DeVito, and P. Dejmek. 2004. Pulsed electric field treatment for solid–liquid extraction of red beetroot pigment. *Journal of Food Engineering* 64: 381–388.

Furuya, T., and H. Yoshi. 2000. Cyclodextrin encapsulation to prevent the loss of 1-menthol and its restriction during drying. *Bioscience, Biotechnology, and Biochemistry* 64: 1608–1613.

Gelli, A., V.J. Higgins, and E. Blumwald. 1997. Activation of plant plasma membrane Ca^{2+} permeable channels by race-specific fungal elicitors. *Plant Physiology* 113: 269–279.

Hart, R.A., and J.E. Bailey. 1991. Purification and aqueous two-phase partitioning properties of recombinant vitreoscilla hemoglobin. *Enzyme and Microbial Technology* 13: 788–795.

Hepler, P.K. 2005. Calcium: a central regulator of plant growth and development. *The Plant Cell* 17: 2142–2155.

Herbach, K.M., F.C. Stintzing, and R. carle. 2006. Betalain stability and degradation-structural and chromatic aspects. *Journal of Food Science* 71: R41–R50.

Joshi, M.S., L.R. Gowda, and S.G. Bhat. 1987. Permeabilization of Yeast cells (*Kluyveromyces fragilis*) to lactose by Cetyltrimethylammonium bromide. *Biotechnology Letters* 9: 549–554.

Joshi, S.C., B.S. Jha, S. Yadav, A. Nerurkar, and A.J. Desai. 2008. Biosurfactant production using molasses and whey under thermophilic conditions. *Bioresource Technology* 99: 195–199.

Kilby, N.J., and C.S. Hunter. 1990. Repeated harvest of vacuole located secondary product from *in vitro* grown plant cells using ultrasound. *Applied Microbiology and Biotechnology* 33: 448–456.

Kilby, N.J., and C.S. Hunter. 1991. Towards optimization of the use of 1.02 MHz ultrasound. *Applied Microbiology and Biotechnology* 4: 283–301.

Kim, J.D., and N.H. Chang. 1990. Enhanced shikonin production from *Lithospermum erythrorhizon* by *in situ* extraction and calcium alginate immobilization. *Biotechnology and Bioengineering* 36: 460–466.

Kino-Oka, M., Y. Hongo, M. Taya, and S. Tone. 1992. Culture of red beet hairy root in bioreactor and recovery of pigment released from cells by repeated treatment of oxygen starvation. *Journal of Chemical Engineering of Japan* 25: 490–495.

Komaraiah, P., R. Naga Amrutha, P.B. Kavi Kishore, and S.V. Ramakrishna. 2002. Elicitor enhanced production of plumbagin in suspension cultures of *Plumbago rosea* L. *Enzyme and Microbial Technology* 31: 634–639.

Levine, A., R.I. Pennell, M.E. Alvarez, R. Palmer, and C. Lamb. 1996. Calcium-mediated apoptosis in a plant hypersensitive disease resistance response. *Current Biology* 6: 427–437.

Lindstedt, G. 1956. Electrophoresis of the red beet pigments. *Acta Chemica Scandinavica* 10: 698.

Loginova, K.V., N.I. Lebovka, and E. Vorobiev. 2011. Pulsed electric field assisted aqueous extraction of colorants from red beet. *Journal of Food Engineering* 106: 127–133.

Martinez, B.C., and C.H. Park. 1993. Characteristics of batch suspension cultures of preconditioned *Coleus blumei* cells: sucrose effect. *Biotechnology Progress* 9: 97–100.

Miranda, M.V., H.M. Fernández-Lahore, J. Dobrecky, and O. Cascone. 1998. The extractive purification of peroxidase from plant raw materials in aqueous two-phase systems. *Acta Biotechnologica* 18: 179–188.

Mukundan, U., E.B. Carvallo, and W.R. Curtis. 1998. Growth and pigment production by hairy root cultures of *Beta vulgaris* L. in a bubble column reactor. *Biotechnology Letters* 20: 469–474.

Muranaka, T., H. Ohakawa, and Y. Yamada. 1992. Scopolamine release into media by *Duboisia leichhardtti* hairy root clones. *Applied Microbiology and Biotechnology* 37: 554–559.

Neelwarne, B., and R. Thimmaraju. 2009. Bioreactor for cultivation of red beet hairy roots and in situ recovery of primary and secondary metabolites. *Engineering in Life Sciences* 9: 227–238.

Nilsson, T. 1970. Studies in to the pigments in beet root (*Beta vulgaris* L. Sp. *vulgaris* var. *rubra* L). *Lantbrukshogsk Annaler* 36: 179–219.

Park, C.H., and B. Martenez. 1994. Growth and production characteristics of permeabilized *Coleus blumei* cells in immobilized fed-batch culture. *Plant Cell Reports* 13: 459–463.

Park, C.H., and B. Martinez. 1992. Enhanced release of rosmarinic acid from *Coleus blumei* permeabilized by dimethyl-sulfoxide (DMSO) while preserving cell viability and growth. *Biotechnology and Bioengineering* 40: 459–464.

Parr, A.J., R.J. Robins, and M.J.C. Rhodes. 1984. Permeabilization of *Cinchona ledgeriana* cells by dimethylsulphoxide. Effects on alkaloid release and long – term membrane integrity. *Plant Cell Reports* 3: 262–265.

Payne, G.F., and M.L. Schuler. 1988. Selective adsorption of plant products. *Biotechnology and Bioengineering* 31: 922–928.

Payne, G.F., N.N. Payne, M.L. Schuler, and M. Asada. 1988. *In situ* adsorption for enhanced production by *Catharanthus roseus*. *Biotechnology Letters* 10: 187–192.

Pedersen, H., C.K. Chin, and A. Dutta. 1999. Yield improvement in plant cell cultures by *in situ* extraction. In *Plant cell and tissue culture for the production of food ingredients*, ed. T.-J. Fu, G. Singh, and W.R. Curtis, 129–138. New York: Kluwer.

Peterson, M. 1999. Biosynthesis and accumulation of rosmarinic acid in plant cell cultures. In *Plant cell and tissue culture for the production of food ingredients*, ed. T.-J. Fu, G. Singh, and W.R. Curtis, 129–138. New York: Kluwer.

Peterson, R.G., and M.A. Joslyn. 1960. The red beet pigment of the root of the beet (*Beta vulgaris*) as a pyrrole compound. *Food Research* 25: 429.

Pradedova, E.V., O.D. Isheeva, and R.K. Salyaev. 2009. Superoxide dismutase of plant cell vacuoles. *Membrane & Cell Biology* 3: 24–32.

Pradedova, E.V., O.D. Isheeva, and R.K. Salyaev. 2011. Antioxidant defense enzymes in cell vacuoles of red beet roots. *Russian Journal of Plant Physiology* 58: 36–44.

Renaudin, J. 1981. [^{14}] Tabernanthine, by cell suspension cultures of *Catharanthus roseus* (L.) G. Don and *Acer pseudoplantanus* L. *Plant Science Letters* 22: 59–69.

Reuffer, M. 1985. The production of isoquinone alkaloids by plant cell cultures. In *The chemistry and biology of isoquinoline alkaloids*, ed. J.D. Philipson, N.F. Roberts, and M.H. Zenk, 265–280. Berlin/Heidelberg/FRG: Springer.

Reznik, H. 1980. Betalains. In *Pigments in plants*, ed. F.C. Czygan, 370–392. Stuttgart/New York: Fischer.

Ruiz-May, E., R.M. Galaz-Ávalos, and V.M. Loyola-Vargas. 2009. Differential secretion and accumulation of Terpene Indole alkaloids in hairy roots of *Catharanthus roseus* treated with methyl jasmonate. *Molecular Biotechnology* 41: 278–285.

Saenz-carbonell, L.A., I.E. Maldoredo-mendoza, O. Moreno-valenzula, R. Ciau-uitz, M. Lopez-meyer, C. Oropeza, and V.M. Loyola-vergas. 1993. Effect of medium pH on the release of secondary metabolites from roots of *Datura stramonium*, *Catheranthus roseas* and *Tagetes patula* cultured *in vitro*. *Applied Biochemistry and Biotechnology* 38: 257–267.

Sahai, O. 1994. Plant tissue culture. In *Bioprocess production of flavour, fragrance and color ingredients*, ed. A. Gabelman, 239–275. New York: Wiley.

Shekinah, P., K. Kadirvelu, P. Kanmani, and V. Subburam. 2002. Adsorption of lead (II) from aqueous solution by activated carbon prepared from *Eichhornia*. *Journal of Chemical Technology and Biotechnology* 77: 458–464.

Shimomura, K., H. Sudo, H. Saga, and H. Kamad. 1991. Shikonin production and secretion by hairy root cultures of *Lithospermum erythrorhizon*. *Plant Cell Reports* 10: 282–285.

Sim, S.J., and H.N. Chang. 1993. Increased shikonin production by hairy roots of *Lithospermum erythrorhizon* in two-phase bubble column reactor. *Biotechnology Letters* 15: 145–150.

Sim, S.J., and H.N. Chang. 1997. Shikonin production by hairy roots of *Lithospermum erythrorhizon* in bioreactors with *in situ* separation. In *Hairy roots: culture and applications*, ed. P.M. Doran, 219–225. The Netherlands: Harwood Academic.

Singh, M. 1994. Production of microbial surfactants along with single cell protein (SCP)/single cell oil (SCO). In: *Microbes for better living*, ed. R. Sankaran, K. S. Manja, 605–612. 35th AMI conference 9–12 Nov 1994.

Sivakumar, G., F. Medina-Bolivar, J.O. Lay, M.C. Dolan, J. Condori, S.K. Grubbs, S.M. Wright, M.A. Baque, E.J. Lee, and K.Y. Paek. 2011. Bioprocess and bioreactor: next generation technology for production of potential plant-based antidiabetic and antioxidant molecules. *Current Medicinal Chemistry* 18: 79–90.

Snyder, L.R. 1968. *Principles of adsorption chromatography: the separation of nonionic organic compounds*, 75–125. New York: Marcel Decker.

Srinivas, N.D., K.R. Rashmi, and K.S.M.S. Raghavarao. 1999. Extraction and purification of a plant peroxidase by aqueous two-phase extraction coupled with gel filtration. *Process Biochemistry* 35: 43–48.

Sticher, L., C. Penel, and H. Greppin. 1981. Calcium requirement for secretion of peroxidase by plant cell suspensions. *Journal of Cell Science* 48: 345–353.

Stintzing, F.C., and R. Carle. 2007. Betalains-emerging prospects for food scientists. *Trends in Food Science & Technology* 18: 514–525.

Sudha, G., and G.A. Ravishankar. 2002. Involvement and interaction of various signaling compounds on the plant metabolic events during defense response, resistance to stress factors, formation of secondary metabolites and their molecular aspects. *Plant Cell, Tissue and Organ Culture* 71: 181–212.

Suresh, B., R. Thimmaraju, N. Bhagyalakshmi, and G.A. Ravishankar. 2004. Polyamine and methyljasmonate influenced enhancement of betalain production in hairy root cultures of *Beta vulgaris* in a bubble column reactor and studies on efflux of pigments. *Process Biochemistry* 39: 2091–2096.

Tanaka, H., C. Hirao, H. Semba, and Y. Tozawin. 1985. Release of intracellularly stored 5′ phosphodiesterase with preserved plant cell viability. *Biotechnology and Bioengineering* 27: 890–892.

Taya, M., K. Mine, M. Kino-Oka, S. Tone, and T. Ich. 1992. Production and release of pigments by culture of transformed hairy root of red beet. *Journal of Fermentation and Bioengineering* 73: 31–36.

Thimmaraju, R., N. Bhagyalakshmi, M.S. Narayan, and G.A. Ravishankar. 2003a. Kinetics of pigment release from hairy root cultures of red beet under the influence of pH, sonication, temperature and oxygen stress. *Process Biochemistry* 38: 1067–1074.

Thimmaraju, R., N. Bhagyalaskshmi, M.S. Narayan, and G.A. Ravishankar. 2003b. Food grade chemical and biological agents permeabilize red beet hairy roots assisting the release of betalains. *Biotechnology Progress* 19: 1274–1282.

Thimmaraju, R., N. Bhagyalakshmi, and G.A. Ravishankar. 2004. *In situ* and *ex situ* adsorption and recovery of betalaines from hairy root cultures of *Beta vulgaris*. *Biotechnology Progress* 20: 777–785.

Thimmaraju, R., V. Kumar, L. Venkatachalam, N. Bhagyalakshmi, and G.A. Ravishankar. 2005. Peroxidase production from hairy root cultures of red beet (*Beta vulgaris*). *Electronic Journal of Biotechnology* 8: 185–196.

Thimmaraju, R., L. Venkatachalam, R.V. Sreedhar, N. Bhagyalakshmi, M.S. Narayan, and G.A. Ravishankar. 2006. Elicitation of peroxidase from hairy root cultures of red beet. *Electronic Journal of Biotechnology* 9: 512–521.

Thimmaraju, R., L. Venkatachalam, K. Roohie, M.S. Narayan, and N. Bhagyalakshmi. 2007. Purification and characterization of an intracellular peroxidase from genetically transformed roots of red beet (*Beta vulgaris* L.). *Food Chemistry* 105: 1312–1320.

Von-Elbe, J.H. 1979. The betalaines. In *Current aspects of food colorants*, ed. T.E. Furuya, 29–39. London: CRC Press.

Wang, L., and C.L. Weller. 2006. Recent advances in extraction of nutraceuticals from plants. *Trends in Food Science and Technology* 17: 300–312.

Weathers, P.T., A. Dilorio, R. Cheetham, and M.R. O'leary. 1990. Recovery of secondary metabolites with minimal loss of cell viability. In *Progress in plant cellular and molecular biology*, ed. A. Nijkamp, L.H.M. Plas, and J. Aartrizk, 582–586. New York: Kluwer.

Wink, M. 1997. Compartmentation of secondary metabolites and xenobiotics in plant vacuoles. In *Advances in botanical research*, The plant Vacuole, 25th ed, ed. R.A. Leigh and D. Sanders, 141–169. San Diego/London: Academic.

Wong, P.L., A.J. Royce, and C.W.T. Lee-Parsons. 2004. Improved ajmalicine production and recovery from *Catharanthus roseus* suspensions with increased product removal rates. *Biochemical Engineering Journal* 21: 253–258.

Wyler, H., and A.S. Dreiding. 1957. Krisellisiertes betanin. *Helvetica Chimica Acta* 40: 191.

Yang, R.Y.K., O. Bayraktar, and H.T. Pu. 2003. Plant-cell bioreactors with simultaneous electropermeabilization and electrophoresis. *Journal of Biotechnology* 100: 13–22.

Zhang, W., C. Curtin, and C. Franco. 2002. Towards manipulation of post-biosynthetic events in secondary metabolism of plant cell cultures. *Enzyme and Microbial Technology* 30: 688–696.

Zhong, J.-J., X.-D. Meng, Y.-H. Zhang, and S. Liu. 1997. Effective release of ginseng saponin from suspension cells of *Panax notoginseng*. *Biotechnology Techniques* 11: 241–243.
Zimmermann, S., T. Nürnberger, J.-M. Frachisse, W. Wirtz, J. Guern, R. Hedrich, and D. Scheel. 1997. Receptor-mediated activation of a plant Ca^{2+} -permeable ion channel involved in pathogen defense. *Proceedings of the National Academy of Sciences USA* 94: 2751–2755.
Zimmermann, U. 1986. Electrical breakdown, electropermeabilization and electrofusion. *Reviews of Physiology, Biochemistry and Pharmacology* 105: 176–250.

Chapter 14
Extraction of Red Beet Pigments

Brijesh K. Tiwari and Patrick J. Cullen

Abstract Red beet is a rich source of water-soluble betalains, which have application in the food and pharmaceutical industries, owing to their visual and health promoting functional properties. Conventional extraction techniques suffer from certain disadvantages, such as low yields, stability problems, solvent contaminations and batch-to-batch variations. Alternative extraction techniques such as permeabilization, pulsed electric field extraction, membrane processing, supercritical extraction, sonication, and nano grinding offer benefits of safety and eco friendly processing steps, higher recovery rates and greater uniformity of the product.

14.1 Introduction

Betalains are water-soluble nitrogen-containing pigments that are synthesised from the amino acid tyrosine into two structural groups: the red–violet betacyanins and the yellow–orange betaxanthins. The term betalain originates from the Latin name of beet root (*Beta vulgaris*), from which betalains were first extracted. Compared with other pigment classes, such as the carotenoids, chlorophylls and anthocyanins, the betalains have been studied to a lesser degree (Stintzing et al. 2006). To date, betalains comprise about 55 structures, including the red–violet betacyanins and the yellow–orange betaxanthins. The betalains in red beet (*Beta vulgaris* L.) consist of

B.K. Tiwari (✉)
Department of Food and Consumer Technology, Manchester Metropolitan University, Manchester M14 6HR, UK
e-mail: b.tiwari@mmu.ac.uk

P.J. Cullen
School of Food and Environmental Health, Dublin Institute of Technology, Dublin, Dublin 1, Ireland
e-mail: pjcullen@dit.ie

B. Neelwarne (ed.), *Red Beet Biotechnology: Food and Pharmaceutical Applications*,
DOI 10.1007/978-1-4614-3458-0_14,

betanin, isobetanin, betanidin, isobetanidin, betaxanthins and some other yellow pigments (Schwartz and von Elbe 1980; Vincent and Scholz 1978). The betalains are natural pigments containing betalamic acid as the chromophore in their structure. Betalamic acid conjugates with *cyclo*-3,4-dihydroxyphenylalanine (DOPA) to produce the red–violet-colored betacyanins, although, if the conjugation is with different amino acids or amines, yellow betaxanthins are produced (Strack et al. 2003; Svenson et al. 2008). The structure of betacyanins consists of betanidin, a glycone that varies with substitutions of acyl groups and sugar moieties. There are two categories of betalains: betacyanins include the reddish to violet betalain pigments and betaxanthins are those betalain pigments that appear yellow to orange. More details regarding the pigments of red beet are available in Chap. 1 of this book. In general, betalains as water-soluble nitrogen-containing pigments are restricted to plants of the Caryophylles family, including *Amaranthaceae* (Cai et al. 2001a). Although these pigments impart high hue value, their stability poses a great challenge during their extraction, storage and incorporation into food and pharmaceutical products. Their health benefits, even after their degradation (see Chap. 7 of this book), is a driving factor behind the food industry's interest in this natural pigment over synthetic colorants. Red beet (*Beta vulgaris*) is the commercial source of betalains. The advantage of betalain color is that the color appearance (to the human eye) does not fluctuate much from pH 3 to pH 7, exhibiting better pH stability (Herbach et al. 2006b; Jackman and Smith 1996) than that of anthocyanins (Tanaka et al. 2008). Similarly, Cai et al. (1998a) showed that the betacyanin pigments from most of the *Amaranthus* genotypes indicated the presence of bright red–violet color and favourable stability in solution at low temperatures in the dark and in the absence of air over the pH range of 5–7, showing good stability at room temperature during long-term storage. These authors also suggested that potential natural colorants are feasible with stability comparable or better than that of red radish anthocyanins. For these characteristics, betalains have emerging interest in food industries and are well reviewed by Stintzing and Carle (2007). This chapter discusses various techniques for the extraction of betalains from natural sources.

14.2 Sources of Betalains

Betalains are found in vacuoles of a wide variety of plants such as red beet roots, cactus fruits, prickly pear, foliages of chards, flowers of bougainvillea and Amaranthus plants. For commercial success, the content of betalains in the source material is very important. Although there are red and yellow beet varieties that purely display betacyanin (reddish purple pigment) or betaxanthins (yellow pigments), respectively, in the common commercially available red beets, the level of betacyanin among red beet varieties varies from 0.44 to 0.60 g of betalain/kg and the yellow pigments ranges from 0.32 to 0.42 g of vulgaxanthin 1/kg (Gasztonyi et al. 2001; von Elbe 1975; Nilsson 1970; Psczczola 1998; Gaertner and Goldman 2005;

Azeredo 2009). Kanner et al. (2001) also determined that red beet contains high amounts of betanin, ranging from 30 to 60 mg/100 g fresh weight (FW), with lower concentrations of isobetanin, betanidin and betaxanthins. However, genetically transformed hairy roots of red beet in cultured conditions showed total betalain contents of >1% on a dry weight (DW) basis (see Chaps. 10, 11 and 13). Among the other sources of betalains, prickly pear cultivars ranging from yellow to red contained total betalains up to 9 mg/100 g fresh pulp (Butera et al. 2002), while the red pitaya cultivars contained concentrations ranging between 0.23 and 0.39 mg/g with the betacyanins predominating. In another study, Kugler et al. (2004) reported that Swiss chard contains about 51.1 μg of betacyanins/g FW and 49.7 μg of betaxanthins/g FW. The red-skinned ulluco tubers are reported to contain mostly betaxanthins, ranging from 41.2 to 70.4 μg/g FW (Svenson et al. 2008). Additionally, the total betacyanin contents of Amaranthus were reported to vary from 46 to 199 mg/100 g FW (Cai et al. 1998a). These water-soluble pigments are generally present at high concentrations (~1% of the total solids) in the vacuole of the plant matrix (see also Chap. 4).

14.3 Extraction Techniques

Extraction of betalains from various pant sources can be executed by employing various conventional techniques using food-grade solvents. Extraction techniques employed for the production of commercial betalains in the food and pharmaceutical industries not only depends on the extraction process parameters but also but also on the availability of sources (Delgado-Vargas et al. 2000; Cai et al. 2005). Red beet (*Beta vulgaris* L.) is widely used for the industrial production of betalains (Stintzing and Carle 2008a, b). Table 14.1 lists some of the techniques employed for the extraction of betalains from a range of plant sources. The major processing methods are of a conventional type, although there is a need to develop non-thermal processing techniques that result in fresher, safer and higher-quality products.

14.3.1 Conventional Extraction Techniques

Conventionally solid–liquid extraction methods are widely used for the extraction of betalains from plants. Extraction methods include maceration, infusion and soxhlet extraction. Extraction of betalains is generally achieved in water; however, in most cases 20–50% v/v methanol or ethanol is used to achieve complete extraction. Maceration for the extraction of betalains is achieved by steeping the betalain source in water or a solvent (e.g., methanol or ethanol) in a closed container that is stirred frequently to increase the rate of extraction from the plant matrix. After the extraction is complete, the plant material is separated from the liquid either by filtration or

Table 14.1 Some of the extraction techniques for betalains

Extraction technique	Plant matrix	Study	Results	Advantages	References
Diffusion-extraction (DE)	Red beetroot	Diffusion apparatus to continuous counter-current extraction to betalains	71% betanine were recovered at in more acid systems at 4.5° incline angle	DE removes water-soluble low molecular weight substances from the beet cossette and juice	Wiley and Lee (1978)
Ultrafilteration (UF) and reverse osmosis (RO)	Red beetroot	Purify and concentrate beet soluble solids in liquid systems and separate non-pigment solids from beet colorants	70–75% of the betacyanins introduced in to the UF processes were recovered in the RO concentrates	Efficient concentrating techniques that remove a substantial portion of water present in betalain-containing beet juices	Lee et al. (1982)
Low DC electrical field versus cryogenic freezing	Red beetroot	Low DC electrical field, 40 V/cm	Electrical field applied to extract betalain pigments from other soluble solids	Electrical treatment causes minor changes, whereas cryogenic freezing ~20 s and thawing cause considerable damage to tissue	Zvitov et al. (2003)
Membrane-based process	Cactus pear fruit juice	Evaluate the potential of ultrafilteration (UF) and osmotic distillation (OD) processes for clarifyingand concentrating	The UF/OD integrated process permits preservation of the nutraceutical and functional importance	Athermal and do not involve phase changes or chemical additives; clarification and concentration of the juice	Cassano et al. (2007)
Aqueous two-phase extraction (ATPE)	Beetroot	ATPE used for differential partitioning of betalains and sugars	70–75% betalain was extracted in top phase and >90% sugar in bottom phase	Alternative for the downstream processing of betalains and removal of sugars in single step	Chethana et al. (2007)

Pulsed electric field (PEF)	Red beetroot	Employed PEF of 27, 54 and 270 pulses	Increased the permeability of red tissues with mass transfer at low levels of PEF treatments (27 pulse)	High- and moderate-intensity PEF may offer a way of disintegrating the cell membrane and an alternative, non-thermal pre-treatment	Chalermchat et al. (2004)
Pulsed electric field (PEF)	Red beetroot	PEF treatment at frequency of 1 Hz at electric field strengths of 0, 1, 3, 5, 7 and 9 kV/cm	Application of PEF treatments at 7 kV/cm increased betanin yield by 4.2 compared with control	Short-duration, high-intensity field strength facilitates betanin diffusion in the extraction from permeabilised cells	López et al. (2009)
Gamma-irradiation	Red beetroot	Doses (2.5, 5.0, 7.5, 10.0 kGy)	Degradation constant increased from 0.050 to 0.079/min with increase in dosage. Diffusion of betanin and ionic components increased with increase in doses	Exposure to gamma-irradiation pre-treatment increases cell wall permeabilisation and pressure, which increase the extractability of betanin from red beetroot	Nayak et al. (2006)
Gamma-irradiation	Red beetroot	Applied dosage of 1 and 2 kGy	Betalain contents sharply decrease at 2 kGy	Irradiation may contribute to higher cell–cell adhesion by increasing calcium cross linking at the middle lamellae regions in addition to an increment of cross-links of polymers to cell walls	Latorre et al. (2010)

Table 14.2 Comparison of effective principles of action of non-thermal membrane permeabilizing methods (Adapted from Knorr 2003)

Non-thermal technique	Driving force
High-pressure processing	Pressure gradient
Pulsed electric field	Electric potential difference
Ultrasound processing	Pressure and temperature gradient
Super critical fluid extraction	Pressure and concentration gradient

membrane processing. During the extraction process, the liquid medium diffuses into the plant matrix, followed by solubilisation of the betalain compounds within the plant matrix and finally diffusion of the betalains rich solvent out of the plant matrix. Although it is a time-consuming process, it is a useful extraction method for heat labile compounds because it is carried out at room temperature. Castellanos-Santiago and Yahia (2008) extracted the betalains from 10 prickly pear cultivars and red beet cultivars using water and buffer solvents, where water was found to extract the highest amount of pigments compared with buffer solvent extraction; this may be due to the high solubility of betalains in water. Similarly, Castellar et al. (2003) also demonstrated the extraction of betacyanin from *Opuntia* fruits using water, 80% ethanol in water and citrate-phosphate buffer pH 5.5, and these authors also found that water extracted the highest quantity of pigments. Water extraction of betalains, although simple, efficient and cost-effective, led to difficulties in the separation of betalains and water-soluble protein components (Cai et al. 1998b). Cai et al. (2001a) reported use of methanol as a solvent, which significantly improved the subsequent separation of betalains and proteins.

14.4 Novel Extraction Techniques

The shortcomings with conventional techniques, such as safety risks with certain solvent systems, time-consuming process steps, solvent contamination of the product and relatively low yields (Wang and Waller 2006), have led to interest in the use of novel processing techniques. Non-thermal techniques, such as high-pressure processing, ultrasound (sonication) processing and pulsed electric field (PEF), have been shown to be very effective for improving the extraction yields of bioactive compounds from biological sources with minimal degradation. These techniques are also known as cold extraction techniques, as temperatures during the extraction process are comparably low and impart minimum effects on the stability betalains and other phenolics, which offer health benefits. Non-thermal techniques improve extraction efficiencies by improving cell membrane permeabilisation by various driving forces, as listed in Table 14.2. These techniques are currently under investigation, with promising preliminary results indicating their potential for industrial applications. These novel extraction techniques aim to improve process efficiency through enhanced mass transfer and more environmentally friendly approaches.

Novel processing techniques such as power ultrasound, high-pressure processing, PEF, dielectric heating (microwave) and supercritical fluid processing techniques can be employed for extraction of betalains from plant matrices. These processes can also be used as a pre-treatment or in combination with environment friendly and safe organic solvents to enhance extraction efficiency. Novel processing techniques typically improve cell membrane permeability, which is considered the governing parameter determining extraction efficiency. Hence the amount of betalains released into the extracting media is characterised by the degree of cell disintegration. Several chemical, physical and mechanical treatments have been suggested to achieve high levels of cellular disintegration to enhance extraction rates (Chalermchat et al. 2004; Rastogi et al. 1999; Ade-Omowaye et al. 2001).

14.4.1 Pulsed Electric Field

PEF is a recognised non-thermal technology that has been extensively investigated in recent years for its applications in food processing and preservation. PEF utilises the influence of a strong electrical field on material located between two electrodes, which leads to cell membrane disruption, thus increasing cell permeability (Zimmermann et al. 1974; Knorr 2003). Fincan et al. (2004) employed high- and moderate-intensity pulsed electric fields to improve the extractability of red pigment from red beetroot in a solid–liquid extraction process. They observed the highest degree of extraction with PEF compared with freezing and mechanical pressing, with relatively low levels of tissue damage and low energy consumption. For example, PEF treatment at 1 kV/cm was shown to be an effective method of permeabilization for the extraction of pigment from beetroots, with a low energy consumption of about 7 kJ/kg (Fincan et al. 2004). Electroporation is reported to be the main mechanism for enhancing the acceleration of the betalain extraction release of colorants through aqueous extraction, as shown in Fig. 14.1 (Loginova et al. 2011). The optical microscopy photos of the red beet tissues indicate that non-treated tissue patterns were a mixture of intensively colored and almost colorless pale cells after some time of extraction (Fig. 14.1a). The intensively colored cells corresponded to undamaged cells and the colorless pale cells corresponded to the damaged cells with lost vacuolar sap and diluted pigment. The PEF pre-treated tissues demonstrated a homogenous distribution of cells having equivalently pale color (Fig. 14.1b) indicating extraction of betalains from the matrix due to electroporation of cells (Loginova et al. 2011).

14.4.2 Gamma Irradiation

Nayak et al. (2006) investigated the potential of gamma-irradiation pre-treatment for enhancement of mass transfer rates during solid–liquid extraction of betalains from beetroot, and observed that the extraction efficiency increased with increases

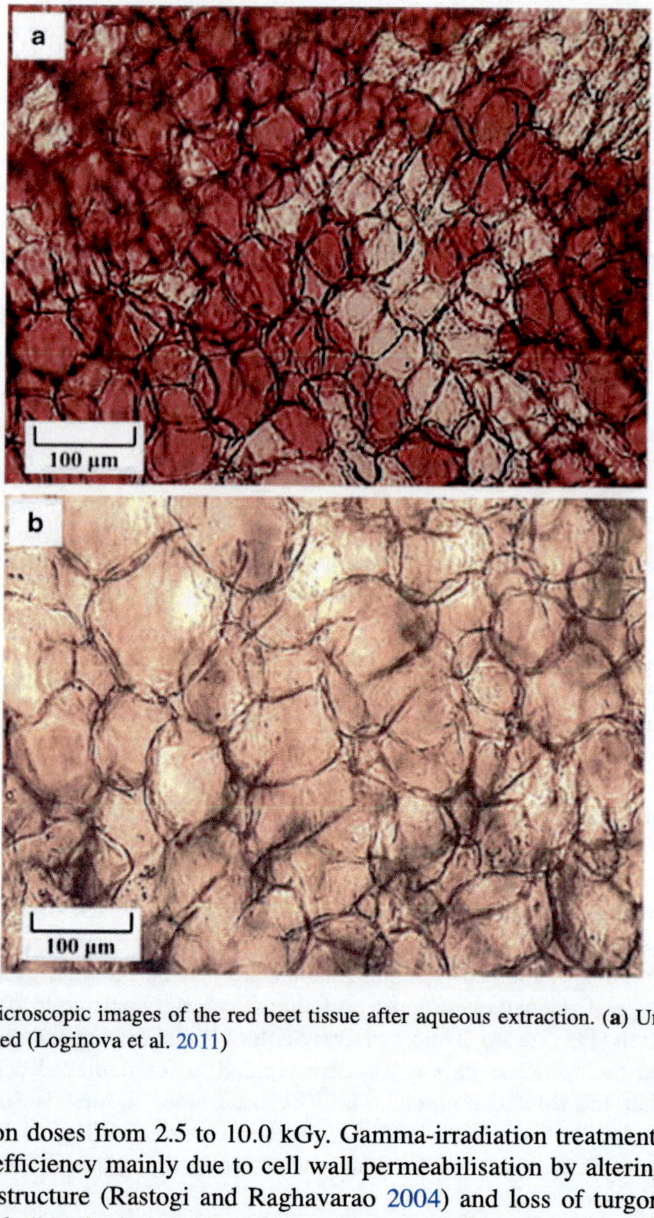

Fig. 14.1 Microscopic images of the red beet tissue after aqueous extraction. (**a**) Untreated; (**b**) PEF pre-treated (Loginova et al. 2011)

in irradiation doses from 2.5 to 10.0 kGy. Gamma-irradiation treatment improves extraction efficiency mainly due to cell wall permeabilisation by altering the interior tissue structure (Rastogi and Raghavarao 2004) and loss of turgor pressure. Latorre et al. (2010) observed that fresh-cut red beetroot exposed to low doses of gamma irradiation of about 1 or 2 kGy caused biochemical changes in the cellular contents and in the cell wall network, as shown in Fig. 14.2. The authors observed that the cell contents of beetroot tissues exposed to gamma irradiation were pushed against the cell membrane along the edges of the cell wall of each neighbouring cell. This study observed that after γ-irradiation at 2 kGy irradiation, the decrease in betacyanin content occurred by 11% and that of betaxanthin by 19% compared with

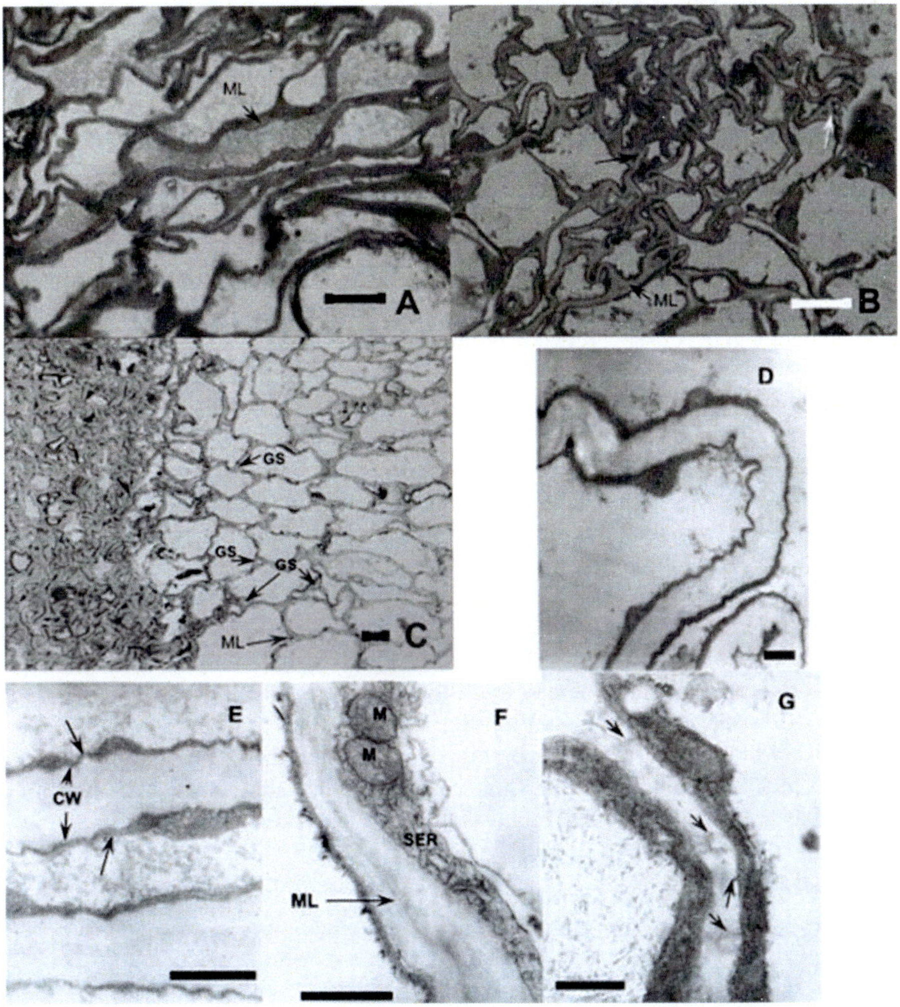

Fig. 14.2 Optical images taken from unstored control (0 kGy) (**a**) and irradiated (**b** 1 kGy, (**c**) 2 kGy) red beet root living tissues. *Arrows* indicate either middle lamellae (*ML*) or gas spaces (*GS*). TEM images got from unstored control (0 kGy) (**d**) and irradiated (**e** 1 kGy; **f**, **g** 2 kGy) red beet root living tissues. (**e**) *arrows* indicate the cell content pushed against the cell membrane along the edges of the cell wall (*CW*) of each neighbour cell. *M* mitochondria, *SER* smooth endoplasmic reticulum. (**g**) Plasmodesmata area between contiguous cells; *arrows* indicate some cytoplasmic bridges. Bar = 10 μm (a–c); bar = 1 μm (d–g). Latorre et al. (2010)

control, whereas pigment concentration was not affected at 1 kGy. There were no changes in the uronic acid and pectic fractions, and an increase in peroxidase activity was suspected because of the observed higher content of ferulic acid and other free phenolics. These observations indicate the suitability of low doses of irradiation for enhancing pigment recovery as well as other nutraceutical biomolecules (Latorre et al. 2010).

14.4.3 Supercritical Extraction

Among novel technologies, use of pressurised solvents has received attention for the extraction of compounds from biological materials, because of the opportunity of low temperatures during processesing. By applying certain pressure and temperature conditions, the physicochemical properties of the solvents, including density, diffusivity, viscosity and dielectric constant, can be controlled. While pre-cooling the solvent is required for maintaining liquid form, application of high pressures and temperatures can improve the extraction of betalains by improving the penetration of solvents into the cellular matrix and efflux of cellular components to the cell exterior. Extraction techniques can be: (1) supercritical fluid extraction, which is called supercritical CO_2 extraction or green technology when carbon dioxide is used as a solvent; or (2) pressurised liquid extraction, when 100% water is used, this technique is called subcritical water extraction or superheated water extraction (Pronyk and Mazza 2009; Mustafa and Turner 2011). When betalains are effluxed from their cellular compartment (the vacuoles), the pigments come in contact with enzymes, mainly peroxidases (POD) and polyphenol oxidases (PPO), which quickly degrade the pigments, causing quality deterioration. Inactivation of POD and PPO in red extract was observed when high pressure carbon dioxide (HPCD) was applied at 37.5 MPa for POD and 22.5 MPa for PPO (55°C, 60 min), where the suppression of enzyme activities occurred at approximately 86% and 95%, respectively. Compared with conventional thermal treatment at the same temperature (55°C), there was a significant reduction in time, from 555.56 min to 74.63 min for POD, and from 161.29 min to 38.31 min for PPO, which followed first-order kinetics. Although there were other energy reduction opportunities in the process, the enzymes were found more sensitive to pressure changes under subcritical rather than under supercritical conditions (Liu et al. 2008).

14.4.4 Ultrasound-Assisted Extraction

Ultrasound-assisted extraction is another technique that offers an inexpensive, environmentally friendly, fast and efficient alternative to conventional extraction techniques. The enhancement in extraction obtained by using ultrasound is mainly attributed to the effect of acoustic cavitations produced in the solvent (Ma et al. 2008; Velickovic et al. 2008). Ultrasound also offers a mechanical effect, allowing greater penetration of solvent into the sample matrix, increasing the contact surface area between the solid and liquid phase, and as a result, the solute more rapidly diffuses from the solid phase into the solvent (Toma et al. 2001; Mason 2003). A sonication-assisted pigment release from cultured red beet hairy roots was reported by Thimmaraju et al. (2003), where pigment release was studied under continuous ultrasound of 0.02 MHz for 15, 30 and 60 s. Only 8% of the total pigments was observed after 60 s and 12% release occurred in hairy roots treated for 120 s,

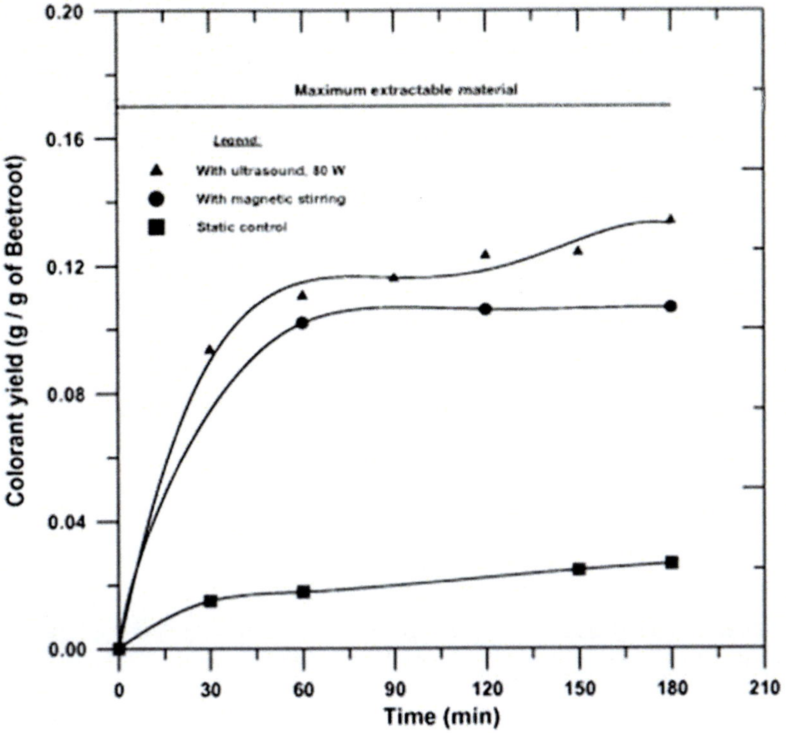

Fig. 14.3 Effect of ultrasound, 80 W, on the extraction of beetroot (1 g in 50 ml water). (Sivakumar et al. (2009))

without causing any loss in the viability of hairy roots. Sivakumar et al. (2009) studied the effect of ultrasound-assisted extraction of betalains from red beetroot and they observed that the ultrasound treatment had a significant improvement in the extraction efficiency. They further observed that the1:1 ethanol–water mixture with 80 W ultrasonic power for 3 h extraction provided a better yield and extraction efficiency, with an 8% increase in yield, as shown in Fig. 14.3.

14.4.5 Nano Grinding

Nano grinding is an innovative technology for ultra-fine sub-micron level grinding/milling applications, producing particles smaller than 10 nm. Tsai et al. (2011) studied the effect of nano grinding on the stability and yield of betanins from djulis (a grain native to Taiwan). They observed that the nanoparticle samples resulted in higher pigment extraction compared with intact granules and microparticles. However, Tsai et al. (2011) also observed reduced pigment stability during storage obtained from nanoparticles probably due to the incorporation of oxygen during the

Table 14.3 Membrane processes and applications (Adapted from Paul and Ohlrogge 1998)

Membrane processes	Phase	Membrane	Application/separation
Dialysis	Liquid/liquid	Dense	Hemicellulose/ steeping solution
Microfiltration	Liquid/liquid	Porous	Particles/solvent
Ultrafiltration	Liquid/liquid	Porous	Paints/solvent
Reverse osmosis	Liquid/liquid	Dense	Pesticide/solvent
Electrodialysis	Liquid/liquid	Porous	Ionic species/water
Pervaporation	Liquid/vapour	Dense	Phenol/water
Membrane distillation	Liquid/liquid/vapour	Dense	Polyglycol/water
Gas separation	Gas/gas	Dense	Nitrogen/oxygen
Vapour permeation	Vapour/vapour	Dense	Chlorohydrocarbon/air

grinding process. Degradation of pigments during storage was also dependent on the storage temperature, for example, after 56 days of storage at 35°C, pigment retention rates were 72.25%, 76.26%, and 63.53%, respectively, for intact granules, microparticles, and nanoparticle samples (Tsai et al. 2011).

14.4.6 Membrane Processing

Membrane processing shows promise in the isolation, enrichment and extraction of functional ingredients by concentrating, fractionation and purification of liquids (Freemantle 2003) without the use of heat. Membranes and membrane-based integrated processes offer practically unlimited selectivity of separation, thereby enabling conservation and the rational use of water and raw materials. Membrane separation processes can be applied in liquid systems for removing salts, water soluble organic compounds and insoluble particles, and in gaseous systems for the separation and recovery and recycling of vapours and gases (Paul and Ohlrogge 1998). Various types of membrane processes can be applied for extraction, purification and concentration of target compounds, as listed in Table 14.3. Membrane-processing techniques, such as microfiltration, ultrafiltration, nanofiltration and reverse osmosis have been reported for various fruit juice applications. Membrane processes have been successfully utilised for apple aroma compounds (Álvarez et al. 1998), clarification of apple juice (Fukumoto et al. 1998), stabilisation of apple juice (Gökmen et al. 1998), concentration of kiwifruit juice (Palmieri et al. 1990; Cassano et al. 2007), production of freshly squeezed orange juice (Todisco et al. 1998) and concentration and purification of cactus peer juice (Cassano et al. 2007). Ultrafiltration techniques are employed in biotechnology to separate compounds with molecular weights ranging from 1 to 1,000 kDa. For successful separation of various components in the feed stream, there should be a tenfold size difference between the components (O'Sullivan et al. 1984; Bayundirli et al. 1988). Lee et al. (1982) employed ultrafiltration and reverse osmosis for the purification and concentration of betalains from beet root. They observed that pectinase enzyme treatment

Table 14.4 Percentage of total betacyanins and soluble solids in the final concentrate as influenced by the different enzymatic treatment on the initial feed using CA-865 membrane in the RO process[a] (Adapted from Lee et al. 1982)

Enzymatic treatment	Pectinase				Invertase and pectinase			
Operating pressure (bar)	30	35	40	Average	30	35	40	Average
Betacyanins	82.2[b]	86.1	89.2	85.8	84.2	84.2	88.9	90.5
Soluble solids	39.0[b]	41.8	44.2	41.6	30.1	31.7	33.9	31.9

[a]Operating conditions: 0.36 m^2 effective membrane area; 20°C and flow rate 8 L/min
[b]Percentage of total betacyanins and soluble solid in the final concentrate, when the amount present in the RO feed is considered as 100%. These values represent the percentage of total betacyanins and soluble solids that can be recovered in the RO concentrate from the RO feed

of juice reduces the viscosity of extracted juices and improves the efficiency of pre-filtration process through diatomaceous earth. They also observed that the addition of invertase enzyme to the pectinase-treated ultrafiltration products increases the concentration of betalains by threefold on a dry weight (DW) basis, with a decrease in soluble solids during the reverse osmosis process, as shown in Table 14.4. The pilot scale reverse osmosis technique to concentrate beet juice for the concentration of betalains, as outlined by Lee et al. (1982) indicates that the juice clarified by ultrafiltration (600–20,000 Da) and concentrated by cellulose acetate (CA) membranes with low salt rejection (30%) was successful, provided that betalains were separated from the majority of soluble solids. The addition of invertase to pectinase-treated juices further decreased flux but yielded a threefold increase in betalain concentration on a dry weight basis with greatly reduced beet-like flavour (Girard and Fukumoto 2000).

Bayundirli et al. (1988) studied the sequential batch ultrafiltration of red beet extract by employing membranes of decreasing molecular weight cut off in sequence and developed a mathematical model. The flux of a filtrate is generally expressed by Darcy's law (O'Sullivan et al. 1984; Bayundirli et al. 1988) as:

$$\frac{dV_p}{dt} = k \cdot \Delta P^m \tag{14.1}$$

where Vp is the volume of the permeate and is the transmembrane pressure, k and m are constants and *t* is time. Equation 14.1 can also be expressed as:

$$\frac{dV_p}{dt} = \frac{\Delta P}{\mu} \cdot A \cdot \left[\frac{1}{R_m + R}\right] \tag{14.2}$$

$$R = \alpha \cdot c \cdot \frac{V}{A} \tag{14.3}$$

where μ is fluid viscosity, α, A and c are the specific cake resistance, effective membrane area and concentration of the solids being collected, respectively and R_m and R are the resistances of the filler and cake, respectively.

Table 14.5 Effect of membrane type on the recovery of soluble solids, total solids and betanins (Based on Bayundirli et al. 1988)

Membrane	MW cut off (kDa)	% Soluble solids	% Total solids	% Betanins
Cellulose nitrate	50	98	89	89
Cellulose acetate	20	84	83	74
Cellulose nitrate	10	79	79	84
Anisotropic	20	93	82	89
Anisotropic	5	78	78	80
Anisotropic	1	32	33	83

Selection of membrane type is critical for the recovery of betanins and soluble solids, as shown in Table 14.5. Bayundirli et al. (1988) observed that the anisotropic membrane system was the best system for the separation of betalains, where 59% of initial betanin was recovered while the percentage of the total and soluble solid was reduced to 21% and 23% of the initial level, respectively.

14.5 Stability of Betalains During Extraction

During extraction of betalains, the stability of a pigment is governed by the presence of enzymes, including peroxidase, polyphenol oxidases, β-glucosidase and betalain oxidase (Martinez-Parra and Munoz 2001; Escribano et al. 2002). These enzymes can cause degradation during the extraction process. The presence of endogenous enzymes responsible for degradation is inactivated by a short heat treatment of the extract (70°C, 2 min). Betalains are thermally unstable and elevated temperatures can accelerate the degradation (Czapski 1990; Herbach et al. 2004a, b, 2006a). Like many other natural pigments, betalains are very sensitive to heat, light and oxidation, especially that caused by peroxidases, which are among the major causes of discolouration of the pigment (Lashley and Wiley 1979; Martinez-Parra and Munoz 2001; Shih and Wiley 1981; Wasserman and Guilfoy 1983; Singer and von Elbe 1980). The stability of betalains during food processing and extraction is influenced by many factors, the most significant of which are temperature, pH, a_w, M^{n+}, O_2, and hυ (Priestley 1979). Figure 14.4 lists the factors that can influence the stability of betalains during extraction. To avoid degradation of betalains during the extraction process, it is advisable to employ cold water for long periods under dark conditions (Delgado-Vargas et al. 2000) or adopt non-thermal extraction processes.

14.6 Conclusion and Future Challenges

One of the most challenging tasks during extraction of betalains is to understand the plant matrix from which a compound is extracted. A new large-scale chromatographic separation method was used by Nemzer et al. (2011) that compared the

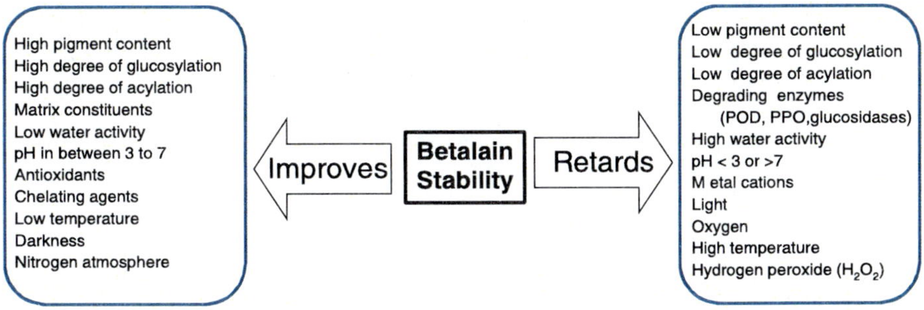

Fig. 14.4 Factors governing the improvement and retardation of the stability of betalains (Adapted with modifications from Herbach et al. 2006b)

quality of the extract powder thus prepared with other methods, and a large number of nutritionally important compounds and minerals were found in the new product along with a very high content of 41% betalain. However, for non-thermal extraction, understanding of the location of betalain compounds within the cellular matrix and their mobility within the cells are important for applying smart pigment effluxing methods and improving extraction efficiency. Non-thermal techniques are known either to improve the penetrability of solvent into the matrix through cell membrane permeabilization or to rupture the cellular structure to release pigments into the extracting medium. Although the potential of novel extraction techniques has been investigated for various phytochemicals, limited research has been carried out on betalains. Industrial adoption of novel extraction techniques for the of pigments have numerous challenges. One of the difficulties encountered in this research area is the non-standardised reporting of methodology and control parameters for novel technologies, which are conducted at experimental scales. These pose further difficulties during scale-up, where re-experimentation and standardisation is crucial. Among the several techniques discussed, PEF seems to be the most promising for betalains; however, further research is required to finely optimise extraction conditions at each step.

References

Ade-Omowaye, B.I.O., A. Angersbach, K.A. Taiwo, and D. Knorr. 2001. Use of pulsed electric field pre-treatment to improve dehydration characteristics of plant based foods. *Trends in Food Science and Technology* 12: 285–295.

Álvarez, S., F.A. Riera, R. Álvarez, and J. Coca. 1998. Permeation of apple aroma compounds in reverse osmosis. *Separation and Purification Technology* 14: 209–220.

Azeredo, H.M.C. 2009. Betalains: properties, sources, applications, and stability – a review. *International Journal of Food Science and Technology* 44: 2365–2376.

Bayundirli, A., F. Yildiz, and M. O'zilgen. 1988. Modeling of sequential batch ultrafiltration of red beet extract. *Journal of Food Science* 53: 1418–1422.

Butera, D., L. Tesoriere, F. Di Gaudio, A. Bongiorno, M. Allegra, A.M. Pintaudi, R. Kohen, and M.A. Livrea. 2002. Antioxidant activities of Sicilian prickly pear (*Opuntia ficus indica*) fruit extracts and reducing properties of its betalains: betanin and indicaxanthin. *Journal of Agricultural and Food Chemistry* 50: 6895–6901.

Cai, Y.Z., M. Sun, and H. Corke. 1998a. Colorant properties and stability of *Amaranthus betacyanin* pigments. *Journal of Agricultural and Food Chemistry* 46: 4491–4495.

Cai, Y.Z., M. Sun, H.X. Wu, R.H. Huang, and H. Corke. 1998b. Characterization and quantification of betacyanin pigments from diverse *Amaranthus* species. *Journal of Agricultural and Food Chemistry* 46: 2063–2070.

Cai, Y.Z., M. Sun, and H. Corke. 2001a. Identification and distribution of simple and acylated betacyanin pigments in the Amaranthaceae. *Journal of Agricultural and Food Chemistry* 49: 1971–1978.

Cai, Y.Z., M. Sun, and H. Corke. 2005. Characterization and application of betalain pigments from plants of the Amaranthaceae. *Trends in Food Science and Technology* 16: 370–376.

Cassano, A., C. Conidi, R. Timpone, M. D'Avella, and E. Drioli. 2007. A membrane-based process for the clarification and the concentration of the cactus pear juice. *Journal of Food Engineering* 80: 914–921.

Castellanos-Santiago, E., and E.M. Yahia. 2008. Identification and quantification of betalains from the fruits of 10 Mexican prickly pear cultivars by high-performance liquid chromatography and electrospray ionization mass spectrometry. *Journal of Agricultural and Food Chemistry* 56: 5758–5764.

Castellar, M.R., J.M. Obon, M. Alacid, and J.A. Fernandez-Lopez. 2003. Color properties and stability of betacyanins from Opuntia fruits. *Journal of Agricultural and Food Chemistry* 51: 2772–2776.

Chalermchat, Y., P. Dejmek, and M. Fincan. 2004. Pulsed electric field treatment for solid–liquid extraction of red beetroot pigment: mathematical modelling of mass transfer. *Journal of Food Engineering* 64: 229–236.

Chethana, S., C.A. Nayak, and K.S.M.S. Raghavarao. 2007. Aqueous two phase extraction for purification and concentration of Betalains. *Journal of Food Engineering* 81: 679–687.

Czapski, J. 1990. Heat stability of betacyanins in red beet juice and in betanin solutions. *European Food Research and Technology* 191: 275–278.

Delgado-Vargas, F., A.R. Jiménez, and O. Paredes-López. 2000. Natural pigments: Carotenoids, anthocyanins, and betalains – Characteristics, biosynthesis, processing, and stability. *Critical Review in Food Science and Nutrition* 40: 173–289.

Escribano, J., F. Gandia-Herrero, N. Caballero, and M.A. Pedreno. 2002. Subcellular localization and isoenzyme pattern of peroxidase and polyphenol oxidase in beet root (*Beta vulgaris L*). *Journal of Agricultural and Food Chemistry* 50: 6123–6129.

Fincan, M., F. De Vito, and P. Dejmek. 2004. Pulsed electric field treatment for solid–liquid extraction of red beetroot pigment. *Journal of Food Engineering* 64: 381–388.

Freemantle, M. 2003. Membrane separations. *Chemical Engineering News* 83(24): 33–35.

Fukumoto, L.R., P. Delaquis, and B. Girard. 1998. Microfiltration and ultrafiltration ceramic membranes for apple juice clarification. *Journal of Food Science* 63: 845–850.

Gaertner, V., and I.L. Goldman. 2005. Pigment distribution and total dissolved solids of selected cycles of table beet from a recurrent selection program for increased pigment. *Journal of the American Society for Horticultural Science* 130: 424–433.

Gasztonyi, M.N., H. Daood, M.T. Hajos, and P. Biacs. 2001. Comparison of red beet (*Beta vulgaris var. conditiva*) varieties on the basis of their pigment components. *Journal of the Science of Food and Agriculture* 81: 932–933.

Girard, B., and L.R. Fukumoto. 2000. Membrane processing of fruit juices and beverages: A review. *Critical Reviews in Food Science and Nutrition* 40: 91–157.

Gökmen, V., Z. Borneman, and H.H. Nijhuis. 1998. Improved ultrafiltration for colour reduction and stabilization of apple juice. *Journal of Food Science* 63: 504–507.

Herbach, K.M., F.C. Stintzing, and R. Carle. 2004a. Impact of thermal treatment on colour and pigment pattern of red beet (*Beta vulgaris L.*) preparations. *Journal of Food Science* 69: C491–C498.

Herbach, K.M., F.C. Stintzing, and R. Carle. 2004b. Thermal degradation of betacyanins in juices from purple pitaya (*Hylocereuspolyrhizus* [Weber] Britton and Rose) monitored by high-performance liquid chromatography-tandem mass spectrometric analyses. *European Food Research and Technology* 219: 377–385.

Herbach, K.M., M. Rohe, F.C. Stintzing, and R. Carle. 2006a. Structural and chromatic stability of purple pitaya (*Hylocereuspolyrhizus* [Weber] Britton and Rose) betacyanins as affected by the juice matrix and selected additives. *Food Research International* 39: 667–677.

Herbach, K.M., F.C. Stintzing, and R. Carle. 2006b. Betalain stability and degradation – structural and chromatic aspects. *Journal of Food Science* 71: R41–R50.

Jackman, R.L., and J.L. Smith. 1996. Anthocyanins and betalains. In *Natural food colourants*, ed. G.F. Hendry and J.D. Houghton, 244–309. Londonpp: Blackie Academic and Professional.

Kanner, J., S. Harel, and R. Granit. 2001. Betalains – a new class of dietary cationized antioxidants. *Journal of Agricultural and Food Chemistry* 49: 5178–5185.

Knorr, D. 2003. Impact of non-thermal processing on plant metabolites. *Journal of Food Engineering* 56: 131–134.

Kugler, F., F.C. Stintzing, and R. Carle. 2004. Identification of betalains from petioles of differently colored Swiss chard (*Beta vulgaris* L. ssp. cicla [L.]Alef. cv. Bright lights) by high-performance liquid chromatography-electrospray ionization mass spectrometry. *Journal of Agriculture and Food Chemistry* 52: 2975–2981.

Lashley, D., and R.C. Wiley. 1979. Abetacyanine decolorizing enzyme found in red beet tissue. *Journal of Food Science* 44: 1568–1569.

Latorre, M.E., P. Narvaiz, A.M. Rojas, and L.N. Gerschenson. 2010. Effects of gamma irradiation on bio-chemical and physico-chemical parameters of fresh-cut red beet (*Beta vulgaris L. var. conditiva*) root. *Journal of Food Engineering* 98: 178–191.

Lee, Y.N., R.C. Wiley, M.J. Sheu, and D.V. Schlimme. 1982. Purification and concentration of betalaines by ultrafiltration and reverse osmosis. *Journal of Food Science* 47: 465–471.

Liu, X., Y. Gao, X. Peng, B. Yang, H. Xu, and J. Zhao. 2008. Inactivation of peroxidase and polyphenol oxidase in red beet (*Beta vulgaris* L.) extract with high pressure carbon dioxide. *Innovative Food Science and Emerging Technologies* 9: 24–31.

Loginova, K.V., N.I. Lebovka, and E. Vorobiev. 2011. Pulsed electric field assisted aqueous extraction of colorants from red beet. *Journal of Food Engineering* 106: 127–133.

López, N., P. Puértolas, I. Hernández-Orte, and R.J. Álvarez. 2009. Effect of a pulsed electric field treatment on the anthocyanins composition and other quality parameters of Cabernet Sauvignon freshly fermented model wines obtained after different maceration times. *LWT- Food Science and Technology* 42: 1225–1231.

Ma, Y., X. Ye, Y. Hao, G. Xu, and D. Liu. 2008. Ultrasound-assisted extraction of hesperidin from Penggan (*Citrus reticulata*) peel. *Ultrasonics Sonochemistry* 15: 227–232.

Martinez-Parra, J., and R. Munoz. 2001. Characterization of betacyanin oxidation catalyzed by a peroxidase from *Beta vulgaris* L roots. *Journal of Agricultural and Food Chemistry* 49: 4064–4068.

Mason, T.J. 2003. Sonochemistry and sonoprocessing: The link, the trends and (probably) the future. *Ultrasonics Sonochemistry* 10: 175–179.

Mustafa, A., and C. Turner. 2011. Pressurized liquid extraction as a green approach in food and herbal plants extraction: A review. *Analytica Chimica Acta* 703: 8–18.

Nayak, C.A., S. Chethana, N.K. Rastogi, and K.S.M.S. Raghavarao. 2006. Enhanced mass transfer during solid–liquid extraction of gamma irradiated red beetroot. *Radiation Physics and Chemistry* 75: 173–178.

Nemzer, B., Z. Pietrzkowski, A. Spórna, P.W. Stalica, W. Thresher, T. Michałowski, and S. Wybraniec. 2011. Betalainic and nutritional profiles of pigment-enriched red beet root (*Beta vulgaris* L.) dried extracts. *Food Chemistry* 127: 42–53.

Nilsson, T. 1970. Studies into the pigments in beetroot (*Beta vulgaris* L. ssp. *vulgaris* var. *rubra* L.). *Lantbrukhogskolans Annaler* 36: 179–219.

O'Sullivan., A.C., S.R. Epstein, S.R. Korchin, and N.C. Beaton. 1984. Applications of ultrafiltration in biotechnology. *Chemical Engineering Progress*, 80 (1): 68–75.
Palmieri, L., M. Dalla Rosa, and C.G. Dall'AglioG. 1990. Production of kiwifruit concentrate by reverse osmosis process. *Acta Horticulturae* 282: 435–439.
Paul, D., and K. Ohlrogge. 1998. Membrane separation processes for clean production. *Environmental Progress* 17: 137–141.
Priestley, R.J. 1979. *Effects of heating on food stuffs*, 307–314. London: Applied Science.
Pronyk, C., and G. Mazza. 2009. Design and scale-up of pressurized fluid extractors for food and bioproducts. *Journal of Food Engineering* 95: 215–226.
Psczczola, D.E. 1998. Natural colors: Pigments of imagination. *Food Technology* 52: 70–82.
Rastogi, N.K., and K.S.M.S. Raghavarao. 2004. Increased mass transfer during osmotic dehydration of irradiated potatoes. *Journal of Food Science* 69: 259–263.
Rastogi, N.K., M.N. Eshtiaghi, and D. Knorr. 1999. Accelerated mass transfer during osmotic dehydration of high intensity electrical field pulse pretreated carrots. *Journal of Food Science* 64: 1020–1023.
Schwartz, S.J., and J.H. von Elbe. 1980. Quantitative determination of individual betacyanin pigments by high-performance liquid chromatography. *Journal of Agricultural and Food Chemistry* 28: 540–543.
Shih, C.C., and R.C. Wiley. 1981. Betacyanine and betaxanthine decolorizing enzymes in the beet (*Beta vulgaris* L) root. *Journal of Food Science* 47: 164–166.
Singer, J.W., and J.H. von Elbe. 1980. Degradation rates of vulgaxanthine-I. *Journal of Food Science* 45: 489–491.
Sivakumar, V., J.L. Anna, J. Vijayeeswarri, and G. Swaminathan. 2009. Ultrasound assisted enhancement in natural dye extraction from beetroot for industrial applications and natural dyeing of leather. *Ultrasonics Sonochemistry* 16: 782–789.
Stintzing, F.C., and R. Carle. 2007. Betalains - emerging prospects for food scientists. *Trends in Food Science and Technology* 18: 514–525.
Stintzing, F.C., and R. Carle. 2008a. N-heterocyclic pigments: betalains. In *Food colorants – Chemical and functional properties*, ed. C. Socaciu, 87–99. Boca Raton FL: CRC Press, Taylor and Francis Group, LLC.
Stintzing, F.C., and R. Carle. 2008b. Betalains in food: occurrence, stability, and postharvest modifications. In *Food colorants: Chemical and functional properties*, ed. C. Socaciu, 277–299. Boca Raton FL: CRC Press/Taylor and Francis/LLC.
Stintzing, F., J. Trichterborn, and R. Carle. 2006. Characterisation of anthocyanin–betalain next term mixtures for food colouring by chromatic and HPLC-DAD-MS analyses. *Food Chemistry* 79: 296–309.
Strack, D., T. Vogt, and W. Schliemann. 2003. Recent advances in betalain research. *Phytochemistry* 62: 247–269.
Svenson, J., B.M. Smallfield, N. Joyce, C.E. Sansom, and N.B. Perry. 2008. Betalains in red and yellow varieties of the Andean tuber crop Ulluco (*Ullucus tuberosus*). *Journal of Agricultural and Food Chemistry* 56: 7730–7737.
Tanaka, Y., N. Sasaki, and A. Ohmiya. 2008. Biosynthesis of plant pigments: anthocyanins, betalains and carotenoids. *Plant* J54: 733–749.
Thimmaraju, R., N. Bhagyalakshmi, M.S. Narayan, and G.A. Ravishankar. 2003. Kinetics of pigment release from hairy root cultures of red beet under the influence of pH, sonication, temperature and oxygen stress. *Process Biochemistry* 38: 1067–1074.
Todisco, S., P. Tallarico, and E. Drioli. 1998. Modelling and analysis of ultrafiltration effects on the quality of freshly squeezed orange juice. *Italian Food and Beverage Technology* 12: 3–8.
Toma, M., M. Vinatoru, L. Paniwnyk, and T.J. Mason. 2001. Investigation of the effects of ultrasound on vegetal tissues during solvent extraction. *Ultrasonics Sonochemistry* 8: 137–142.
Tsai, P.J., Y.N. Chen, C.H. Sheu, and C.H. Chen. 2011. Effect of nanogrinding on the pigment and bioactivity of Djulis (*Chenopodium formosanum* Koidz.). *Journal of Agricultural and Food Chemistry* 59: 1814–1820.

Velickovic, D.T., D.M. Milenovic, M.S. Ristic, and V.B. Veljkovic. 2008. Ultrasonic extraction of waste solid residues from the Salvia sp. essential oil hydrodistillation. *Biochemical Engineering Journal* 42:97–104.

Vincent, K.R., and R.G. Scholz. 1978. Separation and quantification of red beet betacyanins and betaxanthins by high-performance liquid chromatography. *Journal of Agricultural and Food Chemistry* 26: 812–816.

von Elbe, J.H. 1975. Stability of betalaines as food colors. *Food Technology* 5: 42–44.

Wang, L., and C.L. Waller. 2006. Recent advances in extraction of naturaceuticals from plants. *Trends in Food Science and Technology* 17: 300–312.

Wasserman, B.P., and P. Guilfoy. 1983. Peroxidative properties of betanindecolorization by cell walls of red beet. *Phytochemistry* 22: 2653–2656.

Wiley, R.C., and Y.N. Lee. 1978. Recovery of betalaines from red beets by a diffusion-extraction procedure. *Journal of Food Science* 43: 1056–1058.

Zimmermann, U., G. Pilwat, and F. Riemann. 1974. Dielectric breakdown of cell membranes. *Biophysical Journal* 14: 881–899.

Zvitov, R., A. Schwartz, and A. Nussinovitch. 2003. Comparison of betalain extraction from beet (*Beta vulgaris*) by low DC electrical field versus cryogenic freezing. *Journal of Texture Studies* 34: 83–94.

Chapter 15
Aqueous Two-Phase Extraction for the Recovery of Beet Pigments and Enzymes

M.C. Madhusudhan and K.S.M.S. Raghavarao

Abstract Liquid–liquid extraction employing aqueous two-phase systems has proved to be a promising separation strategy for many biological products such as proteins, enzymes, viruses, cells and other biological materials. The major advantages of aqueous two-phase extraction (ATPE) include high capacity, biocompatible environment, low interfacial tension, high yield, lower process time and energy and high selectivity. ATPE can be designed such that the desired biomolecule selectively partitions to one of the phases in a concentrated form, with considerable reduction in the volume of the stream to be handled during the subsequent purification steps. Therefore, it is relatively easier to scale up partitioning steps with greater precision in enzyme/protein isolation and purification when compared with conventional steps. Recovery of natural color pigments (betalains) and enzymes (peroxidases, polyphenol oxidases) from beet root are presented in this chapter, bringing out the versatility of ATPE for their purification and concentration.

15.1 Introduction

Downstream processing is an integral part of any biological product development and the final cost of the product depends largely on the cost incurred during downstream processing (DSP) for its recovery. Scale-up problems are considerable during DSP using conventional methods like centrifugation. Even modern methods such as electrophoresis or column chromatography, make them uneconomical unless the product is of high value. The separation of many biomolecules is still performed

M.C. Madhusudhan • K.S.M.S. Raghavarao (✉)
Department of Food Engineering, Central Food Technological Research Institute (CFTRI), Council of Scientific and Industrial Research (CSIR) India, Mysore 570 020, India
e-mail: fed@cftri.res.in

B. Neelwarne (ed.), *Red Beet Biotechnology: Food and Pharmaceutical Applications*, DOI 10.1007/978-1-4614-3458-0_15, © Springer Science+Business Media New York 2013 in batch mode in small-scale processes, such as column chromatography, salt/solvent-induced precipitation and electrophoresis. These unit operations have scale-up problems and are expensive at larger scale and also often result in low product recovery (Diamond and Hsu 1992; Raghavarao et al. 1998). The downstream processing of biological materials requires purification techniques that are both economically feasible and delicate enough to preserve their biological activity. The problems traditionally faced by the biotechnologist in conventional separation operations pale in comparison with the requirements often placed on separation of biomolecules and recombinant proteins. Various unit operations are required for the downstream processing of biomolecules. While developing a large-scale isolation procedure, it is mandatory to consider processing time, energy, manpower, good manufacturing practices, recycling of chemicals, sterilization and cleaning-in-place (CIP) of equipment, in addition to separation efficiency (Diamond and Hsu 1992; Raghavarao et al. 1995; Rito-palomares 2004). Scaling up of laboratory scale processes is crucial for industrial exploitation. Hence, there is a need to develop simple, efficient, economical, environmentally benign downstream processing methods for the recovery of biomolecules with flexibility for continuous operation. Liquid–liquid extraction using aqueous two-phase systems (ATPS) is one such method, popularly known as aqueous two-phase extraction (ATPE). ATPE has been successful to a large extent in overcoming the drawbacks of even the conventional extraction processes, such as low solubility and denaturation of biomolecules in organic solvents.

Most biotechnological products, soluble molecules and particles are obtained in very dilute solutions. ATPE is able to carry out their concentration simultaneously during the extraction, provided that it is designed in such a way that most of the desired substances are transferred to a phase with a small volume compared with the original solution (Albertsson 1986). The particles may concentrate also at the interface. During ATPE, impurities may be concentrated to a certain extent while a concomitant purification is also achieved. The concentration and purification of viruses could be also reported using ATPE. A one-step or multi-step procedure may be applied, depending on the partitioning of the product and the contaminants. Compared with traditionally used techniques, the main advantage of ATPE is the decrease in the process time, resulting in considerable savings in energy input and manpower (Kroner et al. 1982).

15.1.1 Aqueous Two-Phase Systems (ATPS)

Liquid–liquid extraction using ATPS has been recognized as a superior and versatile technique for the downstream processing of biomolecules (Albertsson 1986).

An ATPS will form by mixing a pair of water-soluble polymers or a polymer and low molecular weight solute with water above critical concentration. Generally, ATPS are of two types, the polymer/polymer type and the polymer/salt type. Some of the commonly used phase systems are listed in Table 15.1. Both components of

Table 15.1 Components for the formation of ATPS

Phase systems	Component 1	Component 2
Polymer/polymer phase systems	Polyethylene glycol, polypropylene glycol, polyvinyl alcohol, polyvinyl pyrrolidone, methyl cellulose, ethyl hydroxyethyl cellulose, hydroxy propyl dextran, ficoll, vinyl-2-pyrrolidone-guar gum	Dextran, ficoll, pullulan, polyvinyl alcohol, Reppal PES 100 (hydroxyl propyl starch), maltodexrin, xanthan, sodium polyacrylate (NaPA), Gemini/SDS, cashew-nut tree gum, mixture of 2-(dimethylamino) ethyl methacrylate, t-butyl methacrylate, methyl methacrylate
Thermo separating polymer-based phase systems	Breox (ethylene oxide–propylene oxide), Ucon 50-HB-5100, poly(ethylene oxide-co-maleic anhydride)	Dextran, Reppal PES 100, potassium phosphate, ammonium sulfate, polyvinyl alcohol
Detergent-based phase systems	Triton X114, Agrimul NRE 1205 (C12–18E5), cetyl trimethyl ammonium bromide (CTAB), Triton X-114, dodecyl trimethyl ammonium bromide (DTAB)	Water, sodium dodecyl sulfonate (AS), polyethylene glycol
Polymer/salt phase systems	Polyethylene glycol, polypropylene glycol, methoxy polyethylene glycol, polyvinyl pyrrolidone, derivatives of PEG (PEG-benzoate (PEG-Bz) PEG-phosphate (PEG-PO_4), PEG-trimethylamine (PEG-tma), PEG-palmitate (PEG-pal), PEG-phenyl acetamide (PEG-paa))	Potassium phosphate, sodium sulfate, sodium formate, sodium potassium tortrate, magnesium sulfate, sodium citrate, ammonium sulfate, ammonium carbamate
Alcohol-based phase systems	Ethanol, 2-propanol, 1-propanol, methanol, acetone	Di-potassium hydrogen phosphate, sodium thiosulfate, magnesium sulfate, ammonium sulfate, sodium di-hydrogen phosphate, cesium carbonate, sodium chloride, tri-potassium phosphate, calcium chloride, sodium carbonate

these systems are separately miscible in water in all proportions and also with each other at low concentrations. As the concentration of these phase components in a common solvent (water) increases above a certain critical value, phase separation occurs. Each ATPS is characterized by an exclusive phase diagram that indicates the equilibrium composition for that particular system and constitutes the most fundamental data for biomolecule extraction involving that system. A model phase diagram is shown in Fig. 15.1. Bamberger et al. (1984) discussed in detail the methods

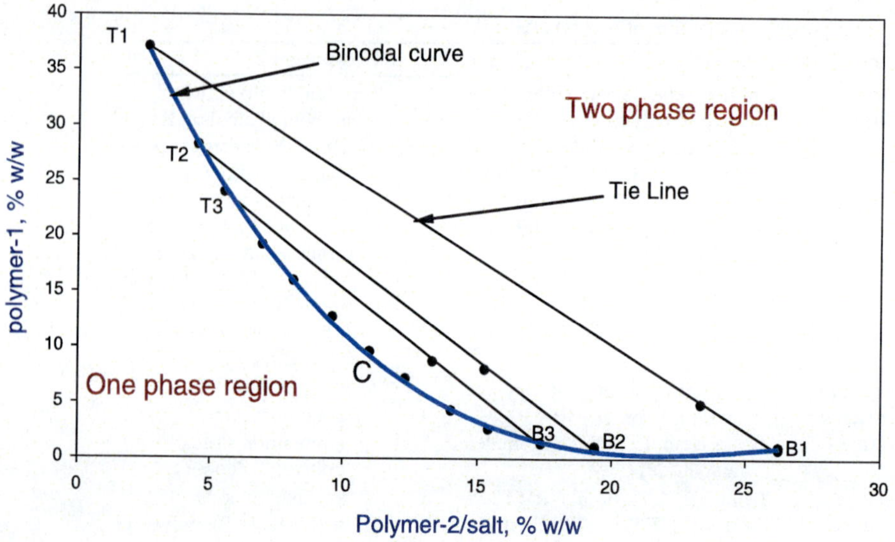

Fig. 15.1 Phase diagram for polymer/salt

for the construction of these phase diagrams. Albertsson (1971), Diamond and Hsu (1989) and Zaslavsky et al. (1982) have compiled phase diagrams for a number of systems. Among these, ATPS formed by polyethylene glycol (PEG)–dextran–water and PEG–salt–water systems are widely used for separation and purification of biomolecules. However, PEG–salt two-phase systems have certain advantages over PEG–polymer systems, such as lower viscosity and cost. Recently new phase systems comprising of alcohol, detergent and ionic liquid-based ATPS have also been reported for the recovery of bioactive components.

ATPS are influenced by many factors like polymer concentration, polymer molecular weight, temperature, hydrophobicity, salt and pH (Albertsson 1986). However, the effects and mechanisms by which they influence phase formation are still not completely understood.

The physical properties of ATPS, such as density, viscosity and interfacial tension, determine the phase demixing and contribute to the biomolecule partition behaviors in a given extraction. The measurement of these physical properties is very important for designing and analyzing the results of extraction employing these phase systems. The information related to the extent of variation of these properties with the tie line length is of prime importance during optimization of extraction.

The protein partitioning in ATPS is influenced by many environmental conditions, such as biomolecule size and concentration, choice of polymers and their molecular weight, composition of the phases, biomaterial surface properties, system pH, temperature, etc. Influence of these parameters on partitioning was collectively explained in terms of relative free volume (Eiteman and Gainer 1989; Grossman and Gainer 1988).

15.1.2 Aqueous Two-Phase Extraction (ATPE) of Biological Products

ATPE has shown its utility in the extraction and purification of biological materials such as proteins/enzymes, nucleic acids, viruses, cell organelles etc. A large number of applications along with thermodynamic properties of phase systems are reported in the literature. (Kula et al. 1982; Walter et al. 1985; Albertsson 1986; Diamond and Hsu 1992; Zaslavsky 1995; Raghavarao et al. 1998; Kaul 2000). Furthermore, ATPE has been recognized as energy efficient and a mild separation technique for product recovery in biotechnology. In some cases ATPE has potential to achieve the desired purification and concentration of the product even in a single step. In others, it is achieved in multiple steps. It reduces the volume of the crude extract after partial purification in order to employ more selective and expensive purification methods for final purification, depending on the need. Thus it is recognized as a primary purification step.

The basis of separation in the ATPE is the selective distribution of different substances in the two phases. The partitioning of small molecules is even in both of the phases and that of macromolecules is extremely variable, whereas the partitioning of these particles is relatively one sided. The distribution of biomolecules is governed by various parameters relating to the properties of the phase system and the substance as well as the interaction between the two. Hence, the prediction of partitioning becomes a difficult task especially for large molecules. The partitioning of proteins/enzymes in ATPS is affected by net charge of each biomolecule and the interaction of water with phase polymers, where buffering salts plays a key role in protein partitioning. The partitioning of biomolecules can be made selective by changing the system properties to make a particular kind of interaction predominant. The multiplicity of factors contributing to partitioning also makes the system very powerful, in contrast to the other conventional separation techniques like centrifugation, electrophoresis etc., allowing the fractionation of molecular or particulate species differing very slightly from each other. Thus, fractionation by partition in ATPS may often substitute for other separation procedures.

15.1.3 ATPE of Enzymes

Generally extraction and purification of enzymes involves a number of steps such as filtration, centrifugation, precipitation, chromatography, electrophoretic techniques, crystallization etc., causing loss of yield at each step, affecting adversely the overall productivity. ATPE has been employed as an alternative for the large-scale

Table 15.2 Enzymes purified by ATPE

Enzymes	References
Peroxidase	Miranda and Cascone (1994); Srinivas et al. (1999, 2002); Silva and Franco (2000)
Alcohol dehydrogenase	Madhusudhan et al. (2008)
Bromelain	Babu et al. (2008)
Lipoxigenase	Lakshmi et al. (2009)
Intracellular glyceraldehyde 3-phosphate dehydrogenase	Rito-Palomares and Lyddiatt (2002)
Lipase	Ooi et al. (2009)
β-glucosidase	Gautam and Simon (2006); Hemavathi and Raghavarao (2011)
Amyloglucosidase	Tanuja et al. (2000)
Lysozyme	Dembczynski et al. (2010)
Plant-esterase	Yanga et al. (2010)
Papain	Nitsawang et al. (2006)
α-amylase	Li et al. (2002)
Endo-polygalacturonase	Pereira et al. (2003)
Phospholipase D	Teotia and Gupta (2004)
Alpha galactosidase	Gautam and Simon (2007)
Pepsin	Imelio et al. (2008)

purification of several products, including recombinant products. In the past three decades, a wealth of information has been reported in the literature on various aspects of ATPE for the isolation of many biological products. A few examples are listed in Table 15.2.

15.1.4 ATPE of Natural Pigments

In recent years, interest in natural colorants has increased considerably, mainly because of the apparent lack of toxicity and their eco-friendliness. Natural and synthetic pigments are used in medicines, foods, clothes and in other products. However, the natural pigments that are permitted for human foods are very limited, and the approval of new sources is often difficult. This is mainly because the US Food and Drug Administration (FDA) considers the pigments as additives, and consequently pigments are under strict regulations. Consumers are aware of the toxicological effects associated with synthetic colors and hence the use of natural colorants has been increasing. Pigments are chemical compounds that absorb light in the wavelength range of the visible region. The colorful appearance is imparted to human eye because of a molecule-specific structure called a chromophore. Pigments are widely distributed in living organisms and a large number of structures have been reported. Most biological pigments are grouped into six kinds of structures: tetrapyrroles, isoprenoids, quinines, benzopyrans, *N*-heterocyclic compounds and metalloproteins. Natural colors have some limitations when compared with the synthetic colors, such as (a) higher sensitivity to light, heat and pH; (b) higher susceptibility to oxidation; (c) lower solubility; (d) lower shelf life and (e) lower pigmenting strength. These limitations should be evaluated and recognized for the food formulation (Attoe and von Elbe 1985).

Some of the most commonly used natural colorants are carotenoids, anthocyanins, betalains, chlorophylls, phycobiliproteins, curcumin from turmeric, etc. The feasibility of ATPE for the purification of natural colorants and synthetic dyes was demonstrated elsewhere (Tong et al. 1999; Wang et al. 1992). Natural colorants such as betalains, C-phycocyanin, carmine etc. were successfully purified using ATPE comprising PEG and inorganic salts (Patil and Raghavarao 2007; Patil et al. 2006, 2008); Chethana et al. 2007; Magestea et al. 2009).

15.2 Extraction of Colorants from Beet Root

As already mentioned, there is a growing interest in the use of natural pigments for food coloring, since synthetic dyes are becoming more and more critically assessed by the consumer. Natural colors are found to have nutritional as well as antioxidant properties and their presence in diet can reduce the risk of cardiovascular diseases, cancer and diseases associated with ageing (Delgado-Vargas et al. 2000). In food processing, betalains are less commonly used than anthocyanins and carotenoids, although these water-soluble pigments, stable between pH 3 and 7, are always well suited for coloring low acid food (Strack et al. 2003).

15.2.1 Pigments in Beet Root

The major pigments present abundantly in beetroot (*Beta vulagaris* L.) are betalains, which are chemically defined as the derivatives of betalamic acids (Delgado-Vargas et al. 2000). Betalains have a large number of applications in food products, such as gelatins, desserts, confectioneries, baked foods, dry powder beverages, poultry products, dairy products (strawberry yogurt, ice creams) and meat products (sausages, cooked ham) (Nilsson 1970; Aparnathi and Borkhatriya 1999). Betalains are also used in powder beverages, water ices, baked goods, biscuit creams, hard candies, jellies and fruit cocktails. Betalains are also blended with other colorants to provide desired color matches. Based on these applications, betalains are used in foods with short shelf lifes, produced by a minimum heat treatment and marketed in a dry state with suitable packaging and to be stored under reduced levels of light, oxygen and humidity. Betalain preparations are water soluble with high tinctorial strength. They are relatively unchanged in color from pH 3 to pH 7 but are violet at pH values below 3 and blue at pH 7.

On the other hand, betalains also have pharmacological applications for mutagenic as well as carcinogenic activity. Recently, betalains have received more attention because betanin (betacyanin) has shown antiviral and antimicrobial activity (Delgado-Vargas et al. 2000). Beet root is reported to be a useful cancer-preventive vegetable (Chap. 7).

The antioxidant activity of betalain pigments from plants was evaluated using the modified 1,1-diphenyl-2-picrylhdrazyl (DPPH) method. All tested betalains exhibited strong antioxidant activity (Cai et al. 2003). Extensive research has already proven that betalains from beetroot impart a high antiradical effect, displaying strong antioxidant activity, representing a new class of dietary cationized antioxidants. Betanin at very low concentration was found to inhibit lipid peroxidation and heme decomposition. Red beet products used regularly in the diet may provide protection against certain oxidative stress-related disorders in humans (Kanner et al. 2001).

15.2.2 ATPE of Betalains

Extraction of betalains was carried out at room temperature (25 ± 1°C) or at slightly lower temperatures. Methanol or ethanol solutions are preferred solvents to achieve complete extraction. The extract obtained by this procedure contains a large quantity of sugars, thus lowering its tinctorial power. Sometimes the extract was heated at 70°C for 2 min, although this may destroy some of the pigments (Delgado-Vargas et al. 2000). The stability of betalains is strongly influenced by sugar, light, oxygen, water activity, pH and temperature (Delgado-Vargas et al. 2000; Nayak et al. 2006). Hence, removal of sugars from beet extract is very much desirable in order to facilitate the application of betalains in food processing. Consequently, a fermentation process was employed to reduce the sugar content. Since the beet extract contains 80% of the fermentable carbohydrates and nitrogenous compounds, yeast fermentation utilizing *Candida utilis* to remove these materials was suggested (Adams and Von Elbe 1977). All betalain pigments are water soluble, and this property is exploited to extract the pigment. On a laboratory scale, betanin can be obtained in crystalline form by employing ion exchange, absorption and/or gel filtration chromatography. Unfortunately, these methods give poor yields and therefore are of very little or no commercial value (Von Elbe and Goldman 1977). The free sugars present in the beet extract causes fermentation of the beet extract and also caramelization during food processing at high temperatures. Hence, removal of sugars from beet extract is very much desirable in order to facilitate the application of betalains in food processing.

ATPE was successfully employed for the purification and concentration of betalains (Chethana et al. 2007). The effects of process parameters such as tie-line length, volume ratio, polymer and salt concentration, pH and neutral salt on partitioning of betalains have been studied. PEG 6000/ammonium sulfate was found to be the most suitable system for the purification of betalains. Differential partitioning of betalains and sugars was achieved in ATPE at a higher tie line (34%), wherein 70–75% of betalains partitioned to the top phase and 80–90% of sugars present in the beet extract partitioned to the bottom phase, thus purifying the betalains.

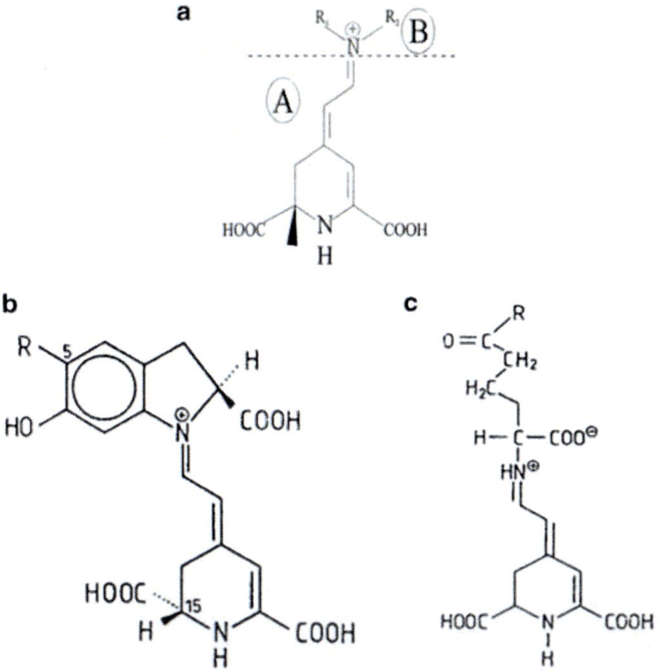

Fig. 15.2 Betalain general formula (**a**); the betalamic acid moiety (*A*) is common in all betalain molecules, and the structure represents a betacyanin or a betaxanthin, depending on the identity of the R_1 and R_2 residues (*B*). Betacyanin (**b**); betaxanthin (**c**)

Betalains have three carboxyl groups and a quaternary nitrogen atom with a weak positive charge, which, together with a carboxyl group in the second position, gives them amphoteric properties. Thus, betalains occur as negatively charged ions at pH >2.0. Hence, betalains preferentially partition to PEG-rich top phase during ATPE. HPLC analysis confirmed that the betalains are stable after ATPE. The PEG was separated from the betalains by organic aqueous extraction; the polymer obtained could be reused for subsequent ATPE (Chethana et al. 2007).

The general structure of the betalains is shown in Fig. 15.2a. Betalains contain two structural groups, namely, the red–purple betacyanins (Fig. 15.2b) and the yellow betaxanthins (Fig. 15.2c), depending on their R_1-N-R_2. Conjugation of a substituted aromatic nucleus to the 1,7-diazaheptamethinium chromophore shifts the absorption maximum from 480 nm in betaxanthins to 540 nm in betacyanins. Betanin and iso-betanin make up 95% of the betacyanins and vulgaxanthin I makes up 95% of the betaxanthins. Because of their high value in purified form, studies on the separation of these closely related pigments have been reported (Wiley and Lee 1978; Stintzing et al. 2002; Fernandez-Lopez and Almela 2001). In all these techniques, the separated betacyanin and betaxanthin are obtained in low quantity and these techniques cannot be easily used for large-scale preparations. ATPE has the potential to separate two closely related pigments, which conventionally requires chromatographic techniques that are more of analytical than preparative methods (Patil and Raghavarao 2007). Differential partitioning of betalains was efficiently carried out employing ATPE comprising PEG6000/ammonium sulfate system. Betacyanin has an aromatic ring in its structure and hence more preferentially partitions to the top phase than betaxanthin. Betaxanthin, which does not contain an aromatic ring, partitions to the bottom phase. Similar results were reported by Kaul (2000), that the aromatic amino acids like tyrosine, tryptophan etc. partition to the top phase. Multiple extractions (triple) were efficiently used for complete removal of betacyanin from the bottom phase. Thus, at the end of the third extraction, the bottom phase had only betaxanthin with no traces of betacyanin (Chethana 2006). Thus, ATPE is a simple and an efficient method for the fractionation of betacyanin and betaxanthin.

15.3 Enzymes from Beetroot

15.3.1 Extraction of Enzymes from Beetroot

Higher cost of production and purification from crude sources limits the use of this versatile enzyme in many industrial applications. Therefore, the constant effort for finding alternative sources for the enzyme has resulted in the exploration of different sources. Beet root is a very good source of sucrose along with many industrially useful enzymes, such as starch-hydrolyzing enzymes (glycolytic enzymes), peroxidases, polyphenol oxidase, acid invertase etc. Further, to improve the productivity, cell cultures of beet root (cell cultures, hairy root cultures) are abundantly used for the production and purification of many enzymes. There are several advantages of hairy roots over cell cultures for the production of high-value secondary metabolites (Mukundan et al. 1998; Georgiev et al. 2008, Suresh et al. 2004).

A few reports indicated that the recovery of enzymes from red beet hairy roots produce copious levels of enzymes that were purified to homogeneity. The purified enzyme (peroxidases) showed better thermal stability than the commercial source from horseradish, indicating the potential for wider applications. An intracellular peroxidase (POD) produced by genetically transformed root cultures of red beet (*Beta vulgaris* L.) was purified using a combination of ammonium sulfate fractionation and ion exchange chromatography, resulting in 15-fold enhancement of activity (Thimmaraju et al. 2007) (Chap. 12).

15.3.2 ATPE of Enzymes from Beetroot

Application of ATPE for the partitioning and recovery of peroxidase enzyme from hairy roots has been reported (Bhagyalakshmi and Thimmaraju 2009). ATPE has often been used for large-scale continuous separation of proteins and natural pigments and removal of contaminants from fermentation broths, as it can form a critical primary purification step. The preliminary set of ATPE experiments used different phase compositions employing various molecular weight PEGs and a constant ammonium sulfate concentration. The increase in recovery of the peroxidase above 100% may be because of the enhancement of the activity of the enzyme due to its stabilization and purification in the salt phase. The enzyme appeared to concentrate in the salt-rich bottom phase, with enhancement in the total activity, especially in low molecular weight PEG (4000). The highest recovery of these biomolecules (~100% enzyme recovery) was observed at low PEG molecular weights (Bhagyalakshmi and Thimmaraju 2009).

Polyphenol oxidase (PPO) is also present abundantly in *Beta vulgaris.* This enzyme has been extracted from beet root in both soluble and membrane fractions. PPO was purified to apparent homogeneity. Conventionally, the soluble PPO purification was achieved by hydrophobic interaction chromatography and gel filtration chromatography. The membrane PPO purification was achieved by anion exchange chromatography and gel filtration (Gandía-Herrero et al. 2004). Purification and characterization of a polyphenol oxidase from red Swiss chard (*Beta vulgaris* spp. *cicla*) has been reported by Gao et al (2009). The purification procedure involved a chromatographic method and resulted in a 39-fold enrichment in specific activity. However, the recovery of total enzyme activity was only 17%. Hence, the application of these methods for large-scale production are limited due to the low yield of the enzyme.

The purification of PPO by employing an integration of ATPE and ultrafiltration (UF) has been explored (Chethana 2006). Influence of the process parameters on partitioning of PPO was studied. It has been shown that polymer molecular weight, tie-line length, phase volume ratio and neutral salt significantly affect the partitioning behavior of PPO. A 4.7-fold purification factor of PPO was obtained at standardized conditions of ATPE. Further, ATPE coupled with membrane process (UF) resulted in an increase in the specific activity of the enzyme, and the purification factor increased by 12.8-fold.

15.4 Simultaneous Recovery of Pigment and Enzyme

ATPE has been explored for simultaneous recovery of the pigment (betalain) and the enzyme (peroxidase) and the results indicated the possibility of partitioning the two products into opposite phases, namely, a bottom and top phase. A system comprising PEG and ammonium sulfate was used for the study. Optimization of process conditions for better differential partitioning of these biomolecules was carried out. The partitioning of pigment to the top PEG phase and a concentration of peroxidase in the bottom salt-rich phase was observed. Among the two phases, peroxidase tends to move towards the salt-rich bottom phase. However, the distribution of the pigment was more or less even in both phases, with a degradation loss of pigment from 30% to 60%. Keeping in mind the separation behavior, more trials were conducted to partition the maximum quantity of the enzyme to bottom phase and the pigment to the top phase. The system comprising 7% PEG and 13% salt appeared best for the partitioning and recovery of betalains and peroxidase simultaneously. The ATPE system resulted in the recovery of only about 25% betalains in the top phase and about 137% peroxidase (Bhagyalakshmi and Thimmaraju 2009).

15.5 Integration of ATPE with Other Methods

In recent years, downstream processing (that is, recovery, purification and concentration of the product) has been faced with a strong demand for intensification and integration of process steps to increase yield, to reduce process time and to cut down running costs and capital expenditure (Schugerl and Hubbuch 2005). Process integration, wherein two or more unit operations are combined into one in order to achieve specific goals which are usually not effectively met by discrete processes, offers considerable potential benefit for the recovery and purification of biological products (Rito-Palomares 2004). Integration of ATPE with other processes (such as fermentation, cell disruption and membrane processes) is one approach that is gaining considerable attention in recent years. Application of ATPE for extractive fermentation is a meaningful approach to overcome low product yield in a conventional fermentation process and, by proper design of the ATPE, it is feasible to obtain the product in a cell-free stream. Recently, extractive fermentations using ATPS have been developed for the recovery of different protein products and has resulted in an increase in productivity (Guan et al. 1996; Sinha et al. 2000). Rito-Palomares and Lyddiatt (2002) have reported the integration of cell disruption and ATPE for the recovery and purification of intracellular proteins. A few research articles on integration of ATPE with membrane processes for the purification and concentration of various biological products are available (Tanuja et al. 2000; Srinivas et al. 2002; Rito-Palomares 2004; Patil and Raghavarao 2007)

15.6 Concluding Remarks

Beet roots are the rich source of many enzymes as well as colorants. Many of them are highly valuable and stable compared with enzymes obtained from other sources. It is very important to identify an economical and scalable method for extraction of these valuable components for industrial success. ATPE is shown to be one such method. A few reports are available for the extraction of pigments and enzymes from beet roots and hence there is large scope to explore efficacy of ATPE for the other enzymes from beetroot for enhanced productivity. It is well known that the betalain pigments are also hydrophilic, like the peroxidase enzyme, and hence extensive study, taking various kinetic parameters into consideration is needed to arriving at a congenial strategy for their simultaneous separation for further improvement of the process. A number of investigators have tried to study and explain the physical interactions influencing two-phase extraction and model the factors affecting biomolecule partitioning. However, the accurate prediction of partition coefficients of proteins is still very difficult. Such difficulties might arise due to the complex and interrelated interactions between the proteins and polymers, salts, buffers and water present in the partition system. Therefore, development of separation processes employing ATPE relies upon extensive experimentation suitably supported by theoretical studies. Optimization methods have been used to discover the best possible combination of factors that can result in maximum productivity.

References

Adams, J.P., and J.H. Von Elbe. 1977. Betanine separation and quantitation by chromatography on gels. *Journal of Food Science* 42: 410–414.

Albertsson, P.A. 1986. *Partition of cells particles and macromolecules*, vol. 3. New York: Wiley.

Albertsson, P.A. 1971. *Partition of cell particles and macromolecules*, 2nd ed. New York: Wiley.

Aparnathi, K.D., and V.N. Borkhatriya. 1999. Improved extraction and stabilization of natural food Colorants. *Indian Food Industry* 18: 164–168.

Attoe, E.L. and J.H. von Elbe. 1985. Oxygen involvement in betanine degradation: effect of antioxidants. *Journal of Food Science* 50: 106–110.

Babu, B.R., N.K. Rastogi, and K.S.M.S. Raghavarao. 2008. Liquid–liquid extraction of bromelain and polyphenol oxidase using aqueous two-phase system. *Chemical Engineering and Processing* 47: 83–89.

Bamberger, S., G.V.F. Seaman, K.A. Sharp, and D.E. Brooks. 1984. The effects of salts on the interfacial tension of aqueous dextran poly(ethylene glycol) phase systems. *Journal of Colloid and Interface Science* 99: 194–200.

Bhagyalakshmi, N., and R. Thimmaraju. 2009. Bioreactor for cultivation of red beet hairy roots and in situ recovery of primary and secondary metabolites. *Engineering in Life Sciences* 9: 227–238.

Cai, Y., M. Sun, and H. Corke. 2003. Antioxidants activity of betalains from plants of the Amaranthaceae. *Journal of Agriculture Food Chemistry* 51: 2288–2294.

Chethana S (2006) Field assisted separation and downstream processing of biomolecules and natural colors. Ph. D. Thesis, Mysore: University of Mysore.

Chethana, S., C.A. Nayak, and K.S.M.S. Raghavarao. 2007. Aqueous two phase extraction for purification and concentration of Betalains. *Journal of Food Engineering* 81: 679–687.

Delgado-Vargas, F., A.R. Jimenez, and O. Pardes-Lopez. 2000. Natural pigments: Carotenoids, anthocyanins and betalains-characteristics, biosynthesis, processing and stability. *Critical Reviews in Food Science and Nutrition* 40: 173–280.

Dembczynski, R., W. Białas, W. Regulski, and T. Jankowski. 2010. Lysozyme extraction from hen egg white in an aqueous two-phase system composed of ethylene oxide–propylene oxide thermoseparating copolymer and potassium phosphate. *Process Biochemistry* 45: 369–374.

Diamond, A.D., and J.T. Hsu. 1992. Aqueous two phase systems for biomolecule separation. *Advances in Biochemical Engineering/Biotechnology* 47: 8–133.

Diamond, A.D., and J.T. Hsu. 1989. Fundamental studies of biomolecule partitioning in aqueous two-phase systems. *Biotechnology and Bioengineering* 34: 1000–1014.

Eiteman, M.A., and J.L. Gainer. 1989. The effect of relative free volume changes on partitioning in magnesium sulphate-poly(ethylene glycol) aqueous two phase systems. *Biochimica et Biophysica Acta* 992: 125–127.
Fernandez-Lopez, J.A., and L. Almela. 2001. Application of high performance liquid chromatography to the characterization of betalain pigments in prickly pear fruits. *Journal of Chromatography. A* 913: 415–420.
Gandía-Herrero, F., F. García-Carmona, and J. Escribano. 2004. Purification and characterization of a latent polyphenol oxidase from beet root (*Beta vulgaris* L.). *Journal of Agricultural and Food Chemistry* 52: 609–615.
Gao, Z.J., X.H. Han, and X.G. Xiao. 2009. Purification and characterisation of polyphenol oxidase from red Swiss chard (*Beta vulgaris subspecies cicla*) leaves. *Food Chemistry* 117: 342–348.
Gautam, S., and L. Simon. 2007. Prediction of equilibrium phase compositions and *β*-glucosidase partition coefficient in aqueous two-phase systems. *Chemical Engineering Communications* 194: 117–128.
Gautam, S., and L. Simon. 2006. Partitioning of β-glucosidase from *Trichoderma reesei* in poly(ethylene glycol) and potassium phosphate aqueous two-phase systems: Influence of pH and temperature. *Biochemical Engineering Journal* 30: 104–108.
Georgiev, V., M. Ilieva, T. Bley, and A. Pavlov. 2008. Betalain production in plant in vitro systems. *Acta Physiologiae Plantarum* 30: 581–759.
Guan, Y.X., Z.Q. Zhu, and L.H. Mei. 1996. Technical aspects of extractive purification of penicillin fermentation broth by aqueous two-phase partitioning. *Separation Science and Technology* 31: 2589–2597.
Grossman, P.D., and J.L. Gainer. 1988. Correlation of aqueous two phase partitioning of proteins with change in free volume. *Biotechnology Progress* 4: 6–11.
Hemavathi, A.B., and K.S.M.S. Raghavarao. 2011. Differential partitioning of α-galactosidase and α-glucosidase using aqueous two phase extraction. *Process Biochemistry* 16: 649–655.
Imelio, N., A. Marini, D. Spelzini, G. Picó, and B. Farruggia. 2008. Pepsin extraction from bovine stomach using aqueous two-phase systems: Molecular mechanism and influence of homogenate mass and phase volume ratio. *Journal of Chromatography B* 873: 133–138.
Kanner, K., S. Harel, and R. Granit. 2001. Betalains – a new class of dietary cationized antioxidants. *Journal of Agriculture and Food Chemistry* 49: 5178–5185.
Kaul, R. 2000. *Aqueous two phase systems: methods and protocols*. Totowa: Humana Press.
Kroner, K.H., H. Schutte, W. Stach, and M.R. Kula. 1982. Scale up of formate dehydrogenase by partition. *Journal of Chemical Technology and Biotechnology* 32: 130–137.
Kula, M.R., K.H. Kroner, and H. Hustedt. 1982. Purification of enzymes by liquid-liquid extraction. *Advances in Biochemical Engineering* 24: 73–118. Springer/New York.
Lakshmi, M.C., M.C. Madhusudhan, and K.S.M.S. Raghavarao. 2009. Extraction and purification of Lipoxygenase from soybean using aqueous two-phase system. *Food and Bioprocess Technology*. doi:10.1007/s11947-009-0278-8.
Li, M., J.W. Kim, and T.L. Peeples. 2002. Amylase partitioning and extractive bioconversion of starch using thermoseparating aqueous two-phase systems. *Journal of Biotechnology* 93: 15–26.
Madhusudhan, M.C., K.S.M.S. Raghavarao, and S. Nene. 2008. Integrated process for extraction and purification of alcohol dehydrogenase from Baker's yeast involving precipitation and aqueous two phase extraction. *Biochemical Engineering Journal* 38: 414–420.
Magestea, A.B., L.R. de Lemosa, G.M.D. Ferreira, M.C.H. Da Silva, L.M.H. da Silva, R.C.F. Bonomob, and L.A. Minim. 2009. Aqueous two-phase systems: An efficient, environmentally safe and economically viable method for purification of natural dye carmine. *Journal of Chromatography. A* 1216: 7623–7629.
Miranda, M.V., and O. Cascone. 1994. Partition behavior of horse radish peroxidase isoenzymes in aqueous two-phase poly (ethylene glycol)/phosphate systems. *Biotechnology Techniques* 8: 275–280.

Mukundan, U., E.B. Carvallo, and W.R. Curtis. 1998. Growth and pigment production by hairy root cultures of *Beta vulgaris* L. in bubble column reactor. *Biotechnology Letters* 20: 469–474.

Nayak, C.A., S. Chethana, N.K. Rastogi, and K.S.M.S. Raghavarao. 2006. Enhanced mass transfer during solid–liquid extraction of gamma irradiated red beetroot. *Radiation Phys and Chemistry* 75: 173–178.

Nilsson, T. 1970. Studies into the pigments in beetroot (*Beta vulgaris* L. ssp. *Vulgaris var. rubra L*). *Lantbrukshoegsk Ann* 36: 179–219.

Nitsawang, S., R. Hatti-Kaul, and P. Kanasawuda. 2006. Purification of papain from *Carica papaya* latex: Aqueous two-phase extraction versus two-step salt precipitation. *Enzyme and Microbial Technology* 39: 1103–1107.

Ooi, C.W., B.T. Tey, S.L. Hii, A. Ariff, H.S. Wu, J.C. Wei, R.S. Juang, S.M.M. Kamal, and T.C. Ling. 2009. Direct purification of *Burkholderia pseudomallei* lipase from fermentation broth using aqueous two-phase systems. *Biotechnology and Bioprocess Engineering* 14: 811–818.

Patil, G., and K.S.M.S. Raghavarao. 2007. Aqueous two phase extraction for purification of C-phycocyanin. *Biochemical Engineering Journal* 34: 156–164.

Patil, G., S. Chethana, M.C. Madhusudhan, and K.S.M.S. Raghavarao. 2008. Fractionation and purification of the phycobiliproteins from *Spirulina platensis*. *Biosource Technology* 99: 7393–7396.

Patil, G., S. Chethana, A.S. Sridevi, and K.S.M.S. Raghavarao. 2006. Method to obtain C-phycocyanin of high purity. *Journal of Chromatography. A* 1127: 76–81.

Pereira, M., Y.T. Wu, A. Venancio, and J. Teixeira. 2003. Aqueous two-phase extraction using thermoseparating polymer: A new system for the separation of endo-polygalacturonase. *Biochemical Engineering Journal* 15: 131–138.

Raghavarao, K.S.M.S., M.R. Guinn, and P. Todd. 1998. Recent developments in aqueous two-phase extraction in bio processing. *Separation and Purification Methods* 27: 1–50.

Raghavarao, K.S.M.S., N.K. Rastogi, M.K. Gowthaman, and N.G. Karanth. 1995. Aqueous two phase extraction for the downstream processing of enzymes/proteins. *Advances in Applied Microbiology* 4: 97–171.

Rito-palomares, M. 2004. Practical application of aqueous two phase partition to process development for the recovery of biological products. *Journal of Chemical Technology and Biotechnology* 807: 3–11.

Rito-Palomares, M., and A. Lyddiatt. 2002. Process integration using aqueous two-phase partition for the recovery of intracellular proteins. *Chemical Engineering Journal* 87: 313–319.

Schugerl, K., and J. Hubbuch. 2005. Integrated bioprocesses. *Current Opinion in Microbiology* 8: 294.

Silva, M.E.D., and T.T. Franco. 2000. Purification of soybean peroxidase (*Glycine max*) by metal affinity partitioning in aqueous two-phase systems. *J Chromatography B: Biomedical Sciences and Applications* 743: 287–294.

Sinha, J., P.K. Dey, and T. Panda. 2000. Extractive fermentation for improved production of endoglucanase by an intergeneric fusant of *Trichoderma reesei/Saccharomyces cerevisiae* using aqueous two-phase system. *Biochemical Engineering Journal* 6: 163.

Srinivas, N.D., R.S. Barhate, and K.S.M.S. Raghavarao. 2002. Aqueous two-phase extraction in combination with ultrafiltration for downstream processing of Ipomoea peroxidase. *Journal of Food Engineering* 54: 1–6.

Srinivas, N.D., K.R. Rashmi, and K.S.M.S. Raghavarao. 1999. Extraction and purification of a plant peroxidase by aqueous two phase extraction coupled with gel filtration. *Process Biochemistry* 35: 43–48.

Stintzing, F.C., A. Schieber, and R. Carle. 2002. Identification of betalains from yellow beet (*Beta vulgaris* L) and cactus pear *(Opuntis ficus indica* (L.) Mill) by high performance liquid chromatography-electro spray ionization mass spectrometry. *Journal of Agricultural and Food Chemistry* 50: 2302–2307.

Strack, D., T. Vogt, and W. Schliemann. 2003. Recent advances in betalain research. *Phytochemistry* 62: 247–269.

Suresh, B., R. Thimmaraju, N. Bhagyalakshmi, and G.A. Ravishankar. 2004. Polyamine and methyl jasmonate-influenced enhancement of betalaine production in hairy root cultures of Beta vulgaris grown in a bubble column reactor and studies on efflux of pigments. *Process Biochemistry* 39: 2091–2096.

Tanuja, S., N.D. Srinivas, M.K. Gowthaman, and K.S.M.S. Raghavarao. 2000. Aqueous two-phase extraction coupled with ultrafiltration for purification of amyloglucosidase. *Bioprocess Engineering* 23: 63–68.

Teotia, S., and M.N. Gupta. 2004. Purification of phospholipase D by two-phase affinity extraction. *Journal of Chromatography. A* 1025: 297–301.

Thimmaraju, R., L. Venkatachalam, K. Roohie, M.S. Narayan, and N. Bhagyalakshmi. 2007. Purification and characterization of an intracellular peroxidase from genetically transformed roots of red beet (*Beta vulgaris L.*). *Food Chemistry* 105: 1312–1320.

Tong, A.J., J.J. Dong, and L.D. Li. 1999. Aqueous two-phase extraction system of sodium perfluorooctanoate and dodecyltriethylammonium bromide mixture and its application to porphyrins and dyes. *Analytica Chimica Acta* 390: 125–131.

von Elbe, J.H., and I.L. Goldman. 1977. The Betalains. In *Current aspects of food colorants*, ed. T.E. Furia, 11–30. Cleveland: CRC press.

Walter, H., D.E. Brooks, and D. Fisher (eds.). 1985. *Partitioning in aqueous two phase systems: Theory, methods, uses and applications to biotechnology*. New York: Academic.

Wang, W.-H., R. Kuboi, and I. Komasawa. 1992. Aqueous two-phase extraction of dehydrogenases using triazine dyes in PEG/phosphate systems. *Chemical Engineering Science* 47: 113–121.

Wiley, R.C., and Y.N. Lee. 1978. Recovery of Betalains from red beets by diffusion extraction procedures. *Journal of Food Science* 43: 1056–1058.

Yanga, L., D. Huoa, C. Houa, K. Hea, F. Lva, H. Fab, and X. Luoa. 2010. Purification of plant-esterase in PEG1000/NaH_2PO_4 aqueous two-phase system by a two-step extraction. *Process Biochemistry* 45: 1664–1671.

Zaslavsky, B.Y. 1995. *Aqueous two phase partitioning: physical chemistry and bioanalytical applications*. New York: Marcel Dekker.

Zaslavsky, B.Y., N.M. Mesetechkina, L.M. Miheeva, and S.V. Rogozhin. 1982. Measurement of relative hydrophobicity of amino-acid side chains by partition in an aqueous two phase system-hydrophobicity scale for non-polar and ionogenic side-chains. *Journal of Chromatography. A* 240: 21–28.

Chapter 16
Techno-commercial Aspects of Relevance to Red Beet

Bhagyalakshmi Neelwarne

Abstract While Chap. 1 of this book provides an overview of what has been researched in red beet, this chapter tries to put red beet into the newer arena of diverting its current progress towards newer technologies and applications. Beets have unlimited applications, from food, fodder, fuel and pharmaceutical to ornamentals and environmental cleansing agents; this means that the allied technologies span from moderate to very high-tech ventures. Scientifically, as a model system, beet has contributed continuously to our knowledge of cellular physiologies of primary metabolism, energy conservation in the form of sugar synthesis and sequestration processes of primary and secondary metabolites. In the field of genetics, their ever-diversifying nature and amazing adaptability to newer environments have mesmerized scientists. Red beet occupies a position among the top ten vegetables for its antioxidant capacity, cancer-prevention activities and instant blood pressure control. However, it is essential to supplement the current knowledge with innovative scientific methods in terms of validation, characterization and further onward integration of processes through advanced mechanotronics to develop mainstream food products. Treating in vitro cultures and normal beet plant with innovative biotechnologies such as metabolic engineering and genetic modification is expected to significantly improve the sustainable production of a wide array of food, pharmaceuticals and bio-energy products. To realize these potentials, the probable technological leaps that are needed are discussed in this chapter.

B. Neelwarne (✉)
Plant Cell Biotechnology Department, Central Food Technological Research Institute
(Council of Scientific and Industrial Research), Mysore 570 020, India
e-mail: bneelwarne@cftri.res.in

B. Neelwarne (ed.), *Red Beet Biotechnology: Food and Pharmaceutical Applications*,
DOI 10.1007/978-1-4614-3458-0_16,

16.1 Introduction

Although historically, cereals occupied a large cultivable surface of the earth and enjoyed a prime place in the food/fuel/fodder markets, the enormous health benefits offered by horticultural commodities coupled with their high yields and the advancement made in processing technologies have culminated in a huge turn-over of a number of horticulture products, resulting in their occupying a larger space in supermarkets than cereals. While ancient people considered red beet as an energy food, the modern drug discovery venture has only recently realized the therapeutic and other such adjunctive effects of red beet, such as its high antioxidant property and as a resource for the alleviation of life-style ailments. Of many groups of plant-derived pharmaceuticals, nitrogen-containing secondary metabolites have contributed the largest number of drugs to the modern pharmacopeia (Schmidt et al. 2007). While red/yellow pigments of beet are nitrogenous betalains, having very high antioxidant and lipoperoxy-radical scavenging efficiencies (see Chaps. 6 and 7), the red beet whole extract, a rich source of nitrate, was found to be very effective in lowering acute blood pressure via vasodilation and hence was also vaso-protective. Moreover, red beet juice showed efficient anti-platelet properties, therefore, foods of this kind hold high promises for treating modern life-style ailments. Such health beneficiary effects were initially thought to occur through the antioxidant route; however, careful insight into the functional aspects of red beet ingredients unravelled that the nitrates are, in fact, responsible for the overall cardio-protection, through the formation of nitric oxide by different mechanisms in each part of the gastro-intestinal tract (Webb et al. 2008). Red beet pigments displayed exceptionally high anticancerous properties in both in vitro and in whole animal models (see Chap. 7). The lipoperoxidative activities rendered by red beet pigments have also confirmed their high efficiency. As indicated in Chap. 6, in an LDL-oxidation assay, betanin was totally consumed before β-carotene, indicating higher effectiveness than β-carotene in the copper-oxidized LDL model. On the processing front, red beet pigment has seen rapid progress in extraction, concentration, encapsulation and packaging technologies. Owing to the great tinctorial value of red beet color and the large array of nutraceutical benefits rendered by red beet extract, efforts are being made to utilize the products of this crop in a commercially feasible manner. For these reasons, and also because of substantial progress made in aseptic cell and organ cultures of red beet, there are ample opportunities for making further progress by the application of modern innovative biotechnologies, leveraging this crop to occupy a prime position in the global nutraceutical market.

16.2 Nutraceutical Market Opportunities for *Beta vulgaris*

The nutraceutical market showed a healthy growth of US$187 billion in 2010 (Global Industry Analyst, Inc.), where functional foods and dietary supplements marked the fastest growth as a consequence of rising health awareness and an increase in the number of people seeking safer ways to keep themselves free from age-related chronic problems. On the other hand, younger generations are also becoming conscious of future ailments of synthetic and obesity/carcinogenicity-related food aspects, driving them to opt for natural and health-promoting ingredients in their food. The innumerable health benefits of red beet and the opportunities for the application of modern biotechnology for improving the production and processing of various red beet products have only started recently, and substantial progress has been made in the recent past. Table 16.1 lists current research developments and the technologies through which better nutraceutical products might enter the global market. An important point to be noted here is that, as shown in Fig. 16.1, the in vitro cultures of red beet and a few other plant species have shown high productivities, to a much higher extent than those produced by field-grown plants. Such productivities offer an opportunity to develop other forward technologies, such as down-stream processing, by economically viable modes. An evaluation made by Sahai (1994) indicated that with a product concentration below 5% and annual production as low as 3,000 kg/year, the product cost would be as high as $1,000/kg, provided the product recovery is 90%. Under such a situation, competing in the market becomes impossible unless the product is a rare and costly material. Although red beet cells/hairy root cultures produce higher quantities of pigments, the production of biomass itself adds enormous cost to the process. However, to realize good economic returns, complete knowledge of the chemistry of the compound is essential to make further changes in its chemical structure to derive newer unique biomolecules. Further monitoring of down-stream processes for maintaining the integrity of the molecule are crucial, as discussed below.

16.3 Red Beet: A Great Source for a Spectrum of Pigments

During plant cell evolution, pigments might have originated with a basic function of an energy trap; they have now diversified to perform various other roles such as internal signalling, to send externally visible signals for attracting fauna for pollination/seed dispersal and, above all, to protect plants from light-induced damage, phytopathogens and heavy metals. Phylogenetic insights have indicated that betalains appear to have evolved at least twice: once within a fungal lineage, the Basidiomycetes (e.g., *Amanita muscaria*), and once among flowering plants of the order Caryophyllales, where they replace the otherwise ubiquitous anthocyanins (Brockington et al. 2011). The fact that betalains originated much later in evolution and exist by mutual replacement of anthocyanins indicate that these plants might have chosen to switch over to betalains because of the wide spectrum of permutations and combinations of colors offered by making small adjustments in the biosynthetic pathway after betalamic acid synthesis and because the pigments are stable through a wide pH range. The clue one may perceive from these biochemical events is that such wide options are also available for the forward applications of betalains. While many normal beets produce more betacyanin than betaxanthin, in several hairy root clones of red beet, this is reversed (Pavlov et al. 2005a, b).

Table 16.1 Applications of innovative biotechnologies for the advancement of current knowledge to economize red beet process technologies

Current knowledge	Innovative biotechnologies	Prospective applications
Red beet extract and pigments	Content improvement in vivo by combinatorial chemistry and chemogenomics	Cost reduction and higher marketability of betalains food color
Phenolics, betacyanin, betaxanthin and their intermediary compounds with scientific data of their biological effects	Cell, tissue and hairy root cultures	High/uniform quality and product availability round the year for onward integration into other products
	Metabolic engineering for higher productivities/sequestration and hyperglycosylated/acylated betacyanins	Process cost reduction and better product stability
	Extraction through permeabilization, electroporation, supercritical and other smart technologies	Higher product recovery and better intactness of molecules
	Post-extraction molecular modification, fermentation, concentration and encapsulation	Improved pigment products with better stabilities, higher therapeutic benefits and wider applications
Morpho-genetic aspects	Chemi-genomics	Field applications for disease resistance, new breeding lines, higher productivities and newer compounds.
Partial information on genetic diversities, gene-flow, pathway biochemistry and ontogenetic–phylogenetic relationships	Combinatorial chemistry	
	Genome-mapping	Eco-friendly technologies with lesser pollutants
Cell cultures and hairy roots	Newer bioreactor designs with continuous production and on-line product recovery and simultaneous fractionations	Improved and newer bioprocess parameters for achieving higher productivities
Typical experimental evidence for kinetics of product formation		
Regulation of pigments by precursor feeding	Genetic engineering for metabolic up-gradation and new proteins	Higher productivities and newer products
	Combinatorial chemistry for newer products	Newer food and pharmaceutical applications
Biochemically proven antioxidant, anti-diabetic anti-cancer and blood pressure-lowering effects	Target-centric biochemical studies, chemogenomics, computational chemistry, micro-arrays, structural evidences with flow cytometry	Better insights into molecular interactions, leading to newer compounds with precisely targeted molecule delivery

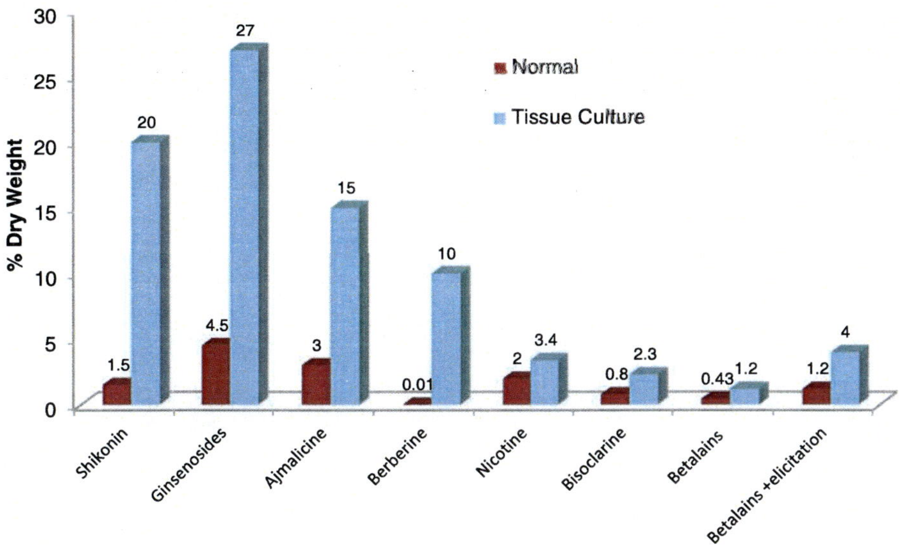

Fig. 16.1 Higher production of metabolites in normal and in vitro cultures. Values of betalains are based on Thimmaraju et al. 2004; Savitha et al. 2006. Other values are obtained from Wang 2008

While betacyanin is a glucosylated molecule, additional glycosylations of the glucose moiety and acylations are possible (Strack et al. 2003). Such additional glycosylation is expected to render additional sweetness to the pigment extract with various advantages, such as products with higher levels of fermentation, better caramelization and more body to the product, with increases in brix as well and improved sensorial properties. Acylation on the other hand is expected to render better stability to the product, as has been observed for anthocyanins (Terahara et al. 2004). Betacyanin acylation is accomplished by respective acyltransferases (Strack et al. 2003), which may also be possible by engineering for targeted gene expression in storage roots, where acylated betacyanin accumulation can be expected. However, the processing needs to be done at carefully maintained pH, since particularly alkaline pH ranges were found to cause acyl-migration (Wybraniec 2008).

Although there are several sources of betalain pigments, red beet has remained the most important source for betalains, where the pigment has both reddish purple betacyanin and yellow betaxanthins, which are abundantly synthesized. Crop yields of 50–70 t/ha yield betanin levels ranging from 0.4 to 20 mg/g of root (Pavokovic and Krsnik-Rasol 2011). Evolutionarily, betalains emerged by suppressing the anthocyanin biosynthetic pathway. Altering this switch partially, either by genetic regulation or combinatorial chemistry, might result in the synthesis of both of the pigments. The high morphogenic plasticity of red beet cells could serve as a model system to explore this possibility. Hairy roots are the right candidates for such genetic engineering because of the ease with which genetic transformation can be accomplished, and the roots thus obtained show consistent productivities.

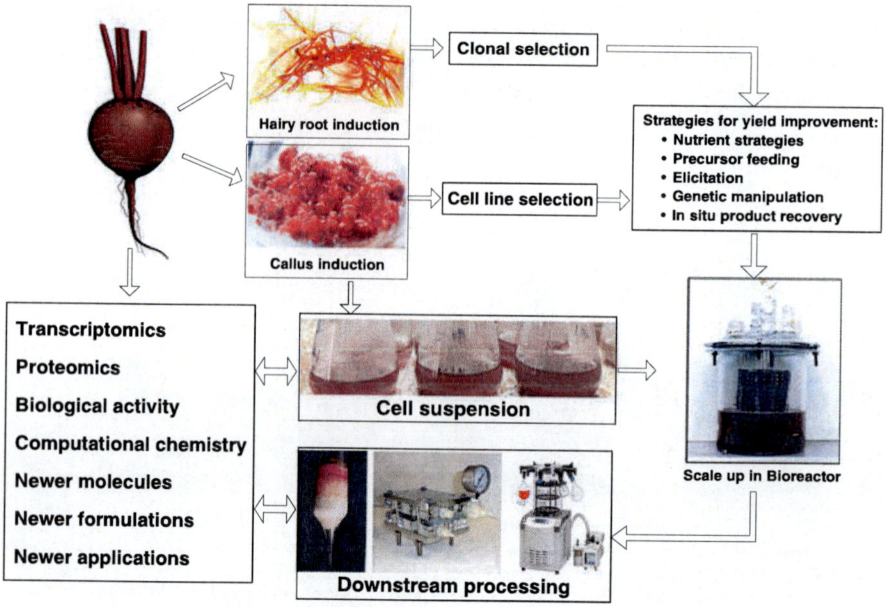

Fig. 16.2 Current in vitro technologies and newer prospects for commercial applications of red beet

Hairy roots and cell cultures produce much higher levels of betalains than in vivo plants (Fig. 16.1) and hence appear feasible for increasing their scale-up cultivation by integrating with technological processes similar to those with proven success in commercializing microbial products. Figure 16.2 depicts current technologies available for product enhancement in cultured cells and hairy roots of red beet, which have great commercial prospects if the hypothetically shown newer methods are integrated into the process. The in vitro cultures come with a package of various advantages, the major ones being the ease of extraction, uniformity in product quality, low earthy aroma and availability around the year. There are several technological issues to be addressed for commercial realization of plant in vitro cultures, as discussed below.

16.4 Commercial Realities of Plant In Vitro Cultures

Unlike purely chemical products, biotechnology products need to stride a long way from research success to reach commercial realities (Table 16.2). A review by Schmidt et al. (2007) highlights steps involved from research success to commercial venture, where a botanical cosmeceutical product may take as little as 3 years, whereas a nutraceutical, being a food ingredient, would take 4 years. Plant extracts

Table 16.2 Time-scale and other requirements for plant-based biotechnology products to reach commercial status[a]

Product type	Technical requirement	Legal requirement	Years to reach market	Investment cost in $	Potential sales/annum in $
Cosmeceutical	Experimentally established cosmetic use	Clinical study for safety	3	100,000	500,000
Nutraceutical	Dietary supplement and history of food use	Clinical study for safety	4	1 million	20 million
GRAS food ingredient	Additional toxicology Scientific publications	Approval by expert panel qualified to evaluate the safety	6	1.5 million	100 million
Botanical new drug that cures/prevents disease	Phase I & II tests for mode of action and actives Study using internationally prescribed protocols	Additional toxicology tests, Pharmacokinetics, Drug-to-drug interactions	10	80 million	1 billion

[a]The values are obtained from reference Schmidt et al. (2007)

with GRAS status that may impart difference to food functionalities need the support of toxicology data and published research information and hence would take about 6 years to reach the market. However, a new drug molecule may take as long as 10 years to enter the market because of the series of toxicology and clinical tests it has to undergo. There are additional corroborative costs involved in the development, nevertheless with good returns.

Biotechnology has utilized red beet initially as a model system for unravelling basic nutritional requirements of cell cultures (Akita et al. 2002) and to study the engineering parameters for scale-up of processes (Bhagyalakshmi et al. 1998; Pavlov et al. 2005a, 2007). Incidentally, red pigment served as an index to keep track, with naked eyes, of the response of cell/organ cultures to different treatments. Since field-grown beets were much cheaper than cultured cells, betalain pigments did not catch the attention of commercial ventures. However, later it was realized that large variations occur in conventionally grown beets and the extraction of compounds from field-grown materials has its own set of problems (Bhagyalakshmi and Ravishankar 1998). Ever since hairy roots were reported, red beet was again the material of choice for research (Taya et al. 1992, 1994; Thimmaraju et al. 2003a, b, 2004). Later Pavlov et al. (2005a, b, 2007), Neelwarne and Thimmaraju (2009) and Suresh et al. 2004) made substantial contributions towards product formation and down-stream recovery of the betalains, substantiating the commercial feasibility of such processes. As an inherent regulatory mechanism, biosynthesis and accumulation of different secondary metabolites are strictly regulated in plant cells, where, after reaching a specific level, no further synthesis and accumulation of the metabolite occurs. When the concentration of a secondary metabolite is increased in plant cells by genetic engineering or elicitation, its intrinsic toxicity has been found to become a limiting factor. For solving such problems and to achieve higher productivities, an alternative option of the removal of metabolite from the site of its synthesis was implemented to eliminate such toxic effects as well as to abolish metabolic feedback inhibition, allowing the cell to shift its metabolic flux further towards product biosynthesis (Wong et al. 2004). Alternatively, removing the metabolite from the site of accumulation within the cell, i.e., recovering the product from the vacuole, increased the chances for further accumulation of the product, causing overall product enhancement. For instance, nicotine, an alkaloid, is tolerated by plant cells up to certain level, beyond which it shows cellular toxicity. However, the over-expression of the yeast ABC transporter PDR5 through genetic transformation allowed transgenic tobacco cells to decrease the cellular toxicity (Goossens et al. 2003). In such cases, it has also been suggested that simultaneously diverting the product to the cell exterior by careful permeabilization would allow for repeated use of biomass for several cycles of product recovery, thereby enhancing overall productivity.

Specific methods of elicitation and application of stress from the cell exterior and through intracellular stress and elicitation, product enhancement and repeated recovery have been enhanced. Therefore, the hypothesized strategies depicted in Fig. 16.3 might further enhance accumulation of betalains in cultured red beet cells and hairy roots. The total product enhancement also depends on the efficient timely

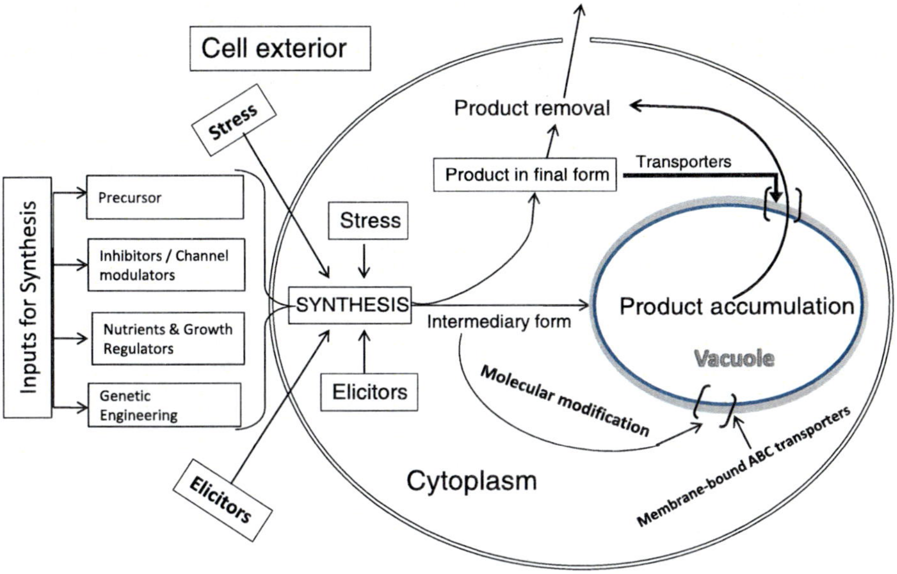

Fig. 16.3 Currently applied methods for product enhancement in cultured cells and organs

repetition in the recovery of products from hairy root biomass, where the recovery process must allow for the good viability of hairy roots and their further performance. The yeast lipid-based permeabilization method demonstrated by Thimmaraju et al. (2004) for the recovery of pigments from in vitro cultures appears efficient and useful for nutraceutical application purpose. Similarly, with external supply of precursors, nutrients and growth regulators, higher productivities were realized (Fett-Neto et al. 1994). Thus the application of biotechnology has already made a mark on quality parameters, nudging towards cost-effective process development. Figure 16.2 also provides probable forward integration of red beet cultures with techniques that are expected to provide better quality products and new biomolecules to the market. As indicated earlier, these technologies are futuristic, with a long way to tread. Alternatively, some of these new technologies may be conveniently integrated to existing conventional methods, as suggested below.

16.5 Value Addition to Conventional Pigment Extraction Technology

Figure 16.4 provides a schematic diagram of various steps involved in conventional pigment extraction (shaded boxes) and the prospective new steps (un-shaded box) that are feasible for the integration to the process of developing newer products with

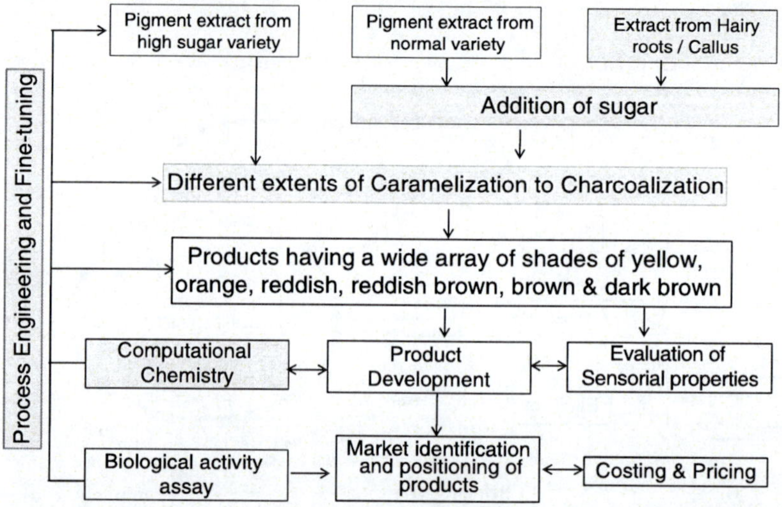

Fig. 16.4 Process engineering and product development by conventional methods where the addition of newer steps such as caramelization and computational chemistry is expected to result in better color stability and identification of new molecules

improved stability. Among the natural pigments used in food industry, caramel is one of the highly used ingredients, where diversity is needed to provide different shades to products. Such diversities are possible by including red beet extract with sugar before caramelization and charcoalization. Further evaluation through computational chemistry is expected to identify the large array of compounds that may be newly formed due to new interactions. Fine-tuning the process parameters on the basis of advanced sensorial analysis methods would then yield market-feasible products, fetching higher prices, since the products are expected to provide additional health benefits imparted by beet extract, although such claims need research back-up, as indicated in Table 16.2. Similarly, red beet polyphenols have also been found useful as anti-diabetic compounds and useful as radical scavengers (Georgiev et al. 2010; Gliszczyǹska-Ŝwiglo et al. 2006). Because of such great health benefits, their processing and delivery to consumers in the right manner has recently attracted great interest in designing functional foods to support nutraceutical industries. The application of encapsulation technology for the delivery of polyphenols, instead of free compounds, has been found to be more effective (Fang and Bhandari 2010). The technologies of encapsulation of polyphenols, including spray drying, complex co-acervation, liposome entrapment, co-complexation, co-crystallization, nano-encapsulation, freeze drying, yeast encapsulation, emulsification are some of the common encapsulation methods (Desai and Park 2005). Kosaraju et al. (2008) demonstrated that natural fruit fibers, with their large surface area, were found useful as potential carriers for spray drying sticky phenolic extracts for a variety of applications in functional foods (Chiou and Langrish 2007).

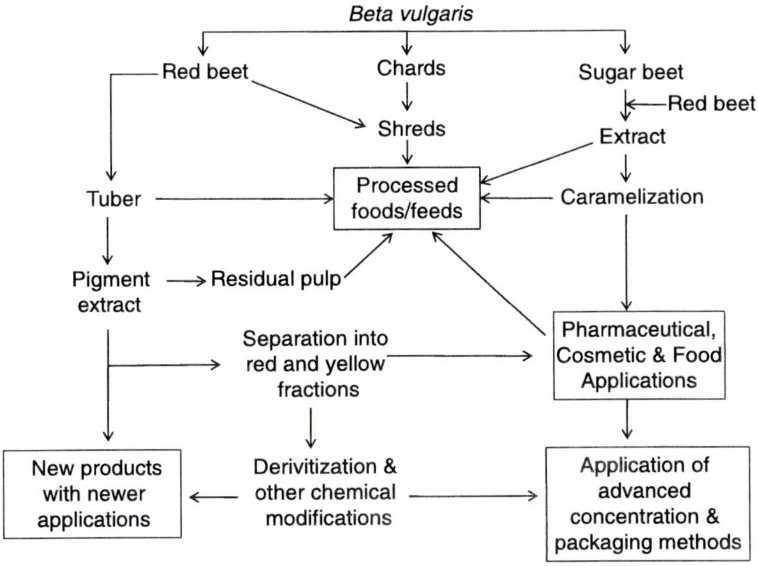

Fig. 16.5 Development of newer products from beets

16.6 Wide Choices and Products of Imagination

Red beet is such a vegetable where no part goes as a waste. Both foliage and storage root are edible, with rich content of nutritious and nutraceutical biomolecules. Both the organs have typical textures that allow their processing as well as incorporation into a large number of product formulations having combinations of leaf and tuber extracts and pulp biomass. The pulp can easily be incorporated in paste-type products such as ginger paste, garlic paste and so on. The rich nutrients in foliage of red/yellow beets, as well as chards, allow several minimally processed food products, similar to those of spinach. Various permutations and combinations of macerated biomass of root pulp and foliage are possible, with an option for the addition of sugar beet extract to allow fermentation as in steps shown in Fig. 16.5, would result in an array of new products. The modern analytical methods and tools such as mechanotronics (Mandenius and Björkman 2010) are also applicable in this context for precisely developing and monitoring of process parameters for each type of product. Fermented vegetable products have several added advantages, the chief ones are improvement in the bio-availability of nutrients, health-promoting effects due to the formation of several nutraceuticals during the fermentation process and long storability (Rakin et al. 2007). Owing to such great versatility, which is topped up with health benefits, there is a continuous addition to the list of beet-based products, not only for human applications but also for pets. Currently a few pet foods and dog chew sticks with red beet pulp are available. As indicated in Fig. 16.5, several areas still need to be strengthened through organized interdisciplinary scientific studies to realize the full benefits offered by beets. However, these products also need toxicological evaluation and certification before they are approved for market.

16.7 New Technical Discovery Strategies

While the societal trend is shifting towards natural medicine and drugs from natural sources (Day et al. 2009), there has been a concomitant increase in the investment in drug research and development, without a great increase in truly innovative drug biomolecules (Munos 2009), probably because of the lack of multidimensional screening and target-based approaches. In the past, the identification of new drug biomolecules was generally based on phenotypic screening and biologic-based approaches. However, newer strategies such as molecular mechanisms of actions have been more efficient in identifying new drug molecules and hence were more successful (Swinney and Antony 2011). Such a giant leap of progress was made possible by the tremendous advancement in the development of new tools to identify target-actions such as specific protein binding (G protein-coupled receptor binding), RNA interference, selective regulations etc. with the aid of high-throughput target-based screening assays. Several structure-based tools are also available for the identification and optimization for certain drug targets, including X-ray crystallography and computational modelling and screening. Apart from such virtual screening methods, computational chemistry and ex situ modification of natural substances has greatly contributed to identifying new drug entities. Red beet is a rich source of several phenolic compounds, and many are yet to be identified. Although computational chemistry is helpful in the identification of such novel chemical entities, the adjunct of combinatorial chemistry, though which further manipulations are possible, would result in striking at some novel biomolecules. There are examples of applying this technique to make alterations in chemical domains that are involved in building the skeletal part of the natural product and generating altogether different metabolites. Such domains can be taken from the same biosynthetic cluster, from different clusters, or even from different organisms to treat the whole plant or cell cultures. Many potentially cytotoxic compounds such as epithilones have been generated by these methods (Muzler and Prantz 2009). Geosmin renders the earthy aroma to red beet and hence is a demerit for the application of beet products. By the application of described methods, this compound can possibly be suppressed. In addition to finding a large variety of hitherto unexplored biomolecules present in red beet, extensive opportunities still exist to tap this crop as a source of several new natural drug candidates.

16.8 Ecological Benefits of Modern Technologies

Product sustainability has been a matter of environmental compliance, encompassing three components: ecological harmony, societal benefits and economic viability. These components need to be perfectly balanced, similar to the legs of a tripod. Therefore product/process sustainability needs to be ascertained with proper evaluation tools. After seeing the enormous growth in chemical technology and its inherent devastating environmental effects, the need for alternative fuels such as bio-fuel has attracted immense research interest in view of the current concerns of environmental sustainability, although economic viability is ascertained. Beet pulp is a source of feed as fresh pulp, silage and silage-dry matter. While sugar beet molasses is presently the most valuable by-product of beet sugar industries, red beet pulp and silage can be used in biogas plants. In addition, there are "energy beet" varieties, evolved as hybrids between fodder beet and sugar beets, which are capable of producing higher biomass (Eggleston 2010). Red beets contain 7% sugars, and hence improved fermentation processes may yield feed biomass, with a simultaneous yield of combustible gases. Since sugar beets have evolved from red beets, re-crossing red beets with sugar beets may produce intermediary types of plants useful for bio-fuel.

Another ecological issue is the use of genetically modified plants, and the problem in beets is the ease with which lateral gene-flow occurs (McGinnis et al. 2010). However, in in vitro cultures such as hairy roots, the cultures are mostly confined to the aseptic environment grown in laboratory conditions, avoiding transgene or pharmacologically active protein dissemination into other organisms or to the environment. While the applications of hairy root technology comes with immense opportunities for producing a large array of products including human proteins such as tissue plasminogens (see Chap. 10) through molecular farming (Graf et al. 2010), their transformed nature is a hurdle for their commercial applications. Such issues have been circumvented through scientifically addressed technological approaches, and now procedures are available to keep these genetically modified (GM) materials under environmentally safe conditions. An assessment of the commercial impact of biotechnology-derived crops showed very high economic returns to the farm sector, with a profound benefit to global environment. The evaluation of four main crops: canola, corn, cotton and soybeans, imparted substantial net economic benefit at the farm level, amounting to $10.1 billion in 2007 and $44.1 billion in a span of 12 years, where 50% of the benefit was reaped by developing countries. At global level, there was a great increase in production levels of food crops; for instance, the cultivation of soybean and corn alone added 68 million tons and 62 million tons, respectively (Brookes and Barfoot 2009). In the environmental perspective, pesticide usage was substantially reduced, to the tune of by 352 million kg (−8.4%), and the associated manufacturing hazards were also reduced by 16.3%. While this assessment was based on the environmental impact quotient of 2008, it also indicated a significant reduction in the release of greenhouse gas emissions from such GM cropping areas, which was equivalent to removing 7.853 million cars from the roads in 2009 (Brookes and Barfoot 2011, see page 19).

Technologically feasible, genetically engineered organs such as hairy roots, if rationally developed for decreased catabolism of the desired compound, with enhanced expression of a rate-limiting enzyme, while simultaneously preventing the feedback inhibition of a key enzyme, are expected to shift the metabolic flux favorably towards the synthesis of target metabolite. This must be accomplished in addition to the newly expressed genes for desired therapeutic protein. Conversion of an existing product into a new product and its further compartmentalization has also been demonstrated in other plant systems. In red beet, the hairy roots were re-transformed to obtain double transformants (Thimmaraju et al. 2008), and such technologies open up large opportunities for using red beet hairy root cultures for molecular farming. However, many biologists are of the opinion that such transformations generally result in marginal net gains because of the notion "gain of function is equivalent to loss of function." Nevertheless, the discovery of transcription factors, which offers the possibility of regulating the entire pathway, has opened new possibilities of better control over the monitoring of secondary metabolites production.

16.9 Red Beet Hairy Roots for Morphophysiological Studies

Red beet hairy roots are ideal candidates for pigment synthesis and a vast array several other applications, including basic research on cellular genomics, described in Chap. 10. Since each root clone originates as a result of genetic transformation of a single cell, each hairy root clone represents a model entity of cellular biochemical environment, where hormone–morphology–secondary metabolism (tri-way response) are precisely balanced by the genes inherited from *Agrobacterium rhizogenes* in harmony with the cell genome (Thimmaraju and Bhagyalakshmi 2002). The fact that these roots also display wide variations in morphologies as a response to different types of sugars (Bhagyalakshmi et al. 2004) is indicative of the fact that they represent complex physiological interactions among the tri-way responses. Such extremely altered tri-way responses herald the occurrence of a great number of genetic interactions and signalling cascades within these organ cultures. Since hairy roots grow under a heterotrophic fully controllable aseptic environment, they represent an ideal system for understanding sugar-sensing mechanisms, which merits more advanced chemo-genomic, transcriptomic and proteomic studies. Large biochemical alterations were noted even in yellow beet hairy roots as a response to sugars (Böhm and Mack 2004). The influence of *rol* genes on carbohydrate metabolism has seldom been studied, even in other plant systems. When *rolB* and *rolC* gene-transformed potato microtubers were analyzed, the structure of the starch granules were very different in each case from those of normal tubers (Aksenova et al. 2010), indicating the commercial importance of such gene interactions. The selective secretion of proteins such as peroxidase enzymes from red beet hairy roots (Thimmaraju et al. 2005, 2007) has demonstrated the inherent secretory nature, which is yet another phenomenon worth characterizing for commercial exploitation in the context that hairy roots ably synthesize human proteins and other animal proteins (Kang et al. 2011; Guillon et al. 2006). Hairy roots offer a better technological edge over shoot cultures, owing to their genetic stabilities and lower requirement of energy inputs (no light requirement and hence low energy light-generated temperature control), presumably making them the most suitable for molecular farming. With the added advantage of their clonal nature, hairy roots are functionally stable with biochemical consistency in product yield.

16.10 Concluding Remarks

Owing to their enormous health benefits, horticultural crops are progressively occupying more acreage globally, with a consequence of an ever-increasing array of products. History has evinced continuous growth in knowledge and technologies, always bringing higher efficiencies and sustainability in crop production and their further utilization through efficient processing, consumption and disposal. Red beet is yet to benefit from modern technologies of multidisciplinary approaches, particularly computational chemistry for the identification of new compounds (formed during breeding and post-harvest processing), combinatorial chemistry for the induction of new pathway controls and chemogenomics for the identification of potential new therapeutics. Applications of these techniques would allow immense opportunities for finding innovative food and pharmaceutical products with red beet-the vegetable has proven itself as an elixir for age-related ailments. However, substantial progress made in genetic analysis of beetroot complements breeding strategies and for developing in vitro cell and hairy roots cultures. These studies have shown the suitability of such cultures for the production of secondary metabolites, enzymes and contributed towards basic knowledge on cellular physiologies. The fact that hairy roots are amenable for re-transformation and that the sugar beet hairy roots produce callus with a great array of pigments (Pavokovic` et al. 2009) prove the vast techno-commercial impact of *Beta vulgaris* in the near future. Backed by substantial efforts made in their characterizations, beet pigments have qualified themselves for modern down-stream processing through smart separation technologies. It is expected that these applications would surely be intercepted by more advanced methods and smarter bioprocesses. Because betalains generate more energy in photocells (Calogero et al. 2010) than other pigments, such applications are expected to benefit from advancements made in nanotechnology for their commercial realization. All in all, the amalgamation of advanced scientific methods for the processing of red beet products would allow all strata of society to reap the health benefits offered by the "marvellous red beet."

References

Akita, T., Y. Hina, and T. Nishi. 2002. New medium composition for high betacyanin production by a cell suspension culture of table beet (*Beta vulgaris* L.). *Bioscience, Biotechnology, and Biochemistry* 66: 902–905.

Aksenova, N.P., L.A. Wasserman, L.I. Sergeeva, T.N. Konstantinova, S.A. Golyanovskaya, A.V. Krivandin, I.G. Plashchina, W. Blaszczak, J. Fornal, and G.A. Romanov. 2010. Agrobacterial *rol* genes modify thermodynamic and structural properties of starch in microtubers of transgenic potato. *Russian Journal of Plant Physiology* 57: 656–663.

Bhagyalakshmi, N., and G.A. Ravishankar. 1998. Application of hairy root cultures for the production of pharmaceutical compounds. In *Role of biotechnology in medicinal and aromatic plants*, ed. I. Khan and A. Khanum, 275–293. Hyderabad: Ukaaz.

Bhagyalakshmi, N., L.V. Venkataraman, and G.A. Ravishankar. 1998. Bioprocess parameters for the scaled-up production of hairy root cultures. In *Tissue culture and molecular biology-applications and prospects*, ed. P.S. Srivastava, 332–358. New Delhi: Narosa Publishing House.

Bhagyalakshmi, N., R. Thimmaraju, and M.S. Narayan. 2004. Various hexoses and di-hexoses differently influence growth, morphology and pigment synthesis in transformed root cultures of red beet (*Beta vulgaris*). *Plant Cell Tissue and Organ Culture* 78: 183–195.

Böhm, H., and G. Mack. 2004. Betaxanthin formation and free amino acids in hairy roots of *Beta vulgaris* var. lutea depending on nutrient medium and glutamate or glutamine feeding. *Phytochemistry* 65: 1361–1368.

Brockington, S.F., R.H. Walker, B.J. Glover, P.S. Soltis, and D.E. Soltis. 2011. Complex pigment evolution in the Caryophyllales. *New Phytologist* 190: 854–864.

Brookes, G., and Barfoot, P. 2009. Global impact of biotech crops: income and production effects, 1996–2007. *AgBioForum* 12: 184–208, http://www.pgeconomics.co.uk/pdf/2009socioeconimpactsagbioforumpaper.pdf

Brookes G, and Barfoot P (2011) *Global impact of biotech crops: income and production effects, 1996–2009*. 1–173. PG Economics Ltd, UK, *Dorchester, UK 2011*.

Calogero, G., G.D. Marco, S. Cazzanti, S. Caramori, R. Argazzi, A.D. Carlo, and C.A. Bignozzi. 2010. Efficient dye-sensitized solar cells using red turnip and purple wild Sicilian prickly pear fruits. *International Journal of Molecular Sciences* 11: 254–267.

Chiou, D., and T.A.G. Langrish. 2007. Development and characterization of novel nutraceuticals with spray drying technology. *Journal of Food Engineering* 82: 84–91.

Day, L., R.B. Seymour, K.F. Pitts, I. Konczak, and L. Lundin. 2009. Incorporation of functional ingredients into foods. *Trends in Food Science and Technology* 20: 388–395.

Desai, K.G.H., and H.J. Park. 2005. Recent developments in microencapsulation of food ingredients. *Drying Technology* 23: 1361–1394.

Eggleston, G. 2010. *Future sustainability of the sugar and sugar – ethanol industries*, ACS symposium series, 1–25. Washington, DC: American Chemical Society.

Fang, Z., and B. Bhandari. 2010. Encapsulation of polyphenols – a review. *Trends in Food Science and Technology* 21: 510–523.

Fett-Neto, A.G., S.J. Melanson, S.A. Nicholson, J.J. Pennington, and F. d'DiCosmo. 1994. Improved taxol yield by aromatic carboxylic acid and amino acid feeding to cell cultures of *taxus cuspidata*. *Biotechnology and Bioengineering* 44: 967–971.

Georgiev, V.G., J. Weber, E. Kneschke, P.N. Denev, T. Bley, and A.I. Pavlov. 2010. Antioxidant activity and phenolic content of betalain extracts from intact plants and hairy root cultures of the red beetroot *Beta vulgaris* cv Detroit Dark Red. *Plant Foods for Human Nutrition (Formerly Qualitas Plantarum)* 65: 105–111.

Gliszczyǹska-Ŝwiglo, A., H. Szymusiak, and P. Malinowska. 2006. Betanin, the main pigment of red beet: Molecular origin of its exceptionally high free radical-scavenging activity. *Food Additives and Contaminants* 23: 1079–1087.

Goossens, A., S.T. Hakkinen, I. Laakso, K.M. Oksman-Caldentey, and D. Inze. 2003. Secretion of secondary metabolites by ATP-binding cassette transporters in plant cell suspension cultures. *Plant Physiology* 131: 1161–1164.

Graf, B.L., I. Raskin, W.T. Cefalu, and D.M. Ribnicky. 2010. Plant-derived therapeutics for the treatment of metabolic syndrome. *Current Opinion in Investigational Drugs* 11: 1107–1115.

Guillon, S., J. Trémouillaux-Guiller, P.K. Pati, M. Rideau, and P. Gantet. 2006. Harnessing the potential of hairy roots: dawn of a new era. *Trends in Biotechnology* 24: 403–409.

Kang, S., H. Ajjappala, H.H. Seo, J.S. Sim, S.H. Yoon, B.S. Koo, Y.H. Kim, and B.S. Hahn. 2011. Expression of the human tissue-plasminogen activator in hairy roots of oriental melon (*Cucumis melo*). *Plant Molecular Biology Reporter*. doi:10.1007/s11105-011-0303-8.

Kosaraju, S.L., D. Labbett, M. Emin, I. Konczak, and L. Lundin. 2008. Delivering polyphenols for healthy ageing. *Nutrition and Dietetics* 65: S48–S52.

Mandenius, C.-F., and M. Björkman. 2010. Mechatronics design principles for biotechnology product development. *Trends in Biotechnology* 28: 230–236.

McGinnis, E.E., M.H. Meyer, and A.G. Smith. 2010. Sweet and sour: a scientific and legal look at herbicide-tolerant sugar beet. *The Plant Cell Online* 22: 1653–1657.

Munos, B. 2009. Lessons for 60 years of pharmaceutical innovation. *Nature Reviews. Drug Discovery* 8: 959–968.
Muzler, J., and K. Prantz. 2009. Total synthesis of epithilones. *Fortschritte der Chemie Organischer Naturstoffe* 90: 55–133.
Neelwarne, B., and R. Thimmaraju. 2009. Bioreactor for cultivation of red beet hairy roots and in situ recovery of primary and secondary metabolites. *Engineering in Life Sciences* 9: 227–238.
Pavlov, A., and T. Bley. 2005. Betalains biosynthesis by *Beta vulgaris* L. hairy root culture in different bioreactor systems. Scientific Works, University of Food Technologies 52: 299–304
Pavlov, A., V. Georgiev, and M. Ilieva. 2005a. Betalain biosynthesis by red beet (*Beta vulgaris* L.) hairy root culture. *Process Biochemistry* 40: 1531–1533.
Pavlov, A., P. Kovatcheva, D. Tuneva, M. Ilieva, and T. Bley. 2005b. Radical scavenging activity and stability of betalains from *Beta vulgaris* hairy root culture in simulated conditions of human gastrointestinal tract. *Plant Foods for Human Nutrition* 60: 43–47.
Pavlov, A., V. Georgiev, and T. Bley. 2007. Batch and fed-batch production of betalains by red beet (*Beta vulgaris*) hairy roots in a bubble column reactor. *Zeitschrift fur Naturforsch C* 62: 439–446.
Pavokovic`, D., and M. Krsnik-Rasol. 2011. Complex biochemistry and biotechnological production of betalains. *Food Technology and Biotechnology* 49: 145–155.
Pavokovic`, D., G. Rusak, V. Besendorfer, and M. Krsnik-Rasol. 2009. Light-dependent betanin production by transformed cells of sugar beet. *Food Technology and Biotechnology* 47: 153–158.
Rakin, M., M. Vukasinovic, S. Silver-Marinkovic, and M. Maksimovic. 2007. Contribution of lactic acid fermentation to improve nutritive quality of vegetable juices enriched with brewer's yeast autolysate. *Food Chemistry* 100: 599–602.
Sahai, O. 1994. Plant tissue culture. In *Bioprocess production of flavour, fragrance and color ingredients*, ed. A. Gabelman, 239–275. New York: Wiley.
Savitha, B.C., R. Thimmaraju, N. Bhagyalakshmi, and G.A. Ravishankar. 2006. Different biotic and abiotic elicitors influence betalain production in hairy root cultures of *Beta vulgaris* in shake flask and bioreactor. *Process Biochemistry* 41: 50–60.
Schmidt, B.M., D.M. Ribnicky, P.E. Lipsky, and I. Raskin. 2007. Revisiting the ancient concept of botanical therapeutics. *Nature Chemical Biology* 3: 360–366.
Strack, D., T. Vogt, and W. Schliemann. 2003. Recent advances in betalain research. *Phytochemistry* 62: 247–269.
Suresh, B., R. Thimmaraju, N. Bhagyalakshmi, and G.A. Ravishankar. 2004. Methyljasmonate and L-Dopa as elicitors of betalaine production in bubble column bioreactor. *Process Biochemistry* 39: 2091–2096.
Swinney, D.C., and J. Antony. 2011. How were new medicines discovered. *Nature Reviews* 10: 507–519.
Taya, M., K. Mine, M. Kino-Oka, S. Tone, and T. Ichi. 1992. Production and release of pigments by culture of transformed hairy root of red beet. *Journal of Fermentation and Bioengineering* 73: 31–36.
Taya, M., K. Yakura, M. Kino-oka, and S. Tone. 1994. Influence of medium constituents on enhancement of pigment production by batch culture of red beet hairy roots. *Journal of Ferment Bioengineering* 77: 215–217.
Terahara, N., I. Konczak, H. Ono, M. Yoshimoto, and O. Yamakawa. 2004. Characterization of acylated anthocyanins in callus induced from storage root of purple-fleshed sweet potato, *Ipomoea batatas* L. *Journal of Biomedicine and Biotechnology* 5: 279–286.
Thimmaraju, R., and N. Bhagyalakshmi. 2002. Application of *rol* genes for enhanced production of bio-active compounds from medicinal plants. In *Role of biotechnology in medicinal and aromatic plants*, VIIth ed, ed. I.A. Khan and A. Khanum, 139–159. Hyderabad: UKAAZ.
Thimmaraju, R., N. Bhagyalakshmi, M.S. Narayan, and G.A. Ravishankar. 2003a. Kinetics of pigment release from hairy root cultures of red beet under the influence of pH, sonication, temperature and oxygen stress. *Process Biochemistry* 38: 1067–1074.

Thimmaraju, R., N. Bhagyalaskshmi, M.S. Narayan, and G.A. Ravishankar. 2003b. Food grade chemical and biological agents permeabilize red beet hairy roots assisting the release of betalains. *Biotechnology Progress* 19: 1274–1282.

Thimmaraju, R., N. Bhagyalakshmi, and G.A. Ravishankar. 2004. Insitu and ex situ adsorption and recovery of betalaines from hairy root cultures of Beta vulgaris. *Biotechnology Progress* 20: 777–785.

Thimmaraju, R., L. Venkatachalam, K. Roohie, M.S. Narayan, and N. Bhagyalakshmi. 2007. Purification and characterization of an intracellular peroxidase from genetically transformed roots of red beet (*Beta vulgaris* L.). *Food Chemistry* 105: 1312–1320.

Thimmaraju, R., L. Venkatachalam, and N. Bhagyalakshmi. 2008. Morphometric and biochemical characterization of red beet (*Beta vulgaris* L.) hairy roots obtained after single and double transformations. *Plant Cell Reports* 27: 1039–1052.

Thimmaraju, R., V. Kumar, L. Venkatachalam, N. Bhagyalakshmi, and G.A. Ravishankar. 2005. Peroxidase production from hairy root cultures of red beet (*Beta vulgaris*). *Electronic Journal of Biotechnology* 8: 185–196.

Wang, Y. 2008. Needs for new plant-derived pharmaceuticals in the post-genome era: an industrial view in drug research and development. *Phytochemistry Reviews* 7: 395–406.

Webb, A.J., N. Patel, S. Loukogeorgakis, M. Okorie, Z. Aboud, S. Misra, R. Rashid, P. Miall, J. Deanfield, N. Benjamin, R. MacAllister, A.J. Hobbs, and A. Ahluwalia. 2008. Acute blood pressure lowering, vasoprotective, and antiplatelet properties of dietary nitrate via bioconversion to nitrite. *Hypertension* 51: 784–90.

Wong, P.L., A.J. Royce, and C.W.T. Lee-Parsons. 2004. Improved ajmalicine production and recovery from *Catharanthus roseus* suspensions with increased product removal rates. *Biochemical Engineering Journal* 21: 253–258.

Wybraniec, S. 2008. Chromatographic investigation on acyl migration in betacyanins and their decarboxylated derivatives. *Journal of Chromatography B* 861: 40–47.

Index

B. Neelwarne (ed.), *Red Beet Biotechnology: Food and Pharmaceutical Applications*,
DOI 10.1007/978-1-4614-3458-0,

Q

R

S

T

U